T0137312

Springer Series in Materials Science

Volume 313

The Springer Series in Materials Science covers the complete spectrum of materials research and technology, including fundamental principles, physical properties, materials theory and design. Recognizing the increasing importance of materials science in future device technologies, the book titles in this series reflect the state-of-the-art in understanding and controlling the structure and properties of all important classes of materials.

More information about this series at http://www.springer.com/series/856

Kamal K. Kar
Editor

Handbook of Nanocomposite Supercapacitor Materials III

Selection

Editor
Kamal K. Kar
Advanced Nanoengineering Materials Laboratory
Department of Mechanical Engineering
and Materials Science Programme
Indian Institute of Technology Kanpur
Kanpur, Uttar Pradesh, India

ISSN 0933-033X ISSN 2196-2812 (electronic)
Springer Series in Materials Science
ISBN 978-3-030-68366-5 ISBN 978-3-030-68364-1 (eBook)
https://doi.org/10.1007/978-3-030-68364-1

This Springer imprint is published by the registered company Springer Nature Switzerland AG
The registered company address is: Gewerbestrasse 11, 6330 Cham, Switzerland

Dedicated to my wife, Sutapa, and my little daughter, Srishtisudha for their loving support and patience, and my mother, late Manjubala, and my father, late Khagendranath

Preface

The global energy scene, which is one of the largest and most diversified fields in the world, is in a state of flux. These include the moving consumption away from non-renewable energy sources, rapid deployment of major renewable energy technologies and deep decline in their costs, and a growing shift towards electricity in energy use across the globe. This power and energy system is experiencing its greatest ever changes and challenges due to the shift from traditional power and energy networks to smart power/energy grids. As long as the energy consumption is intended to be more economical and more environment-friendly, electrochemical energy production is under serious consideration as an alternative energy/power source. In other words, a large amount of electricity can be generated from natural sources like solar, wind and, tidal energy and it is imperative to stock the produced energy since man has constrained control over these natural wonders. Batteries, fuel cells, and supercapacitors belong to the same family of energy storage devices, which are ubiquitous in our day-to-day life. But the supercapacitor is a step-up device in the field of energy storage and has a lot of research and development scope in terms of design, parts fabrication, and energy storage mechanisms.

Various types of supercapacitors have been developed such as electrochemical double-layer capacitors (EDLC), pseudocapacitors (or, redox capacitors), and capacitors. They store charges electrochemically and exhibit high power densities, moderate to high energy densities, high rate capabilities, long life, and safe operation. The electrode, electrolyte, separator, and current collectors are the key parts for the supercapacitors for energy storage to determine the electrochemical properties, energy storage mechanism, and mechanical properties of the supercapacitor devices. Therefore, many significant breakthroughs for a new generation of supercapacitors have been reported in recent years through the development of these materials and novel device designs. But the performance of devices is still challenging in terms of capacitance, flexibility, cycle life, etc. These deciding factors depend on the characteristics of materials used in the devices. The key objective is to select the right materials with new technologies and developments for the electrodes, electrolytes, separators, and current collectors, which are the essential components of supercapacitors with an aim to enhance the performance of supercapacitors.

The book *Handbook of Nanocomposite Supercapacitor Materials with a theme of Selection* focuses on the various characteristics of prospective materials and how to select the right materials with the concept of various material indices using Ashby's chart. This book provides a comprehensive study on several architectural carbon materials, transition metal oxides, conducting polymers, and their binary and ternary composite electrodes that are using in the current era of supercapacitors. Finally, it highlights the advantages, challenges, applications, and future directions of the supercapacitors. Therefore, this book will provide the readers with a complete and composed idea about the fundamentals of supercapacitors, the recent development of electrode materials for supercapacitors, and the design of their novel flexible solid-state devices. This book will be useful to graduate students and researchers from various fields of science and technology, who wish to learn about the recent development of supercapacitor and to select the right material for high-performance supercapacitor.

Chapter 1 discusses about the differences between other energy storage devices and supercapacitor, historical developments of supercapacitors, Faradaic and non-Faradaic processes, types of supercapacitors i.e., electric double layer capacitors, pseudocapacitors, asymmetric supercapacitors, hybrid supercapacitors, quantum supercapacitors, on-chip supercapacitors, hybrid energy storage systems, etc., various components of supercapacitor i.e., electrodes, electrolytes, separators and current collectors, various materials used in these components i.e., activated carbon, graphene/reduced graphene oxide, carbon nanotubes, carbon nanofibers, conducting polymers, transition metal oxides, metal-organic frameworks, covalent organic frameworks, MXenes, black phosphorous, aqueous electrolytes, non-aqueous electrolytes, organic electrolytes, ionic liquids, solid state or quasi-solid state electrolytes, solid/dry polymer electrolytes, gel polymer electrolytes, inorganic solid state electrolytes, polyelectrolytes, redox active electrolytes, etc., and various electrochemical characterization techniques.

As supercapacitors deliver excellent electrochemical performances such as high capacitance, high power density, and long cyclic stability at low cost. In contrast with other energy storage devices, its charge storage mechanism is simple, which makes its charging and discharging process highly reversible. Based on the charge storage mechanism, its electrode material can be categorized as EDLC and pseudocapacitor. EDLC capacitor stores charge electrostatically whereas, reversible redox reaction occurs in pseudocapacitance. The charge is stored via the Faradaic process. The further improvement in the performance is done by the formation of composite electrode material, the introduction of nanostructure electrode, assembling a hybrid capacitor by introducing battery electrode material, and assembly of an asymmetric supercapacitor. Various combinations of electrode and electrolyte material in different types of configuration provide a synergistic effect of both types of charge storage mechanism and wide operating potential range. The main aim is to obtain a high energy density device without compromising other parameters such as power density, rate capability, and cyclic stability. Chapter 2 extensively deals the various materials used in supercapacitors, types of charge storage mechanisms, types of supercapacitor assembly i.e., symmetric supercapacitors, asymmetric supercapacitors, battery-supercapacitor hybrid devices, etc., and their performance to the type of electrode material.

Nowadays flexible solid-state supercapacitors (FSSCs) are the most emerging energy storage devices in modern miniatured technologies. With increasing the use of micro and flexible electronic devices such as, wearable electronic suits, microsensors, and biomedical equipment, the demand for FSSCs is increasing exponentially. These electronic devices focused on the integration of many components in a single compact system that must be flexible in nature, lightweight, smaller in dimension, unbreakable and should be available at a competitive price. Chapter 3 mainly focuses on the advancement of FSSCs devices with its all components such as current collector, electrode materials, and electrolytes. This chapter also discusses the strategies of fabrication techniques, types of FSSCs, design, evaluation of performance, and applications of FSSCs device step-by-step.

CPs are known for their astonishing electrical and electrochemical properties. Characteristic features such as tunable conductivity, structural flexibility, mild synthesis and processing conditions, chemical and structural diversity make them an excellent candidate for different fields of interest. Chapter 4 aims to revisit the journey and recent advancements of CPs in the field of energy storage systems like supercapacitors. CPs have been considered as one of the excellent candidates for supercapacitor as they show miscellaneous redox nature, amazing electrical conductivity, good flexibility, and many others. Therefore, a substantial discussion is required to discuss the supercapacitors and their advantages and disadvantages, recent advancements, future challenges, and new possibilities. Chapter 4 focuses on the synthesis, processing, and chemical modifications of various CPs with various interesting properties and their electrodes used for the advancements of supercapacitors which is the need of the hour.

Among all the components i.e., electrode, electrolyte, separator, and current collector, the electrode plays a major role to store a large amount of charge at its surface. So, characteristics of the electrode such as porosity, surface morphology, surface area, electrical conductivity, etc., are taken into account for selecting suitable electrode material for supercapacitor. Activated carbon, CNT, graphene, carbon aerogel, metal compounds, conducting polymers, and their composites are among various materials, which have been commonly used as electrodes and discussed in detail to select the best material with the concept of various material indices using Ashby's chart in Chap. 5.

Electrode stores charges, electrolyte provide necessary ions, current collector transfer the charge from the electrode to external circuit and separator acts as a membrane, which prevents the device from short-circuit. The separator's main function is to separate cathode and anode electrode material in supercapacitors to prevent short circuits and mainly present in the form of a porous membrane in order to provide easy ion transfer. The common material used as separator includes glass fiber, cellulose, ceramic fibers, or polymeric film materials. Chapter 6 mainly describes functions served and characteristics required for separators and their materials, respectively, which are chosen according to those functions. Finally, the selection of separator material is justified with the help of various material indices using Ashby's chart.

A lot of research is being done to improve the efficiency and performance of supercapacitors by making the right choice for electrodes, electrolytes, separators, and current collectors. Among all the components, electrolytes serve the purpose of balancing charge in supercapacitor and provide necessary ions to form an electrical connection between electrodes. The electrolyte materials used in supercapacitor can be classified as organic, aqueous, ionic liquids, solid-state, and redox-active electrolytes and are chosen according to their properties, ultimate applications, and physical state of the supercapacitor. Chapter 7 explains the functions of electrolytes, classification of electrolytes i.e., aqueous electrolytes, organic electrolytes, ionic electrolytes, etc., characteristics required for electrolytes i.e., conductivity, viscosity, ionic concentration, electrochemical stability, thermal stability, dissociation, toxicity, volatility and flammability, cost, etc., performance of various electrolytes, performance metrics and their relationships, selection of electrolyte material in detail with the support of various material indices using Ashby's chart.

The main function of the current collector is to collect and conduct electric current from electrodes to power sources. It also provides mechanical support to electrodes. To meet the required properties of the current collector materials should have minimum contact resistance, high electric conductivity, and good bonding capacity with electrodes. Most commonly used conventional metals like copper, aluminum, nickel, etc. are being replaced by advanced materials such as nanostructured or composite materials. In addition to this, the demand for flexible electronics is growing rapidly nowadays, these devices require a material with enhanced properties. Different types of materials used for the current collector are thoroughly discussed in Chap. 8, where the selection of materials depends upon the cost of materials and their suitability toward particular applications. Comparative study of properties for various current collector materials has been done to suggest suitable material for supercapacitor applications. The selection of the current collector is discussed with the help of various material indices using Ashby's chart.

Supercapacitor management systems have been developed for supercapacitor usage during demand within safe operating limits. Supercapacitors and batteries are used together with the help of hybrid energy management configurations. Rule-based, optimization-based, and artificial intelligence-based energy management strategies for hybrid energy storage systems are discussed in Chap. 9. The main parameters are adaptability, reliability, and robustness. Computational complexity is a driving parameter for using these techniques in online or offline mode.

The global supercapacitor market is expected to grow at a rapid rate in the coming years owing to the rising demand for supercapacitors in various applications. These supercapacitors are available in varying sizes, capacitances, voltage ranges, etc., and are sometimes tailor-made for certain applications. At present, the market is currently dominated by a few major players such as Murata Technology, Maxwell Technologies, Eaton Corporation, Nippon Chemi-Con, Nesscap among others. Chapter 10 deals with the trends in the supercapacitor market and also sheds light on the properties of supercapacitor cells and modules manufactured by key market players.

The last Chap. 11 provides a brief insight into the commercially available supercapacitors and the applications in various fields like wearable electronics, portable electronic devices, transportation, industrial applications, military, defense, and national security, renewable energy sector, power electronics, communication, artificial intelligence, internet of things, cyber-physical system, soft robotics, complementary metal-oxide-semiconductor, very-large-scale integration, memory, medical and healthcare, buildings, gas sensors, and futuristic applications along with few selected manufactures for the said applications around the globe.

The editor and authors hope that readers from materials science, engineering, and technology will be benefited from the reading of these high-quality review articles related to the characteristics of materials and their selections used in supercapacitor. This book is not intended to be a collection of all research activities on composites worldwide, as it would be rather challenging to keep up with the pace of progress in this field. The editor would like to acknowledge many material researchers, who have contributed to the contents of the book. The editor would also like to thank all the publishers and authors for permitting us to use their published images and original work. I also take this opportunity to thank Viradasarani, Zachary, and Viradasarani Natarajan, and Adelheid Duhm, and the editorial team of Springer Nature for their helpful advice and guidance.

There were lean patches when I felt that 1 would not be able to take time out and complete the book, but my wife Sutapa, and my little daughter Srishtisudha, played a crucial role to inspire me to complete it. I hope that this book will attract more researchers to this field and that it will form a networking nucleus for the community. Please enjoy the book and please communicate to the editor/authors any comments that you might have about its content.

Kanpur, India

Kamal K. Kar

Contents

Editor and Contributors

About the Editor

Prof. Kamal K. Kar, Ph.D. Champa Devi Gangwal Institute Chair Professor; Professor, Department of Mechanical Engineering, Materials Science Programme, Indian Institute of Technology Kanpur, India.

Professor Kar pursued higher studies from the Indian Institute of Technology Kharagpur, India, and Iowa State University, USA, before joining as a Lecturer in the Department of Mechanical Engineering and Interdisciplinary Programme in Materials Science at IIT Kanpur in 2001. He was a BOYSCAST Fellow in the Department of Mechanical Engineering, Massachusetts Institute of Technology, USA in 2003. He is currently holding the Champa Devi Gangwal Chair Professor of the Institute. Before this, he has also held the Umang Gupta Institute Chair Professor (2015–2018) at IIT Kanpur. He is the former Head of the Interdisciplinary Programme in Materials Science from 2011 to 2014, and Founding Chairman of the Indian Society for Advancement of Materials and Process Engineering Kanpur Chapter from 2006 to 2011.

Professor Kar is an active researcher in the broad areas of nanostructured carbon materials, nanocomposites, functionally graded materials, nano-polymers, and smart materials for structural, energy, water, and biomedical applications. His research works have been recognized through the office of the Department of Science and Technology, (ii) Ministry of Human

Resource and Development, National Leather Development Programme, Indian Institute of Technology Kanpur, Defence Research and Development Organisation, Indian Space Research Organization, Department of Atomic Energy, Department of Biotechnology, Council of Scientific and Industrial Research, Aeronautical Development Establishment, Aeronautics Research and Development Board, Defence Materials and Stores Research and Development Establishment, Hindustan Aeronautics Limited Kanpur, Danone research and development department of beverages division France, Indian Science Congress Association, Indian National Academy of Engineering and many more from India.

Professor Kar is the Editor-in-Chief of *Polymers and Polymeric Composites: A Reference Series* published by Springer Nature, and Members in the Editorial Board of *SPE Polymers* published by Wiley, *Advanced Manufacturing: Polymer and Composites Science* published by Taylor & Francis Group, *International Journal of Plastics Technology* published by Springer Nature and many more.

Professor Kar has more than 250 papers in international refereed journals, 135 conference papers, 10 books on nanomaterials and their nanocomposites, 3 special issues on polymer composites, 80 review articles/book chapters, and more than 55 national and international patents to his credits, some of these have over 200 citations. He has guided 18 doctoral students and 80 master students so far. Currently, 17 doctoral students, 10 master students, and few visitors are working in his group, Advanced Nanoengineering Materials Laboratory.

Contributors

Sandeep Ahankari School of Mechanical Engineering, VIT University, Vellore, India

Pallab Bhattacharya Functional Materials Group, Advanced Materials & Processes Division, CSIR-National Metallurgical Laboratory (NML), Burmamines, East Singhbhum, Jamshedpur, Jharkhand, India

Souvik Ghosh Surface Engineering and Tribology Division, Council of Scientific and Industrial Research-Central Mechanical Engineering Research Institute, Durgapur, India;
Academy of Scientific and Innovative Research (AcSIR), Ghaziabad, India

Alka Jangid Advanced Nanoengineering Materials Laboratory, Materials Science Programme, Indian Institute of Technology Kanpur, Kanpur, India

Kamal K. Kar Advanced Nanoengineering Materials Laboratory, Materials Science Programme, Department of Mechanical Engineering, Indian Institute of Technology Kanpur, Kanpur, India

Tapas Kuila Surface Engineering and Tribology Division, Council of Scientific and Industrial Research-Central Mechanical Engineering Research Institute, Durgapur, India;
Academy of Scientific and Innovative Research (AcSIR), Ghaziabad, India

Dylan Lasrado School of Mechanical Engineering, VIT University, Vellore, India

Ravi Nigam Advanced Nanoengineering Materials Laboratory, Materials Science Programme, Indian Institute of Technology Kanpur, Kanpur, India

Ariful Rahaman Department of Manufacturing Engineering, School of Mechanical Engineering, Vellore Institute of Technology, Vellore, Tamil Nadu, India

Prakas Samanta Surface Engineering and Tribology Division, Council of Scientific and Industrial Research-Central Mechanical Engineering Research Institute, Durgapur, India;
Academy of Scientific and Innovative Research (AcSIR), Ghaziabad, India

Prerna Sinha Advanced Nanoengineering Materials Laboratory, Materials Science Programme, Indian Institute of Technology Kanpur, Kanpur, India

M. S. Sreekanth Department of Manufacturing Engineering, School of Mechanical Engineering, Vellore Institute of Technology, Vellore, Tamil Nadu, India

T. P. Sumangala Department of Physics, School of Advanced Sciences, Vellore Institute of Technology, Vellore, Tamil Nadu, India

Harish Trivedi Advanced Nanoengineering Materials Laboratory, Materials Science Programme, Indian Institute of Technology Kanpur, Kanpur, India

Kapil Dev Verma Advanced Nanoengineering Materials Laboratory, Materials Science Programme, Indian Institute of Technology Kanpur, Kanpur, India

Chapter 1
Introduction to Supercapacitors

Ravi Nigam, Prerna Sinha, and Kamal K. Kar

Abstract Supercapacitors are energy storage devices, which display characteristics intermediate between capacitors and batteries. Continuous research and improvements have led to the development of supercapacitors and its hybrid systems and supercapacitors, which can replace traditional batteries. The comparison among different energy storage devices has been introduced in the present chapter. The timeline for the development of supercapacitors is also mentioned along with the introduction of different charge storage mechanisms in supercapacitors. Supercapacitors mainly consist of four components electrodes, electrolyte, separator, and current collector. The different types of supercapacitors have been introduced including the novel quantum supercapacitor. For hybrid energy management configurations, supercapacitors and batteries are used together to mask their limitations of the low energy density and power density, respectively. For miniaturized devices, on-chip supercapacitors and on-chip energy management systems are also discussed. The principles of the most widely used electrochemical characterization techniques and parameters have been incorporated in the chapter.

R. Nigam · P. Sinha · K. K. Kar (✉)
Advanced Nanoengineering Materials Laboratory, Materials Science Programme, Indian Institute of Technology Kanpur, Kanpur 208016, India
e-mail: kamalkk@iitk.ac.in

R. Nigam
e-mail: ravinigam09@gmail.com

P. Sinha
e-mail: findingprerna09@gmail.com

K. K. Kar
Advanced Nanoengineering Materials Laboratory, Department of Mechanical Engineering, Indian Institute of Technology Kanpur, Kanpur 208016, India

K. K. Kar (ed.), *Handbook of Nanocomposite Supercapacitor Materials III*,
Springer Series in Materials Science 313,
https://doi.org/10.1007/978-3-030-68364-1_1

1.1 Introduction

In the present scenario, all the present electronic systems require continuous energy supply to function properly. Renewable energy sources such as sunlight, wind, and tide are intermittent, and non-renewable sources like coal and petroleum products are dwindling day by day. Energy storage devices play a vital role in providing a continuous supply of energy. The supercapacitor is an electrochemical energy storage device that is categorized into various types based on charge transfer or storage mechanisms. There are various types of materials that are used to make different components for the devices. The characteristic properties of the materials are studied by various techniques. The device has practical utility when it is incorporated with energy management systems for end-user applications.

Supercapacitors are also known as ultracapacitors or double-layer capacitors. They bridge the gap between capacitors and batteries. Supercapacitors display higher energy density than a conventional capacitor and higher power density than batteries. They have high cyclic stability, high power density, fast charging, and good rate capability. Supercapacitors are even replacing batteries or integrating with batteries to be used as a hybrid system [1, 2]. There are various types of supercapacitors based on charge storage mechanisms and components.

During charging/discharging of a supercapacitor, opposite charges are accumulated or adsorbed/desorbed on the electrode surface on the application of potential. This is also known as an electric double-layer formation. In pseudocapacitor, reversible redox reactions at the electrode–electrolyte interface occur, which leads to the generation of pseudocapacitance [3].

1.1.1 Differences Between Other Energy Storage Devices and Supercapacitors

The energy storage devices are used in various applications based on their properties. Fuel cell requires a continuous supply of fuel which is not needed in the capacitor, battery, or supercapacitor. The other three devices are to be charged as they discharge on usage. Supercapacitors have medium energy density and high power density when compared to the capacitor and other devices [4]. They have very high capacitance due to the electric double layer and pseudocapacitance. The cycle life of supercapacitor is high, as no chemical changes occur during the charging and discharging process, so there is no degrading effect. Table 1.1 lists the differences between the different types of energy storage devices.

Ragone plot is used to compare the performance of various energy storage devices. Power density (W kg^{-1}) is expressed on the *y*-axis, and energy density (Wh kg^{-1}) is expressed on the *x*-axis. The energy density of capacitors is the lowest, but it has the highest power density. Fuel cells have a higher energy density but undergo complex working mechanism to store charge. Batteries have high energy density but

Table 1.1 Differences between fuel cell, capacitor, battery, and supercapacitors [5, 6]

Fuel cell	Capacitor	Battery	Supercapacitor
Fuel needs to be provided	Electrostatic charge storage	Chemical storage	Surface charge storage
High energy density	Negligible energy density	High energy density	Low energy density
Varies on the type of fuel cell	High power density	Reactant diffusion, low power density	Medium power density
No capacitance	Capacitance present	Non-capacitive faradaic storage	Very high capacitance present
N.A. (produce electricity directly on fuel)	Highest cycle life	Low cycle life due to degradation	High cycle life

low power density. Supercapacitors have properties intermediate between capacitors and batteries. The performance characteristics of different types of supercapacitors are compared in the Ragone plot (Fig. 1.1).

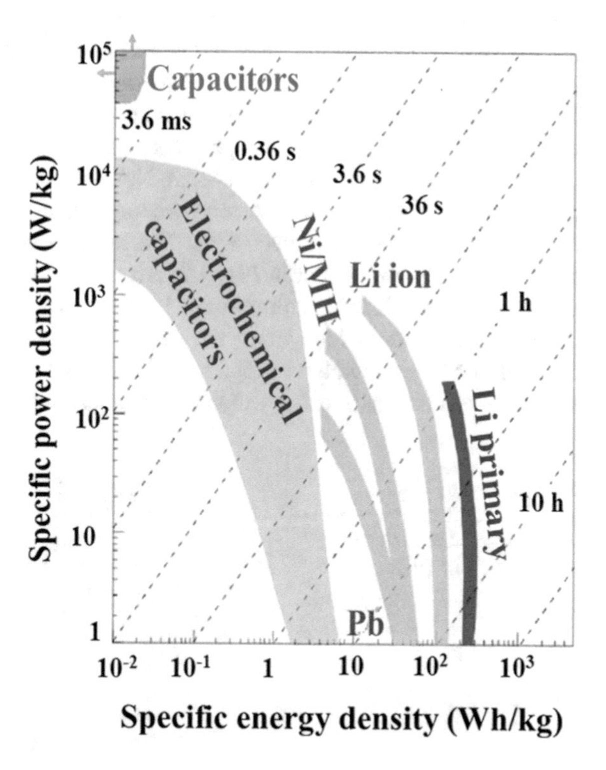

Fig. 1.1 Ragone plot of supercapacitors and other energy storage devices (redrawn and reprinted with permission from [7])

1.2 History of Supercapacitors

1740: Ewald Georg von Kleist constructed the first capacitor

1879: Hermann von Helmholtz predicts capacitive performance at solid and electrolyte interface

1957: Electric double-layer carbon capacitor patented by Becker, General Electric Corporation

1969: Sohio Corporation, Cleveland manufactured commercial electric double-layer capacitor (EDLC)

1978: NEC, Japanese MNC marketed double-layer capacitor as "Supercapacitor"

Early 1990: Conway group developed pseudocapacitance with an increase in energy density.

Further, there have been efforts on different types of supercapacitors with asymmetric electrodes, hybrid systems, or quantum supercapacitors [8].

1.3 Faradaic and Non-faradaic Processes

Non-faradaic process: The non-faradaic energy storage process involves no transfer of electronic or ionic charge in or at the surface of electrodes. During this process, ions undergo physical adsorption at the electrode surface. The non-faradaic process is highly reversible as no chemical change is involved during adsorption and desorption of ions. Non-faradaic charge storage takes place in an electric double-layer capacitor (EDLC) and intercalation between pores. Non-faradaic is the only process that takes place in an ideally polarized electrode, where no charge transfer takes place between the electrode interface and electrolyte. The various energy storage processes taking place in devices are shown in Fig. 1.2. An electric double-layer formation by the non-faradaic mechanism is represented in Fig. 1.3.

Faradaic process: It involves the transfer of charge from the electrode to the external circuit as depicted in Fig. 1.4. Mainly, redox reaction occurs between the electrode–electrolyte interface, and the reactants and products are stored in the bulk

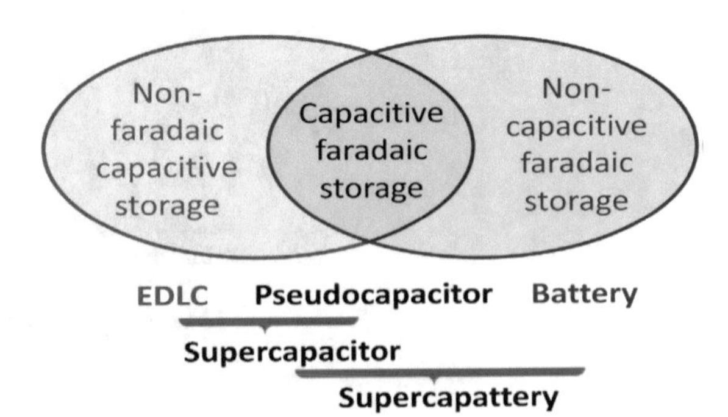

Fig. 1.2 Energy storage processes in different devices (redrawn and reprinted with permission from [9])

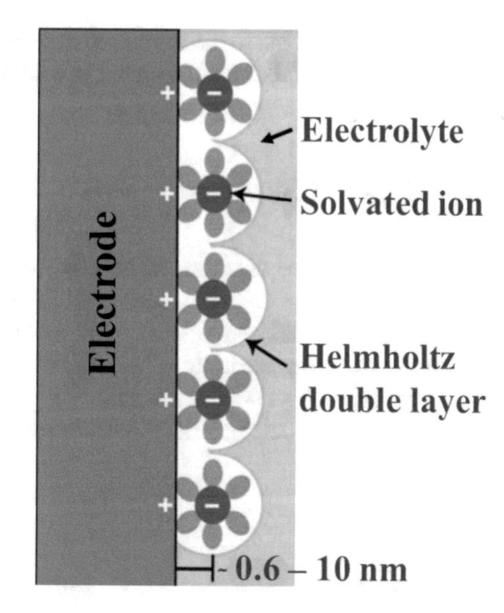

Fig. 1.3 Electrochemical double-layer capacitance due to the non-faradaic process (redrawn and reprinted with permission from [10])

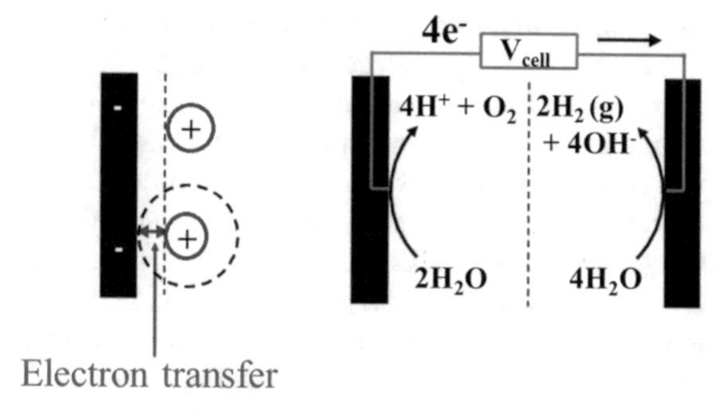

Fig. 1.4 Faradaic process showing charge transfer (redrawn and reprinted with permission from [12])

phases around the electrode acting as reservoirs. The charge in the faradaic process is not stored at the electrode but taken out of the electrode. Energy is stored in a supercapacitor as pseudocapacitance involves electrons transfer reactions due to oxidation and reduction at the electrode–electrolyte interface. In pseudocapacitor, capacitance is due to the capacitive faradaic process. There is also a non-capacitive faradaic process that is observed in batteries, where energy is stored by redox reactions and change in phase. The non-capacitive faradaic process occurs due to localized electrons transfer reactions at redox sites. However, the electrons get delocalized in capacitive storage as in pseudocapacitor electrode materials similar to semiconductors. This is explained by broadband formation due to the superposition of overlapping faradaic redox couples [9]. However, another study has stated that the origin of capacitance despite charge transfer in the faradaic process is due to the formation of an electrical double layer at the electrode–electrolyte interface. This indicates that faradaic reactions lead to the formation of conduction bands which give rise to ohmic conduction potential domain from an insulating state [11].

1.4 Types of Supercapacitors

Supercapacitors are categorized into five categories based on the type of energy storage mechanism or component used (a) EDLC stores energy at the electrode–electrolyte interface due to electrostatic forces, (b) pseudocapacitor utilizes faradaic processes, (c) asymmetric supercapacitors have the electrodes of two different types, (d) hybrid supercapacitors have a battery-type electrode and other a normal electrode, and (e) quantum supercapacitor is the latest research devices based on quantum effects (Fig. 1.5).

1.4.1 Electric Double-Layer Capacitor (EDLC)

In EDLC, charge storage occurs due to the non-faradaic process. There is a formation of a double layer at the electrode–electrolyte interface comprising layers of opposite charges. The formation of an electric double layer is due to electrostatic forces. The charge is stored by simple physisorption, and EDLC delivers high rate capability and power density. However, only the exposed electrode surface is involved to store charge, and EDLC exhibits poor energy density. To understand the phenomena of bilayer formation, different models are proposed as shown in Fig. 1.6. Helmholtz model proposed that a charged electrode in the electrolyte repels the same charged ions and attracts counter ions. This leads to the formation of an electric double layer. There are an inner Helmholtz plane and an outer Helmholtz plane as in Fig. 1.7. Gouy–Chapman's model took into account the mobility of ions and introduced the diffusion layer concept in the double layer. The Stern model combined the Helmholtz model and Gouy–Chapman model explains that the double layers consist of an adsorbed compact immobile ions layer referred to as the Stern layer and another diffusion layer [13].

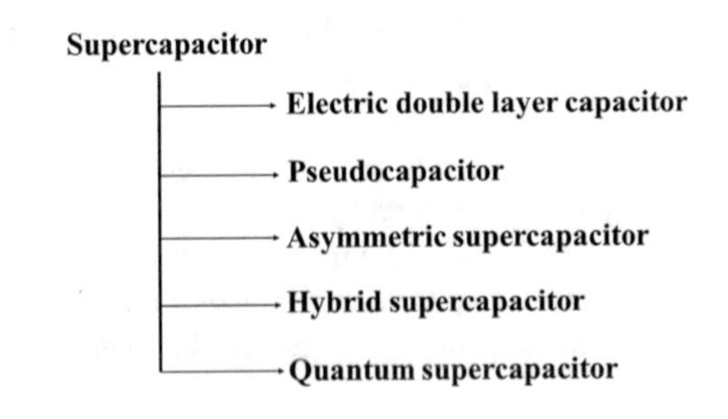

Fig. 1.5 Types of supercapacitors

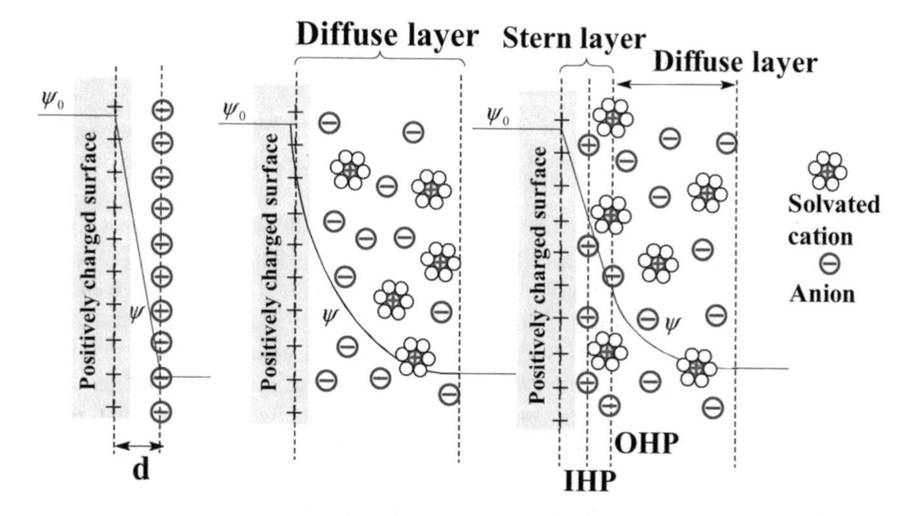

Fig. 1.6 Electric double-layer models. **a** Helmholtz model, **b** Gouy–Chapman model, **c** Stern model (redrawn and reprinted with permission from [14])

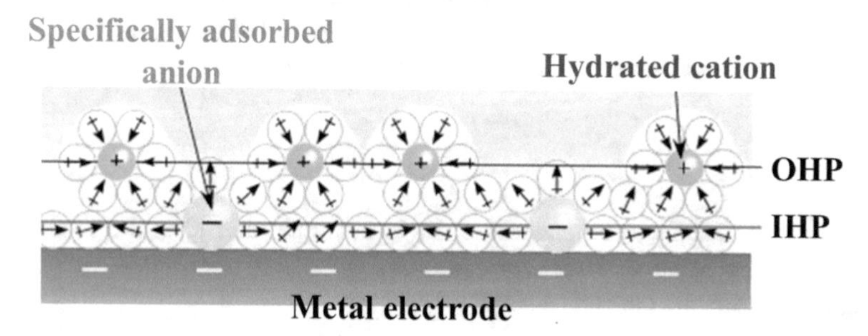

Fig. 1.7 Helmholtz planes. The spheres with arrows show water molecule, and the arrows show the direction of water dipole [IHP-inner Helmholtz plane (solvent layer) and OHP-outer Helmholtz plane (solvated ions)] (redrawn and reprinted with permission from [15])

D. C. Grahame named three regions in double layer as the inner Helmholtz plane (IHP), outer Helmholtz plane (OHP), and diffuse layer. The inner Helmholtz plane is formed by specifically adsorbed ions on the electrode. The outer Helmholtz plane passes through the center of nearest solvated ions, i.e., solvated ions at a distance closest approached to the surface of the electrode [14–16].

Gouy–Chapman or diffuse theory that takes into account the kinetic energy of the ions decides the thickness of the diffuse layer. It follows Boltzmann distribution for ions, where ion concentration is an important parameter. The Boltzmann distribution expresses the relative probability of a subsystem, a part of a physical system, in thermal equilibrium that has a certain energy. Equation 1.1 is the relation between the distribution of ions, the number density of ions, the energy to bring from infinity to 'r', and temperature.

$$n(r) = N \exp\left(-\frac{V(r)}{k_b T}\right) \quad (1.1)$$

where $n(r)$ is the distribution of ions, N is the number density of ions, $V(r)$ is the energy to bring from infinity to r, k_b is the Boltzmann constant, and T is the temperature.

This model assumes that the ions are point charges with electrostatic interaction, no polarizability, no chemical adsorption, IHP, and OHP which will not be present as they require finite ion size for polarization, metal has a planar surface with a surface charge density and ions follow Maxwell Boltzmann statistics. Stern modified the diffuse layer model with finite ion size which can form an electric double layer [17]. A drawback of these models is that the real charge distribution in a hierarchical pore structure in which varying pore size and shapes are not explained. The small pores block ions adsorption and prevent electric double-layer formation and capacitance decreases.

The working of an electric double-layer supercapacitor is simple. The electric double layer is formed on charging on positive and negative electrodes. Opposite charges get aligned to electrodes. On the application of load, the current flows, and potential in double layer decreases. The electrostatic force decreases between charges and the ions get randomly distributed in the electrolyte. This is shown in Fig. 1.8.

The electric double-layer capacitance is calculated by (1.2)

$$C = \frac{Q}{V} \quad (1.2)$$

where C is the EDL capacitance of a single electrode and Q is the total charge transferred at potential V.

Equation 1.3 is the current relation.

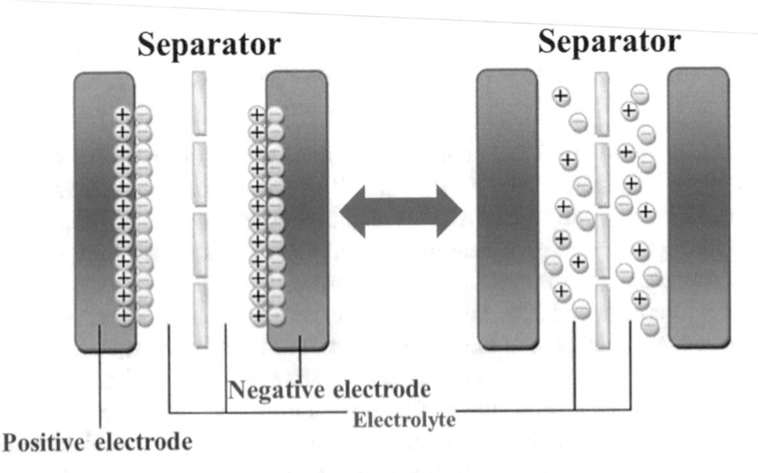

Fig. 1.8 Charge and discharge mechanism of EDLC supercapacitor (redrawn and reprinted with permission from [16])

$$I = \left(\frac{\mathrm{d}Q}{\mathrm{d}t}\right) = C\left(\frac{\mathrm{d}V}{\mathrm{d}t}\right) \tag{1.3}$$

where I is the response current and t is the charge time.

1.4.2 *Pseudocapacitors*

Pseudocapacitor store energy by fast and reversible charge transfer reactions at or near the electrode–electrolyte surface leading to pseudocapacitance. Pseudocapacitance is a faradaic process that involves the reduction-oxidation of electro-active species. The energy density of the pseudocapacitor is higher than the electric double-layer capacitor (EDLC).

The charge storage mechanism in pseudocapacitor is categorized into three types:

- underpotential deposition
- redox pseudocapacitance
- intercalation pseudocapacitance

1.4.2.1 Underpotential Deposition

This type of reaction is due to adsorption and reduction of ions to form a monolayer on the different electrode surfaces. An example of this process is the deposition of lead on gold. The interaction between Pb–Au is more favorable than Pb–Pb, so the deposition of lead over the gold surface can occur.

A monolayer of metal can be electrochemically deposited on a different metal substrate at a higher potential than required for bulk deposition [4]. The reason is the binding energy of adsorbed monolayer on the different substrates is more than the binding energy for depositing the same metal. This is underpotential deposition. Figure 1.9i allows bulk deposition as the required applied potential is less than required for dissociation in order to maintain a minimum energy state. The monolayer deposition on the different substrates is also allowed due to higher binding energy. Figure 1.9ii, iii shows no bulk deposition on the substrate as dissociation of metal is preferred due to low potential. The binding energy of monolayer deposition on the different substrates is also low as shown in Fig. 1.9ii. Figure 1.9iii shows that binding energy (Metal (Me)-Substrate (S)) is more, and monolayer formation is favorable.

1.4.2.2 Redox Pseudocapacitance

It involves the reduction and oxidation of redox-active species under certain potential. The ions are electrochemically adsorbed on or near the surface of the material. It is accompanied by faradaic charge transfer. Transition metal oxides and conducting polymers undergo a faradaic redox reaction to provide pseudocapacitance.

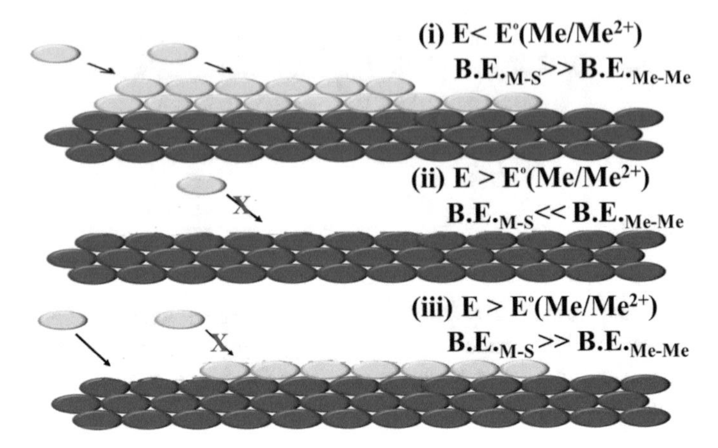

Fig. 1.9 Illustrations showing cases of monolayer deposition (reprinted and redrawn with permission from [18])

1.4.2.3 Intercalation Pseudocapacitance

This process occurs due to the intercalation of the ions into the layers and pores of electrode materials. There is a faradaic charge transfer process with no crystallographic phase change. The intercalation changes the metal valency to maintain the electrical neutrality of the device.

The quantitative study of pseudocapacitive material is studied by the amount of charge stored by the material.

The state of charge (q) depends on the quantitative value of the faradaic charge transferred on charging or discharging (Q) and the electrode potential (V). Pseudocapacitance is defined as the change in Q with respect to potential as in (1.4).

$$C_{\mathrm{Pseudo}} = \frac{\mathrm{d}Q}{\mathrm{d}V} \tag{1.4}$$

The current response at a fixed potential is due to the surface capacitive effect and diffusion-controlled insertion process shown in (1.5) [19].

$$I_{\mathrm{total}} = i_c + i_d = k_c v + k_d v^{0.5} \tag{1.5}$$

where $k_c v$ corresponds to the surface capacitive current and $k_d v^{0.5}$ is due to the diffusion-controlled intercalation process. The two terms indicate the contribution of different effects on the total current.

$$I_{\mathrm{total}} = a v^b \tag{1.6}$$

where a is an adjustable parameter and b is a variable term dependent on contributions from different processes. Equation 1.6 is the relation for total current based on diffusion or surface capacitive effect. $b = 0.5$ or 1 if the diffusion process or surface-capacitive effect dominates correspondingly. An important point is that $b = 1$ in the case of the battery may represent a fast redox kinetics comparable to pseudocapacitive materials instead of surface capacitive effects like in EDLC or pseudocapacitor. b value depends on the potential, sweep rate, charge storage mechanism, and type of material. $b > 0.5$ when the redox process is no longer limited by ion diffusion. The linear dependence of the charge storage over the potential window is an important method to differentiate between pseudocapacitive charge storage from battery behavior.

Ion insertion characteristics in pseudocapacitor can be classified into two categories, i.e., intrinsic and extrinsic materials. The charge storage properties of the intrinsic pseudocapacitor are not related to crystalline grain size and morphology. Intrinsic pseudocapacitive materials have capacitive charge storage characteristics irrespective of crystalline properties, morphology, or particle size [20]. Examples are MnO_2, RuO_2, polypyrrole, polyaniline, α-MoO_3. For extrinsic pseudocapacitor, it uses battery-type material which shows battery-type behavior in bulk, but on decreasing particle size to the nanolevel, these materials demonstrate pseudocapacitive behavior. So, depending on the particle size, extrinsic material can exhibit battery-type or pseudocapacitive charge storage mechanism. The overall process of ion intercalation can be divided into three parts: (i) faradaic contributions from the bulk solid-state diffusion dominated by ion intercalation, (ii) fast ion diffusion dynamics by faradaic charge transfer process at the surface of active material, and (iii) electrostatic adsorption and desorption of ions depicting non-faradaic EDLC contribution. Nanostructured materials have short ion and electron transport paths. An example is bulk $LiCoO_2$ which shows redox peaks battery-type behavior, but as the particle size is reduced to 6 nm, intercalation increases, and there is a transition to pseudocapacitive behavior. Capacitive behavior dominates with decreasing crystallite size.

1.4.3 Asymmetric Supercapacitors

Supercapacitor displays poor energy density. One major limitation is the low working potential window. Asymmetric supercapacitors are designed to utilize both types of electrode materials in order to expand the operating potential window. It is well known that electric double-layer capacitors have high power density, but poor energy density and pseudocapacitors have better capacitance and energy density but lack high power density and long cycle life. The asymmetric assembly utilizes two dissimilar types of electrodes. Preferably, an electrode is an electric double-layer capacitor type, and the other one is pseudocapacitor type. The potential difference between two electrodes is utilized to expand the operating potential of the device.

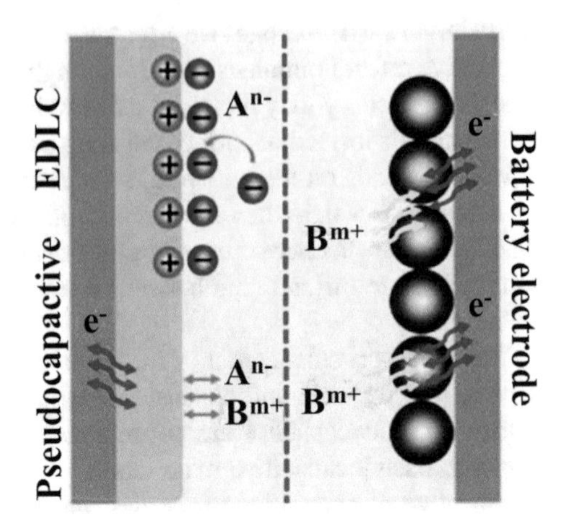

Fig. 1.10 Energy storage mechanism in hybrid asymmetric supercapacitor (redrawn and reprinted with permission from [23])

1.4.4 Hybrid Supercapacitors

Hybrid supercapacitors came into existence to focus on enhancing energy density without compromising its power density. The difference between asymmetric supercapacitor and hybrid supercapacitor is that latter uses a battery (faradaic)-type electrode with a capacitive (non-faradaic, EDLC)-type electrode as shown in Fig. 1.10 [21]. The working potential range is also increased in hybrid supercapacitors than symmetric supercapacitors. Lithium-ion batteries are the most widely used devices in the electronics sector. They have exceptional energy density. Combining the lithium-ion battery electrode with the capacitor-type electrode has both the advantages of higher energy density and power density. Sodium-ion hybrid supercapacitors are also an important area of research. Composite electrode-based supercapacitors are also classified as hybrid supercapacitors [22]. These can be symmetric or asymmetric electrodes. The composite electrode is a combination of EDLC and pseudocapacitor electrode materials. The composite electrode aims to address the limitation of the individual electrode materials in order to provide enhanced energy density, power density, and longer cycle life.

1.4.5 Quantum Supercapacitors

Low-dimensional materials exhibit quantum effects such as exchange and correlation energy with long-range coulombic interactions. The capacitance provided by an electronic system to a mesoscopic device (C_e) is shown in (1.7).

$$\frac{1}{C_e} = \left(\frac{1}{C_g}\right) + \left(\frac{1}{C_q}\right) \tag{1.7}$$

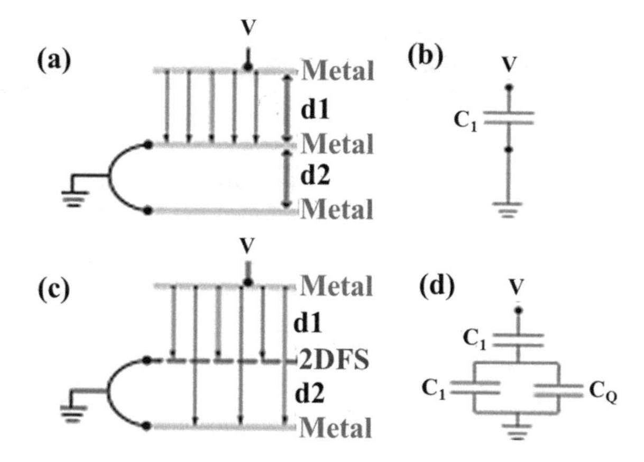

Fig. 1.11 **a** A parallel plate capacitor with **a** metal plate inserted, **b** the circuit model for (**a**), **c** a parallel plate capacitor with a 2D charge (fermion, 2DFS) system inserted, and **d** equivalent circuit model for (**c**) having a quantum capacitor C_Q (redrawn and reprinted with permission from [25])

where C_g is the normal geometric capacitance and C_q is the quantum capacitance.

The total capacitance decreases due to positive quantum capacitance. Some materials exhibit negative exchange and correlation energy, which is more than positive kinetic energy. This leads to quantum capacitance as a negative value and the total capacitance increases greater than geometric capacitance. The enhancement in capacitance has been found in the interface between $LaAlO_3/SrTiO_3$, two-dimensional (2D) monolayers of WSe_2, and graphene-MoS_2 heterostructures [24]. Figure 1.11 shows a schematic of normal and quantum capacitors.

1.5 Hybrid Energy Storage Systems (HESS)

Batteries have higher energy density but low power density. Supercapacitors have low energy density and higher power density. The drawbacks of these two energy storage devices are compensated by integrating them together known as hybrid energy storage systems. The life of a battery gets lowered by the large fluctuating current in and out of the battery due to the generation of heat and an increase in the internal resistance of the battery. The hybrid energy storage management system has two important functions (a) to minimize the variations of the current and their magnitude while charging or discharging and (b) to reduce the energy loss of the connected supercapacitors. The batteries and supercapacitors are connected to hybrid energy storage systems in various configurations. The various battery supercapacitor hybrid energy storage system topologies are shown in Fig. 1.12.

(i) Passive HESS

Passive topology is simple, least expensive, and requires no control system as in Fig. 1.13. The usable power capability of the energy storage devices depends on the state of charge, which depends upon voltage level and charge/discharge current limit. The batteries generally in these systems have a relatively low depth of discharge.

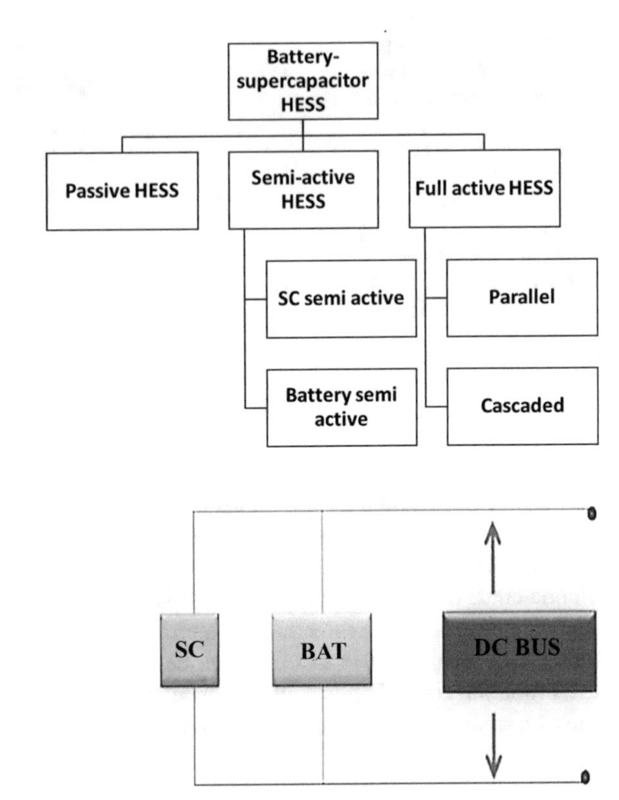

Fig. 1.12 Various battery supercapacitor hybrid energy storage system topologies

Fig. 1.13 Passive topology (redrawn and reprinted with permission from [27])

The battery and supercapacitors are directly connected to the DC bus with the same voltage. The battery state of charge defines the voltage [26]. The supercapacitor is underutilized due to the restrained voltage limit. The high power density of supercapacitors is compromised as the capability to easily charge and discharge is hindered.

(ii) Semi-active HESS

A semi-active hybrid energy storage system is most widely used due to better energy utilization than passive HESS and at a lower cost than active HESS. The semi-active topology is of two types depending on the controlled or interfaced energy storage device with DC bus [28]. Only one storage device is connected to the DC bus via unidirectional or bidirectional DC-to-DC converter. DC-to-DC converter is a device that converts one voltage level to another. Supercapacitor semi-active HESS topology has supercapacitor connected to DC bus via DC-to-DC converter. The supercapacitor is isolated from the battery, and the DC-to-DC converter allows a wider range of permissible operating voltage levels for the supercapacitor. The supercapacitor is efficiently utilized. The battery is directly connected to the DC bus that results in a

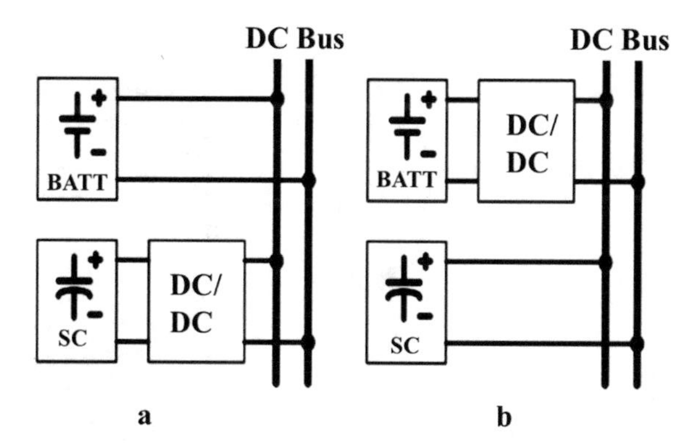

Fig. 1.14 **a** supercapacitor semi-active HESS and **b** battery semi-active HESS (redrawn and reprinted with permission from [29])

stable voltage level. The direct connection of the battery to the DC bus may result in exposure to changing currents in the circuit decreasing the lifespan of batteries.

Battery semi-active HESS topology has DC-to-DC converter connected between the battery and DC bus. The supercapacitor is passively directly connected to the DC bus. The advantage is that the battery is protected against fluctuating currents. The supercapacitor has reduced efficiency due to the restricted operating voltage range. The charging and discharging current from the supercapacitor leads to fluctuations in DC bus possibly resulting in system instability and performance degradation. Both semi-active topologies are illustrated in Fig. 1.14.

(iii) Active HESS

The active HESS topology consists of both controlled battery and supercapacitor [29]. There are cascaded full active topology and parallel topology with different configurations as in Fig. 1.15. The full active HESS increases system performance and stability. The drawbacks are higher cost due to more converters, bigger size, and an increase in energy management complexity. The full active HESS leads to an increase in battery life by protection against current fluctuations. The freedom of a wider operating voltage range leads to the efficient use of supercapacitors due to a wider state of charge. The battery has higher energy density, and energy storage management can be programmed to meet low-frequency power exchange between battery and DC bus. Supercapacitors can be controlled to respond to high-frequency power increase and regulate DC bus voltage. The system efficiency decreases with an increase in the DC-to-DC converter due to losses.

An example of the active use of hybrid HESS is in electric vehicles as shown in Fig. 1.16. Vehicle to grid or vehicle to home is an emerging concept by which electric cars are built. These cars can be integrated with solar panels or piezoelectric energy harvesting. The harvested energy can be stored or used for charging the vehicle, which can be consumed on driving. But in an ideal state, this excess stored energy can be transferred to the grid or households. The magnitude of energy is large if we consider

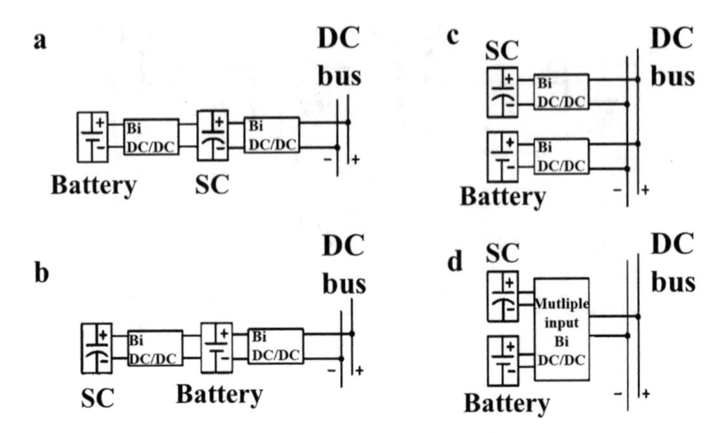

Fig. 1.15 **a** cascaded full active topology 1, **b** cascaded full active topology 2, **c** parallel full active topology, and **d** multiple input full active topology (redrawn and reprinted with permission from [30])

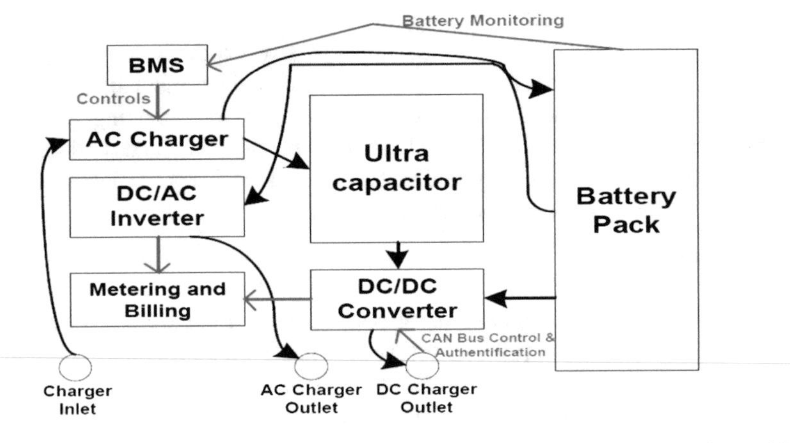

Fig. 1.16 Hybrid energy storage system based on the above-mentioned topology for mobile charging station (reprinted and redrawn with permission from [32])

the lakhs of electric cars that are projected in the market. There is a technology that allows the energy storage system to have a requisite amount of energy stored for driving, and the rest is excess energy that can be transmitted. The charging station is the main point, where the vehicle-to-grid or vehicle-to-home transmission is made possible. There are fixed charging stations and a mobile charging station. There is an off-grid charging mode and an on-grid charging mode. During on-grid charging mode, the charging station is directly connected to the grid. The off-grid charging uses the stored energy platform for electric vehicle charging [31].

1.6 On-Chip Supercapacitors

Texas Instruments have developed a supercapacitor manager, which is an integrated single-chip solution. It can manage charge control, monitor, and protect supercapacitors with individual and collective monitoring and cell balancing capability. These types of miniaturized energy management systems coupled with on-chip supercapacitors can act as a revolutionary stage in electronics, wireless devices, security sensors, and implantable medical devices.

The components of a micro-supercapacitor are the same as a normal supercapacitor. The left schematic in Fig. 1.17a is the stacked configuration. A substrate is taken, and the current collector is deposited on it. Subsequently, first electrode material, then solid electrolyte to prevent leakage or any other electrolyte as per requirements is deposited. Next counter electrode material and again current collector is deposited. The right schematic of Fig. 1.17b is a planar configuration. A current collector is deposited on the insulator substrate in a pattern form. The positive and negative electrodes are deposited on the current collector in an alternating fashion. A liquid electrolyte submerging the electrodes or a solid electrolyte over the electrodes is used.

The fabrication of micro-supercapacitors is a complex process that requires no short circuit between electrodes, compatibility with existing technology, and high-resolution patterning. The fabrication of planar on-chip micro-supercapacitors is widely assembled by two methods. The first method includes a powder form of electrode material that is converted into a usable viscous paste. This paste is deposited on the patterned current collectors using precision patterning. The patterned current collector may be formed using lithography techniques and selective etching. The second method is the deposition of a material on the current collector and chemical transformation of this coated material. An electrolytic deposition is also possible,

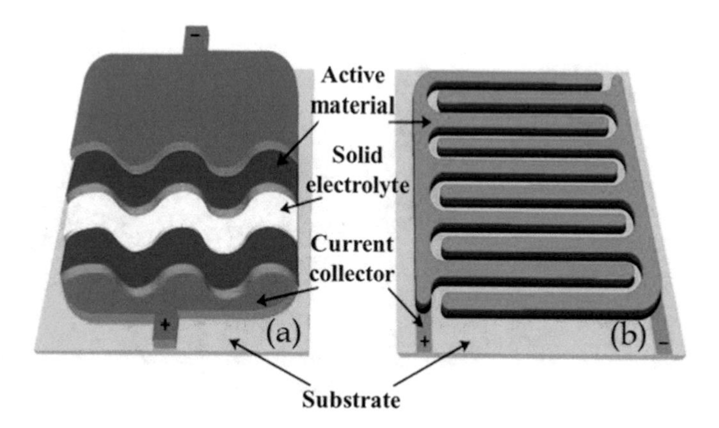

Fig. 1.17 Micro-supercapacitors configuration for on-chip integration, **a** stacked configuration and **b** planar configuration (redrawn and reprinted with permission from [33])

where electrode formation is from solution constituents. The encapsulation of on-chip supercapacitors is an important process.

The drawback of 2D limited planar or stack micro-supercapacitors is that energy density is very low due to thin electrodes that hold less charge. 3D micro-supercapacitors have been developed to increase energy density. Nanofiber, nanotube, and the nanowire morphology-based material can be used as electrodes and serve as 3D device components with a high surface area. Pseudocapacitors can also be utilized by depositing capacitive redox-active materials between arrays of deposited vertical structures like nanotubes or nanowires on the substrate. The gravimetric capacitance of micro-supercapacitors is not used as the weight of the active electrode material is low in micro-supercapacitors, so a real or volumetric capacitance is calculated for efficiency and performance comparison [34].

The fabrication of multiple micro-supercapacitors is done by a laser scribe method. In this process, a laser is used to make interdigitated planar electrode structures on the substrate. Copper is the commonly used current collector. Kapton tape is used as electrical insulation and protective layer. The electrolyte is deposited between the electrodes. After the deposition of all the components, the micro-supercapacitor is sealed or encapsulated for use as an on-chip component.

1.7 Components of Supercapacitors

A supercapacitor consists of four primary components as shown in Fig. 1.18. Among all the components, electrodes and electrolytes are active components, which are involved directly in the charge storage mechanism. The other essential components include separator and current collector. The following section discusses each component.

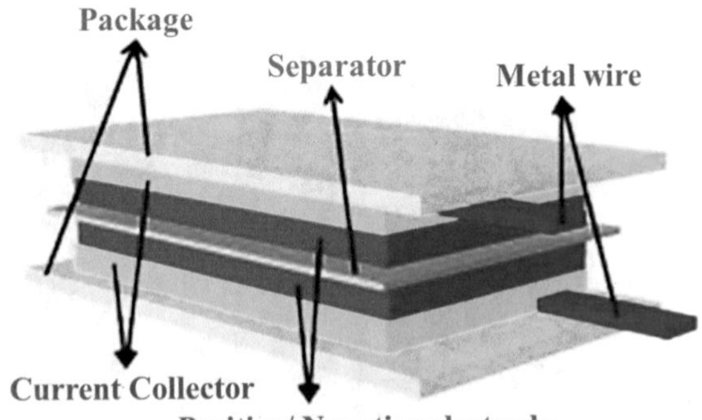

Fig. 1.18 Components of a supercapacitor (redrawn and reprinted with permission from [35])

1.7.1 Electrodes

Electrodes form an electric double layer or undergo redox reactions due to applied potential. The energy is stored electrostatically, electrochemically, or both. The electrodes should have good electrical conductivity, large surface area for better electrolyte interaction and increased specific capacitance, and wide pore size distribution as it increases surface area and redox activity [36]. In order to obtain a high surface area, nanostructured materials are used as electrodes. The different types of material used in supercapacitor electrodes are mentioned below. These materials are also used in combination with each other and referred to as nanocomposite materials [37]. The thickness of the electrode at higher mass loading is an important parameter. The difference in the volume of charged and discharged states develops mechanical strain in the thicker electrode. This disintegrates the electrode and reduces cyclic performance. Some of the widely used electrode materials are discussed in the next section.

1.7.1.1 Carbon-Based Materials

Activated carbon, carbon nanotubes, carbon nanofibers, carbon aerogel, template-mediated porous carbon materials, and graphene are chemically stable, electrical conducting, large surface area, low fabrication cost, environmentally friendly, and can withstand high temperatures. These materials can be easily functionalized, which lead to the incorporation of heteroatoms and functional group in carbon structure for attaining pseudocapacitance.

Activated Carbon

The properties of activated carbon are determined by various parameters:

- Raw materials: Activated carbon is generally synthesized from organic materials. The raw material should have high carbon content, low inorganic or ash content, high density, low cost, and stability on storage.
- Surface area and pore size distribution: This factor depends on the temperature and activating agent. The raw material is decomposed at high temperatures and increases crystallinity. However, a large amount of crystallinity decreases surface area due to densification.
- Graphitization: Annealing happens at higher temperatures reducing defects and increase in graphitization. This leads to a reduction of heteroatom content and a decrease in carrier concentration and pseudocapacitance. Graphitization increases electrical conductivity.
- The surface concentration of heteroatoms.
- Type of functional groups.

- Activation time: An increase in activation time reduces the yield of activated carbon due to the volatility of organic material, but the BET surface is enhanced.
- Activation temperature: The BET surface area increases with an increase in activation temperature due to the liberation of volatile matter. This results in new pores and the widening of existing pores. There is a reduction in the amount of activation carbon produced with an increase in temperature due to the reduction of volatile raw material.

Physical activation involves the pyrolysis of carbon in steam, CO_2, and air at 700–1200 °C. Chemical activation is done at a temperature range of 400–1000 °C under the influence of phosphoric acid, KOH, NaOH, and zinc chloride. Chemical activation provides higher carbon yield, higher surface area, and microporosity than physical activation.

Activated carbon doped with nitrogen or oxygen has better ion adsorption and transportation. The reason is the electronegativity difference between doped and carbon atoms, which leads to polarization and increases the wettability of the electrode surface.

The porosity of activated carbon depends on the activating agent, activation temperature and time, gas flow rate, gases used during pyrolysis and activation, solution or mechanical mixing, and heating rate. Macropores act as electrolyte buffering reservoirs providing ion transport into the interior of the active material, and mesopores provide a diffusion channel for electrolyte ions in their interior and micropores. Micropores have capacitive contribution due to charge accumulation as a result of controlled diffusion and molecular sieve effect.

Activated carbon is used in supercapacitor because of its high surface area, high electrical conductivity, easy availability, and moderate cost. Activated carbon displays a three-dimensional porous structure with a high specific surface area and excellent electronic conductivity at a moderate cost. The electrochemical performances of recently synthesized biomass-based activated carbons are presented elsewhere [38]. However, by tailoring the synthesis route to (a) reduce electrode dead volume by modifying morphology, (b) enhance conductivity, (c) incorporate heteroatom content, and (d) increase surface area, the electrochemical properties can be greatly enhanced [8, 39]. In this vein, peach-gum-derived 3D continuous porous activated carbon network shows a high surface area of 1535 $m^2\ g^{-1}$ displaying specific capacitance of 406 $F\ g^{-1}$ [40]. However, ginger-derived carbon nanosheet displays a specific surface area of 720 $m^2\ g^{-1}$ that delivers excellent specific capacitance of 456 $F\ g^{-1}$ [41].

Graphene/Reduced Graphene Oxide

Graphene is one atomic thick sp^2 hybridized carbon arranged in a honeycomb sheet-like structure. It has high electrical conductivity, excellent chemical stability, high surface area, and light weight. It finds wide application in modern electronic devices due to its high conductivity and flexibility [42]. In a supercapacitor device, graphene has been widely used as an electrode material due to its large surface area and extraordinary high electrical conductivity, and excellent chemical and thermal stability. Graphene is synthesized using mechanical exfoliation, chemical vapor deposition, and chemical derived methods. Depending upon the synthesis technique, the morphology of graphene can be tuned to different dimensions, such as 0D graphene quantum dots to 1D graphene nanofiber to 3D graphene aerogel, hydrogel, and sponges. Since, graphene stores charges via a capacitive mechanism, it shows excellent cyclic stability. The capacitance value ranges from 210 F g^{-1} at a scan rate of 1 mV s^{-1} in particle morphology [43] to 441 F g^{-1} at a current density of 1 A g^{-1} in 3D hydrogel [44]. Nowadays, graphene-based hybrid electrode materials have emerged as promising electrode material for assembling high-performance devices. The further details of graphene as discussed elsewhere [45].

Carbon Nanotubes (CNTs)

CNTs are carbon nanomaterials, displaying tube-shaped carbon materials with tube diameter in the nanometer range. The CNT's structure is formed by rolling graphitic layer, known as MWCNTs or graphene sheet, known as SWCNTs. CNTs have high electrical and thermal conductivity, good mechanical strength and porosity, and high surface area. For the supercapacitor application, CNT based electrode shows excellent performance due to the high surface area, excellent conductivity, and resistance-free electron transfer. The tube structure enables high rate capability since it acts as a highway, which provides fast charge transportation. Due to its high mechanical resilience, CNTs are widely integrated with pseudocapacitive material, where it not only undergoes charge storage, but also provides mechanical support and conductive pathway to the composite electrode. CNT with a specific surface area of 430 m^2 g^{-1} displays a specific capacitance of 102 F g^{-1} [46]. The details of the CNT are discussed elsewhere [47, 48].

Carbon Nanofibers

Carbon nanofibers are one-dimensional carbon material. It can be classified into three categories as (a) ribbon or tubular carbon nanofiber, (b) herringbone carbon nanofibers, and (c) platelet carbon nanofibers [49]. Carbon nanofibers are synthesized in many ways, but chemical vapor deposition and electrospinning are the preferred methods. The synthesis methods lead to various configurations of carbon nanofiber. The graphite layers are stacked in various arrangements that results in different carbon nanofibers. There are weak Van der Waals forces between stacked graphite layers. Carbon nanofibers are weaker than carbon nanotubes. The nanofiber has a high surface area and porosity making it suitable for electrode applications. It undergoes the EDLC charge storage mechanism. High surface area, excellent conductivity, and tunable porosity make it a suitable electrode for supercapacitor devices. Zinc glycolate fibers derived carbon nanofibers exhibit a specific surface area of 1725 m^2

g^{-1}, with the dominance of mesopores delivering specific capacitance of 280 F g^{-1} at a current density of 0.5 A g^{-1} [50]. The details of the carbon nanofiber are discussed elsewhere [51].

1.7.1.2 Conducting Polymers

Besides carbon materials and metal oxides, conducting polymers possess many advantages for electrode materials. Conducting polymers have high conductivity in the doped state, high operating potential window, and adjustable redox activity via chemical modification. Many conducting polymers have low cost and low environmental impact. The charge storage mechanism of conducting polymers is via a redox reaction. During the oxidation reaction, ions get transferred to the polymer backbone. When the reduction process occurs, ions get released from the polymer backbone into the electrolyte. The capacitance value achieved is much higher than other electrode materials because redox reaction occurs throughout the material, not just concentrate on the surface. Charging and discharging in conducting polymer do not involve any structural changes, and these processes are highly reversible, which increase the rate capability of the device. Depending on the type of ion insertion, conducting polymers can be charged positively or negatively. The conductivity of conducting polymer can be altered during reduction and oxidation processes that generates delocalized n electrons into the polymeric chain. During the oxidation process, polymers get positively charged and termed as p-type, while negatively charged polymers are prepared via reduction reactions and termed as n-type.

Commonly used conducting polymers as supercapacitor electrode material are polyaniline (PANI), polythiophene (PTh), polypyrrole (PPy), and their corresponding derivatives. p-doped PANI and PPy are used as its n-doped state show very less potential in an electrolyte solution. Oxidation induces doping (p-doping) ions from the electrolyte which are transferred to the polymer backbone. De-doping/undoping happens on a reduction, where the ions are released back into the solution [52]. Generally, the stability of p-doped polymers is more than n-doped polymers possibly due to low electronic repulsion. Equation 1.8 is the doping/de-doping reaction mechanism of polypyrrole.

$$\mathrm{PPy^{+}A^{-}} + \mathrm{e^{-}} \leftrightarrow \mathrm{PPy} + \mathrm{A^{-}} \tag{1.8}$$

Another polymer, PTh, and its derivatives can be doped in both states. In general, conducting polymers show poor conductivity in the n-doped state, i.e., in negative working potential. However, conducting polymers can work in a defined potential window, and beyond its working potential, the polymer may degrade or switch to an insulating and un-doped state. So, the selection of suitable potential is essential for conducting polymers. The details of the conducting polymers are discussed elsewhere [53, 54].

1.7.1.3 Transition Metal Oxides

Metal oxides provide higher energy density than the EDLC and better electrochemical stability than polymers. Metal oxides are cheap and easy to synthesize. Various nanostructured morphology of single metal oxides can be obtained by varying the synthesis condition. It stores charges electrostatically and faradaically. Pseudocapacitive metal oxide should have [52]:

- High electronic conductivity: For uninterrupted transportation of electrons
- Multiple oxidations state: Metal can exist in two or more oxidation states due to its multiple valences. This enables them to coexist over a continuous range with no phase change.
- Facile interconversion of $O^{2-} \leftrightarrow OH^-$: to facilitate free intercalation/deintercalation of protons into the oxide lattice on reduction/oxidation.

Various metal oxides have been studied for pseudocapacitive electrode material among which RuO_2, MnO_2, $NiO_2/Ni(OH)_2$, $CoxO_{1-x}$, and $NiCo_2O_4$ are well known. The theoretical capacitance value of transition metal oxides is much higher than carbon-based materials. However, poor electrical conductivity and low cyclic stability limit their advanced application. In this regard, various composite electrodes using metal oxides have been prepared, which provide a synergistic contribution to deliver high-performance electrode. The details of the transition metal oxides are discussed elsewhere [55, 56].

1.7.1.4 Metal-Organic Frameworks

These are compounds consisting of metal ions or clusters coordinated to organic ligands forming 1D, 2D, or 3D structures. The various synthesis routes are diffusion method, hydrothermal, electrochemical, mechanochemical, and microwave-assisted method [57]. These are porous, molecular-level tunable frameworks, lightweight, and high surface area. But they have low electronic conductivity. The details of the metal-organic frameworks are discussed elsewhere [58].

1.7.1.5 Covalent Organic Frameworks

The framework is composed of lightweight elements like carbon, boron, oxygen with covalent bonds. The synthesis methods are solvothermal, ionothermal, microwave synthesis, sonochemical, mechanochemical, and light-promoted synthesis. These have a high surface area, controllable pore size, and flexibility. The details of the covalent organic frameworks are discussed elsewhere [59].

1.7.1.6 MXenes

This is a 2D few atomic layers thick material structure composed of transition metal carbide, nitride, and carbonitrides [60]. They have high electronic conductivity, good mechanical strength, and hydrophilicity. The synthesis of MXenes involves etching A layers from the MAX phases as M-A bonds are weaker compared to M-X bonds. The general chemical formula of MAX phases is $M_{n+1}AX_n$, *where M is an early transition metal, A is a* group IIIA or IVA element, X is C and/or N, and $n = 1, 2, 3$. MAX phases alternately stacked $M_{n+1}X_n$ units and the A layers. After etching, the left layer is weakly bonded $M_{n+1}X_n$ layers. These layers are known as MXenes that can be functionalized or delaminated by sonication. The surfaces of $M_{n+1}X_n$ units have functional groups such as oxygen, hydroxyl, or fluorine. MXenes chemical formula is $M_{n+1}X_nT_x$, where T_x represents the surface functional groups. This is shown in Fig. 1.19. The details of the MXenes are discussed elsewhere [61].

Metal nitrides

Molybdenum nitride, titanium nitride, vanadium nitride, niobium nitride, ruthenium nitride, chromium nitride, cobalt nitride, lanthanum nitride, nickel nitride, gallium nitride, iron nitride, and tungsten nitride are used as electrodes. The synthesis method involves the formation of metal oxide nanostructures from metal oxide precursors. The nanostructures are annealed in an ammonia or nitrogen environment to yield metal nitride nanostructures. The sputtering of metal source with ammonia or nitrogen also synthesizes metal nitride. These are electronic conductive and serve as the mechanical backbone. The details of the metal nitrides are discussed elsewhere [63].

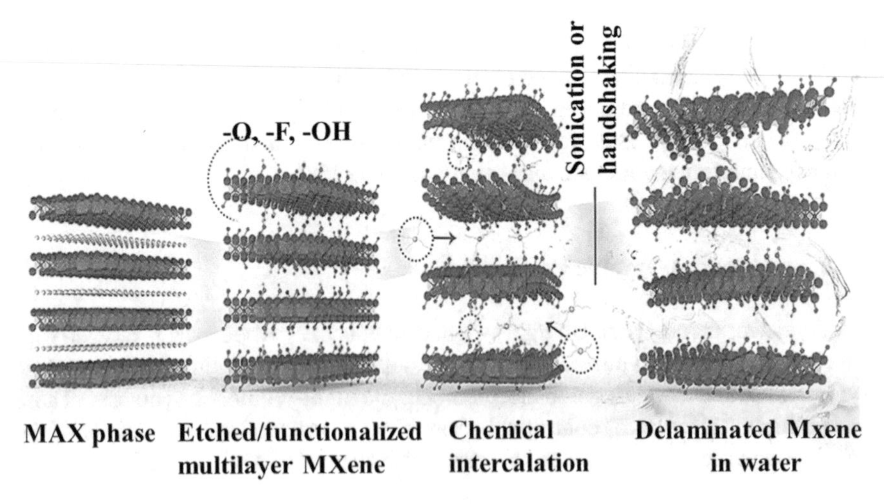

Fig. 1.19 MXene formation and delamination (redrawn and reprinted with permission from [62])

1.7.1.7 Black Phosphorous

It has 2D puckered structure with high surface area, electronic mobility, and good mechanical strength. The structure of black phosphorous is similar to graphite. Black phosphorous is synthesized from white phosphorous. It has a high theoretical capacitance. The details of the black phosphorous are discussed elsewhere [64].

1.7.2 Electrolytes

The electrolyte is a compound that produces solvated ions when dissolved in a solution. The electrolyte is responsible for transferring and balancing the charges between the two electrodes. They can enhance the energy density of supercapacitors, improve cycle life, minimize internal resistance, control operational temperature, and decrease the self-discharge process. The power density of supercapacitors depends on the resistance of the electrolyte, if the resistance is high, the power density will be low.

The electrolyte should have a high potential window, high ionic conductivity, and chemical stability, compatible with the electrode materials, high operating temperature range, non-toxic, optimum viscosity, and low cost [65]. The performance of electrolytes depends on the ion size, hydrated ion size, corrosive behavior, mobility, and ion conductivity. The conductivity of the electrolyte depends on the mobility of ions, concentration of ions, elementary charge, and valence of mobile ion charges. The number of free ions determines the ionic conductivity, which implies different salt concentration in the same solvent has different conductivity. The electrochemical stability depends on the interaction of electrolyte with electrodes and electrolyte composition [52]. Figure 1.20 shows the different types of electrolytes.

1.7.2.1 Aqueous Electrolytes

Aqueous electrolytes are highly conducting, which enhances the power density of the device. It is cost-effective as water is used as a solvent, ease of handling, and simplified preparation procedure. The major limitations are low potential window as the water molecules start to decompose on the potential above 1.23 V, narrow operating temperature range, and possible leakage. Aqueous electrolytes are mainly categorized as acidic (H_2SO_4), alkaline (KOH, NaOH, LiOH), and neutral (Na_2SO_4, K_2SO_4, Li_2SO_4, $NaNO_3$, KCl). Acid and alkaline electrolytes give high capacitance in comparison with the neutral electrolyte due to the high mobility of H^+ and OH^-. However, neutral electrolytes undergo the faradaic process. Due to higher ionic concentration and lower resistance, the aqueous electrolyte solution may have a higher capacitance and more power density than an organic electrolyte-based supercapacitor [67].

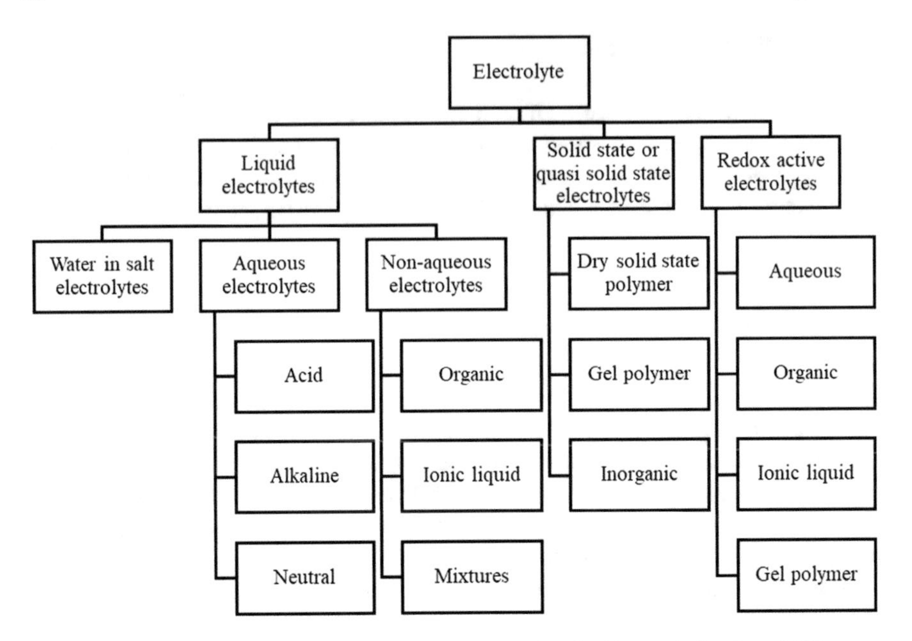

Fig. 1.20 Different types of electrolyte (redrawn and reprinted with permission from [66])

1.7.2.2 Non-aqueous Electrolytes

These are higher working potential window, robust but high cost, toxic, and low conductivity electrolytes. There are different types of non-aqueous electrolytes like organic electrolytes, ionic liquid, solid-state or quasi solid-state electrolyte, solid/dry polymer electrolyte, gel polymer electrolyte, or inorganic solid-state electrolytes, polyelectrolytes, and redox-active electrolytes, as discussed below.

Organic Electrolytes

These are formed by dissolving a conducting salt in an organic solvent. Organic electrolytes have a high cost, lower specific capacitance, higher internal resistance, low ionic conductivity, highly flammable, toxic, difficult fabrication, degradation in capacitance value, and self-discharge. Tetraethyl ammonium tetrafluoroborate and triethylmethyl ammonium tetrafluoroborate in acetonitrile/propylene carbonate solvents are commonly used organic electrolytes [68]. The electric double-layer capacitor performance of these electrolytes is not good due to large solvated ions and low dielectric constant.

Generally, low viscosity solvent results in more conducting electrolytes than a high viscous solvent. But sometimes, deviation from this behavior is noted because viscosity is associated with dipole moment and molecular interaction of the solvent. The dielectric constant (ε) of the solvent is also associated with these two parameters.

A higher dielectric constant reduces ion-pairing that improves the conductivity of a given salt. Lower dielectric constant if associated with low viscosity solvents that are due to weak intermolecular interactions tends to reduce the conductivity. Mixed solvents optimize the conductance.

Ionic Liquids

Ionic liquids are made of salts that are liquid at room temperature or slightly above, i.e., less than 100 °C. They usually have an asymmetric organic cation weakly coordinated with an inorganic/organic anion. They have a low melting point or are liquid at low temperatures because of the large size of organic molecules that prevents close packing. The conformational flexibility of the ions changes due to external factors and prevents rigid structure and low melting point [69]. Molten salts have a higher melting point.

Ionic liquids are dissolved in an organic solvent or taken in pure form. The cations are phosphonium, sulfonium, pyrrolidinium, and ammonium, and anions are tetrafluoroborate, dicyanamide, and bis(trifluoromethanesulfonyl)imide. These have higher ionic conductivity, wider cell voltage, but low capacitance, high cost, and high viscosity. The larger ion size limits the interaction with electrodes. Ionic liquids are usually neutral at room temperature, but the electrolytes have free or paired ions. The high potential window results in high energy and power density. Heterocyclic cations form thermally stable ionic liquid due to resonance stabilization. The thermal stability of ionic liquid is governed by intermolecular interactions like hydrogen bonds, coulombic interactions between anionic and cationic charges, and van der Waals interactions between the side chains.

1.7.2.3 Solid-State or Quasi Solid-State Electrolytes

These have high electrochemical stability but low ionic conductivity. The electrolyte contains a polymer matrix embedded with a liquid electrolyte. PVA is widely used as a polymer matrix as it is hydrophilic in nature, involves simple synthesis, non-toxic, and cost-effective. Polythiophene, polyaniline, and polypyrrole are also used as the polymer matrix.

Solid/Dry Polymer Electrolytes

The solid polymer matrix holds salt, and there is no solvent. Conductivity depends on the transfer of ions of the salt to the polymer [70]. Ions move in the polymer phase in place of solvent represented in Fig. 1.21.

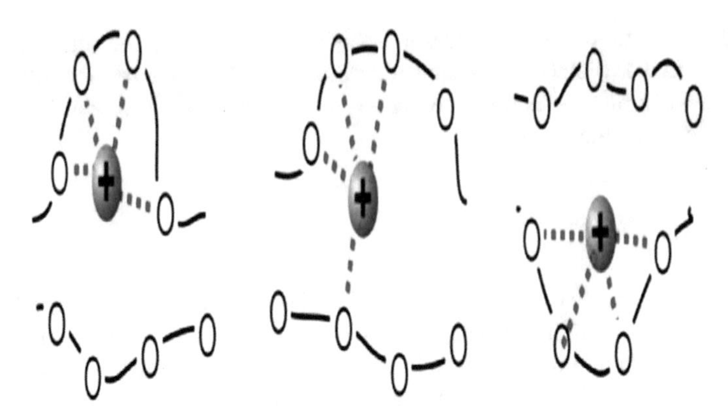

Fig. 1.21 Cation transport in a solid polymer matrix (redrawn and reprinted with permission from [71])

Gel Polymer Electrolytes

A gel polymer electrolyte is composed of a polymer matrix (host polymer) and a liquid electrolyte. The liquid electrolyte can be an aqueous, organic solvent containing conducting salt and ionic liquid. Hydrogel electrolytes are those having aqueous electrolytes with polymer hosts [72].

Inorganic Solid-State Electrolytes

These are neither flexible nor stretchable. They have high mechanical strength and thermally stable at a wide temperature range. Li_2-S-PS_5 is a solid-state inorganic electrolyte, which is ion conductive that can also act as a separator [73].

Polyelectrolytes

Polymer electrolytes are linear macromolecular chains when dissolved in a suitable solvent have several charged or chargeable groups. Macro-ions are polymer molecules having one or a few ionic groups. These are active polymers, wherein the polymer chain grows as long as monomers are supplied. The ionic charge of the macro-ion is transferred to the added monomer. The macro-ion remains charged by this mechanism leading to the addition of further monomers. Ionomers are polymers having a considerable number of ionic groups and a nonpolar backbone. Polyelectrolytes are water-soluble polymers with several ionic groups. Polyelectrolytes having both positive and negative charges like proteins are referred to as polyampholytes [74]. When polyelectrolytes dissolve or dissociate in water, polymers retain one type of charge (polyions). Their counterions diffuse in the solution. This is different from salts such as NaCl and HCl, where both the ions dissociate uniformly in water.

1.7.2.4 Redox-Active Electrolytes

Redox-active electrolytes: pseudocapacitance is not only contributed by electrode material but also from the electrolyte. Redox-active aqueous electrolytes are hydroquinones, m-phenylenediamine, KI, and lignosulfonates. Redox-active non-aqueous electrolytes are polyfluorododecaborate cluster ions into an organic mixture solvent propylene carbonate (PC) and dimethyl carbonate (DMC), tetraethylammonium undecafluorododecaborate with PC-DMC, 5 wt% 1 ethyl 3 methylimidazolium iodide into EMIM-BF_4, N-ethyl N-methylpyrrolidinium fluorohydrogenate.

Heteropoly acids are phosphotungstic acid and silicotungstic acid. These act as redox mediators for redox-active electrolytes. Polymetalate acid or polyacid is the polymerized form of the weak acids of amphoteric metals like vanadium, niobium, tantalum, chromium, molybdenum, and tungsten. The anions of these poly acids are made up of several molecules of the acid anhydride. Polysalts are the corresponding salts. An acid anhydride is a chemical compound formed by the removal of water molecules from an acid. Isopoly acids have only one type of acid anhydride as mentioned in (1.9)–(1.11).

$$2H_2CrO_4 \rightarrow H_2Cr_2O_7 + H_2O \tag{1.9}$$

$$3H_2CrO_4 \rightarrow H_2Cr_3O_{10} - 2H_2O \tag{1.10}$$

$$4H_2CrO_4 \rightarrow H_2Cr_4O_{13} - 3H_2O \tag{1.11}$$

Heteropolyanions are expressed as $[XM_{12}O_{40}]^{n-}$, M is Mo or W, and X represents the heteroatom possibly P, Si, As, Ge, or Ti.

1.7.3 Separators

Separators are electrical insulators and ion-conductive membrane. They prevent contact or provide electrical insulation between two electrodes of different polarizations. Separators also allow the passage of ions from one electrode to another. They are chemically inert, high mechanical strength, optimal thick, and porosity along good electrolyte wettability [75, 76]. Polypropylene, PVDF, PTFE, cellulose polymer membrane (cellulose nitrate, cellulose acetate), glass fiber, Celgard, Nafion, graphene oxide, etc., are used as separators.

The study of separators is carried out by MacMullin number, which is the ratio of the resistance of the electrolyte filled separator to the resistance of the electrolyte alone. Gurley value defines air permeability. It is the time required for a specified amount of air to pass through a specified area of the separator at a specified pressure. Equation 1.12 is the relation of porosity.

$$\text{Porosity}(P) = \left(\frac{W - W_0}{\rho_e \times V}\right) \times 100\% \tag{1.12}$$

where W is the weight of the separator immersed in the electrolyte, W_o is the weight of the separator before immersing in the electrolyte, ρ_e is the density of the electrolyte, and V is the volume of the separator.

Tortuosity is the ratio of the length of the streamlined path between two points in the separator to the straight-line distance connecting the same points.

1.7.4 Current Collectors

The current collector has desired properties of high electrical and thermal conductivity, low contact resistance, high chemical and electrochemical stability, low corrosion resistance, compatible with electrode material, and light material [77]. It is a good electronic conductor that collects electrons from the electrodes and transport to the external load [78]. Corrosion-resistive metal foil-like gold is used in acid-based electrolytes; Ni-based materials, stainless steel, alloy (Inconel 600), and carbon-based material are used in case of alkaline electrolyte; ITO, stainless steel, Ni, CNT, and titanium oxynitride are used in neutral electrolytes, and aluminum is used in non-aqueous electrolyte.

The most important is that the time constant of a supercapacitor is the measure of how quickly it can be charged or discharged. This time constant is RC, i.e., circuit resistance and circuit capacitance product, which should be low such that the resistance of the device is as low as possible. The low resistance of the current collector increases energy density [79].

Dimensionally stabilized anodes are the materials that can be used as both electrodes and current collectors. It is made up of a valve metal as an electrode with a film of platinum group metal deposited on the electrode material. Valve metals show rectification properties due to oxide coating in most electrolytes. There are asymmetric current–voltage characteristics because anodic current under reverse bias flows in high fields due to the formation of surface oxide film by ionic transport.

1.8 Electrochemical Characterization Techniques

Specific capacitance, energy density, and power density are important parameters for the evaluation of supercapacitor device performance. The specific capacitance of the supercapacitor is the capacitance with respect to a known entity like mass, area, and volume. Energy density is the amount of energy stored in the supercapacitor with respect to the active mass of the material of the electrode or volume of the supercapacitor and given by (1.13).

$$E = \frac{CV^2}{2} \tag{1.13}$$

where E is the energy density, C is the specific capacitance, and V is the operating voltage of the supercapacitor

The power density of a supercapacitor is the amount of energy available per unit of time and given by (1.14).

$$P = \frac{V^2}{4R_S} \tag{1.14}$$

where P is the power density, V is the operating voltage, and R_s is the equivalent series resistance of the supercapacitor.

The techniques involved in the study of supercapacitor device performance are as follows:

1.8.1 *Electrochemical Impedance Spectroscopy*

The impedance of an electrochemical system is measured by applying a low amplitude alternative voltage ΔV to a steady-state potential V_s as in (1.15).

$$\Delta V(\omega) = \Delta V_{\max} e^{j\omega t} \tag{1.15}$$

where $\Delta V_{\max}$ is the signal amplitude and ω is the pulsation.

The applied voltage gives rise to alternating output current ΔI given by (1.16).

$$\Delta I(\omega) = \Delta I_{\max} e^{j(\omega t+\varphi)} \tag{1.16}$$

where φ is the phase angle of the current with respect to voltage and $\Delta I_{\max}$ is the signal amplitude.

Equation 1.17 represents impedance

$$Z(\omega) = \Delta V(\omega)/\Delta I(\omega) = |Z(\omega)|e^{-i\varphi} = Z' + iZ'' \tag{1.17}$$

where $Z(\omega)$ is the electrochemical impedance

$$Z'^2 + Z''^2 = |Z(\omega)|^2 \tag{1.18}$$

The Nyquist plot is also used for the evaluation of impedance data as in (1.18). The real and imaginary parts of the impedances are displayed on a complex plane. Electrochemical impedance spectroscopy can be used to measure impedance, specific capacitance, charge transfer, mass transport, and charge storage mechanisms [80].

1.8.2 Cyclic Voltammetry

This technique involves an applied linearly changed electric potential between two electrodes of a supercapacitor in a two-electrode cell configuration. The instantaneous current is measured during the anodic and cathodic sweeps. The current profile is used to determine the electrochemical reactions involved in the supercapacitor. Anodic sweep is increasing. The reductive current is negative, and the oxidative current is positive [81].

CV testing is used to determine the operating voltage or potential window for supercapacitor materials. The reversal potential in a three-electrode system is adjusted, and simultaneously the reversibility of the charge and discharge processes is studied. Specific capacitance, energy performance, and total cell capacitance can also be evaluated.

1.8.3 Galvanostatic Charge/Discharge

The constant current charge/discharge (CCCD) measurement involves repetitive charging and discharging of the supercapacitor or working electrode at a fixed current level. The dwelling period is the time between charging and discharging keeping constant peak voltage applied or not in the measurement. Cycling stability can be measured by running repeated charge–discharge cycles. The specific capacitance, reversibility, and potential window for the material of the supercapacitors can be calculated by using a three-electrode setup and CCCD test [83]. Figure 1.22 shows a potential profile on the applied current.

The parameters obtained are:

Cell capacitance is given by (1.19).

$$C = \frac{It}{\Delta E} \tag{1.19}$$

where I is the charging current, t is the discharging time, and ΔE is the operating potential window.

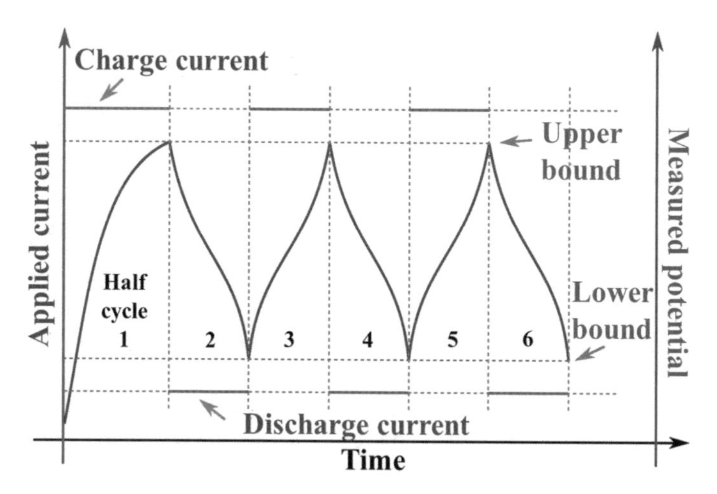

Fig. 1.22 Example of a potential profile on applied current (redrawn and reprinted with permission from [82])

Areal capacitance = Cell capacitance/(Total geometric area of two supercapacitor electrodes (i.e., two times the area of the single electrode))

$$\text{Areal energy density: Areal capacitance } X = \frac{(\Delta E)^2}{2 \times 3600} \tag{1.20}$$

Rate capability: The supercapacitor generating a considerable amount of power at high current loads has high rate capability. The rate characteristics of the supercapacitor can be measured by comparing the results of galvanostatic charge/discharge measurement at different current densities [84].

Flexibility: The CCCD test is carried out at different bending angles to measure flexibility. The capacitor is bent from 0° to 180° (or other degrees) and then flattened again to 0° in each bending cycle [85].

Cycle life: The charging/discharging or cyclic voltammetry test is carried out at a particular current density or scan rate for the "n" number of cycles. The performance parameters obtained from tests at particular no. of cycles are compared to determine cycle life.

1.8.4 Electrode System

There are three types of electrodes, a working electrode that is under evaluation, a counter electrode, which does not take part in chemical reactions and act as a current source or sink completing the circuit, and a reference electrode whose potential is constant during the reaction is acting as a reference point in electrochemical cells and no current passes through it [86].

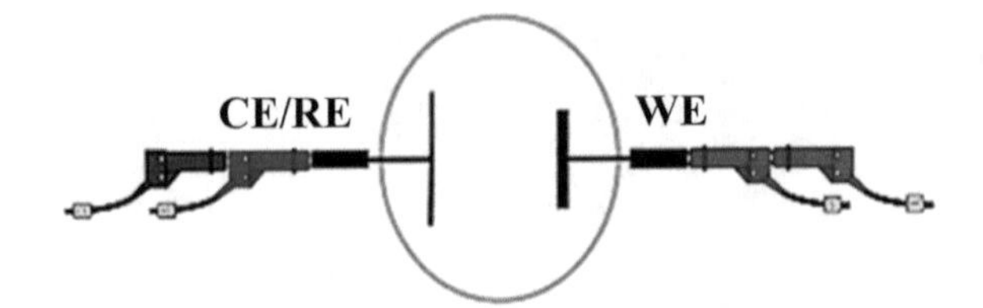

Fig. 1.23 Two-electrode configuration. S is sense electrode shorted with a working electrode (redrawn and reprinted with permission from [87])

Potentiostatic mode: The potential of the counter electrode (CE) is controlled against the working electrode (WE), so that the potential difference between the working electrode and the reference electrode (RE) is at a constant user-specified value. The current between the WE and RE is measured over time.

Galvanostatic mode: The current flow between the WE and CE is controlled. The potential difference between the RE and WE is monitored.

1.8.4.1 Two-Electrode Configuration

The drawbacks are constant counter electrode potential which cannot be maintained while the current is flowing. There is a lack of compensation for the voltage drop across the solution between the two electrodes. A two-electrode configuration is shown in Fig. 1.23.

1.8.4.2 Three-Electrode Configuration

The three-electrode setup is a widely used configuration as shown in Fig. 1.24. It has a working electrode, counter/auxiliary electrode, and a reference electrode. Current flows between the working and counter electrode. The corresponding potential is measured between the working and reference electrode. No current flows between the working and reference electrode that results in negligible potential drop. Thus, the compensation for the voltage drop due to the solution is done. Counter electrode

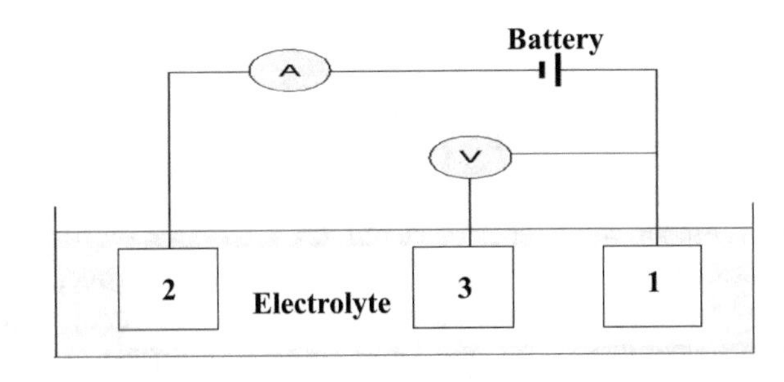

Fig. 1.24 Three-electrode setup ((1) working electrode; (2) auxiliary electrode; (3) reference electrode)

use is to balance all the current observed at the working electrode [88]. The counter electrode has wide potential fluctuations.

1.9 Concluding Remarks

Energy storage is an important aspect of national security to fulfill the uninterruptible energy demand of the country by utilizing renewable sources or other emerging concepts. The chapter discusses the difference between other energy storage devices and supercapacitors. Supercapacitors have energy density more than capacitors and power density more than batteries. These devices are replacing batteries with continuous improvement. The energy storage mechanism in supercapacitors is the non-faradaic and capacitive faradaic process. There are different types of supercapacitors depending on the charge storage mechanisms and components. Supercapacitor management systems increase the reliability and efficient use of supercapacitors. The supercapacitors are used with battery in various circuit configurations like passive, semi-active, or fully active with each having its advantages and disadvantages. The four main components of a supercapacitor are electrodes, electrolyte, separator, and current collector. The different materials used for electrodes are carbon-based, conducting polymers, transition metal oxide, and novel materials. The electrolytes can be aqueous or non-aqueous based on organic or ionic liquid or mixture electrolytes. Electrolytes can contribute to pseudocapacitance by redox reactions. The separator prevents short-circuit, and the current collector facilitates current flow. The various electrochemical characterization techniques mostly used are electrochemical impedance spectroscopy, cyclic voltammetry, galvanostatic charge/discharge, and electrode setups.

Acknowledgements The authors acknowledge the financial support provided by the Science and Engineering Research Board, Department of Science and Technology, India (SR/WOS-A/ET-48/2018), for carrying out this research work.

References

1. R. Nigam, K.D. Verma, T. Pal, K.K. Kar, in *Handbook of Nanocomposite Supercapacitor Materials II Performance*, ed. by K.K. Kar (Springer, 2020), p. 463. https://doi.org/10.1007/978-3-030-52359-6_17
2. S. Banerjee, P. Sinha, K.D. Verma, T. Pal, B. De, J. Cherruseri, P.K. Manna, K. K. Kar, in *Handbook of Nanocomposite Supercapacitor Materials I Characteristics*, ed. by K.K. Kar (Springer, 2020), p. 53. https://doi.org/10.1007/978-3-030-43009-2_2
3. P. Sinha, S. Banerjee, K.K. Kar, in *Handbook of Nanocomposite Supercapacitor Materials I Characteristics*, ed. by K.K. Kar (Springer, 2020), p. 125. https://doi.org/10.1007/978-3-030-43009-2_4
4. P. Sinha, K.K. Kar, in *Handbook of Nanocomposite Supercapacitor Materials II Performance*, ed. by K.K. Kar (Springer, 2020), p. 1. https://doi.org/10.1007/978-3-030-52359-6_1

5. S.I. Yoon, D.J. Seo, G. Kim, M. Kim, C.Y. Jung, Y.G. Yoon, S.H. Joo, T.Y. Kim, H.S. Shin, ACS Nano **12**, 10764 (2018)
6. J. Tahalyani, M.J. Akhtar, J. Cherruseri, K.K. Kar, in *Handbook of Nanocomposite Supercapacitor Materials I Characteristics*, ed. by K.K. Kar (Springer, 2020), p. 1. https://doi.org/10.1007/978-3-030-43009-2_1
7. H.S. Yi, S. Cha, Int. J. Precis. Eng. Manuf. Green Technol. **7**, 913 (2020)
8. Y. Shao, M.F. El-Kady, J. Sun, Y. Li, Q. Zhang, M. Zhu, H. Wang, B. Dunn, R.B. Kaner, Chem. Rev. **118**, 9233 (2018)
9. L. Guan, L. Yu, G.Z. Chen, Electrochim. Acta **206**, 464 (2016)
10. J.W. Long, D. Bélanger, T. Brousse, W. Sugimoto, M.B. Sassin, O. Crosnier, MRS Bull. **36** (2011)
11. C. Costentin, J.M. Savéant, Chem. Sci. **10**, 5656 (2019)
12. J. Dykstra, Desalination with porous electrodes, Dissertation, Wageningen University, 2018
13. L.L. Zhang, X.S. Zhao, Chem. Soc. Rev. **38**, 2520 (2009)
14. Y. Kumar, S. Rawal, B. Joshi et al., J. Solid State Electrochem. **23**, 667 (2019)
15. M. Nakamura, N. Sato, N. Hoshi, O. Sakata, ChemPhysChem **12**, 1430 (2011)
16. X. Li, B. Wei, Nano Energy **2**(2), 159 (2013)
17. X. Zhao, K.J. Aoki, J. Chen, T. Nishiumi, RSC Adv. **4**, 63171 (2014)
18. B.A. Rosen, Tel Aviv University, Lecture 8, Electrochemistry for Engineers. https://dokumen.tips/documents/05815271-electrochemistry-for-engineers-lecture-5-lecturer-dr-brian-rosen.html
19. Y. Jiang, J. Liu, Energy Environ. Mater. **2**, 30 (2019)
20. V. Augustyn, P. Simon, B. Dunn, Energy Environ. Sci. **7**, 1597 (2014)
21. A. Vlad, N. Singh, J. Rolland, S. Melinte, P.M. Ajayan, J.-F. Gohy, Sci. Rep. **4**, 4315 (2015)
22. B. Zhao, D. Chen, X. Xiong, B. Song, R. Hu, Q. Zhang, B.H. Rainwater, G.H. Waller, D. Zhen, Y. Ding, Y. Chen, C. Qu, D. Dang, C.P. Wong, M. Liu, Energy Storage Mater. **7**, 32 (2017)
23. W. Liu, L. Li, Q. Gui, B. Deng, Y. Li, J. Liu, Acta Phys. Chim. Sin. **36**(2), 1904049 (2020)
24. D. Ferraro, G.M. Andolina, M. Campisi, V. Pellegrini, M. Polini, Phys. Rev. B **100**, 075433 (2019)
25. B. Nabet (ed.), *Photodetectors Materials, Devices and Applications* (Elsevier, 2015)
26. W. Jing, C. Hung Lai, S.H.W. Wong, M.L.D. Wong, IET Renew. Power Gener. **11**, 4 (2017)
27. L. Kouchachvili, W. Yaïci, E. Entchev, J. Power Sources **374**, 237 (2018)
28. Z. Song, H. Hofmann, J. Li, X. Han, X. Zhang, M. Ouyang, J. Power Sources **274**, 400 (2015)
29. Z. Song, J. Li, J. Hou, H. Hofmann, M. Ouyang, J. Du, Energy **154**, 433 (2018)
30. K.V. Singh, H.O. Bansal, D. Singh, J. Mod. Transp. **27**(2), 77 (2019)
31. K.W. Hu, P.H. Yi, C.M. Liaw, IEEE Trans. Ind. Electron. **62**, 8 (2015)
32. T.D. Atmaja, Amin, in *Energy Procedia*, vol. 68, ed. by R.B.A. Bakar, C. Froome (2015), pp. 429–437
33. N.A. Kyeremateng, T. Brousse, D. Pech, Nat. Nanotechnol. **12** (2017)
34. C. Shen, S. Xu, Y. Xie, M. Sanghadasa, X. Wang, L. Lin, J. Microelectromech. Syst. **26**, 5 (2017)
35. S. Shi, C. Xu, C. Yang, J. Li, H. Du, B. Li, F. Kang, Particuology **11**(4), 371 (2013)
36. K.D. Verma, P. Sinha, S. Banerjee, K.K. Kar, in *Handbook of Nanocomposite Supercapacitor Materials I Characteristics*, ed. by K.K. Kar (Springer, 2020), p. 269. https://doi.org/10.1007/978-3-030-43009-2_9
37. S. Kumar, R. Nigam, V. Kundu, N. Jaggi, J. Mater. Sci.: Mater. Electron. **26**, 3268 (2015)
38. P. Sinha, S. Banerjee, K.K. Kar, in *Handbook of Nanocomposite Supercapacitor Materials II Performance*, ed. by K.K. Kar (Springer, 2020), p. 113. https://doi.org/10.1007/978-3-030-52359-6_5
39. P. Sinha, A. Yadav, A. Tyagi, P. Paik, H. Yokoi, A.K. Naskar, T. Kuila, K.K. Kar, Carbon **168**, 419 (2020)
40. Y. Lin, Z. Chen, C. Yu, W. Zhong, A.C.S. Sustain, Chem. Eng. **7**, 3389 (2019)
41. A. Gopalakrishnan, T.D. Raju, S. Badhulika, Carbon **168**, 209 (2020)

42. B. De, S. Banerjee, T. Pal, K.D. Verma, P.K. Manna, K.K. Kar, in *Handbook of Nanocomposite Supercapacitor Materials II Performance*, ed. by K.K. Kar (Springer, 2020), p. 271. https://doi.org/10.1007/978-3-030-52359-6_11
43. Q. Ke, Y. Liu, H. Liu, Y. Zhang, Y. Hu, J. Wang, RSC Adv. **4**, 26398 (2014)
44. X. Huang, X. Qi, F. Boey, H. Zhang, Chem. Soc. Rev. **41**, 666 (2012)
45. P. Chamoli, S. Banerjee, K.K. Raina, K.K. Kar, in *Handbook of Nanocomposite Supercapacitor Materials I Characteristics*, ed. by K.K. Kar (Springer, 2020), p. 155. https://doi.org/10.1007/978-3-030-43009-2_5
46. C. Niu, E.K. Sichel, R. Hoch, D. Moy, H. Tennent, Appl. Phys. Lett. **70**, 1480 (1997)
47. S. Banerjee, K.K. Kar, in *Handbook of Nanocomposite Supercapacitor Materials I Characteristics*, ed. by K.K. Kar (Springer, 2020), p. 179. https://doi.org/10.1007/978-3-030-43009-2_6
48. B. De, S. Banerjee, K.D. Verma, T. Pal. P.K. Manna, K.K. Kar, in *Handbook of Nanocomposite Supercapacitor Materials II Performance*, ed. by K.K. Kar (Springer, 2020), p. 229. https://doi.org/10.1007/978-3-030-52359-6_9
49. B. De, S. Banerjee, K.D. Verma, T. Pal. P.K. Manna, K.K. Kar, in *Handbook of Nanocomposite Supercapacitor Materials II Performance*, ed. by K.K. Kar (Springer, 2020), p. 179. https://doi.org/10.1007/978-3-030-52359-6_7
50. W. Li, F. Zhang, Y. Dou, Z. Wu, H. Liu, X. Qian, D. Gu, Y. Xia, B. Tu, D. Zhao, Adv. Energy Mater. **1**, 382 (2011)
51. R. Sharma, K.K. Kar, in *Handbook of Nanocomposite Supercapacitor Materials I Characteristics*, ed. by K.K. Kar (Springer, 2020), p. 215. https://doi.org/10.1007/978-3-030-43009-2_7
52. M. Kumar, P. Sinha, T. Pal, K.K. Kar, in *Handbook of Nanocomposite Supercapacitor Materials II Performance,* ed. by K.K. Kar (Springer, 2020), p. 29. https://doi.org/10.1007/978-3-030-52359-6_2
53. T. Pal, S. Banerjee, P.K. Manna, K.K. Kar, in *Handbook of Nanocomposite Supercapacitor Materials I Characteristics*, ed. by K.K. Kar (Springer, 2020), p. 247. https://doi.org/10.1007/978-3-030-43009-2_8
54. S. Banerjee, K.K. Kar, in *Handbook of Nanocomposite Supercapacitor Materials II Performance*, ed. by K.K. Kar (Springer, 2020), p. 333. https://doi.org/10.1007/978-3-030-52359-6_13
55. A. Tyagi, S. Banerjee, J. Cherusseri, K.K. Kar, in *Handbook of Nanocomposite Supercapacitor Materials I Characteristics*, ed. by K.K. Kar (Springer, 2020), p. 91. https://doi.org/10.1007/978-3-030-43009-2_3
56. B. De, S. Banerjee, K.D. Verma, T. Pal, P.K. Manna, K.K. Kar, in *Handbook of Nanocomposite Supercapacitor Materials II Performance*, ed. by K.K. Kar (Springer, 2020), p. 89. https://doi.org/10.1007/978-3-030-52359-6_4
57. M. Safaeia, M.M. Foroughi, N. Ebrahimpoor, S. Jahani, A. Omidi, M. Khatami, TrAC Trends Anal. Chem. **118**, 401 (2019)
58. A.E. Baumann, D.A. Burns, B. Liu, V.S. Thoi, Commun. Chem. **2**, 86 (2019)
59. R.K. Sharma, P. Yadav, M. Yadav, R. Gupta, P. Rana, A. Srivastava, R. Zboril, R.S. Varma, M. Antonietti, M.B. Gawande, Mater. Horiz. **7**, 411 (2020)
60. Y. Gogotsi, B. Anasori, ACS Nano **13**(8), 8491 (2019)
61. M. Hu, H. Zhang, T. Hu, B. Fan, X. Wang, Z. Li, Chem. Soc. Rev. **49**, 6666 (2020)
62. B. Anasori, Y. Gogotsi (ed.), *2D Metal Carbides and Nitrides (MXenes)* (Springer, 2019)
63. S. Ghosh, S.M. Jeong, S.R. Polaki, Korean J. Chem. Eng. **35**(7), 1389–1408 (2018)
64. J. Yang, Z. Pan, Q. Yu, Q. Zhang, X. Ding, X. Shi, Y. Qiu, K. Zhang, J. Wang, Y. Zhang, A.C.S. Appl, Mater. Interfaces **11**(6), 5938 (2019)
65. K.D. Verma, S. Banerjee, K.K. Kar, in *Handbook of Nanocomposite Supercapacitor Materials I Characteristics*, ed. by K.K. Kar (Springer, 2020), p. 287. https://doi.org/10.1007/978-3-030-43009-2_10
66. B. Pal, S. Yang, S. Ramesh, V. Thangadurai, R. Jose, Nanoscale Adv. **1**, 3807 (2019)

67. F. Barzegar, D.Y. Momodu, O.O. Fashedemi, A. Bello, J.K. Dangbegnona, N. Manyala, RSC Adv. **5**, 107482 (2015)
68. Z. Cheng, Y. Deng, H. Wenbin, Q. Jinli, Z. Lei, Z. Jiujun, Chem. Soc. Rev. **44**(21), 7484 (2015)
69. I. Krossing, J.M. Slattery, C. Daguenet, P.J. Dyson, A. Oleinikova, H. Weingärtner, J. Am. Chem. Soc. **128**(41), 13427 (2006)
70. S.B. Aziz, T.J. Woo, M.F.Z. Kadir, H.M. Ahmeda, J. Sci.: Adv. Mater. Devices **3**(1), 1 (2018)
71. P. Judeinstein, D. Reichert, E.R.D. Azevedo, T.J. Bonagamba, Acta Chim. Slov. **52**, 349 (2005)
72. N.A. Choudhury, S. Sampath, A.K. Shukla, Energy Environ. Sci. **2**, 55 (2009)
73. M. Calpa, N.C.R. Navarro, A. Miura, K. Tadanaga, Inorg. Chem. Front. **5**, 501 (2018)
74. A.V. Dobrynin, R.H. Colby, M. Rubinstein, J. Polym. Sci.: Part B: Polym. Phys. **42**, 3513 (2004)
75. K.D. Verma, P. Sinha, S. Banerjee, K.K. Kar, M.K. Ghorai, in *Handbook of Nanocomposite Supercapacitor Materials I Characteristics*, ed. by K.K. Kar (Springer, 2020), p. 315. https://doi.org/10.1007/978-3-030-43009-2_11
76. C. Man, P. Jiang, K. Wong, Y. Zhao, C. Tang, M. Fan, W. Lau, J. Mei, S. Li, H. Liu, D. Huib, J. Mater. Chem. A **2**, 11980 (2014)
77. K.D. Verma, P. Sinha, S. Banerjee, K.K. Kar, in *Handbook of Nanocomposite Supercapacitor Materials I Characteristics*, ed. by K.K. Kar (Springer, 2020), p. 327. https://doi.org/10.1007/978-3-030-43009-2_12
78. S.A. Kazaryan, V.P. Nedoshivin, V.A. Kazarov, G.G. Kharisov, S.V. Litvinenko, S.N. Razumov, U.S. Patent 7,446,998 B2, 2008
79. S. Fletcher, I. Kirkpatrick, R. Dring, R. Puttock, R. Thring, S. Howroyd, J. Power Sources **345**, 247 (2017)
80. S. Zhang, N. Pan, Adv. Energy Mater. **5**, 1401401 (2015)
81. N. Elgrishi, K.J. Rountree, B.D. McCarthy, E.S. Rountree, T.T. Eisenhart, J.L. Dempsey, J. Chem. Educ. **95**, 197 (2018)
82. T. Dobbelaere, P.M. Vereecken, C. Detavernier, HardwareX **2**, 34 (2017)
83. C. Tang, Z. Tang, H. Gong, J. Electrochem. Soc. **159**(5), A651 (2012)
84. D. Zhang, X. Zhang, Y. Chen, C. Wang, Y. Ma, H. Dong, L. Jiang, Q. Meng, W. Hu, Phys. Chem. Chem. Phys. **14**, 10899 (2012)
85. D.W. Kim, S.M. Jung, H.Y. Jung, J. Mater. Chem. A **8**, 532 (2020)
86. K.K. Kar (ed.), *Handbook of Nanocomposite Supercapacitor Materials I Characteristics* (Springer, 2020). https://doi.org/10.1007/978-3-030-43009-2
87. Metrohm Autolab, Basic overview of the working principle of a potentiostat/galvanostat (PGSTAT)—electrochemical cell setup. Application Note EC08 (2011), p. 4. https://www.ecochemie.nl/download/Applicationnotes/Autolab_Application_Note_EC08.pdf
88. W. Wang, Y. Fu, Q. Lv, H. Bai, H. Li, Z. Wang, Q. Zhang, Sens. Actuators B: Chem. **297**, 126719 (2019)

Chapter 2
Supercapacitor Devices

Prerna Sinha and Kamal K. Kar

Abstract Supercapacitors are electrochemical energy storage devices that can be used to store a large amount of energy. It delivers excellent electrochemical performances such as high capacitance, high power density, and long cyclic stability at low cost. In contrast with other energy storage devices, its charge storage mechanism is simple, which makes its charging and discharging process highly reversible. Based on the charge storage mechanism, its electrode material can be categorized as EDLC and pseudocapacitor. EDLC capacitor stores charge electrostatically whereas, reversible redox reaction occurs in pseudocapacitance. Here, the charge is stored via the Faradaic process. The further improvement in the performance is done by the formation of composite electrode material, the introduction of nanostructure electrode, assembling a hybrid capacitor by introducing battery electrode material, and assembly of an asymmetric supercapacitor. Various combinations of electrode and electrolyte material in different types of configuration provide a synergistic effect of both types of charge storage mechanism and wide operating potential range. The main aim is to obtain a high energy density device without compromising other parameters such as power density, rate capability, and cyclic stability. This chapter extensively deals the various materials used in supercapacitors, types of charge storage mechanisms, types of supercapacitor assembly i.e., symmetric supercapacitors, asymmetric supercapacitors, battery-supercapacitor hybrid devices, etc., and their performance to the type of electrode material.

P. Sinha · K. K. Kar
Advanced Nanoengineering Materials Laboratory, Materials Science Programme, Indian Institute of Technology Kanpur, Kanpur 208016, India
e-mail: findingprerna09@gmail.com

K. K. Kar (✉)
Advanced Nanoengineering Materials Laboratory, Department of Mechanical Engineering, Indian Institute of Technology Kanpur, Kanpur 208016, India
e-mail: kamalkk@iitk.ac.in

K. K. Kar (ed.), *Handbook of Nanocomposite Supercapacitor Materials III*,
Springer Series in Materials Science 313,
https://doi.org/10.1007/978-3-030-68364-1_2

2.1 Introduction

Electrochemical capacitors, ultracapacitors, or commonly known as supercapacitors are designed to bridge the gap between conventional capacitors and batteries [1]. Supercapacitor consists of electrodes, electrolyte, separator, and current collector as shown in Fig. 2.1. It has been considered as fast-charging energy storage device, which can deliver higher power density than batteries and higher energy density than

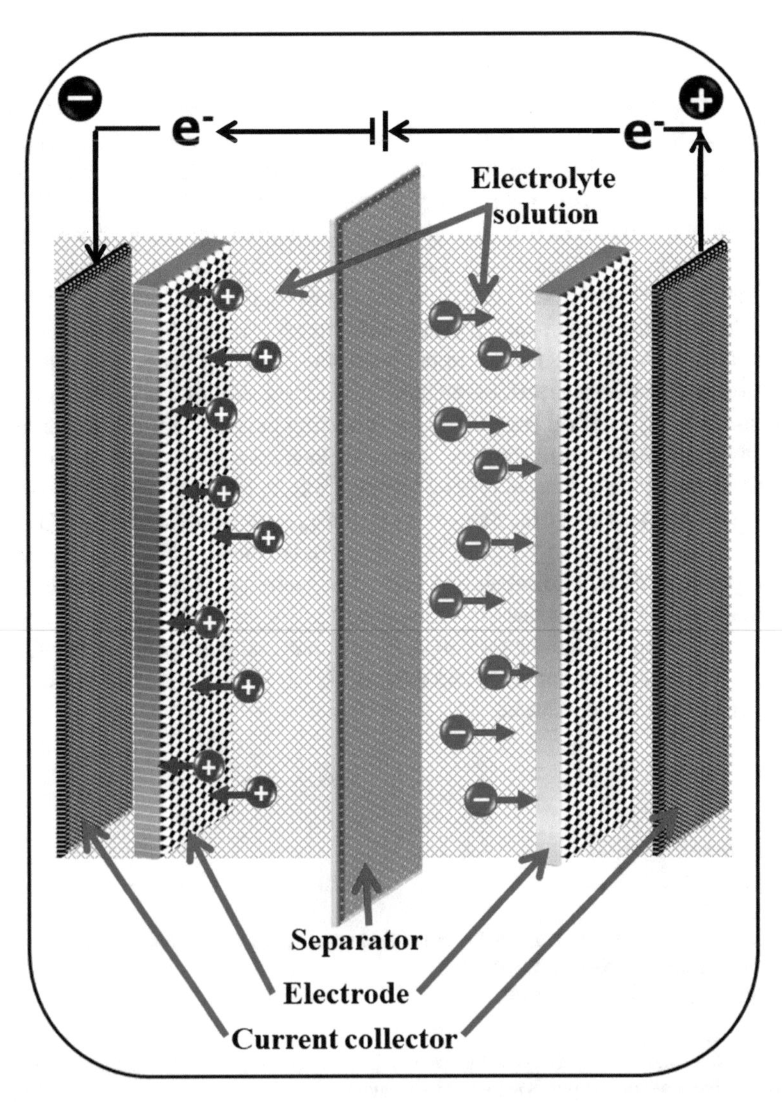

Fig. 2.1 Schematic diagram of a supercapacitor device assembly with its components (electrode, electrolyte, separator, and current collector)

the conventional capacitors [2, 3]. The fundamental of capacitor and characteristics of the capacitor to supercapacitor is reported elsewhere [4, 5]. At present, lithium-ion batteries are widely used in consumer electronics and lead-acid batteries find application as emergency lighting, powerful electric motors, etc., [6–9]. However, due to sluggish electron transfer during charging/discharging, batteries undergo heat generation and dendrite formation, which lead to serious safety issues. Also, the materials used to fabricate batteries are expensive, toxic, and detrimental to the environment. On the other hand, supercapacitors provide a sustainable approach to material selection during fabrication. It can safely provide rapid charging, and high power density with incredibly long cyclic stability. This has boosted a growing research interest to develop high cycle life, fast charging, and high power delivery supercapacitor devices. Till now, supercapacitors have been used to complement batteries in the electric component but new advancement may allow its independent utilization [10–13].

The major challenge associated with supercapacitor is its poor energy density. But, the growing demand for electronic devices requires the application of a high energy density electrochemical energy storage system [1]. To overcome this limitation, extensive research efforts have been devoted to increase the energy density of supercapacitor without compromising its other characteristics [14, 15]. In an electrochemical capacitor, the fabrication of the individual component is relatively easy. The main effort has been required to promote the synergistic effect of all the components as a device. For instance, it is well known that the pore size of activated carbon should be comparable with the size of the electrolytic ion used [16, 17]. So, an extensive study of each component is necessary before designing a supercapacitor device.

2.2 Materials Used in Supercapacitors

Varieties of approach have been undertaken to design high-performance supercapacitor. This includes

(a) tailoring the properties of electrode material [18, 19] with highly effective surface area of carbon structure [20–29], electroactive transition metal oxide (TMO) [30, 31], and conducting polymer [32, 33],
(b) introduction of different types of electrolyte solutions [34], use of speciality current collectors [18], use of high-performance separator [35],
(c) formation of composite electrode material using TMO and nanostructured carbon [25, 27, 36, 37], TMO and conducting polymers [38, 39], and ternary composites [40–42],
(d) assembly of asymmetric supercapacitor design, and
(e) fabrication of hybrid devices including battery type electrode [3, 11, 25, 27, 36, 37, 39, 41, 43–50].

This approach provides a novel hybrid electrode material, which can deliver a high energy and power density with high capacitance, low resistance, and high cycle life

[51]. Hybrid supercapacitors provide superior electrochemical properties because of the synergistic effect of individual material present in the composite. This effort paves a new path to develop a highly efficient device [49, 52, 53]. The supercapacitor is a versatile device; it can be molded into numerous shapes. So, there are some emerging supercapacitor energy storage devices, which can be designed for future generation electronics [54–57]. The current chapter provides an overview of all kinds of supercapacitor devices, in terms of assembly and electrode materials [58].

2.3 Types of Charge Storage Mechanisms

Electrochemical energy storage devices undergo different types of charge storage mechanism. Supercapacitor stores charges via electrostatic adsorption and desorption of ions commonly known as electric double layer charge storage mechanism. There is another class of supercapacitor material, which store charge via reversible redox reaction and deliver pseudocapacitance. However, there are several confusions regarding the occurrence of redox reaction in pseudocapacitors and batteries. As it is well known that batteries also undergo redox reactions to store charge, batteries and pseudocapacitors also share a similar type of charge storage mechanism, which is intercalation/de-intercalation of cationic species within the electrode structure. In order to design supercapacitor devices, the deep knowledge about different types of charge storage mechanism is required. This helps in the selection of ideal electrode material for a particular application [59, 60]. Figure 2.2 shows various charge storage mechanisms of rechargeable battery, supercapacitor, and intercalation pseudocapacitor.

The capacitance in supercapacitor devices arises from the electrostatic adsorption and desorption of ions at the surface of the electrode material. This results in the formation of an electric double layer at the electrode-electrolyte interface indicating capacitive behavior [1, 61, 62]. Carbon-based materials such as carbon nanotubes, activated carbon, and graphene are widely used as electrode material for EDLC. However, capacitive charge storage is limited until the exposed surface of carbon materials. So, nanostructured carbon materials with high surface area and tunable pore size enhance the capacitance of the electrode material [63].

Pseudocapacitance and batteries both undergo redox reactions during charge storage mechanism. In batteries, charge storage depends upon cation diffusion, which is limited by the type of crystalline framework. However, pseudocapacitance is not limited to cation diffusion. Cyclic voltammetry (CV) is used as an efficient electrochemical technique to distinguish the kinetics, which differences between both types of charge storage mechanism. The voltammetric response at different scan rate can be expressed as [59]:

$$i = a\nu^b \tag{2.1}$$

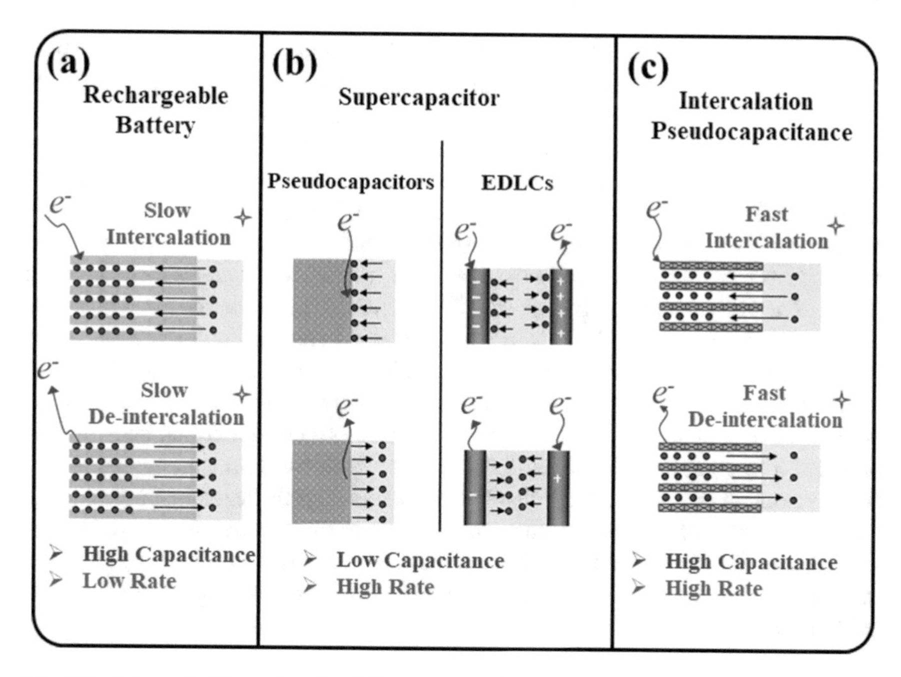

Fig. 2.2 Schematic illustration for different types of charge storage mechanisms. **a** rechargeable battery, **b** supercapacitor, and **c** intercalation pseudocapacitance (redrawn and reprinted with permission from [59])

Here, i is the measured current at a fixed potential, which obeys a power-law relationship with the potential scan rate ν.

Intercalation of batteries is a diffusion-controlled process so the corresponding voltammetric response can be expressed as [59]:

$$i = a\nu^{0.5} \tag{2.2}$$

In contrary to the battery behavior, supercapacitor behavior does not depend upon diffusion. So, 2.1 is expressed as [59]:

$$i = C_d A\nu \tag{2.3}$$

Here, C_d is the capacitance and A is the effective surface area of the active material.

Pseudocapactior electrode displays linear dependence of stored charge within the charging potential. Here, the origination of charge arises from the electron transfer mechanism [59]. The deviation of linear behavior in the galvanostatic charge-discharge (GCD) curve arises due to the presence of redox-active species. On the contrary, battery-based electrode material, charge storage occurs at a special potential window. The increase in redox activity in batteries is depicted as a plateau in the charge-discharge curves [59].

Besides, capacitive and Faradaic charge storage process, there is another type of charge storage mechanism exhibited by both batteries and the pseudocapacitor electrode. This charge storage mechanism depends upon the intercalation and de-intercalation of cations such as Li^+, Na^+, K^+, and H^+ into bulk electrode material [59]. For batteries, the intercalation and de-intercalation are controlled by cation diffusion with a crystalline framework. For example, in aqueous Ni-MH batteries, H^+ ions are de-intercalated from the crystalline structure of $Ni(OH)_2$, which acts as a cathode to form NiOOH. During the discharge process, the proton is again inserted into the $Ni(OH)_2$ crystalline framework [59]. Here, charge-discharge takes place via reversible redox reaction of Ni^{2+} to Ni^{3+} ions [59, 64]. In rechargeable batteries, phase transformation or alloying mechanism takes place to store charge at the electrode material. In supercapacitor, pseudocapacitor intercalation and de-intercalation are not limited only to the diffusion-controlled process. Pseudocapacitance intercalation provides a synergistic effect of utilization of bulk electrode for intercalation and de-intercalation and electrostatic charge storage, which is diffusion independent. As a result, the intercalation pseudocapacitive electrode shows a hybrid charge storage mechanism. So, the peak current is the combination of two separate charge storage mechanism, (a) capacitive effect ($k_1\nu$) and (b) diffusion controlled insertion and deinsertion ($k_2\nu^{1/2}$), which is expressed:

$$i(\nu) = k_1\nu + k_2\nu^{1/2} \tag{2.4}$$

Here, k_1 and k_2 values distinguish the fraction of current arising from capacitive or cation intercalation at specific potential [59].

2.4 Types of Supercapacitor Assembly

Based on the difference in the composition of electrode materials, supercapacitor devices can be classified as symmetric, asymmetric supercapacitor, and battery supercapacitor hybrid device [59]. Symmetric supercapacitors are those in which the same electrode material is used. The asymmetric supercapacitor involves the assembly of two different types of electrode material in a single device assembly. Hybrid supercapacitors with asymmetric assembly show better electrochemical performance than the similar symmetric assembly. One major advantage of asymmetric assembly over symmetric is the enhanced capacitance value and expansion of the operating voltage window. In the present time, the commercially available symmetric supercapacitor comprises activated carbon in organic electrolytes having a potential range of 2.7 V [65]. For asymmetric assembly, the commercially available supercapacitors are activated carbon and MnO_2 along with activated carbon-$Ni(OH)_2$ [66]. Currently, carbon-based supercapacitors dominate the market due to their technical maturity.

2.4.1 Symmetric Supercapacitors

Assembly of symmetric supercapacitor involves two identical positive and negative electrode materials. The electrode material can be carbon, metal oxides, conducting polymer, and the composite of the capacitive and pseudocapacitive electrode material. Based on the different charge storage mechanisms, symmetric supercapacitor can be purely capacitive or purely Faradaic. However, by forming a composite electrode by using EDLC and pseudocapacitive material, deliver synergistic charge storage. This yields better properties than its individual material. Symmetric assembly is simple and reduces charge imbalance between the two electrodes, which may lead to the poor performance of the device.

2.4.1.1 EDLC-EDLC Symmetric Supercapacitors

Electric double layer capacitors (EDLC) contain various forms of carbon material possessing a high surface area assembled in the electrolyte. Carbon is commonly used due to its easy processing, abundance, non-toxic, appreciable electronic conductivity, thermal and chemical stability [67]. Carbon material in aqueous electrolyte finds the major advantage of high ionic conductivity, low cost, ease in handling. However, symmetric EDLC-based supercapacitors only utilize electrostatic charge formed at the electrode surface, which limits its capacitance value [14]. To increase the performance of the device, carbon used as an EDLC must possess high surface area, hierarchical pores for electrolyte accessibility, and good intra/inter particle conductivity [59]. Interconnected pores facilitate ion transportation, which increases rate capability and power density. Hierarchical pore structure with abundant micropores and mesopores offers a large surface area for the interaction of electrolytic ions, which provide high capacitance and energy density [59]. In this vein, various strategies have been presented to enhance the capacitance of the device:

(a) introduction of redox-active species (heteroatom) into carbon skeleton;
(b) addition of redox-active species into conventional electrolyte; and
(c) development of nanostructured carbon materials with high surface area and tunable pore size.

The most widely used form of carbon is activated carbon [68]. It displays a high surface area, tunable pore size, and excellent electrical conductivity. Also, its synthesis technique is simple, which can be scaled for bulk production. Most commercially available devices are assembled using activated carbon as electrode material soaked in organic electrolytes. The operating cell voltage of 2.7 V is achieved with a specific capacitance of 100–200 F g^{-1} [69–73]. The presence of a high surface area enables the charge accumulation at the electrode surface by forming an electrode-electrolyte interface. However, there is no linear relation between surface area and specific capacitance. The charge accumulation mainly depends upon the tunable pore size distribution. Activated carbon possesses a wide range of pore size distribution with micropores, macropores, and mesopores. The pore smaller than electrolytic ions does not participate in charge storage. Also, inaccessible pores slower the ion transport channel, lowering the energy and power density of the EDLC system. Figure 2.3 depicts the effect of pore size on the adsorption of electrolytic ions at the pore surface. Here, smaller pores block the larger electrolytic ion to fully utilize the electrode surface. Activated carbon is cheap and widely available electrode material. It shows excellent physicochemical properties along with high cyclic stability and chemical inertness. Over the past decade, biomass-derived activated carbon is widely used. The synthesis involves simple carbonization and activation processes [23]. Qian et al. derived porous activated carbon flakes from waste human hair [74]. The symmetric assembled device in 1 M $LiPF_6$ EC/DEC electrolyte shows a specific capacitance of 107 F g^{-1} at a current density of 2 A g^{-1}. The device also delivers a high energy density of 29 Wh kg^{-1} with a high power density of 2243 W kg^{-1} [74]. However, activated carbon has relatively low specific capacitance. This is due to the inaccessibility of electrolytic ion at the pore. This issue is attributed to the mismatch of the electrolytic ion size and pore size of the electrode. In order to address the issue of the wide pore size distribution of activated carbon, Zhou et al. have prepared self-ordered mesoporous carbon [75]. Carbon material with narrow pore size has been developed using the hexagonal SiO_2 template using sucrose as a carbon source. The narrow pore diameter of 3.90 nm is reported with a specific surface area of 900 m^2 g^{-1}. The specific capacitance of 60–95 F g^{-1} is reported at a scan rate of 0.5–5 mV s^{-1}. Self-ordered mesoporous carbon acts as a promising electrode material, which provides narrow pore size distribution [75].

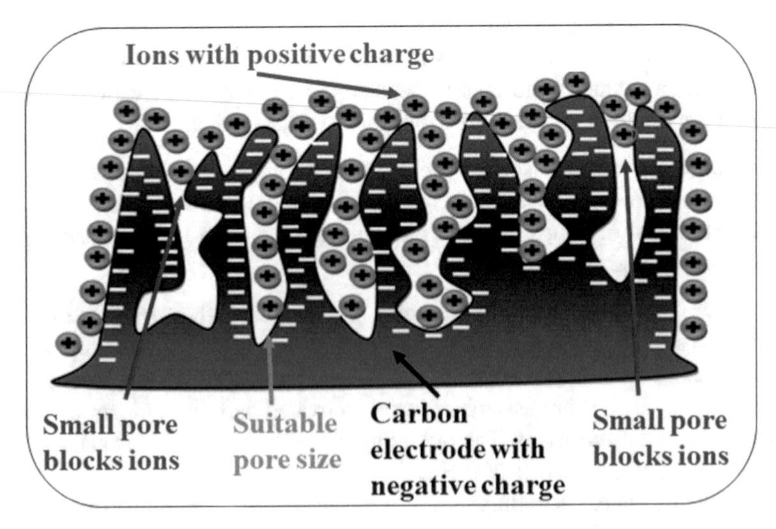

Fig. 2.3 Schematic of the ion size and pore size effect on nanoporous carbon electrode (redrawn and reprinted with permission from [76])

Carbide-derived carbon (CDCs) is another class of carbon material, which provides better control over the pore size development. CDCs are synthesized via the extraction of metals from carbides at high temperatures. Carbide precursors provide fine-tuning of pores. This allows improved control over the surface functional groups and surface area than activated carbon [73]. Cambaz et al. have synthesized silicon carbide CDC show specific capacitance of 126 F cm^{-3} in aqueous and 72 F cm^{-3} in organic electrolyte [77]. The presence of micropores increases the capacitance of the device and the uniform distribution of mesopores increases the rate capability. To study this phenomenon, Liu et al. have used titanium carbide/carbon composite to develop ordered hierarchical mesoporous/microporous carbon (OHMMC). OHMMC depicts the high surface area of 1917 m^2 g^{-1} having mesopores and micropore distribution [78]. The as-assembled device exhibits a specific capacitance of 146 F g^{-1} in the organic electrolyte with high rate capability [78]. The mesoporous channel provides a favorable path for electrolyte ions. Also, an increase in pore volume provides more surface-active sites to store charge. The above studies show that the properties of the CDC can be tuned by changing the synthesis condition.

Carbon materials have also been explored at the nano-level. The commonly used nanostructured carbon as electrode material includes carbon nanotubes (CNT), carbon nanofibers (CNF), and graphene or reduced graphene oxide (rGO) [22]. These materials are produced by catalytic decomposition of hydrocarbon [71, 79]. CNT has attractive structural, electrical, and mechanical properties, which if used wisely can yield high-performance electrode material. It can be single-walled CNT (SWCNT) or multi-walled CNT (MWCNT) [80–82]. It offers high surface area, high stability, size in nano distribution range, and low resistivity. Futaba et al. have assembled bulk SWCNT supercapacitors in organic electrolytes [80]. The specific capacitance of 80 F g^{-1} has been reported with a high surface area of 1000 m^2 g^{-1} [80]. CNTs are generally grown on a substrate, which minimizes the contact resistance. Zhou et al. have fabricated another high-performance symmetric supercapacitor using a nanocomposite of CNT/graphite nanofiber (GNF) via chemical vapor deposition [82]. As assembled supercapacitor shows excellent cyclic stability of 96% capacitance retention after 10,000 cycles. It delivers a specific energy density of 72.2 Wh kg^{-1} at a power density of 686 W kg^{-1}. The high performance of CNT/GNF is due to a large accessible site for charge storage and efficient charge transport due to CNT skeleton [81, 82]. CNTs are the primary choice for composite formation with another electrode material because of its

(a) efficient percolation of active particles within nanotubes;
(b) the tubular structure of CNT allows percolated particles to undergo mechanical changes during the charge-discharge process, which is widely beneficial for conducting polymers and intercalation pseudocapacitors materials; and
(c) open mesopores tubes allow easy diffusion of the electrolytic ion into the active surface [61, 83].

Apart from tube and fiber morphology, carbon material also exists in a two-dimensional sheet structure exhibiting high charge transport mobility and high electrical conductivity. This two-dimensional, one atomic thick sheet with sp^2 hybridized carbon atoms is called graphene. Graphene displays excellent physicochemical properties, high rate capability. Besides superior properties, graphene limits its use due to restacking tendency. This tremendously decreases the properties of graphene material [84]. In this regard, Zhu et al. have synthesized oxygen-enriched crumpled graphene-based symmetric supercapacitor. The device delivers excellent gravimetric and volumetric energy density of 20.4 Wh kg^{-1} and 26.1 Wh L^{-1} in aqueous electrolyte [85]. This shows that modification of graphene structure by the addition of functionalities can deliver high electrochemical performance.

Apart from tuning the pore structure and introducing nanostructured carbon, surface functionalization is another effective approach for the fabrication of high performance of EDLC-based carbon electrode material [59]. Surface functionalization is the introduction of active species or heteroatoms at the carbon skeleton. Various literature reports that the introduction of surface functionalities has increased the capacitance and rate capability of all forms of carbon materials, from activated carbon, CNT, graphene to CNF. The commonly used heteroatoms are nitrogen, oxygen, boron, and sulfur. The presence of nitrogen in the carbon matrix generates redox-active sites, increases electron donor capability, and improves the wettability of electrolytic ions onto the carbon surface. Nitrogen atoms bonded with carbon skeleton can be present in three states

(a) pyridinic-type nitrogen, where sp^2 bonded to two carbon atoms, which donates a *p*-type electron to the aromatic system;
(b) pyrrolic nitrogen, it is associated with carbonyl or phenolic group with five-membered carbon atom ring; and
(c) quaternary type nitrogen, here nitrogen atom is surrounded by three carbon atoms in the central or valley position [59].

Pyridinic and pyrrolic nitrogen group induce pseudocapacitance, whereas quaternary nitrogen promotes electron transfer throughout the carbon matrix. Besides nitrogen, oxygen is another electroactive atoms, which tailor the surface properties of the carbon structure. Oxygen functionalities mainly exist in quinone (C=O), phenol group (C–OH), ether (C–O–C), carboxylic (COOH), and chemisorbed oxygen or water molecules [59]. Among all oxygen functionalities, quinone displays electroactive properties in acidic medium. Zhao et al. have synthesized oxygen-doped hierarchical porous carbon via HNO_3-activation [86]. The presence of oxygen functionalities provides additional pseudocapacitance. The electrode material displays a high specific capacitance of 369 F g^{-1} in 1 M H_2SO_4 [86]. However, most of the carbon material possess both nitrogen and oxygen functionalities. During charge-discharge cycles, heteroatoms induce a Faradaic reaction, which can be clearly observed in cyclic voltammetry and charge-discharge profile. Further discussion for the introduction of heteroatoms can be extended by the use of cheap and abundant biomass precursors for the synthesis of functionalized carbon. Recently, tremendous studies have been carried out to utilize biomass as the primary precursor for the preparation of carbon. This includes the carbonization of biomass in a controlled atmosphere mainly in argon and nitrogen followed by the activation process. During carbonization, biomass decomposes into carbon blocks by evaporating all its inorganic and volatile species. The activation process uses activating agents such as KOH, $ZnCl_2$, CO_2, air, and steam to generate porosity inside the carbon blocks. The extraction of carbon from biomass provides two major advantages

(a) inherent heteroatom doping, which is present as inorganic molecules in biomass; and
(b) bio-inspired nanoarchitecture of resultant carbon obtained from biomass, which otherwise requires extensive efforts to obtain the desired carbon morphology [87–89].

Liu et al. have prepared oxygen, nitrogen, and sulfur co-doped hierarchical porous carbon via direct pyrolysis of kraft lignin [90]. The reported surface area is 1307 m^2 g^{-1} along with 19.91 wt% of heteroatom doping. The schematic illustration for the preparation of O–N–S co-doped hierarchical porous carbon is shown in Fig. 2.4a. The symmetric assembled device in 6 M KOH electrolyte exhibits near rectangular CV curves indicating capacitive behavior as shown in Fig. 2.4b. Also, the GCD plot in Fig. 2.4c depicts symmetric and linear charge-discharge plot at different current densities, which indicates that the charges are mainly stored by electrostatic forces. The symmetric assembled device delivers a high specific capacitance of 244.5 F g^{-1} at a current density of 0.2 A g^{-1}, having an excellent rate capability of 81.8% at a high current density of 40 A g^{-1}. The device also shows an energy density of 8.5 Wh kg^{-1} at a power density of 100 W kg^{-1} [90]. The study reveals that biomass-derived functionalized carbon provides a green and sustainable approach at a low cost. Also, the major advantage of utilizing biomass precursors for the synthesis of the carbon material is its bulk production, which however is the major drawback with CNT and graphene type nanostructured carbon materials [59].

2.4.1.2 Pseudocapacitor-Pseudocapacitor Based Symmetric Supercapacitors

Metal oxides based symmetric supercapacitors

After carbon-based electrode material, transition metal oxides have been explored as an alternative electrode material for supercapacitors. Metal oxide possesses diverse crystal structure and morphology, high theoretical capacitance value, and large surface area [91]. Also, metal oxides are abundant in nature; some are environmentally safe and easily approachable. Metal oxide stores charges via Faradaic redox reaction and adsorption/desorption ions at electrode surface [30]. Similar to batteries, metal oxide also undergo Faradaic redox reaction to store charge. However, their diffusion kinetics and electrochemical profiles are entirely different [59]. Various metal oxides such as RuO_2, MnO_2, IrO_2, NiO, Co_3O_4, and metal hydroxides such as $Co(OH)_2$, $Ni(OH)_2$ are widely investigated as pseudocapacitive electrode material. RuO_2 is the first redox material, which has been explored as high-performance electrode material for supercapacitor [92]. It shows high electrical conductivity [73]. The variable oxidation state delivers a large amount of electron, which increases the

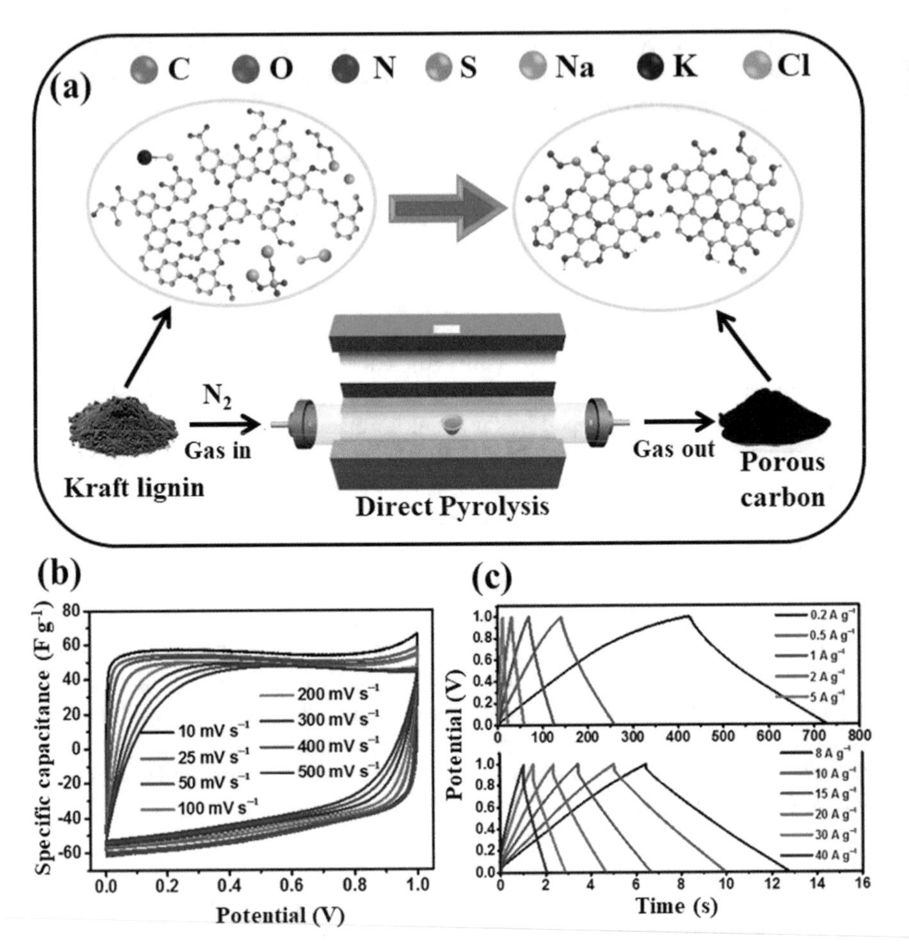

Fig. 2.4 **a** Schematic illustration of the direct conversion of kraft lignin into the heteroatom-doped hierarchical porous carbons, **b** CV curves at different scan rates, and **c** GCD profiles at different current densities of symmetrically assembled porous carbon in 6 M KOH electrolyte (redrawn and reprinted with permission from [90])

capacitance. However, RuO_2 is an expensive material, which limits wide application as electrode material. Alternatively, several other metal oxides such as MnO_2, NiO, Co_3O_4, and mixed transition metal oxides such as $NiCo_2O_4$ and $NiFe_2O_4$ have been explored. However, these metal oxides suffer from poor electrical conductivity. Also, repeated chemical reaction during charging and discharging decreases its cyclic stability [14, 93]. The design of metal oxide by fabrication of novel nanostructure, tailoring oxygen vacancies, and introduction of ternary metal oxides can boost electro-conductivity and enhance the chemical and physical performance. Few strategies that improve the performance of the metal oxide electrode are as follows:

(a) Ternary metal oxides

Metal oxides possess a group of promising pseudocapacitor electrode materials, which have improved the supercapacitive performance than single-component metal oxides. These are called ternary metal oxide with the general formula of $A_xB_yO_z$, which includes AB_2O_4, ABO_4, and $A_3B_2O_8$. Here, A and B elements represent the metal elements with high and low oxidation states and participate in the energy storage process. The co-existence of two cations generates more electron than the single metal oxides. The presence of a variable oxidation state enables multiple redox reactions and improves the electrical conductivity, which is however one of the major limitations associated with single-component metal oxide. The nanostructure with a higher surface to volume ratio improves the electrochemical activity during the charge-discharge process. Spinel structure with a formula of AB_2O_4 has emerged as a high energy density electrode material. The commonly used electrode is $NiCo_2O_4$, $MnCo_2O_4$, and $MnFe_2O_4$. Nickel cobaltite possesses much better electrical conductivity and superior electrochemical activity as compared to their monometallic oxides [94]. Lu et al. have synthesized vertically aligned nickel cobalt oxide porous nanosheet structure as shown in Fig. 2.4a [95]. The unique interconnected porous morphology provides penetration and diffusion of electron and ions. Figure 2.5b,c shows the SEM image of nickel cobalt oxide porous nanosheet at different magnification. The electrochemical studies were carried out in 1 M KOH electrolyte, wherein three-electrode system nickel cobalt oxide delivers 405.6 F g^{-1}. In the two-electrode system, the CV profile as shown in Fig. 2.5d shows rectangular-like shape indicating ideal capacitive behavior. The GCD curve (Fig. 2.5e) also resembles capacitive behavior with a triangular-shape curve. The maximum specific capacitance reported is 89.2 F g^{-1} at a current density of 0.17 A g^{-1} [95].

Another class of ternary metal oxide includes ABO_4, in which $AMoO_4$ (A=Ni, Co, Mn) is used as electrode material. Ghosh et al. have carried out the electrochemical performance of $MnMoO_4$ in 1 M Na_2SO_4 electrolyte [96]. Here, major contribution arises from Mn atom, whereas Mo does not take part directly in the charge storage process. However, the presence of Mo enhances the conductivity of $MnMoO_4$ to 4.27×10^{-3} S cm^{-1} [96]. The other class contains the ternary metal oxide with the formula of $A_3B_2O_8$, in which $M_3V_2O_8$ (M=Ni, Co) is the material used as an electrode. Here, the ratio of A/B atom is 1.5:1, which is higher than other classes of ternary metal oxides. Theoretical expectations reveal the strong redoxomorphism between M^{2+} ions and the electrolyte, which can result in higher capacitance. Zhang et al. have studied the electrochemical performance of $Co_3V_2O_8$ nanoplates obtained via hydrothermal synthesis [97]. The electrode delivers a high specific capacitance of 739 F g^{-1} at a current density of 0.5 A g^{-1} with excellent cyclic stability of 95% after 2000 cycles [97].

(b) Fabrication of high surface area nanostructured metal oxides

Nano-level metal oxide provides a high surface area to volume ratio. This offers a large contact area between the active electrode and electrolyte. Nanostructure

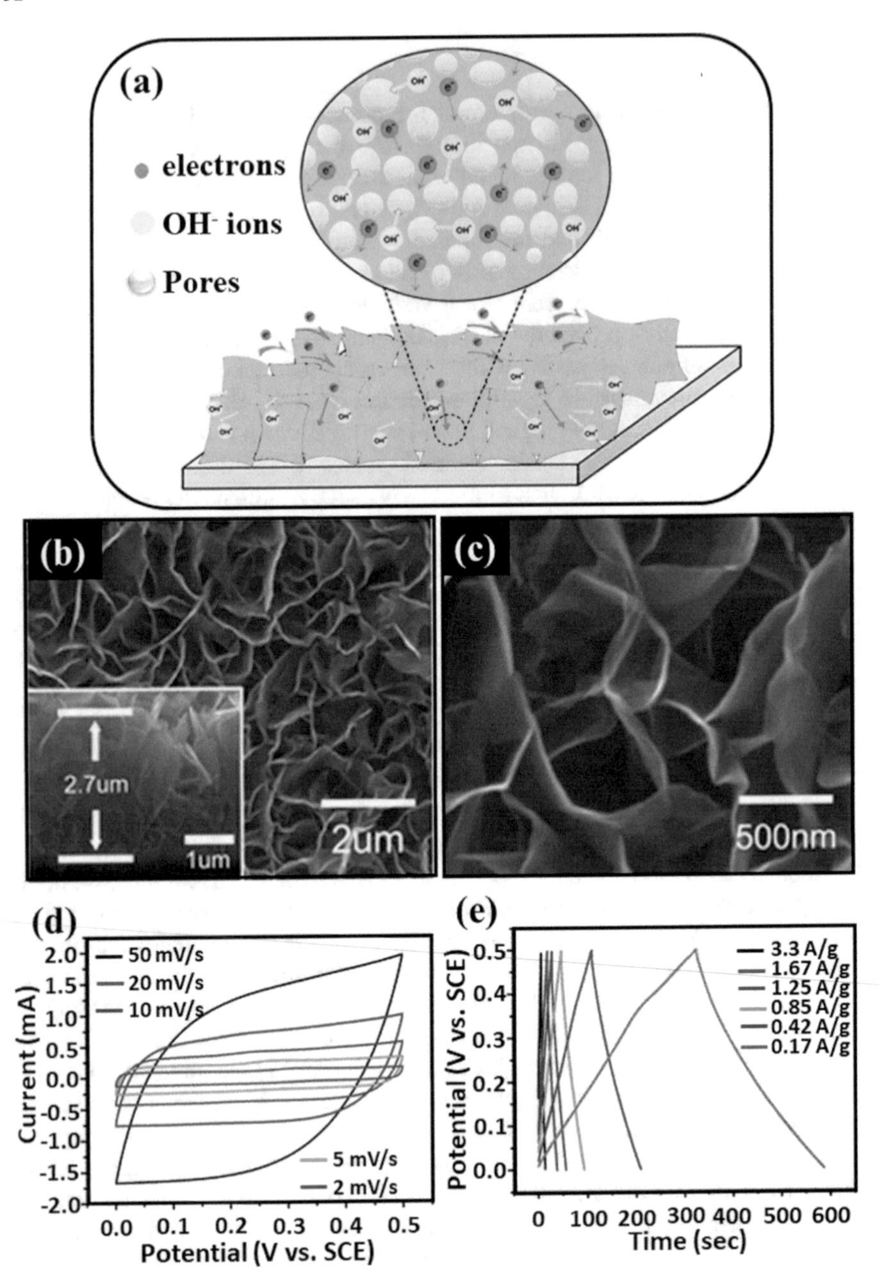

Fig. 2.5 **a** Schematic illustration of the transportation of species on nickel cobalt oxide porous nanosheet in 1 M KOH solution. **b** and **c** SEM images of nickel cobalt oxide porous nanosheet, **d** and **e** CV and GCD profile of symmetric assembled nickel cobalt oxide porous nanosheet as an electrode in 1 M KOH electrolyte (redrawn and reprinted with permission from [95])

reduces the ion diffusion length and enables the full utilization of material during the charge-discharge process. A large number of electroactive surface sites provide high capacitance and high rate capability. Different nanostructure with different dimensions possesses their inherent unique advantage. Figure 2.6 depicts the schematic representation of different dimensional nanostructured transition metal oxide. One-dimensional nanostructure such as nanowires, nanotubes, nanobelts and nanorods acts as a highway for charge transportation, which reduces the ion diffusion path. This increases the rate capability and power delivery of the electrode material. Two-dimensional nanostructure includes nanoflakes and nanosheets that exhibit superior cyclic performance. These structures enable efficient electron and ion transportation. Ultrathin nanosheet with interconnected porous morphology provides better ion and electron accommodation of volume variation. However, three-dimensional nanostructure consists of micro-spheres, micro-flowers, and hierarchical 3D porous network.

(c) Introduction of oxygen vacancies into metal oxides

The absence of oxygen atom in metal oxide crystal structure widens the interlayer spacing. This enables fast charge storage kinetics with intercalation pseudocapacitive behavior [91]. Figure 2.7a shows the schematic illustration for the charge-discharge mechanism of $MoO_{3-x.}$ Xiao et al. have studied the hydrogenation effect on creating oxygen vacancies in MoO_3 [101]. Hydrogenation leads to the partial reduction and introduction of oxygen vacancies without changing the shape and size of MoO_3. The conductivity gets enhanced from MoO_3 to the MoO_{3-x} compound [101]. Similarly, Kim et al. have synthesized α-MoO_3 through the introduction of oxygen vacancies [102]. This approach improves the electrical conductivity as oxygen vacancies act as shallow donors that further increase the carrier concentration of the electrode material. The oxygen vacancies induce large interlayer spacing, which enables to retain its structure during charging and discharging. Figure 2.7b shows that reduced MoO_{3-x} has higher charge storing capacity than fully oxidized MoO_3. The capacitor like charge storage mechanism of reduced MoO_{3-x} and fully oxidized MoO_3 are shown in the CV curve, Fig. 2.7c,d. The CV curve indicates that reduced MoO_{3-x} displays better capacitive charge storage capability. It also improves the cycle stability of electrode and fast charge storage capability [102].

Transition metal oxide can provide high specific capacitance than carbon materials and improved electrochemical stability than polymer materials. However, low electronic conductivity and poor cyclic stability limit its practical application. The introduction of a different class of supercapacitive electrode material into the matrix can provide a synergistic approach toward the development of ideal material. For proper utilization of metal oxide properties, these materials are a couple with carbon and conducting polymer-forming composite. By this process, the high-performance electrode material can be achieved with excellent electrochemical properties.

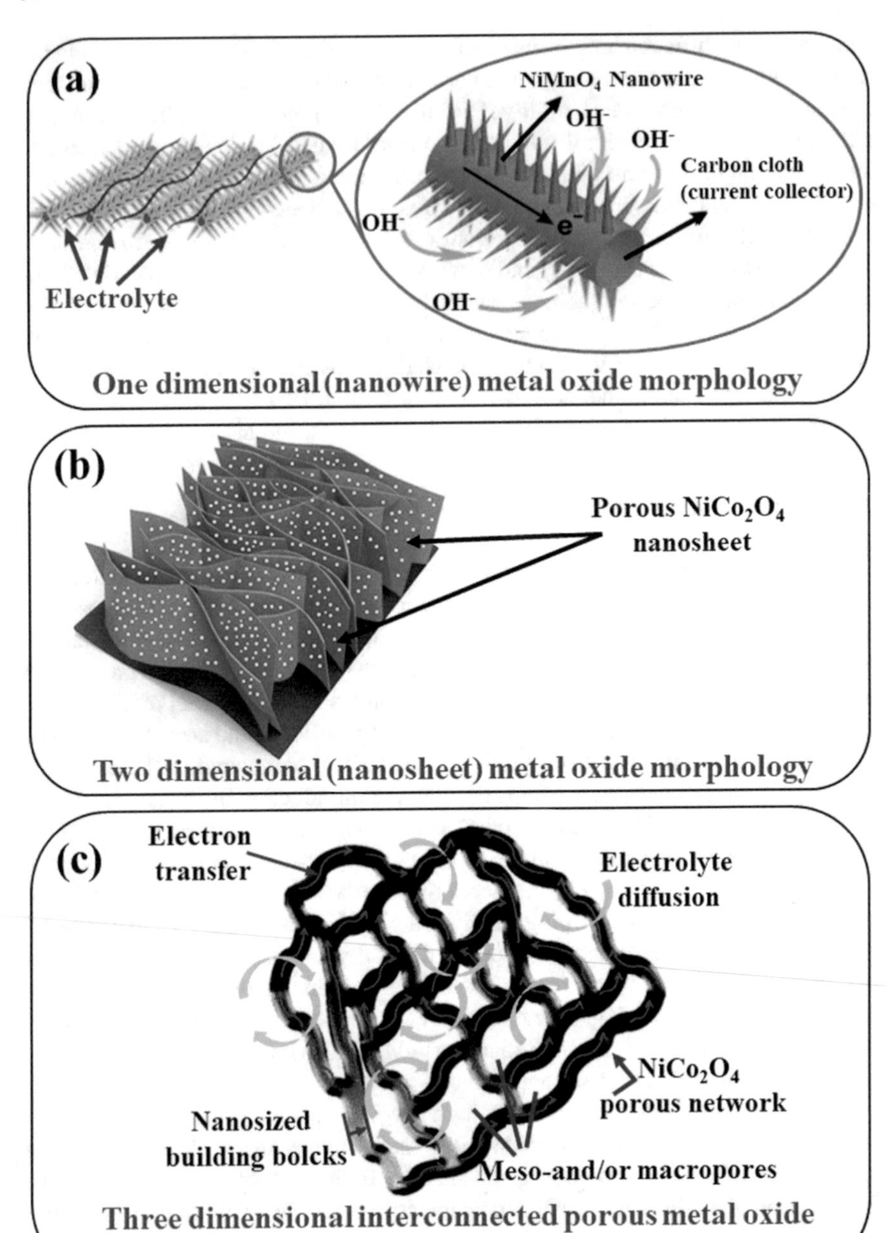

Fig. 2.6 Schematic representation of nanostructure metal oxide as electrode material. **a** one-dimensional $NiMoO_4$ nanowire on carbon cloth [98], **b** two-dimensional porous $NiCo_2O_4$ nanosheet arrays on flexible carbon fiber [99], and **c** three-dimensional interconnected hierarchical porous network-like $NiCo_2O_4$ framework electrode [100] (redrawn and reprinted with permission)

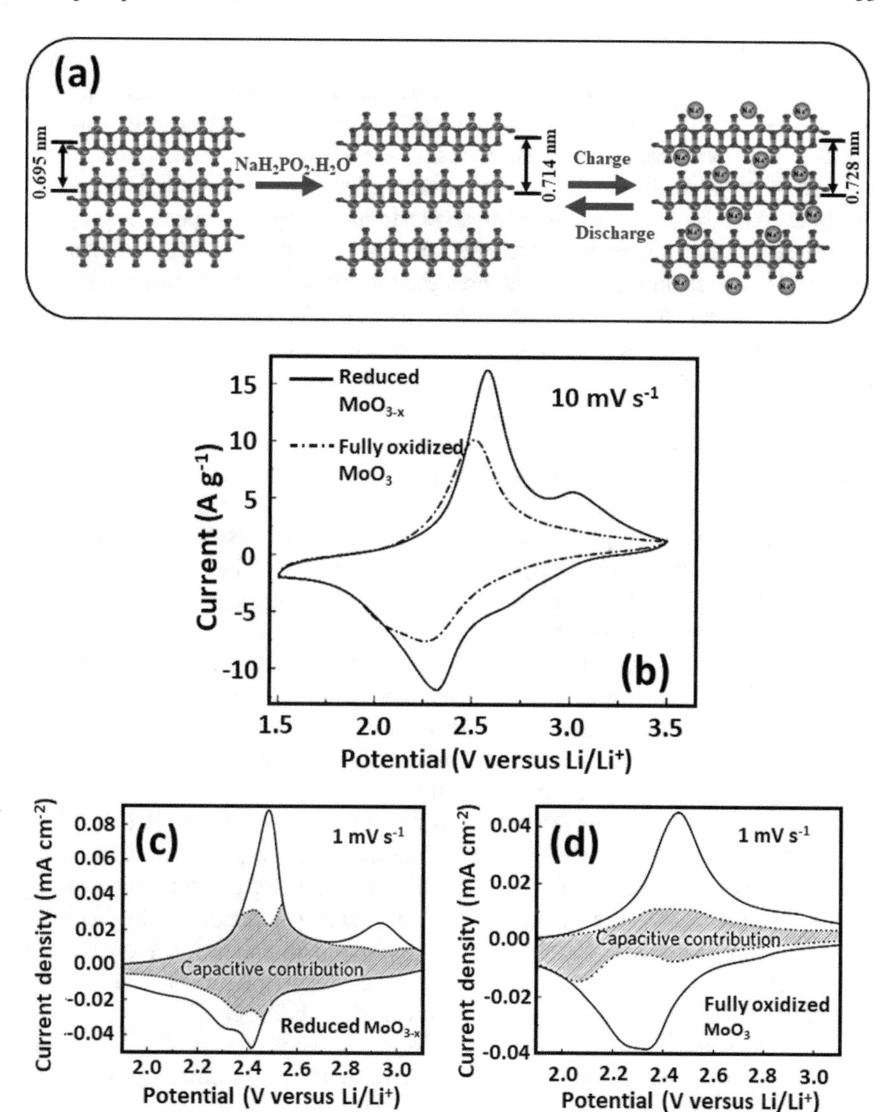

Fig. 2.7 **a** Schematic illustration of intercalation and de-intercalation mechanism of MoO_{3-x} [103], **b** voltammetric sweep at 10 mV s^{-1} for R-MoO_{3-x}, and F-MoO_3, **c** the contribution of capacitive charge storage for R-MoO_{3-x}, and **d** the contribution of capacitive charge storage for F-MoO_3 [102] (redrawn and reprinted with permission)

Conducting polymer-based symmetric supercapacitors

Conducting polymers gained considerable attention as they provide high specific capacitance, rapid charge-discharge, and low resistance. This high capacitance is due to the participation of the whole electrode material for the charge storage mechanism [32]. The presence of a conjugated bond along the polymer backbone renders the movement of ions during the charge storage process. The redox process is highly reversible as no phase transformation occurs. The conducting polymers are easy to synthesize, environment friendly, and low-cost material. The *n*- and *p*-type conducting polymers are present, where *p*-type are widely used as positive electrode material for supercapacitor applications. *n*-doped conducting polymers such as polyaniline and polypyrrole have very low reduction potential in a common electrolyte. Conducting polymer requires the transfer of proton during the charge-discharge process; hence, improved performances have been obtained in acidic, protic, and aprotic ionic liquid solution electrolyte. Unfortunately, during the repeated charge-discharge process, conducting polymer swells and shrinks. This leads to the mechanical degradation of the material. Wang et al. have studied conducting polymer-based supercapacitor by surface-modified nanocelluse fiber (NCF) [104]. The modified nanocelluse fiber aims to display high gravimetric and volumetric capacitance. The modification process takes place by the introduction of carboxylate or quaternary amine groups, which results in anionic (a-NCF) and cationic (c-NCF) surface charge species, respectively, where for comparison unmodified NCF is denoted as u-NCF. NCF has been further polymerized with polypyrrole. The electrochemical characterization of symmetric assembled devices of u-NCF, a-NCF, and c-NCF has been reported in 2 M NaCl electrolyte. Figure 2.8a presents the CV curve of the symmetric assembled devices depicting near-symmetrical and rectangular shape. Figure 2.8b,c shows the gravimetric and volumetric capacitance over various current density, respectively. However, all the samples deliver similar gravimetric capacitance, whereas their volumetric capacitance found to differ substantially. PPy@c-NCFs exhibits the highest volumetric capacitance of 173 F cm^{-3} at a current density of 1 mA cm^{-2}. Figure 2.8d shows the Ragone plot of all the samples, where PPy@c-NCFs deliver an energy density of 3.6 mWh cm^{-3} at a power density of 3.5 W cm^{-3} [104]. The performance of conducting polymer can be improved by controlling microstructure, surface morphology, and enhancing the degree of crystallinity. Also, the electrode performance of conducting polymer depends upon dopant concentration, oxidation state, and processing abilities [105].

However, conducting polymer owns a lot of interesting properties as an electrode for supercapacitor. They are not appropriate to be solely utilized as electrodes for supercapacitors. Polyaniline offers large specific capacitance, easy fabrication, high doping level, and high theoretical capacitance with controllable conductivity. However, it fails to deliver material stability over repeated cycles. Also, polyaniline limits its application only for proton type electrodes [106]. After polyaniline, polypyrrole is the most common polymer, which provides easy fabrication, high cyclic stability and can be used for neutral electrolyte. The major drawbacks associated with polypyrrole as an electrode are low specific capacitance, difficulty in doping and de-doping and can only be used as a cathode material. Polythiophene offers favorable cyclic stability and it is environment friendly. Low specific capacitance and poor conductivity limit its practical application as an electrode for supercapacitor [105]. In order to fully utilize its electrochemical properties and over its limitation, it is mixed with another electrode material to form a composite. This provides support

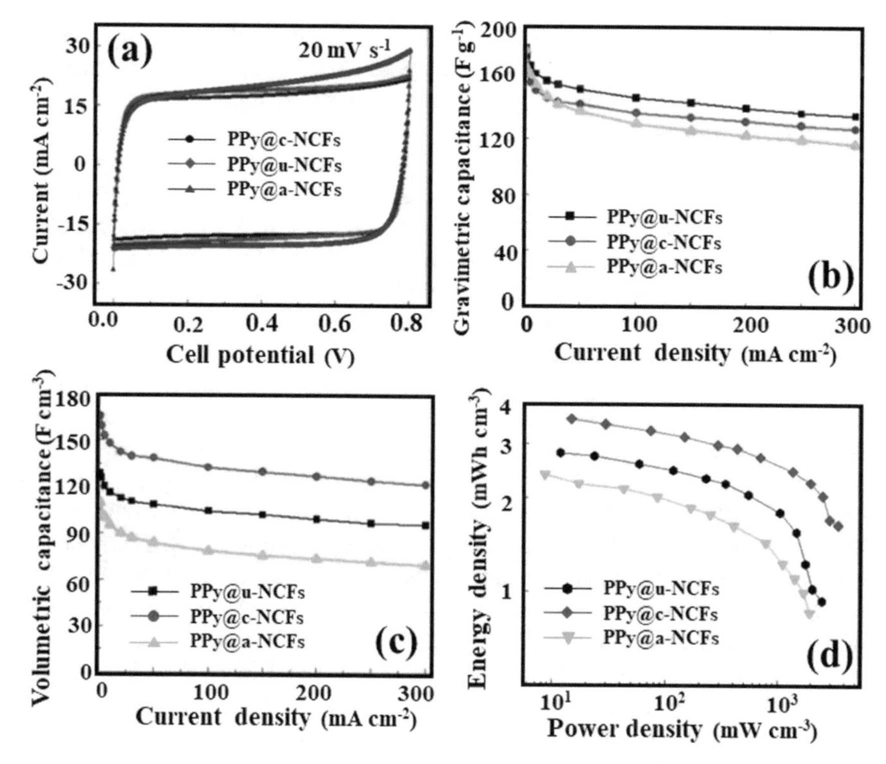

Fig. 2.8 Electrochemical performance of composite electrodes in symmetric assembly. **a** CV curves at a scan rate of 20 mV s^{-1}, **b** gravimetric and **c** volumetric capacitances for different charging/discharging current densities and **d** Ragone plots for symmetric supercapacitor devices containing the composite upon charging the devices to 0.8 V (redrawn and reprinted with permission from [104])

to the conducting polymer and increases its cyclic stability. Conducting polymer coating in metal oxide and carbon material provides good material stability and high capacitance value. Other conducting polymers include polypyrrole, polythiophene, etc., [105, 107].

2.4.1.3 Hybrid Composite-Composite Based Symmetric Supercapacitors

The hybrid composite electrode eliminates the inherent limitation of the individual electrode. Metal oxides and conducting polymer suffers from poor material stability and low charge-discharge rate capability. The introduction of porous carbon material as reinforcement aims to improve the cyclic stability and rate capability of the composite [59]. The porous three-dimensional pathway efficiently transfer charge and micro to nanochannels serve as reservoirs for electrolytic ions that provide a short diffusion path, which minimizes the resistance. Carbon material acts as a filler for metal oxide and conducting polymer-based hybrid electrode material [105]. Also, nanoarchitecture carbon provides mechanical support and participates in charge storage mechanisms resulting in additional capacitance [108]. Kashani et al. have synthesized three-dimensional nanoporous graphene-polypyrrole composite by electrochemical deposition of polypyrrole into graphene nanoparticle [108]. The symmetric assembly of graphene-polypyrrole composite in PVA/H_2SO_4 gel electrolyte delivers a rectangular-shaped CV curve, which remains even at a high scan rate of 100 mV s^{-1}. The identical CV curve depicts the efficient charge propagation across the two symmetric electrodes. The linear GCD plot exhibits the specific capacitance of 514 F g^{-1} at a current density of 0.2 A g^{-1} along with a 63% rate capability at a current density of 20 A g^{-1}. The superior electrochemical performance suggests the efficient utilization of porous nanotubular composite, which enhances the contact interface between polypyrrole and electrolyte. The assembled device delivers ultra-high energy and power density of 21.6 Wh kg^{-1} and 60.1 kW kg^{-1} [108]. Another literature by Chen et al. has reported a composite electrode of the sandwich-like structure of Holey graphene/polyaniline/graphene nanohybrid [109]. The structure of the nanohybrid and its charge storage mechanism is shown in Fig. 2.9a. The resultant nanohybrid structure promotes fast ionic diffusion and electronic transportation and also contributes to prominent rate capability. The electrochemical studies of symmetric assembled nanohybrid have been reported in 1 M H_2SO_4. Figure 2.9b shows the CV curve of nanohybrid measured at different scan rates. The reduction-oxidation peaks are noticed due to the pseudocapacitance arising from polyaniline. GCD plot as shown in Fig. 2.9c shows symmetric charge and discharge curve with a maximum specific capacitance of 437 F g^{-1} at a current density of 1 A g^{-1}. The device also delivers a high energy density of 55 Wh kg^{-1} at a power density of 900 W kg^{-1}, along with excellent stability of 84% capacitance retention after 5000 cycles [109].

It is well known that metal oxides and carbon composite provide a synergistic charge storage mechanism [59]. The high capacitance of metal oxide and excellent rate capability and cyclic stability of carbon material helps to achieve high-performance electrode material. Pandit et al. have studied hexagonal VS_2 anchored multi-walled CNT as an electrode for the assembly of symmetric supercapacitor [110]. VS_2 undergoes structural degradation during the charge-discharge process. The addition of multi-walled CNT into VS_2 undergoes rapid ion diffusion and fast charge-discharge electrochemical kinetics. The symmetric assembled device in PVA-$LiCl_4$ electrolyte offers a working potential window of 1.6 V. The device delivers a specific capacitance of 182 F g^{-1} at a scan rate of 2 mV s^{-1}. The energy density of 42 Wh kg^{-1} with a power density of 2.8 KW kg^{-1} has been reported with excellent cyclic stability of 93.2% capacitance retention after 5000 cycles [110]. The utilization of a hybrid electrode offers a new class of electrode material, which utilizes a synergetic charge storage mechanism.

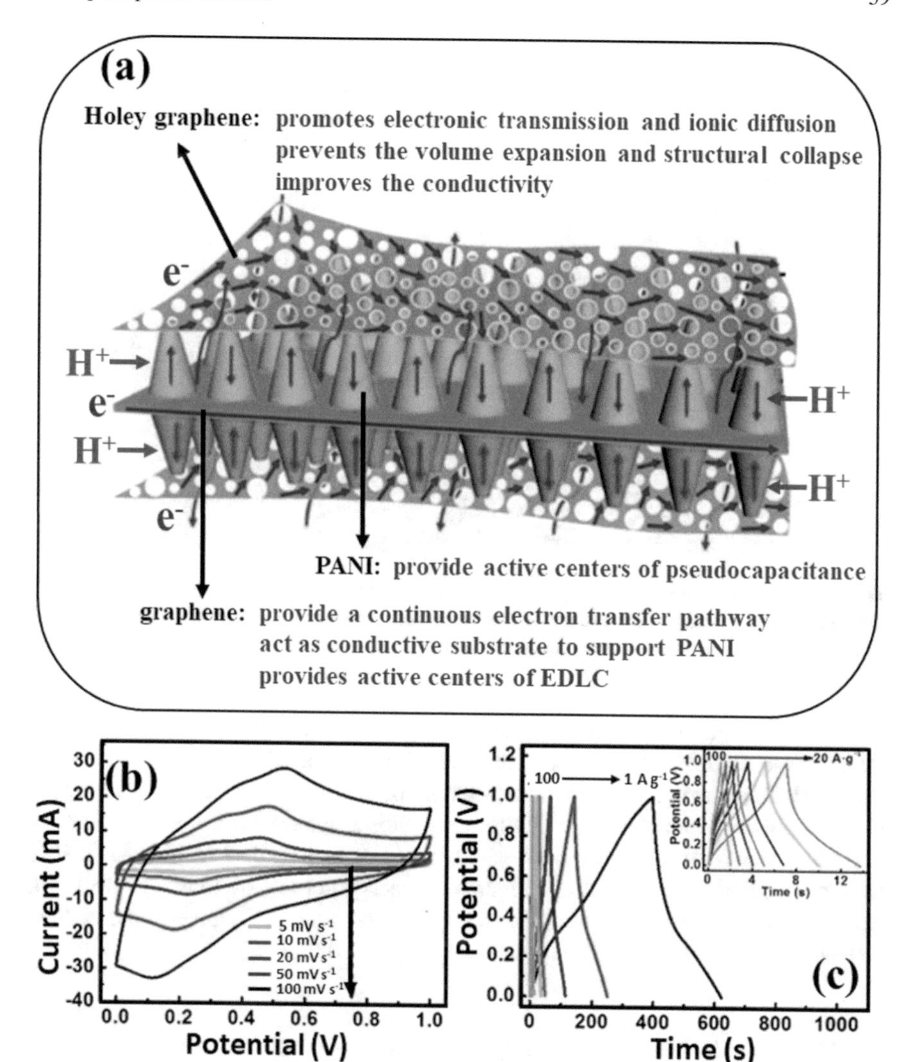

Fig. 2.9 **a** Schematic representation of the energy-storage mechanism of holey graphene/PANI/graphene nanohybrid composite, **b** CV curves at different scan rates, and **c** GCD curves at different current densities of symmetric assembled nanohybrid composite (redrawn and reprinted with permission from [109])

2.4.2 Asymmetric Supercapacitors

The design of an asymmetric supercapacitor is a remarkable approach for increasing the power density of the supercapacitor. The combination of high energy and power density can be successfully obtained from asymmetric hybrid electrode assembly [91]. This assembly covers a wide range of device configurations including diverse

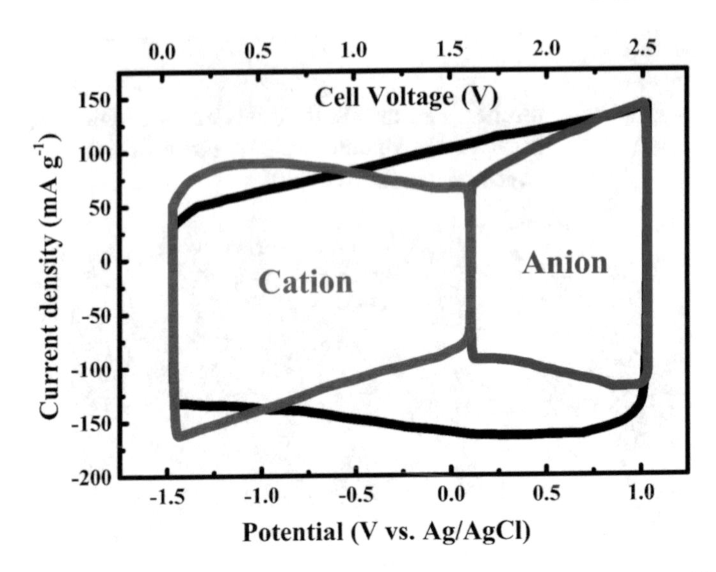

Fig. 2.10 Example of corresponding CV curves for the entire cell in asymmetric assembly (black), three-electrode CV curves of the negative electrode (red), and a positive electrode (blue) (redrawn and reprinted with permission from [112])

electrode material and electrolyte solutions. During the charge-discharge process, the asymmetric device takes full advantage of each electrode potential as shown in Fig. 2.10. This maximizes the overall potential of the full cell. For example, aqueous electrolyte-based supercapacitor limits its working potential till 1–1.2 V. But, by designing asymmetric supercapacitor it can get extended till 2 V [84, 111]. It is well known that working potential increases energy density by fourfolds [59]. So, considerable efforts have been made to increase the working voltage of the device. One such effort is the design of asymmetric supercapacitor [6].

2.4.2.1 EDLC—Metal Oxide and Its Composite-Based Asymmetric Supercapacitors

Transition metal oxide and carbon-based material store charge by different charge storage mechanism. Both materials suffer from inherent limitations, which can be addressed by the formation of composite and designing asymmetric assembly. In addition to this, Hwang et al. have studied the behavior of metal oxide and carbon material in asymmetric supercapacitors [113]. As a positive electrode, laser-scribed graphene (LSG)/hydrous RuO_2 composite has been used and activated carbon as a negative electrode material. The electron microscopy of the composite electrode shows that RuO_2 nanoparticles are well dispersed onto the surface of graphene. This prevents the agglomeration of RuO_2 nanoparticle. The composite delivers excellent pseudocapacitive properties, where RuO_2 nanoparticles undergo fast reversible redox reaction and graphene provides a resistance-free pathway for electron transportation. The CV curve of each electrode material in the three-electrode system is used to obtain a mass balance ratio. The assembled asymmetric supercapacitor provides the maximum output voltage of 1.8 V [113]. In addition to this, Wang et al. have reported NiO/graphene foam (GF)-based hybrid electrode, which delivers high capacitance of 1225 F g^{-1} at a current density of 2 A g^{-1} [114]. The asymmetric device assembly with a hybrid electrode as positive and CNT as cathode shows an energy density of 32 Wh kg^{-1} at a power density of 700 W kg^{-1} in aqueous electrolyte. The device displays excellent cyclic stability with 94% capacitance retention after 2000 cycles [114]. Another study utilizes V_2O_5-CNT composite as a positive electrode and carbon fiber as a negative electrode. The asymmetric assembly delivers an energy density of 46 Wh kg^{-1} along with a power density of 5.26 W kg^{-1} [115]. Fe_2O_3 and Fe_3O_4 have gained significant attention, owing to their low cost, environment friendliness, and high specific capacitance. However, poor conductivity and large volume expansion during charging and discharging limit its practical performance. In this vein, Sheng et al. have designed asymmetric assembly, where Fe_3O_4 nanosphere decorated in graphene nanosheet act as negative electrode and graphene MnO_2 composite as positive electrode [116]. The working potential of asymmetric assembly extends up to 2 V without any polarization in aqueous 1 M Na_2SO_4 electrolyte. The device delivers energy density 87.6 Wh kg^{-1} at a power density of 394 W kg^{-1}, along with excellent cyclic stability of 93.1% after 10,000 cycles. The work demonstrates that the combination of novel graphene structure and metal oxide enhances the electrical conductivity and prevents agglomeration of metal oxide [116]. Another literature demonstrates the utilization of nanostructured MnO_2//graphene as an electrode for asymmetric devices. Gao et al. have demonstrated asymmetric devices with three-dimensional interconnected porous graphene hydrogel as a cathode and vertically aligned MnO_2 nanoplates in nickel foam as an anode in aqueous Na_2SO_4 electrolyte as shown in Fig. 2.11a [117]. Figure 2.11b depicts the three electrodes CV measurement of both the electrodes. The stable potential window of 0–2 V has been reported due to the asymmetry assembly of both the material as shown in Fig. 2.11c. The asymmetric device delivers near rectangular CV curves even at a high scan rate of 100 mV s^{-1} as shown in Fig. 2.11d. The GCD tests have been performed at different current density shown in Fig. 2.11e, which shows a symmetric plot. The maximum specific capacitance of 41.7 F g^{-1} has been reported at a current density of 1 A g^{-1} which retains up to 26.8 F g^{-1} at a high current density of 10 A g^{-1} as shown in Fig. 2.11f. Figure 2.11g shows the Ragone plot of comparative value for the asymmetric and symmetric devices. The energy density of asymmetric supercapacitor graphene hydrogel//MnO_2-Nickle foam delivers a high energy density of 23.2 Wh kg^{-1} at a power density of 1 kW kg^{-1} along with the stable cyclic performance of 83.4% capacitance retention after 5000 cycles [117].

The combination of carbon and metal oxide can provide enhance energy density, power density, rate capability, cyclic stability, and capacitance value. Due to their simple synthesis technique and abundance, these materials can be used to fabricate a high-performance supercapacitor [47].

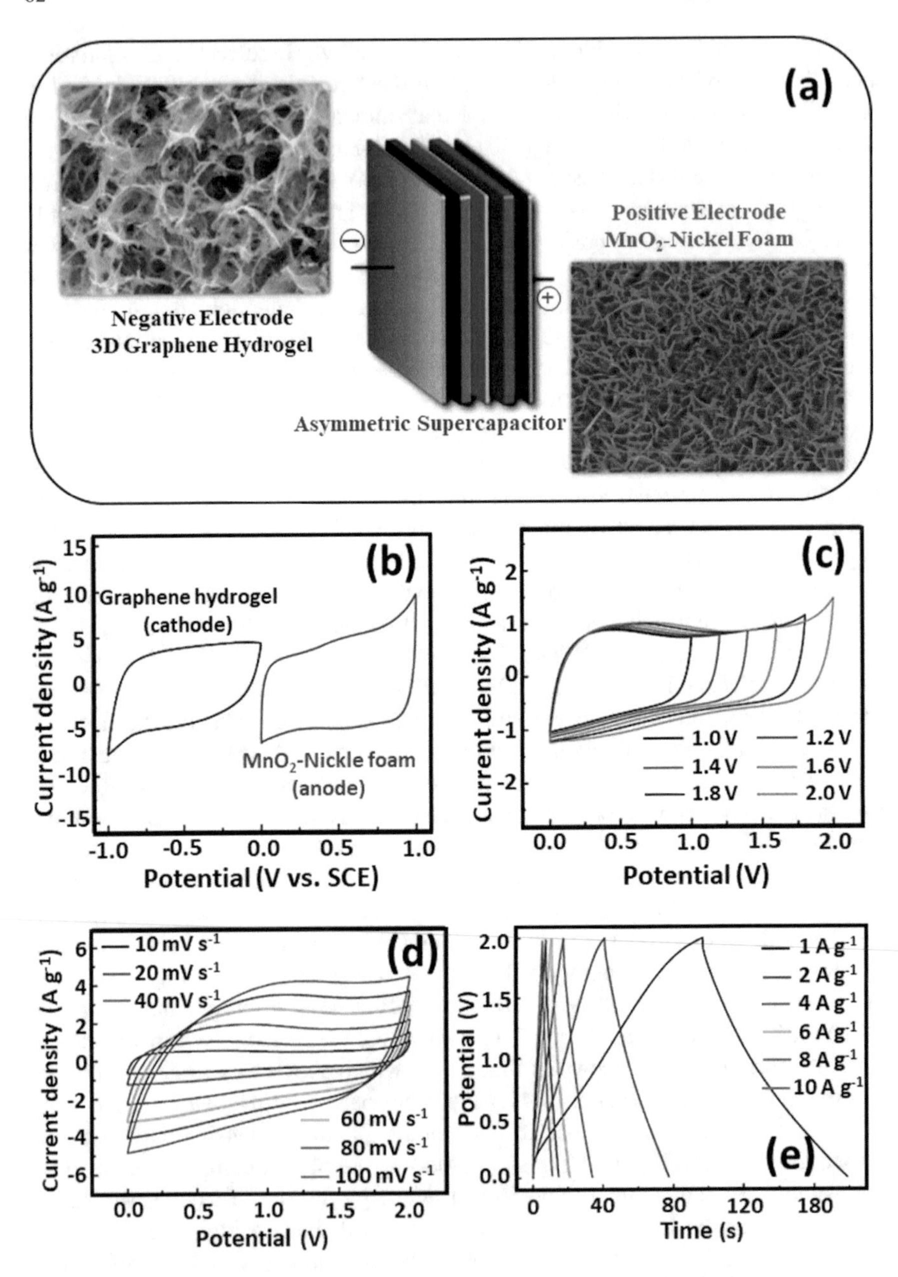

Fig. 2.11 **a** Schematic representation of asymmetric assembly, **b** comparative cyclic voltammograms of graphene hydrogel and MnO_2-nickel foam in a three-electrode system at a scan rate of 20 mV s^{-1}, **c** CV at various potential windows at a scan rate of 20 mV s^{-1}, **d** CV at different scan rates, **e** GCD at various current densities, **f** capacitance retention ratio as a function of discharge currents of the asymmetric supercapacitor of graphene hydrogel//MnO_2-nickel foam measured, and **g** comparative Ragone plots of the asymmetric supercapacitor of graphene hydrogel//MnO_2-nickel foam, and symmetric supercapacitors of MnO_2-nickel foam//MnO_2-nickel foam and graphene hydrogel//graphene hydrogel (redrawn and reprinted with permission from [117])

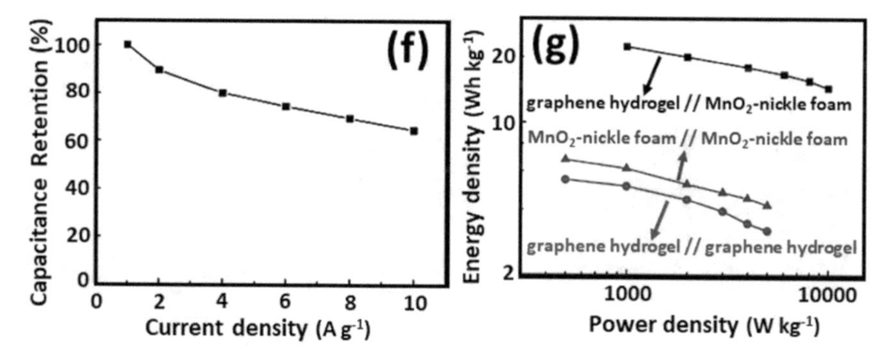

Fig. 2.11 (continued)

2.4.2.2 EDLC—Conducting Polymer and Its Composite-Based Asymmetric Supercapacitors

Carbon material as a capacitive electrode has gained a lot of attention in the field of the supercapacitor. But its low capacitance has paved the path for alternative electrode material. On the other hand, it is well known that carbon material exhibits many outstanding properties such as high electronic conductivity, high surface area, high mechanical stability, and most important excellent cyclic stability. The amalgamation of conducting polymer and carbon material shows better electrochemical properties. In conducting polymer, the charges are stored via a redox reaction. During oxidation, the ions get transferred to the polymer backbone, while during reduction ion gets released back to the electrolyte solution. The charging and discharging process occurs along the bulk volume of the conducting polymer. This shows the major advantage of conducting polymer over carbon materials, where surface participates in charge storage mechanism [50, 107]. It possesses a high charge density and low cost. This makes it possible to construct a low resistance device and high energy and power density device. Lota et al. assembled asymmetric supercapacitor with 80% PEDOT-20% acetylene black, which acts as a positive electrode and activated carbon as a negative electrode [118]. The assembled device delivers excellent specific capacitance of 160 F g^{-1} with minimal loss of capacitance after 2000 cycles [118]. Another literature by Shen et al. has assembled graphene nanosheet/carboxylated multi-walled carbon nanotube/polyaniline (sGNS/cMWCNT/PANI) nanocomposites as anode and activated porous graphene (aGNS) as cathode for asymmetric supercapacitor as shown in Fig. 2.12a [119]. Hierarchical porous graphene/CNT/PANI composite facilitates charge transportation. The well-ordered composite structure also increases the electrochemical performance of PANI. Figure 2.12b shows the CV curve of the as-fabricated asymmetric device at the different potential windows. The device exhibits good capacitive behavior with a quasi-rectangular curve till potential 1.6 V. After 1.6 V, the curve deviates from rectangularity due to the decomposition of PANI. Also, the CV curve at different scan rates in Fig. 2.12c shows a pair of redox peaks suggesting Faradaic reactions; however, the curve dominates the capacitive type

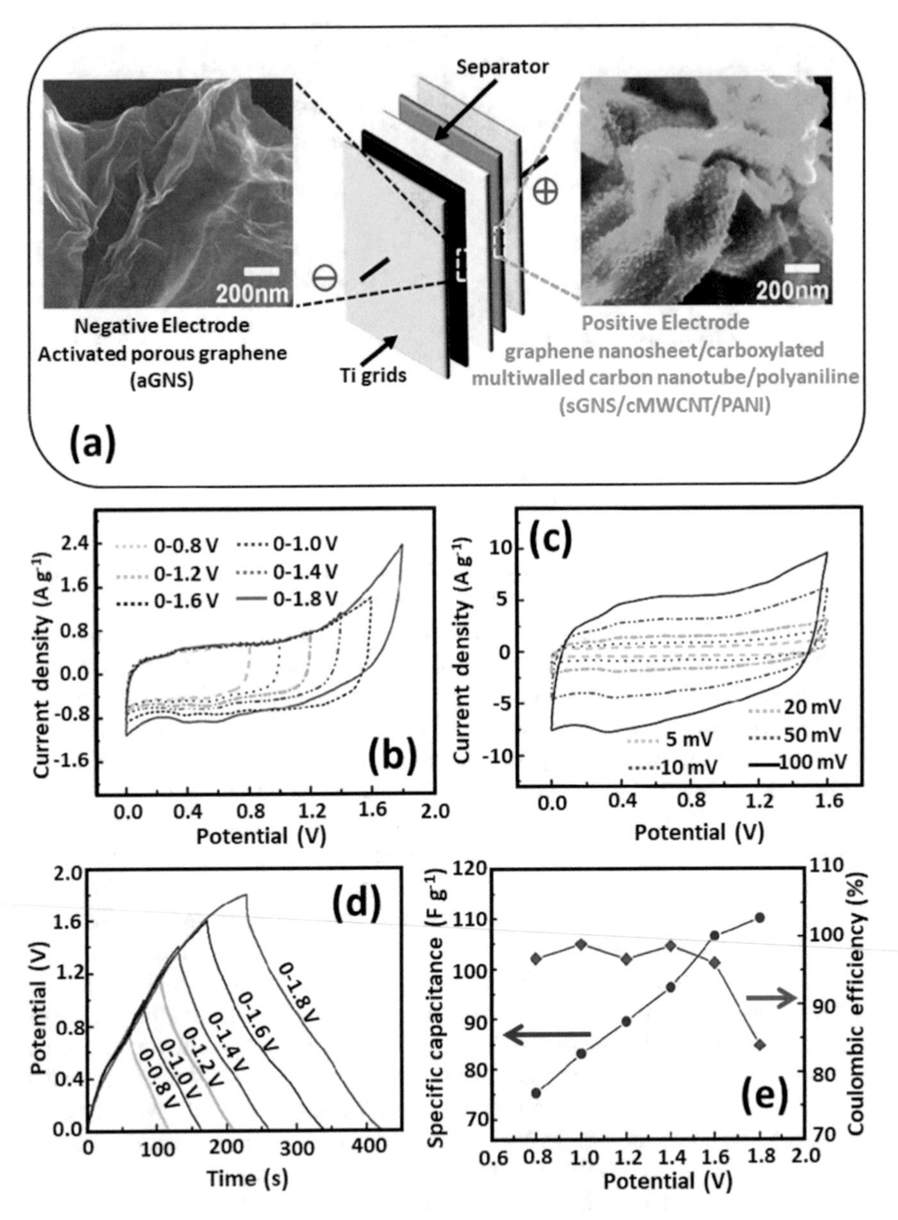

Fig. 2.12 **a** Schematic illustration of the asymmetric device based on an aGNS as the negative electrode and sGNS/cWMCNT/PANI composite as the positive electrode, electrochemical performance of the fabricated asymmetric device in 1 M H_2SO_4, **b** CV curves measured at various potential windows at a scan rate of 5 mV s^{-1}, **c** CV curves at different scan rates with a maximum potential window of 1.6 V, **d** GCD plot at the various potential window at a current density of 1 A g^{-1}, and **e** specific capacitances and coulombic efficiencies as a function of the various potential windows (redrawn and reprinted with permission from [119])

charge storage mechanism. Figure 2.12 shows the GCD plot at a current density of 1 A g^{-1} in the various potential windows. The charge and discharge curve remains in symmetry until the potential of 1.6 V, indicating excellent reversibility. Figure 2.12e shows the specific capacitance and coulombic efficiency of the asymmetric device at the different potential ranges. The capacitance increases from 75 to 17 F g^{-1} from 0.8 to 1.6 V. Moreover, the coulombic efficiency remains above 95% throughout the potential range till 1.6 V. After 1.6 V the coulombic efficiency decreases due to prominent polarization by decomposition of PANI. The device delivers a maximum energy density of 41.5 Wh kg^{-1} and a power density of 1667 W kg^{-1}. In addition to this, the fabricated asymmetric device depicts 91.4% capacitance retention after 5000 cycles [119].

Applications of conducting polymer as anode and carbon-based material as a cathode are the most promising candidate for the fabrication of high efficient, flexible, and wearable supercapacitor devices. Asymmetric configuration delivers a large cycle life and small voltage fluctuation. A range of composite material can be fabricated to get promising electrode material [47, 107].

2.4.2.3 Metal Oxides—Conducting Polymer and Its Composite-Based Asymmetric Supercapacitors

Metal oxide and conducting polymer are pseudocapacitive electrode materials. The combination of both material either as composite or asymmetric assembly can provide high energy densities. Since both materials undergo fast and reversible redox reactions, they offer superior electrochemical performance [120]. In this vein, Peng et al. have synthesized template confined growth of poly(4-aminodiphenylamine) (P(4-ADPA)) as a positive electrode and $W_{18}O_{49}$ as a negative electrode for the assembly of asymmetric supercapacitor [121]. P(4-ADPA) provides an integral nano-framework structure with intertwined and uniform nanosheets. Additionally, $W_{18}O_{49}$ possesses multiple oxidation states due to the intercalation of electrons and protons in the monoclinic structure. The as-assembled P(4-ADPA)//$W_{18}O_{49}$ asymmetric supercapacitor in H_2SO_4 electrolyte offers a wide operating voltage of 1.5 V. Figure 2.13a shows the working potential of the positive and negative electrodes in the three-electrode system. The CV curve of P(4-ADPA)//$W_{18}O_{49}$ device (Fig. 2.13b) exhibits a distorted semi-rectangular shape with obvious redox peaks. The symmetric GCD plot as shown in Fig. 2.13c peak indicates reversible and fast pseudocapacitance behavior. The device delivers an energy density of 24.4 Wh kg^{-1} at the power density of 1491 W kg^{-1}. Figure 2.13d depicts the cyclic stability with 92% capacitance retention after 10,000 cycles. However, the specific capacitance initially increases till 2500 cycles owing to the improvement of electrode activation and surface wettability. The outstanding performance of assembled asymmetric supercapacitor is due to a reasonable matching of the positive and negative electrodes [121].

It is well known that the composite electrode of metal oxide and conducting polymer provides superior electrochemical performance. Another literature by Zhang et al. has fabricated an asymmetric assembly supercapacitor device, where

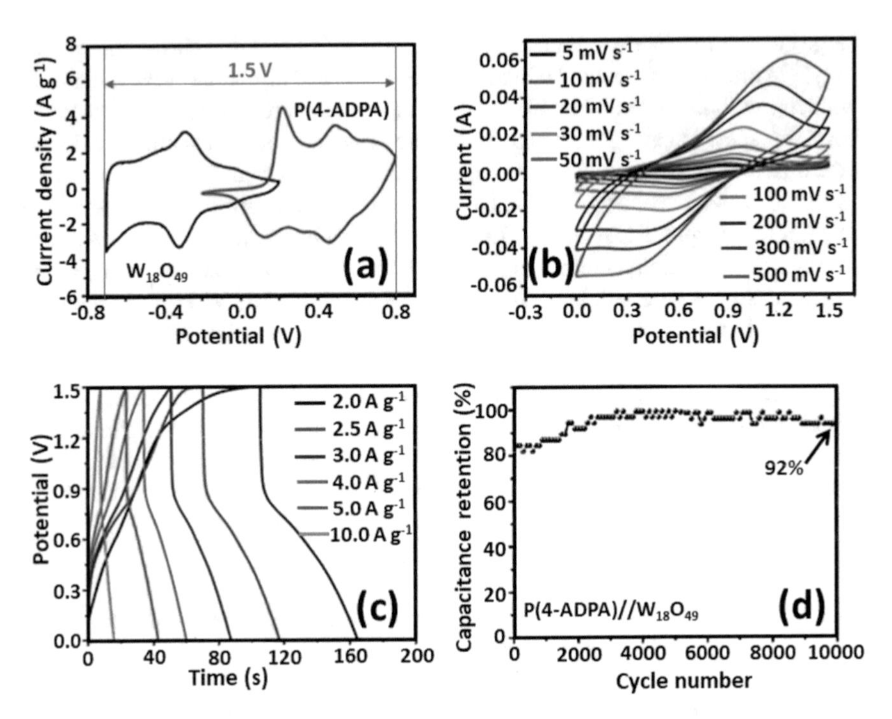

Fig. 2.13 **a** Comparative CV curves of P(4-ADPA) and $W_{18}O_{49}$ electrodes in a three-electrode cell attested in 1 M H_2SO_4 electrolyte at a scan rate of 5 mV s^{-1}, **b** CV curves at various scan rates, **c** GCD curves at different current densities, and **d** cycling stability of P(4-ADPA)//$W_{18}O_{49}$ asymmetric supercapacitor (redrawn and reprinted with permission from [121])

$NiCo_2O_4/MnO_2$ servers as the positive electrode and MoO_3/PPy composite as a negative electrode [122]. The fabrication of both the electrode materials is depicted in Fig. 2.14a,b. The heterostructure electrode results in better cyclic stability and resistance-free charge transportation. $NiCo_2O_4/MnO_2$ structure provides large charge storage capacity than the pristine $NiCo_2O_4$, indicating that the composite possesses high areal capacity. The thin coating of MnO_2 at $NiCo_2O_4$ provides additional capacitance, where MnO_2 can absorb electrolyte ion at the electrode surface and easily intercalate and de-intercalated electrolytic ions near the electrode surface. As a negative electrode material, MoO_3/PPy, polypyrrole serves as a conducting medium, which enhances the electrical conductivity of the MoO_3 metal oxide. Evaluation of asymmetric assembly of the two electrodes enables the expansion of electrochemical stable potential window as shown in Fig. 2.14c. The operating potential of the device expands until 1.6 V in 1 M Na_2SO_4 aqueous electrolyte. Figure 2.14d shows the CV curve of asymmetric assembly at various scan rates. It clearly depicts that the CV curve remains undistorted even at a high scan rate of 100 mV s^{-1}, indicating good rate capability. The symmetrical charge-discharge behavior as shown in Fig. 2.14e implies good capacitance property. The specific capacitance of 156.5 F g^{-1} is reported at an areal current density of 4 mA cm^{-2}. A high energy density of

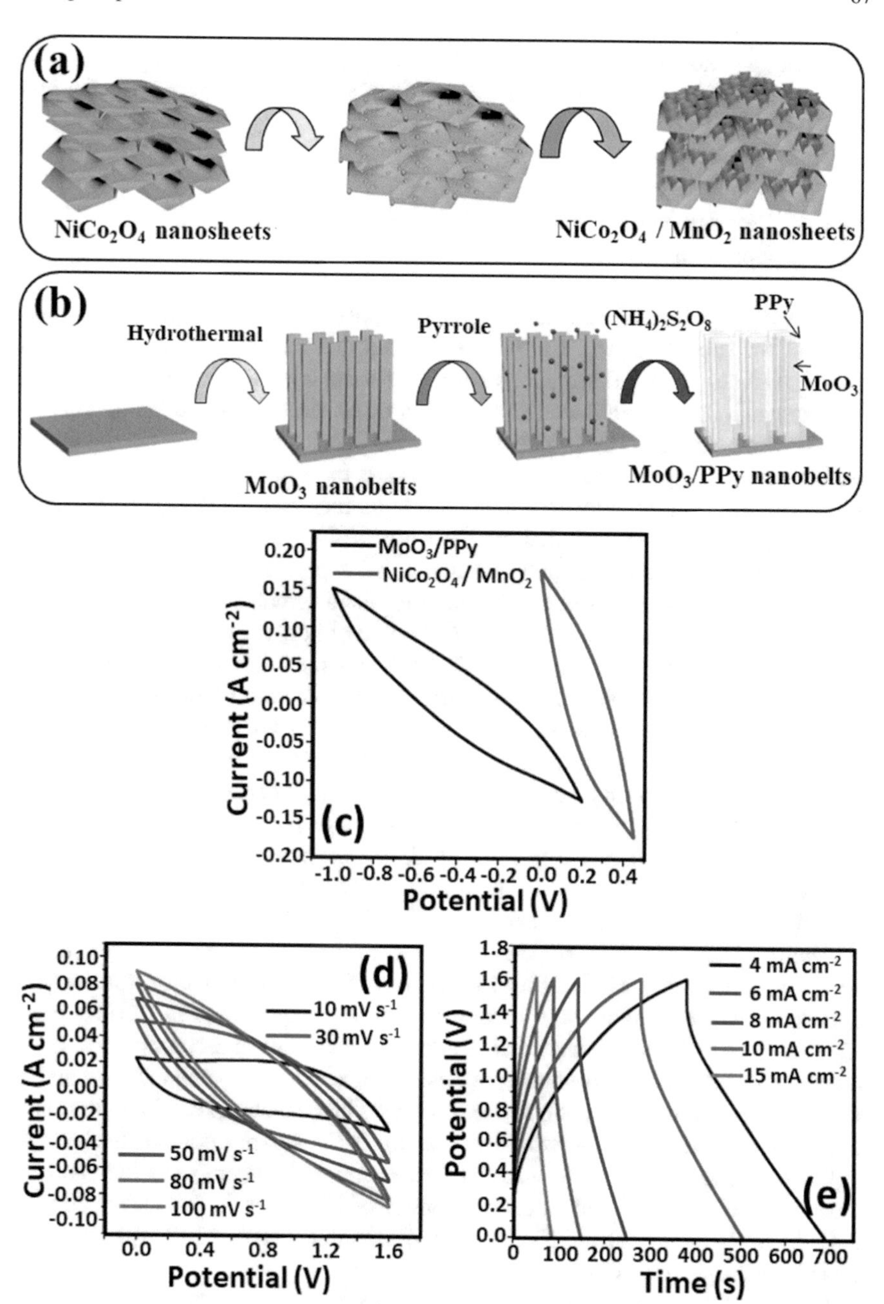

Fig. 2.14 Schematic diagram representing the synthesis procedure of **a** $NiCo_2O_4/MnO_2$ and **b** MoO_3/PPy, **c** comparative CV curves of $NiCo_2O_4/MnO_2$ nanoflakes and MoO_3/PPy nanobelts electrodes performed in a three-electrode cell, **d** CV curves at different scan rates, and **e** GCD curves various current densities of asymmetric assembled MoO_3/PPy//$NiCo_2O_4/MnO_2$ supercapacitor (redrawn and reprinted with permission from [122])

60.4 Wh kg^{-1} and a maximum power density of 2400 W kg^{-1} are reported. Also, the assembled device delivers a high cycle life of 88.2% capacitance retention after 10,000 cycles [122].

2.4.2.4 Ternary Nanocomposites Electrode-Based Asymmetric Supercapacitors

Till now carbon, metal oxide, and conducting polymer have emerged as electrode material for supercapacitor. Each class of electrode material shows some of the excellent electrochemical properties but is also associated with inherent limitation. Ternary nanocomposite allows the utilization of all class of electrode material in a single device. The synergistic effect of different species would enhance the overall performance of diverse electrode material, where the combined effect of diffusion, ion adsorption, and ion transportation leads to an increment in device performance [91]. He et al. have demonstrated the $FeCo_2O_4$@PPy hybrid electrode material via hydrothermal reaction and oxidative polymerization process [123]. The hybrid electrode acts as a positive electrode and porous activated carbon as a negative electrode. Figure 2.15a shows the three-electrode CV curve of both the electrodes, exhibiting an overall stable potential window of 0–1.6 V. Figure 2.15b shows the CV curve of $FeCo_2O_4$@PPy//activated carbon asymmetric device with prominent redox peak indicating the Faradaic process. The near-symmetrical GCD curve in Fig. 2.15c shows good reversibility during the charge and discharge processes. The device delivers a specific capacitance of 194 F g^{-1} at a current density of 1 A g^{-1}, which retains up to 146 F g^{-1} at a current density of 20 A g^{-1}. This shows a high rate capability of 75% as depicted in Fig. 2.15d. The asymmetric device shows a high energy density of 68.8 Wh kg^{-1} at a power density of 0.82 kW kg^{-1}, which retains up to 52 Wh kg^{-1} at a power density of 15.5 kW kg^{-1} [123].

Another literature displays the ternary combination of the electrode in the asymmetric assembly. Nallappan et al. have studied the asymmetric assembly of CeO_2/polyaniline as anode and carbon as cathode material [124]. Cerium oxide offers multiple oxidation states and it is one of abundant material on earth crust. However, its intrinsic nature shows poor electronic conductivity and low mechanical stability. On the other hand, polyaniline possesses high electrical conductivity, and it is a low-cost material. The attempt has been made to prepare a composite electrode of CeO_2/polyaniline as anode material. The asymmetric assembly delivers a high capacitance of 60 F g^{-1} and an energy density of 18.75 F g^{-1}. The introduction of CeO_2 finds beneficial material for the development of low-cost and high-performance hybrid supercapacitor [124].

Additionally, Zhou et al. demonstrate asymmetric supercapacitor using 3D CoO@polypyrrole nanowire array as anode material [125]. The well-aligned CoO nanowire array has been grown on three-dimensional nickel foam with polypyrrole as depicted in Fig. 2.16a. Each nanowire of CoO@polypyrrole boosts the pseudocapacitive performance. The high electronic conductivity of polypyrrole and short diffusion path of nanowire increases the electrochemical activity of the composite material. The

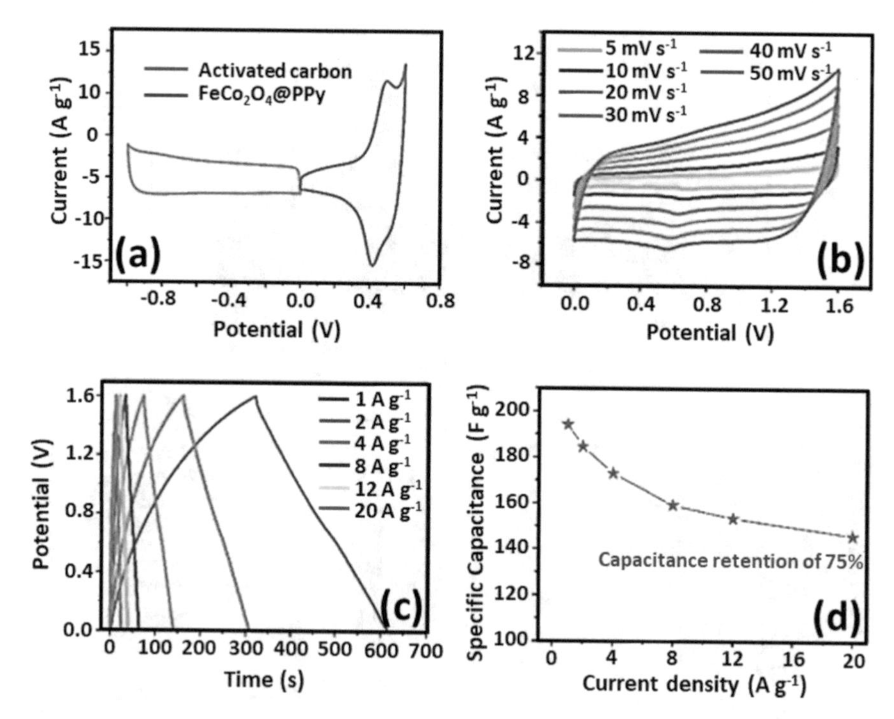

Fig. 2.15 **a** Comparative CV curves in the three-electrode system of the $FeCo_2O_4$@PPy and activated carbon electrodes at a scan rate of 5 mV s^{-1}, **b** CV curves at various scan rates, **c** GCD curve at various current densities, and **d** capacitance retention over different current densities of the $FeCo_2O_4$@PPy//activated carbon asymmetric supercapacitor (redrawn and reprinted with permission from [123])

electrochemical performance of asymmetric assembly of CoO@polypyrrole as anode and activated carbon a cathode material has been tested in 3 M NaOH electrolyte as shown in Fig. 2.16b. The CV curve of asymmetric device displays quasi-rectangular behavior with broad hump representing the Faradaic charge storage mechanism. The cell offers a wide operating potential of 1.8 V. The discharge curves at various current densities have been illustrated in Fig. 2.16c. From the discharge curve, the energy density has been calculated as 43.5 Wh kg^{-1} at a power density of 87.5 W kg^{-1} [125]. The ternary assembly opens up the possibility to engineer promising pseudocapacitive and electric double layer electrode material to design hybrid electrode architecture for energy storage devices.

2.4.3 Battery Supercapacitor Hybrid Devices

Battery supercapacitor-type assembly of hybrid device displays the amalgamation of two different types of the electrode, one having capacitive mechanism and other

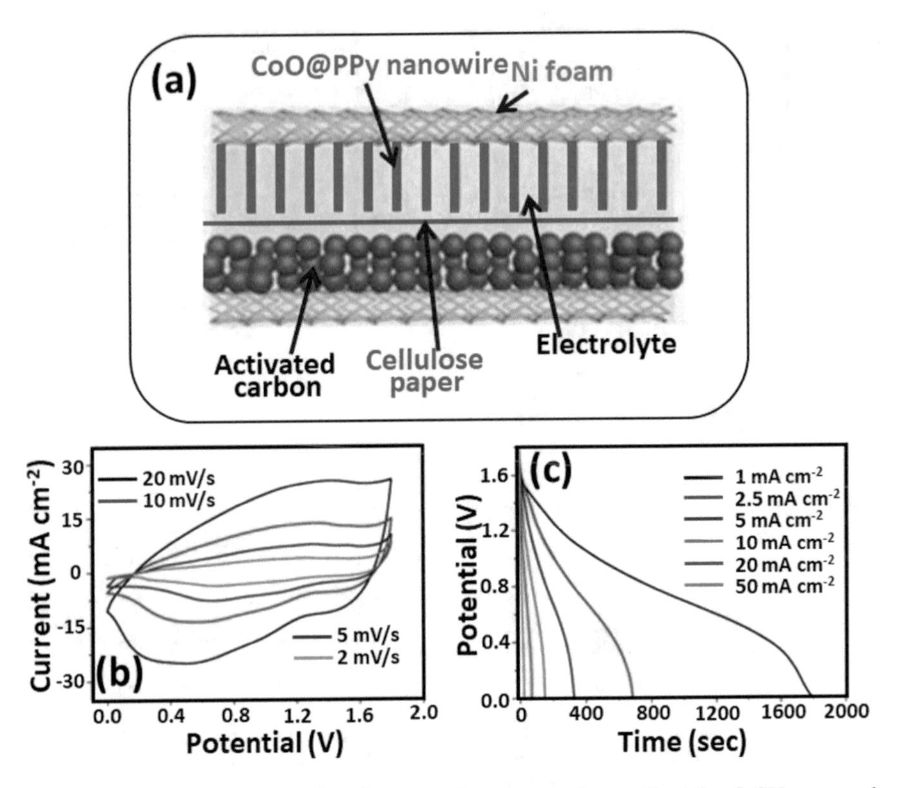

Fig. 2.16 **a** Schematic representation of asymmetric supercapacitor configuration, **b** CV curve, and **c** GCD plots of the CoO@PPy//activated carbon asymmetric supercapacitor(redrawn and reprinted with permission from [125])

as battery mechanism. This assembly aims to accelerate the energy feature of the supercapacitor electrode and enhances the power characteristics of the battery electrode. Battery supercapacitor hybrid device has attracted significant attention due to its application in smart electric grids, hybrid electric vehicles, and miniaturized electronic and optoelectronic devices [49]. It has the advantage of the high electrochemical performance, low cost, safe to handle, and environment friendly [49, 50]. During charge and discharge process, anion and cation move toward the electrode. Ion adsorption and rapid charge transfer occur at the capacitive electrode and bulk redox reaction occurs at the battery-type electrode [50]. Battery supercapacitor-type hybrid device provides several combinations to design a variety of energy storage devices, using diverse electrode and electrolyte material, and device configuration. These devices can eliminate the energy density of supercapacitors due to the presence of high capacity battery-type electrode and surpass low power density of batteries because of the capacitive electrode [49, 91]. Also, the hybrid design of using two entirely diverse electrodes enables fast electrochemical kinetics. The energy density in these devices is improved by two approaches:

(a) improvement in capacitance due to the introduction of battery type electrode and
(b) expansion of electrochemical stable potential window, because of the utilization of asymmetric assembly.

Battery supercapacitor hybrid can be classified as Li-ion capacitor and Na-ion capacitor. Organic electrolytes such as $LiPF_6$, $LiClO_4$, $NaPF_6$, and $NaClO_4$ are widely used electrolytes. Aqueous electrolyte H_2SO_4, CH_3SO_3OH, KOH, NaOH, and $LiSO_4$ are also used in hybrid assembly [49].

2.4.3.1 Li-Ion Battery Supercapacitor Hybrid

Li-ion battery is a widely used electrochemical energy storage device. For the development of the battery supercapacitor device, the use of Li-ion is the primary choice [49, 60, 126]. Nanostructured metal oxide electrode and their hybrid with carbon material are widely used electrode material. Hemmati et al. have studied the nitrogen-doped T-Nb_2O_5/tubular carbon hybrid electrode [127]. The structure possesses robust frameworks, which consist of nanostructure active material that is encapsulated in an electronic conducting framework. The hollow structure and a large number of voids promote fast electron/ion transport. The Li-ion hybrid system has been assembled, where the prepared composite acts as anode and activated carbon as a cathode. A high energy density of 86.6 Wh kg^{-1} is achieved with a high power density of 6.09 kW kg^{-1}. The excellent performance is due to the wide operating potential window of 0–3 V [127]. Naoi et al. have used the ultracentrifuge technique to synthesize the nano-$LiTi_5O_{12}$/carbon composite electrode as anode material [111]. The electrode shows a high capacity of 78 mAh g^{-1} at 1200 C. The full cell can deliver the energy density of 28 Wh L^{-1} at a power density of 10 kW L^{-1} [111]. This shows that the synthesis of the nanostructure electrode can provide extremely high capacity, which can be efficiently utilized as battery supercapacitor hybrid electrode material. Another study carried out by Brandt et al. assembled high power Li-ion-based battery supercapacitor hybrid device [128]. $LiNi_{0.5}Mn_{1.5}O_4$ and activated carbon are used as electrode material. The cell voltage of 0–3.3 V is reported along with 89% capacitance retention after 4000 cycles. Also, the specific energy is 50 Wh kg^{-1} at a power density of 1100 W kg^{-1}. High working voltage allows a high chance of intercalation and de-intercalation reaction at $LiNi_{0.5}Mn_{1.5}O_4$ [128].

Yang et al. have fabricated high power Li-ion hybrid supercapacitor using ultra three-dimensional structure [129]. The maximum surface exposure facilitates a fast interfacial reaction. The assembled device demonstrates excellent energy power integration. Well-designed MnO@graphene composite acts as anode and three-dimensional hierarchical porous *N*-doped carbon nanosheet (HNC) acts as a cathode. Prelithiated MnO@GNS anode keeps anodic potential below the lithiation plateau. This allows for high operation voltage. Figure 2.17a illustrates Li-MnO@GNS//HNC hybrid device that can achieve a wide potential range of 1–4 V. The electrochemical performance has been studied by CV and GCD. The CV curves of Li-ion hybrid

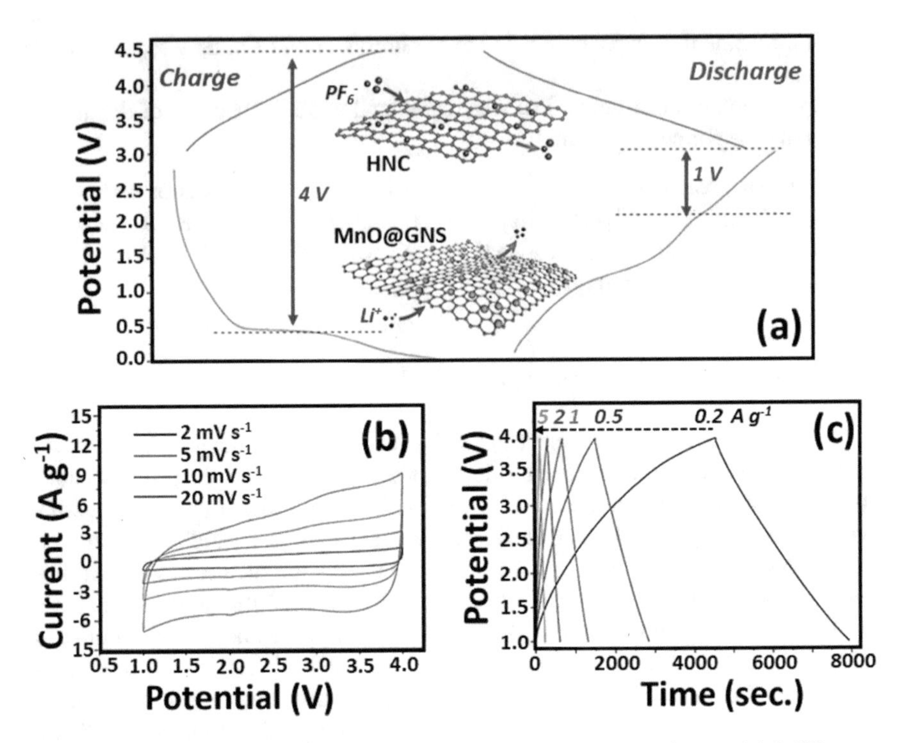

Fig. 2.17 **a** Voltage range representation of positive and negative electrode material, **b** CV curves at various scan rates, **c** GCD plots of at different current densities of MnO@GNS//HNC devices (redrawn and reprinted with permission from [129])

device demonstrate near rectangular shape, indicating fast energy storage kinetics with interfacial electrochemical reactions as depicted in Fig. 2.17b. However, broad humps are also present indicating the lithiation/delithiation plateau of MnO demonstrating intercalation pseudocapacitance behavior of MnO. The linear behavior of redox peak current over different scan rates confirms that electrochemical reactions are not diffusion controlled. The active high surface area also participates in charge storage. Figure 2.17c displays the GCD curve of a hybrid device, where near-linear behavior demonstrates capacitive behavior. The MnO@GNS//HNC hybrid device delivers high capacitance of 244 F g^{-1} at a current density of 0.2 A g^{-1} with an excellent rate capability of 66% at a high current density of 40 A g^{-1}. The hybrid Li-ion system achieves an attractive energy density of 127 Wh kg^{-1} with a high power density of 25 kW kg^{-1} with rapid charging within 8 s. The coherent design can mark a new strategy for the fabrication of a new Li-ion battery supercapacitor hybrid device [129].

2.4.3.2 Na-Ion Battery Supercapacitor Hybrid

Li-ion-based energy storage devices have been widely used in the electronic system. But due to their limited abundance, Na-ion-based energy devices have proven the best alternative. Also, sodium is an abundant material and display similar characteristics as lithium [49]. The device configuration of Na-ion-based battery supercapacitor hybrid is a similar Li-ion-based hybrid device. In this vein, Ding et al. have assembled carbon//carbon Na-ion battery supercapacitor hybrid [130]. The carbon used here is flaky, porous, and thin morphology. Carbon for anode has the morphology of thin graphene domain and cathode carbon has a large surface area. The as-assembled device shows an excellent energy density of 201 Wh kg^{-1} at power density of 285 W kg^{-1}. The device also delivers high thermal stability by widening its working temperature from 0 to 65° [130]. Battery-type electrodes such as V_2O_5, Na_xMnO_2, and $Na_4Mn_9O_{18}$ show poor electrical conductivity. Chen et al. have addressed this issue by synthesizing the porous composite architecture of V_2O_5 nanowire and carbon nanotube (CNT) [131]. CNT provides an effective electron transfer network and V_2O_5 nanowire undergoes a fast pseudocapacitive charge storage process. The as-assembled Na-ion AC//V_2O_5-CNT device delivers an energy density of 39 Wh kg^{-1} at a power density of 140 W kg^{-1} [130, 131]. Zhang et al. have studied the Na-ion battery supercapacitor hybrid using Na_xMnO_2. The nanowire structure of $Na_{0.35}MnO_2$ shows better insertion ability than a rod-like structure [132]. The assembled $Na_{0.35}MnO_2$//AC device delivers a high energy density of 42.6 Wh kg^{-1} at a working potential of 1.8 V. In order to further improve the battery supercapacitor hybrid device precise design at the micro- and nano-level structure form electrode material is most important criteria. Fast intercalation of ions and high surface area electrode can greatly enhance the device performance [132].

However, manganese oxide has dominated the energy storage device due to its diversity in oxidation state (Mn^{2+}, Mn^{3+}, Mn^{4+}) and varieties of crystal structure and morphology. Another report by Karikalan et al. has assembled three-dimensional fibrous $NaMnO_2$ network as the cathode and reduced graphene oxide (rGO) as anode for Na-ion supercapacitor battery hybrid [133]. The asymmetric assembly offers a wide potential window of 2.7 V in aqueous 2 M Na_2SO_4 electrolyte. $NaMnO_2$ and rGO both exhibit high over potential toward water oxidation and reduction process, respectively. The presence of strong Na-ion enables high overpotential during intercalation and de-intercalation into rGO and $NaMnO_2$. Also, Mn-based compounds exhibit high overpotential due to strong bonding of Mn ions with adsorbed OH^- ions. Figure 2.18 shows the schematic of the charge-discharge process taking place in $NaMnO_2$//rGO-Na-ion hybrid device. During charging, Na-ions are de-intercalated from $NaMnO_2$ and move toward rGO, where Na-ion gets intercalated via reduction of carbonyl group present in rGO (–C=O to –C–O–Na). Na-ion can also get stored by non-covalent interaction between rGO layers. During the discharge process, Na-ions reversibly get intercalated with MnO_2. The charging and discharging equations of $NaMnO_2$//rGO-Na-ion hybrid device are as follows:

$$NaMnO_2 \rightarrow MnO_2 + Na^+ + e^- \text{(cathode while charging)} \tag{2.5}$$

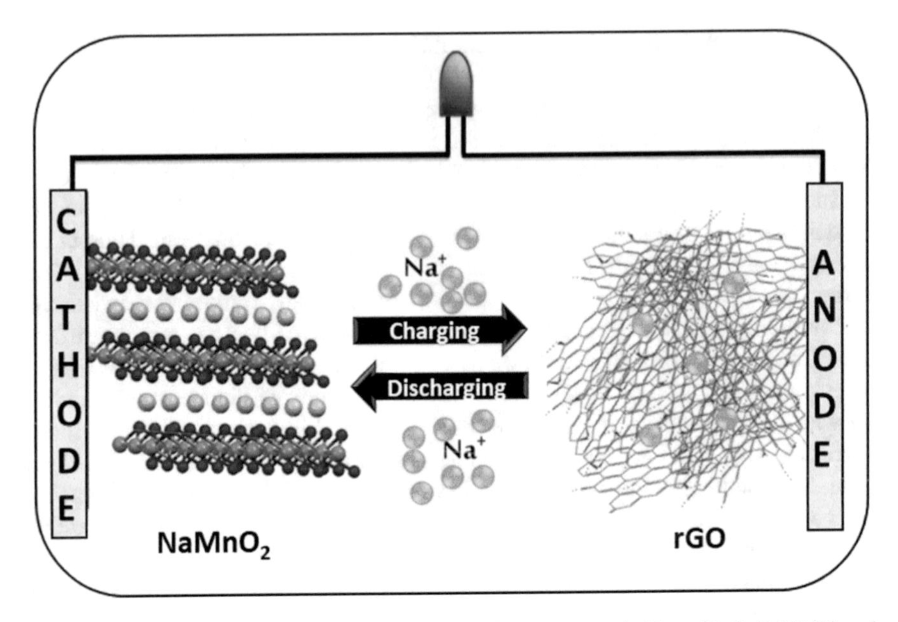

Fig. 2.18 Schematic illustration of the mechanism of the asymmetric $Na_{0.21}MnO_2$//rGO SC and its Na-ion intercalation/de-intercalation (redrawn and reprinted with permission from [133])

$$\mathrm{rGO} + \mathrm{Na}^+ + \mathrm{e}^- \rightarrow \mathrm{rGO} - \mathrm{Na}^+ (\text{anode while charging}) \tag{2.6}$$

$$\mathrm{NaMnO_2} + \mathrm{rGO} \rightarrow \mathrm{MnO_2} + \mathrm{rGO} - \mathrm{Na}^+ (\text{overall charging reaction}) \tag{2.7}$$

$$\mathrm{MnO_2} + \mathrm{Na}^+ + \mathrm{e}^- \rightarrow \mathrm{NaMnO_2} (\text{cathode while discharging}) \tag{2.8}$$

$$\mathrm{rGo} - \mathrm{Na}^+ \rightarrow \mathrm{rGO} + \mathrm{Na}^+ + \mathrm{e}^- (\text{anode while discharging}) \tag{2.9}$$

$$\mathrm{MnO_2} + \mathrm{rGO} - \mathrm{Na}^+ \rightarrow \mathrm{NaMnO_2} + \mathrm{rGO} (\text{overall discharging reaction}) \tag{2.10}$$

The device offers a high energy density of 31.8 Wh kg^{-1} at a power density of 1.35 kW kg^{-1}, which retains up to 7.11 Wh kg^{-1} at 8 kW kg^{-1}. It also delivers high cyclic stability of 86.7% capacitance retention after 1000 cycles [133].

However, the progress in the battery supercapacitor hybrid device is still in its initial stage as a lot of complex reactions are involved during charge storage. Simulations and theoretical calculation of electrode and electrolyte are required to study possible electrochemical mechanism involved. Also, in situ spectroscopy and microscopy can be helpful to upgrade the understanding of the mechanism involved in charge storage in a supercapacitor battery hybrid system [91].

2.5 Concluding Remarks

Tremendous studies have been carried out in the field of the supercapacitor to obtain a highly efficient supercapacitor. Electrochemical supercapacitors have bridged the gap between batteries and conventional capacitors. However, the key challenge is to address its poor energy density. Several efforts have been carried out to obtain a high energy density device without hampering other parameters. The efforts include (a) synthesis of composite electrode materials, which can store a large number of charge and deliver high cycle stability. The composite electrodes include the combination of carbon-based EDLC electrode and metal oxide and conducting polymer-based pseudocapacitor electrode. Various combinations of these materials are used to synthesize high-performance electrodes, where pseudocapacitive electrode provides high capacitance and carbon provides electrostatic capacitance, conductivity throughout the electrode material, and high cyclic stability, (b) assembly of the asymmetric supercapacitor, which uses different types of the electrode material. This design stretches the working potential of the device by utilizing the stable electrochemical window of each electrode. This assembly enhances the energy density of the whole device, (c) introduction of battery type electrode to design battery supercapacitor hybrid device. This asymmetric device utilizes the capacitive behavior of the supercapacitor electrode at one electrode and battery-type bulk redox reaction at another electrode. The simple charge storage phenomena of electrochemical supercapacitors have made it possible to be a highly promising candidate for a high-performance energy storage device.

Acknowledgements The authors acknowledge the financial support provided by the Science and Engineering Research Board, Department of Science and Technology, India (SR/WOS-A/ET-48/2018) for carrying out this research work.

References

1. B.E. Conway, *Electrochemical Supercapacitors* (Springer, US, Boston, MA, 1999)
2. M. Vangari, T. Pryor, L. Jiang, J. Energy Eng. **139**, 72 (2013)
3. K.K. Kar (ed.), *Handbook of Nanocomposite Supercapacitor Materials II* (Springer International Publishing, Cham, 2020)
4. J. Tahalyani, M.J. Akhtar, J. Cherusseri, K.K. Kar, in *Handbook of Nanocomposite Supercapacitor Materials I Characteristics*, ed. by K.K. Kar (Springer International Publishing, 2020), pp. 1–51
5. S. Banerjee, P. Sinha, K.D. Verma, T. Pal, B. De, J. Cherusseri, P.K. Manna, K.K. Kar, in *Handbook of Nanocomposite Supercapacitor Materials I Characteristics*, ed. by K.K. Kar (Springer International Publishing, 2020), pp. 53–89
6. Y. Shao, M.F. El-Kady, J. Sun, Y. Li, Q. Zhang, M. Zhu, H. Wang, B. Dunn, R.B. Kaner, Chem. Rev. **118**, 9233 (2018)
7. B.K. Kim, S. Sy, A. Yu, J. Zhang, in *Handbook of Clean Energy Systems*, ed. by R.-S. Liu, L. Zhang, X. Sun, H. Liu, J. Zhang (Wiley, Ltd, Chichester, UK, 2015), pp. 1–25
8. A. Burke, J. Power Sources **91**, 37 (2000)

9. R. Kumar, S. Sahoo, E. Joanni, R.K. Singh, W.K. Tan, K.K. Kar, A. Matsuda, Prog. Energy Combust. Sci. **75**, 100786 (2019)
10. J. Li, K. Liu, X. Gao, B. Yao, K. Huo, Y. Cheng, X. Cheng, D. Chen, B. Wang, W. Sun, D. Ding, M. Liu, L. Huang, ACS Appl. Mater. Interfaces **7**, 24622 (2015)
11. S.M. Chen, R. Ramachandran, V. Mani, R. Saraswathi, Int. J. Electrochem. Sci. **9**, 4072 (2014)
12. S. Banerjee, B. De, P. Sinha, J. Cherusseri, K.K. Kar, in *Handbook of Nanocomposite Supercapacitor Materials I Characteristics*, ed. by K.K. Kar (Springer International Publishing, 2020), pp. 341–350
13. R. Nigam, K.D. Verma, T. Pal, K.K. Kar, in *Handbook of Nanocomposite Supercapacitor Materials II Performance*, ed. by K.K. Kar (Springer International Publishing, 2020)
14. A. González, E. Goikolea, J.A. Barrena, R. Mysyk, Renew. Sustain. Energy Rev. **58**, 1189 (2016)
15. M. Zhi, C. Xiang, J. Li, M. Li, N. Wu, Nanoscale **5**, 72 (2013)
16. X. Lang, A. Hirata, T. Fujita, M. Chen, Nat. Nanotechnol. **6**, 232 (2011)
17. W. Deng, X. Ji, Q. Chen, C.E. Banks, RSC Adv. **1**, 1171 (2011)
18. K.D. Verma, P. Sinha, S. Banerjee, K.K. Kar, in *Handbook of Nanocomposite Supercapacitor Materials I Characteristics*, ed. by K.K. Kar (Springer International Publishing, 2020)
19. P. Sinha, K.K. Kar, in *Handbook of Nanocomposite Supercapacitor Materials II Performance*, ed. by K.K. Kar (Springer International Publishing, 2020)
20. J. Cherusseri, R. Sharma, K.K. Kar, Carbon **105**, 113 (2016)
21. S. Banerjee, K.K. Kar, in *Handbook of Nanocomposite Supercapacitor Materials I Characteristics*, ed. by K.K. Kar (Springer International Publishing, 2020)
22. R. Sharma, K.K. Kar, in *Handbook of Nanocomposite Supercapacitor Materials I Characteristics*, ed. by K.K. Kar (Springer International Publishing, 2020), pp. 215–245
23. P. Sinha, S. Banerjee, K.K. Kar, in *Handbook of Nanocomposite Supercapacitor Materials I Characteristics*, ed. by K.K. Kar (Springer International Publishing, 2020), pp. 125–154
24. P. Chamoli, S. Banerjee, K.K. Raina, K.K. Kar, in *Handbook of Nanocomposite Supercapacitor Materials I Characteristics*, ed. by K.K. Kar (Springer International Publishing, 2020)
25. P. Sinha, S. Banerjee, K.K. Kar, in *Handbook of Nanocomposite Supercapacitor Materials II Performance*, ed. by K.K. Kar (Springer International Publishing, 2020)
26. B. De, S. Banerjee, K.D. Verma, K.K. Kar, in *Handbook of Nanocomposite Supercapacitor Materials II Performance*, ed. by K.K. Kar (Springer International Publishing, 2020)
27. B. De, S. Banerjee, K.D. Verma, T. Pal, P.K. Manna, K.K. Kar, in *Handbook of Nanocomposite Supercapacitor Materials II Performance*, ed. by K.K. Kar (Springer International Publishing, 2020)
28. B. De, S. Banerjee, T. Pal, K.D. Verma, P.K. Manna, K.K. Kar, in *Handbook of Nanocomposite Supercapacitor Materials II Performance*, ed. by K.K. Kar (Springer International Publishing, 2020)
29. J. Cherusseri, K.K. Kar, J. Mater. Chem. A **3**, 21586 (2015)
30. A. Tyagi, S. Banerjee, J. Cherusseri, K.K. Kar, in *Handbook of Nanocomposite Supercapacitor Materials I Characteristics*, ed. by K.K. Kar (Springer International Publishing, 2020), pp. 91–123
31. B. De, S. Banerjee, K.D. Verma, T. Pal, K.K. Kar, in *Handbook of Nanocomposite Supercapacitor Materials II Performance*, ed. by K.K. Kar (Springer International Publishing, 2020)
32. T. Pal, S. Banerjee, P.K. Manna, K.K. Kar, in *Handbook of Nanocomposite Supercapacitor Materials I Characteristics*, ed. by K.K. Kar (Springer International Publishing, 2020), pp. 247–268
33. S. Banerjee, K.K. Kar, in *Handbook of Nanocomposite Supercapacitor Materials II Performance*, ed. by K.K. Kar (Springer International Publishing, 2020)
34. K.D. Verma, S. Banerjee, K.K. Kar, in *Handbook of Nanocomposite Supercapacitor Materials I Characteristics*, ed. by K.K. Kar (Springer International Publishing, 2020), pp. 287–314

35. K.D. Verma, P. Sinha, S. Banerjee, K.K. Kar, M.K. Ghorai, in *Handbook of Nanocomposite Supercapacitor Materials I Characteristics*, ed. by K.K. Kar (Springer International Publishing, 2020), pp. 315–326
36. B. De, S. Banerjee, T. Pal, A. Tyagi, K.D. Verma, P.K. Manna, K.K. Kar, in *Handbook of Nanocomposite Supercapacitor Materials II Performance*, ed. by K.K. Kar (Springer International Publishing, 2020)
37. B. De, P. Sinha, S. Banerjee, T. Pal, K.D. Verma, A. Tyagi, P.K. Manna, K.K. Kar, in *Handbook of Nanocomposite Supercapacitor Materials II Performance*, ed. by K.K. Kar (Springer International Publishing, 2020)
38. J. Cherusseri, K.K. Kar, RSC Adv. **6**, 60454 (2016)
39. B. De, S. Banerjee, T. Pal, K.D. Verma, A. Tyagi, P.K. Manna, K.K. Kar, in *Handbook of Nanocomposite Supercapacitor Materials II Performance*, ed. by K.K. Kar (Springer International Publishing, 2020)
40. J. Cherusseri, K.K. Kar, Phys. Chem. Chem. Phys. **18**, 8587 (2016)
41. B. De, S. Banerjee, T. Pal, K.D. Verma, A. Tyagi, P.K. Manna, in *Handbook of Nanocomposite Supercapacitor Materials II Performance*, ed. by K.K. Kar (Springer International Publishing, 2020)
42. P. Sinha, B. De, S. Banerjee, K.D. Verma, T. Pal, P.K. Manna, K.K. Kar, in *Handbook of Nanocomposite Supercapacitor Materials II Performance*, ed. by K.K. Kar (Springer International Publishing, 2020)
43. J. Kim, R. Kumar, A.J. Bandodkar, J. Wang, Adv. Electron. Mater. **3**, 1600260 (2017)
44. Y.K. Penke, A.K. Yadav, P. Sinha, I. Malik, J. Ramkumar, K.K. Kar, Chem. Eng. J. **390**, 124000 (2020)
45. A. Davies, A. Yu, Can. J. Chem. Eng. **89**, 1342 (2011)
46. L. Feng, Y. Zhu, H. Ding, C. Ni, J. Power Sources **267**, 430 (2014)
47. K. Poonam, A. Sharma, S.K. Arora, J. Tripathi, Energy Storage **21**, 801 (2019)
48. J. Jiang, Y. Li, J. Liu, X. Huang, C. Yuan, X.W.D. Lou, Adv. Mater. **24**, 5166 (2012)
49. W. Zuo, R. Li, C. Zhou, Y. Li, J. Xia, J. Liu, Adv. Sci. **4**, 1600539 (2017)
50. A. Muzaffar, M.B. Ahamed, K. Deshmukh, J. Thirumalai, Renew. Sustain. Energy Rev. **101**, 123 (2019)
51. E. Lim, H. Kim, C. Jo, J. Chun, K. Ku, S. Kim, H.I. Lee, I.-S. Nam, S. Yoon, K. Kang, J. Lee, ACS Nano **8**, 8968 (2014)
52. C. Peng, X. Bin Yan, R.T. Wang, J.W. Lang, Y.J. Ou, Q.J. Xue, Electrochim. Acta **87**, 401 (2013)
53. J. Cherusseri, K.K. Kar, J. Mater. Chem. A **4**, 9910 (2016)
54. P. Yang, W. Mai, Nano Energy **8**, 274 (2014)
55. F. Wang, X. Wu, X. Yuan, Z. Liu, Y. Zhang, L. Fu, Y. Zhu, Q. Zhou, Y. Wu, W. Huang, Chem. Soc. Rev. **46**, 6816 (2017)
56. H. Jiang, J. Ma, C. Li, Adv. Mater. **24**, 4197 (2012)
57. J. Cherusseri, K.K. Kar, RSC Adv. **5**, 34335 (2015)
58. R. Kumar, S. Sahoo, E. Joanni, R.K. Singh, K. Maegawa, W.K. Tan, G. Kawamura, K.K. Kar, A. Matsuda, Mater. Today (2020)
59. Y. Wang, Y. Song, Y. Xia, Chem. Soc. Rev. **45**, 5925 (2016)
60. C. Zhong, Y. Deng, W. Hu, J. Qiao, L. Zhang, J. Zhang, Chem. Soc. Rev. **44**, 7484 (2015)
61. P. Simon, Y. Gogotsi, *Materials for Electrochemical Capacitors* (Co-Published with Macmillan Publishers Ltd, UK, 2009)
62. H.A. Andreas, B.E. Conway, Electrochim. Acta **51**, 6510 (2006)
63. G. Wang, L. Zhang, J. Zhang, Chem. Soc. Rev. **41**, 797 (2012)
64. Y. Jiang, J. Liu, Energy Environ. Mater. **2**, 30 (2019)
65. M. Mastragostino, Solid State Ionics **148**, 493 (2002)
66. Y. Wang, L. Yu, Y. Xia, J. Electrochem. Soc. **153**, A743 (2006)
67. T. Ogoshi, K. Yoshikoshi, R. Sueto, H. Nishihara, T. Yamagishi, Angew. Chemie Int. Ed. **54**, 6466 (2015)
68. E. Frackowiak, F. Béguin, Carbon **39**, 937 (2001)

69. K. Jurewicz, C. Vix-Guterl, E. Frackowiak, S. Saadallah, M. Reda, J. Parmentier, J. Patarin, F. Béguin, J. Phys. Chem. Solids **65**, 287 (2004)
70. J.A. Fernández, T. Morishita, M. Toyoda, M. Inagaki, F. Stoeckli, T.A. Centeno, J. Power Sources **175**, 675 (2008)
71. S.K. Singh, H. Prakash, M.J. Akhtar, K.K. Kar, ACS Sustain. Chem. Eng. **6**, 5381 (2018)
72. A. Tyagi, A. Yadav, P. Sinha, S. Singh, P. Paik, K.K. Kar, Appl. Surf. Sci. **495**, 143603 (2019)
73. M. Aulice Scibioh, B. Viswanathan (eds.), *Materials for Supercapacitor Applications* (Elsevier, 2020)
74. W. Qian, F. Sun, Y. Xu, L. Qiu, C. Liu, S. Wang, F. Yan, Energy Environ. Sci. **7**, 379 (2014)
75. H. Zhou, S. Zhu, M. Hibino, I. Honma, J. Power Sources **122**, 219 (2003)
76. X. Li, B. Wei, Nano Energy **2**, 159 (2013)
77. Z.G. Cambaz, G.N. Yushin, Y. Gogotsi, K.L. Vyshnyakova, L.N. Pereselentseva, J. Am. Ceram. Soc. **89**, 509 (2006)
78. H.-J. Liu, J. Wang, C.-X. Wang, Y.-Y. Xia, Adv. Energy Mater. **1**, 1101 (2011)
79. A.G. Pandolfo, A.F. Hollenkamp, J. Power Sources **157**, 11 (2006)
80. D.N. Futaba, K. Hata, T. Yamada, T. Hiraoka, Y. Hayamizu, Y. Kakudate, O. Tanaike, H. Hatori, M. Yumura, S. Iijima, Nat. Mater. **5**, 987 (2006)
81. S.K. Singh, M.J. Akhtar, K.K. Kar, ACS Appl. Mater. Interfaces **10**, 24816 (2018)
82. Y. Zhou, P. Jin, Y. Zhou, Y. Zhu, Sci. Rep. **8**, 9005 (2018)
83. L.L. Zhang, R. Zhou, X.S. Zhao, J. Mater. Chem. **20**, 5983 (2010)
84. F. Wang, S. Xiao, Y. Hou, C. Hu, L. Liu, Y. Wu, RSC Adv. **3**, 13059 (2013)
85. J. Zhu, S. Dong, Y. Xu, H. Guo, X. Lu, X. Zhang, J. Electroanal. Chem. **833**, 119 (2019)
86. Y. Zhao, W. Ran, J. He, Y. Song, C. Zhang, D.-B. Xiong, F. Gao, J. Wu, Y. Xia, ACS Appl. Mater. Interfaces **7**, 1132 (2015)
87. H.M. Coromina, D.A. Walsh, R. Mokaya, J. Mater. Chem. A **4**, 280 (2016)
88. M. Sevilla, A.B. Fuertes, Chemsuschem **9**, 1880 (2016)
89. H. Chen, D. Liu, Z. Shen, B. Bao, S. Zhao, L. Wu, Electrochim. Acta **180**, 241 (2015)
90. F. Liu, Z. Wang, H. Zhang, L. Jin, X. Chu, B. Gu, H. Huang, W. Yang, Carbon **149**, 105 (2019)
91. C. An, Y. Zhang, H. Guo, Y. Wang, Nanoscale Adv. **1**, 4644 (2019)
92. D. Majumdar, T. Maiyalagan, Z. Jiang, ChemElectroChem **6**, 4343 (2019)
93. C. Liu, F. Li, L.-P. Ma, H.-M. Cheng, Adv. Mater. **22**, E28 (2010)
94. D. Chen, Q. Wang, R. Wang, G. Shen, J. Mater. Chem. A **3**, 10158 (2015)
95. X. Lu, X. Huang, S. Xie, T. Zhai, C. Wang, P. Zhang, M. Yu, W. Li, C. Liang, Y. Tong, J. Mater. Chem. **22**, 13357 (2012)
96. D. Ghosh, S. Giri, M. Moniruzzaman, T. Basu, M. Mandal, C.K. Das, Dalt. Trans. **43**, 11067 (2014)
97. Y. Zhang, Y. Liu, J. Chen, Q. Guo, T. Wang, H. Pang, Sci. Rep. **4**, 5687 (2015)
98. D. Guo, Y. Luo, X. Yu, Q. Li, T. Wang, Nano Energy **8**, 174 (2014)
99. J. Du, G. Zhou, H. Zhang, C. Cheng, J. Ma, W. Wei, L. Chen, T. Wang, ACS Appl. Mater. Interfaces **5**, 7405 (2013)
100. C. Yuan, J. Li, L. Hou, J. Lin, X. Zhang, S. Xiong, J. Mater. Chem. A **1**, 11145 (2013)
101. X. Xiao, Z. Peng, C. Chen, C. Zhang, M. Beidaghi, Z. Yang, N. Wu, Y. Huang, L. Miao, Y. Gogotsi, J. Zhou, Nano Energy **9**, 355 (2014)
102. H.-S. Kim, J.B. Cook, H. Lin, J.S. Ko, S.H. Tolbert, V. Ozolins, B. Dunn, Nat. Mater. **16**, 454 (2017)
103. J. Yang, X. Xiao, P. Chen, K. Zhu, K. Cheng, K. Ye, G. Wang, D. Cao, J. Yan, Nano Energy **58**, 455 (2019)
104. Z. Wang, D.O. Carlsson, P. Tammela, K. Hua, P. Zhang, L. Nyholm, M. Strømme, ACS Nano **9**, 7563 (2015)
105. Q. Meng, K. Cai, Y. Chen, L. Chen, Nano Energy **36**, 268 (2017)
106. A. Eftekhari, L. Li, Y. Yang, J. Power Sources **347**, 86 (2017)
107. G.A. Snook, P. Kao, A.S. Best, J. Power Sources **196**, 1 (2011)
108. H. Kashani, L. Chen, Y. Ito, J. Han, A. Hirata, M. Chen, Nano Energy **19**, 391 (2016)

109. N. Chen, L. Ni, J. Zhou, G. Zhu, Q. Kang, Y. Zhang, S. Chen, W. Zhou, C. Lu, J. Chen, X. Feng, X. Wang, X. Guo, L. Peng, W. Ding, W. Hou, ACS Appl. Energy Mater. **1**, 5189 (2018)
110. B. Pandit, S.S. Karade, B.R. Sankapal, ACS Appl. Mater. Interfaces **9**, 44880 (2017)
111. K. Naoi, W. Naoi, S. Aoyagi, J. Miyamoto, T. Kamino, Acc. Chem. Res. **46**, 1075 (2013)
112. K.L. Van Aken, M. Beidaghi, Y. Gogotsi, Angew. Chemie Int. Ed. **54**, 4806 (2015)
113. J.Y. Hwang, M.F. El-Kady, Y. Wang, L. Wang, Y. Shao, K. Marsh, J.M. Ko, R.B. Kaner, Nano Energy **18**, 57 (2015)
114. H. Wang, H. Yi, X. Chen, X. Wang, J. Mater. Chem. A **2**, 3223 (2014)
115. S.D. Perera, B. Patel, N. Nijem, K. Roodenko, O. Seitz, J.P. Ferraris, Y.J. Chabal, K.J. Balkus, Adv. Energy Mater. **1**, 936 (2011)
116. S. Sheng, W. Liu, K. Zhu, K. Cheng, K. Ye, G. Wang, D. Cao, J. Yan, J. Colloid Interface Sci. **536**, 235 (2019)
117. H. Gao, F. Xiao, C.B. Ching, H. Duan, ACS Appl. Mater. Interfaces **4**, 2801 (2012)
118. K. Lota, V. Khomenko, E. Frackowiak, J. Phys. Chem. Solids **65**, 295 (2004)
119. J. Shen, C. Yang, X. Li, G. Wang, ACS Appl. Mater. Interfaces **5**, 8467 (2013)
120. M.A.A. Mohd Abdah, N.H.N. Azman, S. Kulandaivalu, Y. Sulaiman, Mater. Des. **186**, 108199 (2020)
121. H. Peng, R. Zhao, J. Liang, S. Wang, F. Wang, J. Zhou, G. Ma, Z. Lei, ACS Appl. Mater. Interfaces **10**, 37125 (2018)
122. S.-W. Zhang, B.-S. Yin, C. Liu, Z.-B. Wang, D.-M. Gu, Chem. Eng. J. **312**, 296 (2017)
123. X. He, Y. Zhao, R. Chen, H. Zhang, J. Liu, Q. Liu, D. Song, R. Li, J. Wang, ACS Sustain. Chem. Eng. **6**, 14945 (2018)
124. M. Nallappan, M. Gopalan, Mater. Res. Bull. **106**, 357 (2018)
125. C. Zhou, Y. Zhang, Y. Li, J. Liu, Nano Lett. **13**, 2078 (2013)
126. T. Aida, K. Yamada, M. Morita, Electrochem. Solid-State Lett. **9**, A534 (2006)
127. S. Hemmati, G. Li, X. Wang, Y. Ding, Y. Pei, A. Yu, Z. Chen, Nano Energy **56**, 118 (2019)
128. A. Brandt, A. Balducci, U. Rodehorst, S. Menne, M. Winter, A. Bhaskar, J. Electrochem. Soc. **161**, A1139 (2014)
129. M. Yang, Y. Zhong, J. Ren, X. Zhou, J. Wei, Z. Zhou, Adv. Energy Mater. **5**, 1500550 (2015)
130. J. Ding, H. Wang, Z. Li, K. Cui, D. Karpuzov, X. Tan, A. Kohandehghan, D. Mitlin, Energy Environ. Sci. **8**, 941 (2015)
131. Z. Chen, V. Augustyn, X. Jia, Q. Xiao, B. Dunn, Y. Lu, ACS Nano **6**, 4319 (2012)
132. B.H. Zhang, Y. Liu, Z. Chang, Y.Q. Yang, Z.B. Wen, Y.P. Wu, R. Holze, J. Power Sources **253**, 98 (2014)
133. N. Karikalan, C. Karuppiah, S.-M. Chen, M. Velmurugan, P. Gnanaprakasam, Chem. Eur **23**, 2379 (2017)

Chapter 3
All Types of Flexible Solid-State Supercapacitors

Souvik Ghosh, Prakas Samanta, and Tapas Kuila

Abstract Nowadays, flexible solid-state supercapacitors (FSSCs) are the most emerging energy storage devices in modern miniatured technologies. With increasing the use of micro- and flexible electronic devices such as wearable electronic suits, microsensors, and biomedical equipment, the demand of FSSCs is increasing exponentially. These electronic devices focused on the integration of many components in single compact system that must be flexible in nature, lightweight, smaller in dimension, unbreakable and should be available at competitive price. Although the FSSCs device fabrication is in the early stage of development, there is a significant effort to investigate the new strategies to fulfill the demand of advanced energy technology. Therefore, exploring the novel approaches always remains an important academic and industrial challenge to the researchers. The chapter mainly focuses on the advancement of FSSCs devices with its all components such as current collector, electrode materials, and electrolytes. The chapter also discusses the strategies of fabrication techniques, types of FSSCs, design, evaluation of performance of FSSCs device step by step. Hence, the chapter gives an idea about the recent progress and challenges of the FSSCs devices.

S. Ghosh · P. Samanta · T. Kuila (✉)
Surface Engineering and Tribology Division, Council of Scientific and Industrial Research-Central Mechanical Engineering Research Institute, Durgapur 713209, India
e-mail: tkuila@gmail.com

Academy of Scientific and Innovative Research (AcSIR), Ghaziabad 201002, India

S. Ghosh
e-mail: sgkssm92@gmail.com

P. Samanta
e-mail: samantaprakas2@gmail.com

K. K. Kar (ed.), *Handbook of Nanocomposite Supercapacitor Materials III*,
Springer Series in Materials Science 313,
https://doi.org/10.1007/978-3-030-68364-1_3

3.1 Introduction

With the rapid growth of electronics technology, energy requirement in our daily life is increasing day by day. In order to fulfill the huge requirement of electricity, consumption of fossil fuels is increasing in alarming rate and causes global warming. In 2015, the total global energy demand was 400 EJ, equivalent to the burning of 9.6 billion tons of oil. It has been estimated that the global energy requirement would reach to ~430 EJ by 2050. The report on global oil reserves shows that there are about 1.688 trillion barrels of crude oil on the earth, which will last for 53.3 years at the current rate of extraction. Therefore, the researchers are trying hard to develop sustainable solution for future energy demand to mitigate the issues of fossil fuel depletion and global warming. In this regard, efficient and eco-friendly energy storage devices could be helpful to meet the future energy demand [1–3]. The Ragon plot shown in Fig. 3.1a reveals that the rechargeable batteries and supercapacitors (SCs) have attracted more research attention as the commonly used energy storage devices. SCs are the most promising modern energy storage devices owing to its fast charge–discharge rates, long cycle life, and wide operating temperature

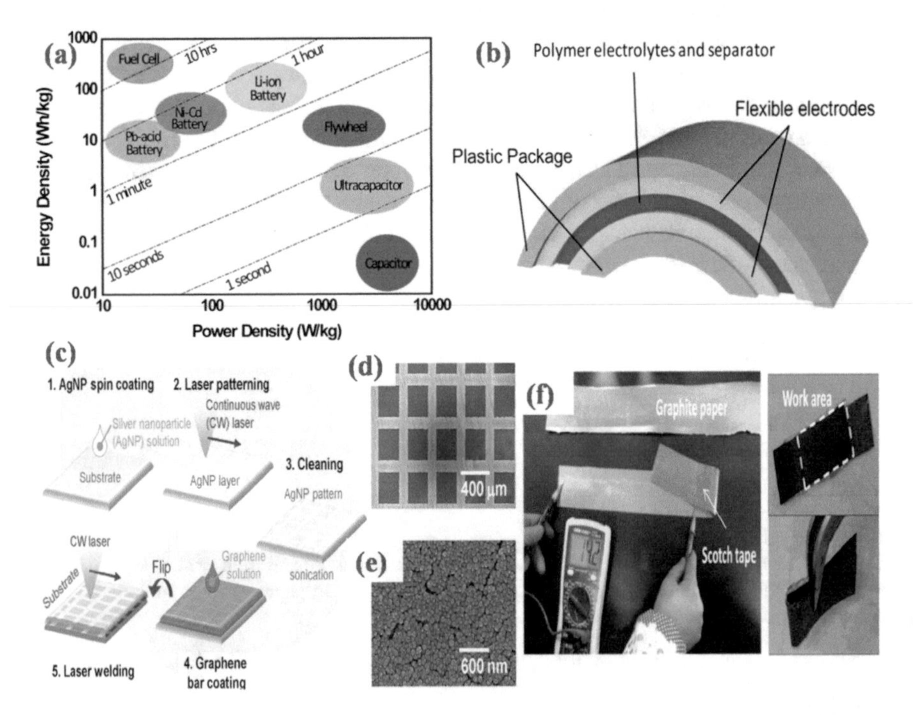

Fig. 3.1 **a** Ragon plots for different energy storage systems [2], **b** different components of FSSCs device [9], **c** schematic of flexible graphene electrode fabrication process, **d**, **e** FE-SEM images of the Ag grid film and the sintered Ag nanoparticles [14], **f** left panel shows the digital image of scotch tape of current collector, upper right panel shows the devices and the bottom right panel demonstrates its flexibility [16]

range. SCs holds a unique and important position as the energy storage device by bridging the gap between the battery and conventional capacitor. SCs can be used as energy backup for mobile phones, digital cameras, implantable heart sensors, notebook computer, smart phone, etc. [4, 5]. With the emergence of increasing demand of wearable and flexible electronics, development of flexible energy storage devices is also highly demanding. Rigid-type SCs cannot meet the requirements for powering flexible electronic devices. Therefore, it is the need of the hour to develop the flexible solid-state SCs (FSSCs) device. It has attracted considerable attention owing to their lightweight, small size, wide range of operating temperatures and ease of handling and outstanding reliability. As FSSCs can be used in bent, folded, compressed, or even stretched electronic devices, it is a challenging job to fabricate such types of SCs devices with high power density (PD) and energy density (ED) along with long life span. Currently, the commercially available SCs consist of rigid electrodes, which severely hinder the feasibility of manufacturing FSSCs. In comparison to the metal foil-based usual electrodes, FSSCs electrodes replace the non-conductive organic binders and heavy current collectors with flexible and bedded current collectors. Therefore, the basic research and technical applications of FSSCs devices have attracted widespread attention [6–8]. Research attention on FSSCs is mainly focus on the design of novel electrode materials or electrolytes or architects the novel device for improving the performance. Thus, a flexible energy storage device with high ED and PD should provide a high-rated current output. For the FSSCs devices, the most important considerations are the utilization of appropriate electrode materials and fabrication strategies to achieve high ED and cyclic stability. FSSCs are a novel emerging branch of SCs, usually oblige thin-film electrode materials with good electrochemical performance and high mechanical stability against bending, folding, or rolling. The main difference of FSSCs with commercial SCs is the variable shape of FSSCs and the key in fabricating different shape FSSCs is the development of flexible electrode materials. So, the choice of electrode material is very important in fabricating FSSCs. Figure 3.1b shows the schematic diagram of a FSSCs device with different components. Several nanostructured materials such as transitional metal oxide (TMO), transitional metal chalcogenides (TMC), various carbonaceous materials (graphene, activated carbon, CNT), and conductive polymers are used as the electrodes in FSSCs [9, 10]. In addition to the choice of electrode materials, selection of electrolyte is also a major factor in FSSCs. Generally, aqueous and non-aqueous electrolytes are used in conventional types of SCs. However, the use of liquid electrolytes in FSSCs device is not preferred due to its toxic and corrosive nature. In addition, there is an issue of packaging with liquid electrolyte due to its tendency toward leakage. In order to mitigate this lacuna, different types of solid-state or gel-polymer electrolytes are used for FSSCs device fabrication. However, further research is required to optimize the structural design with superior flexibility rather than focusing on the electrode materials or electrolytes [11, 12].

This chapter discusses the aforementioned important topics of recent trends on the fabrication strategies of high-performance solid-state FSCs devices, electrode materials, electrolytes, and current collectors. The chapter also discusses the basic working principle and device configuration of FSSCs, which provides necessary background information including the current collector, electrode materials, and electrolytes. Different types of fabrication strategies, their advantages, and disadvantages are discussed in detail. Different types of FSSCs devices along with their electrochemical performance are also discussed. Special emphasis is given on the applications of all types of FSSCs devices.

3.2 Flexible SCs Device Configuration

3.2.1 Current Collector

Carbon fibers and carbon fabrics are mostly used as current collectors in FSSCs device due to their excellent mechanical strength and flexibility. In addition, metal-coated polymers and metal wires are also used as FSSCs current collectors. Although nickel foil, stainless steel mesh, titanium foil, and copper foam shows high electrical conductivity, low flexibility and poor stability in acidic/alkaline media limit their applications in FSSCs. However, the mechanical flexibility and long-term stability in acidic/alkaline medium are the essential parameters for the fabrication of FSSCs devices [13]. In order to address these issues, graphene-based materials can be coated onto the metallic substrate through a selective laser sintering process to form the microgrid. Figure 3.1c shows the schematic diagram of the flexible graphene electrode fabrication on the Ag nanoparticle layers. Electrical conductivity and adhesion between the metal mesh with electrode materials are improved in presence of GO [14]. Figure 3.1d, e is the FE-SEM images of the Ag grid film and the sintered Ag nanoparticles, respectively. The FE-SEM image shows that the width of the Ag nanoparticles is ~100 μm and the spacing between the two patterns is ~300 μm. Highly conductive current collector also decreases the equivalent series resistance of the FSSCs device. Highly conductive, non-toxic and extremely flexible graphite foil (25 and 150 μm), and graphite ink-coated aluminum foil show remarkable decrease in ESR value (~80%) [15]. Graphite foil is the best alternative than the metal current collector, as it is free from corrosion and shows large flexibility. Highly conductive scotch tape also acts as a good current collector in FSSCs. Nitrogen-doped-graphene/MnO_2 nanosheet composite materials deposited on scotch tape acts as the positive and activated carbon as the negative electrode materials in a flexible SCs device (Fig. 3.1f). The device can be operated reversibly up to ~1.8 V potential range and exhibits a maximum ED of 3.5 mWh cm^{-3} at the PD of 0.019 W cm^{-3} [16].

3.2.2 Electrodes

SCs can store the energy by accumulating the charge and formation of a double layer or through fast redox reaction on the surface of the electrodes. Therefore, the electrode material plays an important role in determining the performance of the SCs. Design of electrode materials is a major challenge in terms of cost effectiveness and capacitive properties. In addition, the electrode materials must be foldable, lightweight, and should have high mechanical properties. The flexible electrode materials used in flexible SCs can be classified into two categories: substrate-supported flexible electrodes and flexible freestanding electrodes.

3.2.2.1 Flexible Substrate-Supported Electrodes

This section emphasizes on the flexible substrates used in the FSSCs device. Thin, lightweight, and mechanically robust plastics substrates are used to support the electrode materials. Stainless steel (SS), nickel (Ni), copper (Cu), titanium (Ti), etc., are used as flexible substrates due to their high mechanical strength, good electrical conductivity, and ease of processing. The direct growth of electroactive materials on the substrate surface may decrease the resistance and increase the electrochemical performance as well as the flexibility of the device. In contrast, the conventional powder-based materials require extra non-conducting polymer binder for binding the materials with the substrate resulting the lowering of electrochemical properties as well as mechanical durability of the device. In addition to the metal substrates, carbon paper, carbon fabric, textile, ordinary paper, etc., are widely used as the supporting substrate in FSSCs.

The superior mechanical properties and high electrical conductivity of the metal substrate make them promising support for the active material of FSSCs [17, 18]. Dubal and co-workers prepared branched MnO_2 nanorods on the SS substrate through an inexpensive and template-free solution-based facile route [19]. Figure 3.2a, b shows the preparation of brunched MnO_2 structure and urchin like MnO_2 nanorods on substrates. The brunched urchin like structure opens up more active sites for electrolytes and improves the electrochemical properties of SCs. Figure 3.2c represents the schematic of symmetric FSSCs device, where two MnO_2 electrodes are sandwiched with a flexible polypropylene separator [19]. Copper foil is also considered as very good flexible substrate for the direct growth of the electroactive materials. 3-D $Cu(OH)_2$ nanoporous structures can be organized on a Cu foil through a facile one-step electrodeposition process. The $Cu(OH)_2$ deposited Cu foil can be used as binder-free electrode and provides a large number of active sites for redox reactions and electrolyte ion transport pathways. Although metallic support provides good electrical conductivity, however, its low corrosive resistance and extra weight decrease the gravimetric performance of the device. In order to mitigate this lacuna, highly conducting, corrosion resistant, lightweight, and flexible carbon-based substrate is used as the electrode support. Carbon fiber paper (CFP) consists of a regular arrangement of microsized fabrics with increased pores and surface area are suitable to support the electrode materials. It forms a good conducting channel with appropriate pores. It can effectively create an effective electron transmission path and efficiently allow the electrolyte to enter the electrochemically active material [20–22].

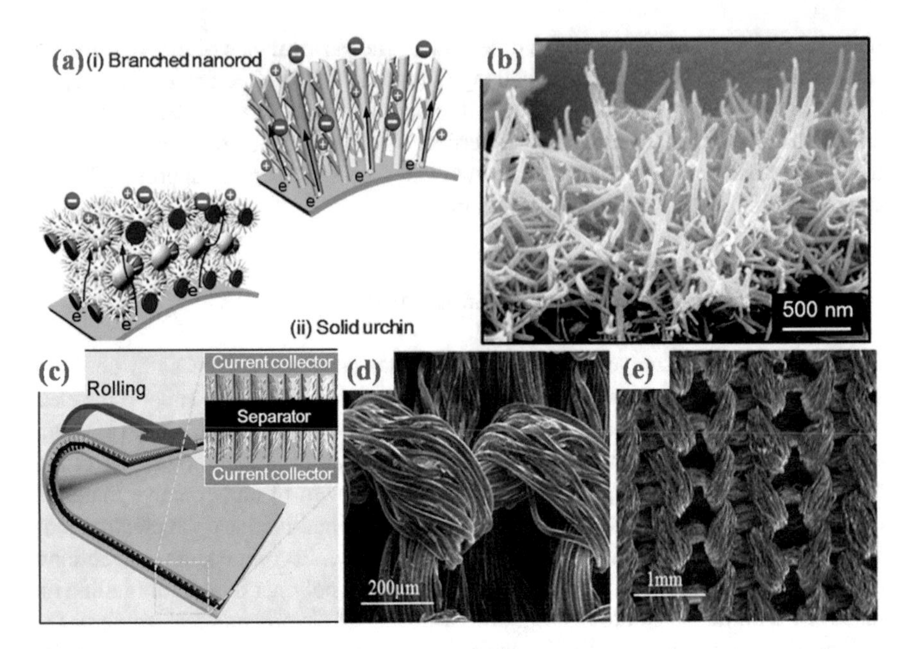

Fig. 3.2 **a** Schematic of the ion diffusion path of MnO_2, **b** FE-SEM image of MnO_2 nanorods, **c** illustration of the stacked MnO_2 electrode with a separator [19], **d**, **e** FE-SEM image of CNT-polyester and CNT-cotton textile [23]

Sawangphruk et al. reported a solid-state FSSCs using MnO_2-rGO coated on flexible CFP by spray-coating technique [23]. MnO_2-rGO/CFP exhibits high specific capacitance of ~393 F g^{-1} at 0.1 A g^{-1}current density. The MnO_2-rGO/CFP FSSCs device shows high performance due to the high porosity created between each MnO_2-rGO-coated porous CFP. Textile or cloth is also considered as the highly flexible porous material, which can be easily scaled up from natural or synthetic fibers. Nowadays, textile is extensively used in FSSCs due to its good flexibility, high stretchability, and lightweight. The 3D open-pore structure of textile as shown in Fig. 3.2d, e is suitable for the conformal coating applied to the entire surfaces [24, 25]. He et al. synthesized a hierarchical ultrathin $Cu_{1.4}S$ nanoplate on the electrochemically active carbon cloth ($Cu_{1.4}S$@CC) and formed a flexible electrode. The oxygen-containing groups of CC show strong interaction with $Cu_{1.4}S$ particles leading to good mechanical performance and cyclic ability of the electrode. The electrode exhibits ~485 F g^{-1} specific capacitance at 0.25 A g^{-1}current density. The retention of specific capacitance is ~80.2% after 1000 charge–discharge cycles [26].

3.2.2.2 Flexible Freestanding Electrodes

The conductive substrate like SS, Ni, and Ti foil, etc., is used as the current collector in FSSCs. However, the corrosive nature of the electrolytes and less flexibility of metal foils restricts the durability of the device. The presence of non-conductive binder increases the weight and resistance parameters of the device. In order to overcome these difficulty, an emerging approach has been introduced to develop freestanding films as active materials in fabricating FSSCs device [2, 27]. Various carbonaceous materials, transition metal oxides, and conducting polymers are used as flexible free-standing electrodes. Among the various nanostructured carbonaceous materials, 1D CNT, 2D graphene, and their hybrids have received a great attention as freestanding FSSCs electrodes owing to the high electrical conductivity, large surface area, and good mechanical properties [28–30]. In comparison to the other carbonaceous materials, CNT possesses high specific surface area (1240–200 m^2 g^{-1}), flexibility, electrical conductivity (104–105 S m^{-1}), long continuous conductive paths, and regular pore size. All these parameters make CNT an attractive freestanding electrode material [31, 32]. Anwer et al. developed freestanding electrode by the growth of $Ni(OH)_2$ nanosheet on the 3D architecture of jointly welded CNT foam (CNTF). Figure 3.3a shows a graphical illustration of the CNTF preparation by chemical vapor deposition (CVD) method and CNT organized them into a cotton-like network. The thin CNT segments and thickened joints are shown in FE-SEM image (Fig. 3.3b). Owing to the high electrical conductivity and flexible nature, CNT accelerates the performance of the flexible electrode, reduces the weight and internal resistance between the FSSCs active materials and current collector [33]. Graphene could be the better choice as electrode materials due to its large effective surface area, van der Waals

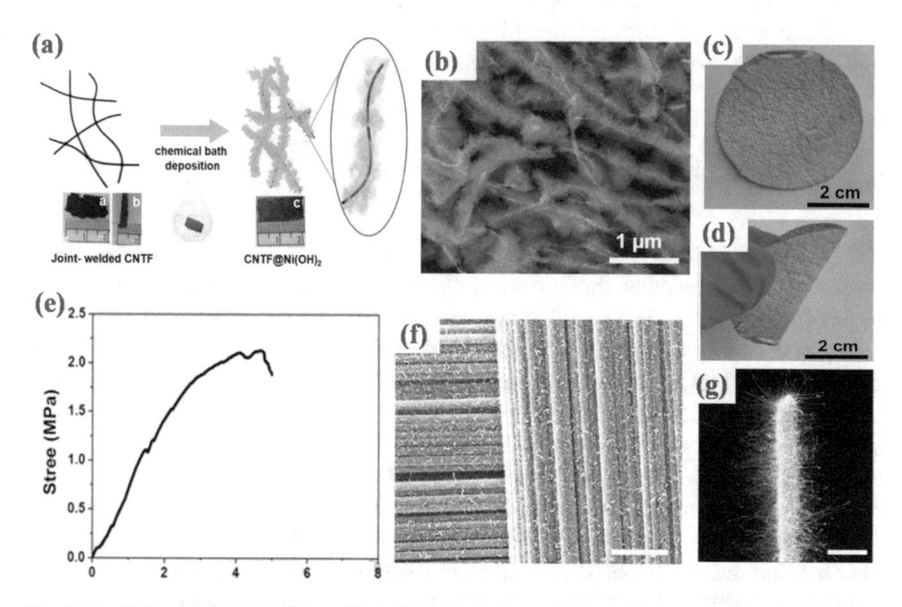

Fig. 3.3 **a** Schematic of the joint welded CNTF@$Ni(OH)_2$ coaxial composite preparation process and corresponding photographic images, **b** FE-SEM image of CNTF@$Ni(OH)_2$ coaxial composite [33], **c**, **d** digital images of flexible cellulose/rGO/AgNPs composite in natural and curved states, respectively. **e** stress–strain curve of cellulose/rGO/AgNPs composite [34], **f**, **g** FE-SEM images of ZTO/CMFs [36]

attractions, and superior electronic and mechanical properties. Zou et al. fabricated cellulose fiber-based flexible self-supporting graphene electrodes by reacting GO and silver ammonia complex. Figure 3.3c, d shows that 400-μm-thick flexible cellulose/RGO/AgNPs composite film can utilized directly as the working electrode and shows good mechanical robustness with ~2.14 MP of tensile strength at a fracture elongation of 4.6% as demonstrated in Fig. 3.3e. However, the electrode does not have any non-conducting binder and active materials resulting in an excellent rate capability and ~99.6% retention in specific capacitance after 10,000 GCD cycles [34]. Flexible freestanding, transparent, and lightweight graphene paper can be prepared by microwave plasma-enhanced chemical vapor deposition (CVD) process, which acts as the good freestanding electrode material [35]. Metal oxides can also be used as the flexible freestanding SCs electrode materials. However, the poor electrical conductivity of metal oxides restricts its wide range of applications as SCs electrode. The conductive polymers like PANI and PPy are widely used to prepare the composite with metal oxides for the improvement of electrochemical properties. Bao et al. fabricated novel flexible freestanding electrode by coating ultrathin MnO_2 films on the highly conducting Zn_2SnO_4 nanowires grown on carbon microfibers (CMF) [36]. Figure 3.3f, g shows the growth of Zn_2SnO_4 nanowires on CMFs provides highly conductive cores and the backbones for MnO_2 coating. Such type of morphological architecture facilitates the transport of electrolytes and shortens the ion diffusion path, resulting good electrochemical performance. The Zn_2SnO_4@MnO_2 core–shell nanowire shows high specific capacitance of ~621.6 F g^{-1} at 2 mV s^{-1} scan rate [36]. The electrochemical performance of a few substrates supported and freestanding electrode materials are summarized in Table 3.1.

3.2.3 Electrolyte

The choice of electrolyte in FSSCs is an important factor as it controls the capacitive properties. Recent studies showed that solid-state electrolytes play a significant role in flexible SCs due to its several advantages, like ion conducting media, cost effectiveness, ease of packaging, leak proof, and perform the role of separator as well. Solid-state electrolytes offer good mechanical stability, which facilitates to assemble various flexible and bendable SCs devices [37]. However, the main disadvantage of solid-state electrolytes is its low ionic mobility. Most of the solid-state electrolytes are based on polymeric electrolytes and limited focus on the inorganic solid materials. Polymer-based electrolytes are further classified into a solid polymer electrolyte (SPEs), polyelectrolyte, and gel-polymer electrolyte (GPEs). Figure 3.4a illustrates the use of methacrylate-based GPE in solid-state SCs device [38]. The SPEs consist of a polymers [like polyethylene oxide (PEO), poly(aryl ether ketone), etc.] and a salt (LiCl, NaCl, KCl, etc.) without any solvents. The use of SPEs in SCs is limited due to the low ionic conductivity of the SPEs. Wang et al. synthesized a solid-polymer electrolyte by using poly(aryl ether ketone)-poly(ethylene glycol) copolymer and $LiClO_4$ as an electrolyte (Fig. 3.4b). The SPEs perform at wide temperature range

Table 3.1 Electrochemical performance of some substrates supported or self-supported FSSCs electrode materials

Electrode	Substrate	Capacitance	Current density	Stability (%)	cycles	Resistance (Ω)	Ref
PPy nanosheet	SS	586 F g^{-1}	5 mA cm^{-2}	81	5000	11.66 (R_{ct})	18
MnO_2 nanorod	SS	578 F g^{-1}	5 mV s^{-1}	86	5000		19
MnO_2 nanorod	SS	578 F g^{-1}	5 mV s^{-1}	67	5000		19
CNT/PANI	Print paper	332 F g^{-1}	1 A g^{-1}	88.6	1000	11.1 (R_{ct})	20
(PEDOT:PSS)-CNT/Ag	Paper	23.6 F cm^{-3}	10 mV s^{-1}	92	10,000		21
Graphene paper/carbon black	Graphene paper	138 F g^{-1}	10 mV s^{-1}	96.15	2000		22
(MnO_2-rGO/CFP)	CFP	393 Fg^{-1}	0.1 A g^{-1}	98.5	2000	2 (R_{ct})	23
CNT-coated textile	Cotton	122.1 F g^{-1}	0.2 A g^{-1}			31.7 (ESR)	24
CNT-coated textile	Polyster	53.6 F g^{-1}	0.2 A g^{-1}				24
Cu1.4S@CC	Carbon cloth (CC)	485 F g^{-1}	0.25A g^{-1}	80.25	1000		26
Carbon nanocages	Self-supported	260 F g^{-1} (2E)	0.1 A g^{-1}	90	10,000		28
Graphene–PANI paper	Self-supported	763 F g^{-1}	1 A g^{-1}	82	1000	4.1 (R_s)	29
CNF/GNS	Self-supported	197 F g^{-1}	1.25 A g^{-1}	84	1500		30
MnO_2/ethylene–vinyl acetate copolymers	Self-supported	169.5 F g^{-1}	10 mV s^{-1}	100	2000	28.23 (R_{ct})	31
CNT Foam @ $Ni(OH)_2$	Self-supported	1272 F g^{-1}	2 A g^{-1}	83.7	3000		33
Cellulose/rGO/Ag nanoparticles	Self-supported	1242.7 mF cm^{-2}	2 mA cm^{-2}	99.6	10,000		34
Graphene paper	Self-supported	3.3 mF cm^{-2}	0.02 mA cm^{-2}	95.4	20,000	98 (ESR)	35
MnO_2/ZTO/CMF	Self-supported	642.4 F g^{-1}	1 A g^{-1}	98.8	1000	4.9 (R_{ct})	36

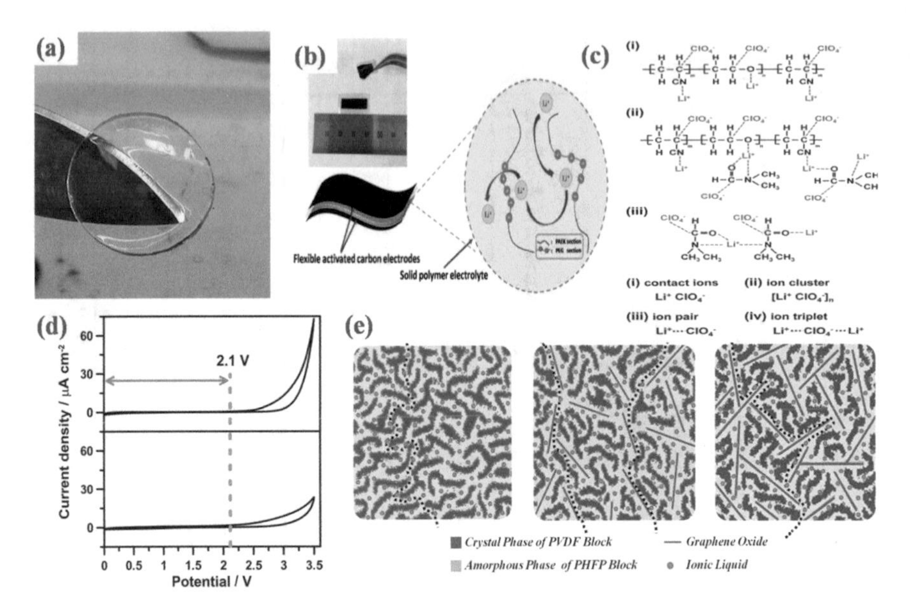

Fig. 3.4 **a** Digital image of methacrylate-based GPE separator [38], **b** upper part is the optical image of FSSCs device and the lower part is the schematic of the FSSCs device [39], **c** three interacting modes of solvated Li^+ and ClO_4^- ions, **d** CV curves of the GPE (upper) and the liquid-phase electrolyte (lower) [45], **e** proposed cross-sectional schematic structures of GO-doped ion gels with different GO doping level [49]

(30–120 °C) in flexible SCs and shows good tensile strength (~12.2 MPa) and elongation at break (~467.4%) [39]. In poly-electrolyte, the charged polymer chains contribute the ionic conductivity. However, it is very difficult to find out the single solid-state electrolyte that should fulfill all the requirements of FSSCs.

On the other hand, GPEs consist of a polymer [PVA, poly(methyl methacrylate), etc.] as host and aqueous electrolyte (H_2SO_4, KOH, H_3PO_4, etc.) or any solvated conducting salts as guests. The polymer serves as the matrix, which can be swollen by the solvent and ion transport through the solvent. The GPEs show comparatively high ionic conductivity (~10^{-4} to 10^{-3} S cm^{-1}) under ambient conditions and are called quasi-solid-state electrolytes [40, 41]. It is generally composed of a host polymer matrix, electrolytic salts, and an organic/aqueous solvent. Various polymer matrices have been used in GPE including poly(polyacrylate) (PAA), poly(vinylalcohol), poly(ethylene oxide), poly(vinylidene fluoride-co-hexafluoropropylene)(PVDFco-HFP) and poly(amine-ester) (PAE), etc. Organic solvents such as tetrahydrofuran (THF), ethylene carbonate (EC), propylene carbonate (PC), and dimethyl carbonate (DMC) are used as commonly used plasticizer [42, 43]. The electrolytic salts should have low dissociation energy and large number of ions to provide more free ions. The ratio of polymers to plasticizers is important in determining the degree of plasticization, which affects the flexibility and stability of an electrolyte. According to the use of electrolytic salts, GPE can be classified into four categories: (1) proton conducting,

(2) alkaline, (3) Li-ion, and (4) other ion GPE [44]. $LiClO_4$, LiCl, etc., are dissolved in an organic solvent at a temperature range of 70–140 °C to achieve Li-ion GPE. This solution is mixed with the polymer matrix and stirred vigorously at ambient temperature and the mixture turned into a gel form. In the $LiClO_4$-based GPEs, interaction of Li^+ and ClO_4^- ions with the copolymer and plasticizer in three different modes with various states is described in Fig. 3.4c,d. This type of GPE provides excellent ionic conductivity (~7×10^{-3} S cm^{-1}) and a wide potential window of ~2 V than the liquid electrolyte. However, organic electrolytes are now replaced with aqueous electrolytes due to their high cost, toxicity, corrosiveness, and flammable in nature. In aqueous electrolytes, use of water increases the stability of the electrode materials by retaining its structure, suppresses the chemical dissolution and reversible electrochemical reaction [45, 46]. Proton conducting polymer-gel electrolyte (PCPE) is considered as the promising charge carriers in charge–discharge process due to the higher mobility of protons in aqueous medium as compared to the lithium ions. Recently, a large number of PCPEs (PVA/HSO_4, PVA/H_3PO_4, PVA/H3PO_4/silicotungstic acid, etc.) are developed for solid-state FSSCs. PCPE shows high ionic conductivity (10^{-4}–10^{-2} S cm^{-1}) at room temperature owing to the higher ionic mobility of H^+ ions. The ionic conductivity of the PCPE electrolyte reaches to a maximum value (34.8 mS cm^{-1}) with the addition of redox-mediated (p-benezenediol) electrolyte [20, 47]. Alkaline GPE exhibits high ionic conductivity of 10^{-3}–10^{-2} S cm^{-1}. This type of polymer-gel electrolyte has attracted more attention due to their higher potential use in all flexible solid alkaline rechargeable energy storage devices [2]. In addition to the above mentioned GPEs, some other types like ionic liquid (1-butyl-2,3-dimethylimidazolium bis-trifluoromethane sulfonylimide) or inorganic salts (Li_2S–P_2S_5, $LiClO_4$–Al_2O_3) are also used in FSSCs. Chang et al. prepared a GPEs by doping 1 wt% GO with PVDF-HFP-1-ethyl-3-methylimidazoliumtetrafluoroborate ($EMiMbF_4$) and is seen that the GO significantly improves the ionic conductivity of the GPEs by ~260% [48]. Figure 3.4e shows the schematic of the GO-doped ion gels with different GO doping mass fraction [48, 49].

3.3 Device Fabrication Technique

The performance of SCs depends to a large extent on the related design and manufacturing technology. Flexible solid-state SCs can be divided into two categories: solid-state symmetrical and solid-state asymmetrical SCs. This section discusses various available methods for the fabrication of high performances FSSCs [50].

3.3.1 Pencil Drawing

Pencils are commonly used graphite products in our daily lives. Nowadays, pencils are also used as writing tools for manufacturing energy storage devices. Pencil is the hybrid of graphite, wax, and clay (such as SiO_2 and Al_2O_3). Handwriting can be thought of as a conductive film made from a network of penetrating graphite particles on paper. It can be used to create arbitrary shapes and patterns and is very stable to chemical, moisture, and ultraviolet radiation. Therefore, pencil drawing electrical equipment has been widely used in zinc oxide ultraviolet sensors, lithium-air batteries, SCs, and other fields [50, 51]. Pencil drawing SCs has several advantages, e.g., the design and geometry can be easily changed by hand drawing on a suitable substrate, eco-friendly nature, controlled thickness, lighter weight, and cost effectiveness [29, 51]. The physicochemical properties of pencil rods are remarkably different from that of pure graphite. It exhibits room temperature ferromagnetism by varying the mechanical and electrical properties due to the disordered grains induced by the clay particles. For the preparation of pencil drawing electrode, a substrate is required for drawing. Paper is the cheapest flexible substrate and thus, widely used as a substrate for SCs. For instance, commonly used papers having varied roughness like Xerox paper, filter paper, weighing paper, and glossy paper are used as substrates. Based on the amount of clay and graphite, pencils are classified in the scale of 9B to 9H. Some reports show the use of graphite rods, graphite paste, and customized "pencil" components to draw and fabricate FSSCs electrodes on paper [29, 52]. Zheng et al. synthesized and fabricated FSSCs by directly drawing the graphite on a cellulose paper and showed a favorable synergy among the components. Figure 3.5a, b shows the schematic of drawing a graphite pencil on cellulose paper to fabricate conductive flexible electrode and a digital photograph of this electrode. Figure 3.5c shows the fabrication of SCs using sandwiched Xerox paper (separator) between

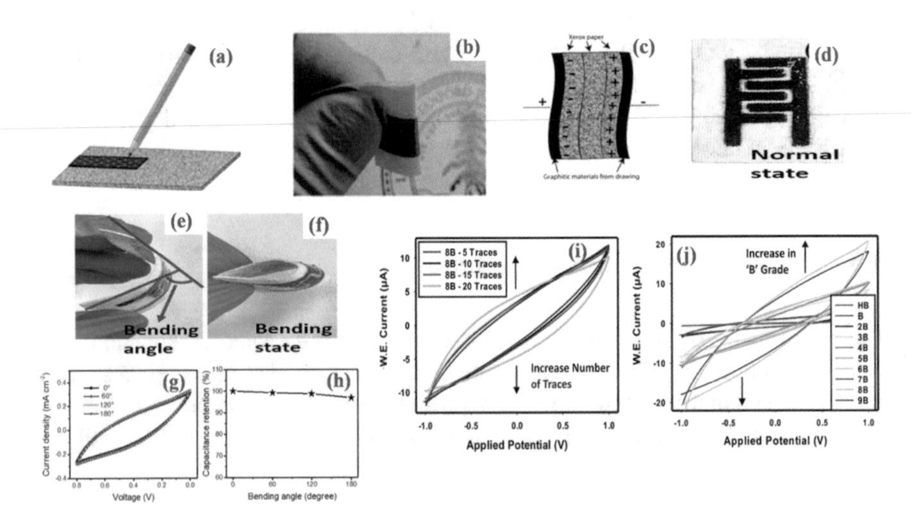

Fig. 3.5 **a** Schematic of pencil drawing electrode, **b** digital image of the pencil drawing paper-based FSSCs device, **c** schematic of the Xerox paper-based SCs device [53], **d**, **e**, **f** optical images of the MnO_2/graphite solid FSSCs device at normal and bending states, **g**, **h** CV curves and capacitance retention curve of the MnO_2/graphite solid FSSCs under various bending states [55], and **i**, **j** CV curves of the 8B pencil drawn electrodes [56]

the two conductive paper electrodes. The fabricated device exhibits ~2.3 mF cm^{-2} areal capacitance in presence of 1 M H_2SO_4 and 100% cycle stability up to 1500 GCD cycles. It has been found that the specific energy and power of the paper-based SCs are comparable to the commercial and other carbonaceous SCs [53]. Yao et al. synthesized graphite/PANI hybrid electrodes for FSSCs by drawing a thin layer of graphite on A4 size printing paper and grows 3D PANI network through electrodeposition method. The device can be fabricated by sandwiching a cellulose membrane separator with PVA/H_2SO_4 electrolyte in between the two electrodes. The FSSCs show nearly unchanged CV diagram after 90° bending suggesting its highly flexible characteristics. The capacitance of the device deteriorates gradually and retains ~83% after 10,000 cycles with respect to the 1st cycle [54]. Li and co-workers fabricated a plaster-like thin-film solid micro-SCs through simple pencil drawing on medical tapes [55]. In addition to the flexibility, pencil drawing micro-SCs exhibits good electrochemical properties by the deposition of MnO_2. The device can be fabricated by sandwiching two electrodes using PVA/H_3PO_4 solid–gel electrolyte. Figure 3.5d–f shows the optical image of the fabricated thin-film solid micro-SCs in different forms. The plaster like micro-SC can be directly mounted on human body by simple uncovering–affixing technique. Figure 3.5g shows 200 CV cycles after bending from 0 to 180° and ~90% retention in specific capacitance is recorded after 200 cycles as shown in Fig. 3.5h [55]. The properties of the pencil drawing FSSCs not only depend on the quality of the pencil but also on the soaking time of the electrolyte and number of traces on the substrate. Figure 3.5i, j illustrates that with increasing the number of traces of the pencil more amount of graphite is deposited on the substrate, which directly affects the cell capacitance of the device. The resistance of the electrode varies with carbon content in pencil and the electrical resistance decreases from 898 to 39 Ω while changing the pencil from HB to 9B [56].

The pencil drawn electrodes provide a good basis for testing hypotheses, electrode design, or electrode application of electrolyte performance. The difficulty in using pencil electrodes is to provide uniformity in the manufactured electrodes. The error range in sheet resistance measurement represents the variation in each device. The optimization of the double-sided pencil drawn electrodes provides brief understanding for the development of flexible paper-based pouch SCs [2, 57].

3.3.2 Deposition

The deposition of thin layer active materials on the substrate is another useful technique for the fabrication of FSSCs. Chemical bath deposition (CBD), electrochemical deposition (ED), spray deposition, and sputtering are equally useful like chemical vapor deposition (CVD) for the fabrication electrodes of FSSCs [58, 59].

CVD technique is employed when the porosity of the electrode materials is playing an important role. The CVD grown electrode does not use binder or surfactant leading to the improved device performance. The process is very simple and is capable to cover a large surface area. Solution-based carbon materials like graphene, CQD,

CNTs, and their composite materials are deposited by CVD techniques. It is the most commonly used technique for the fabrication of graphene or graphene-based composites [58, 60]. A group of researchers showed the scalable preparation of multi-layered graphene-based flexible micro-SCs device, by direct laser writing of stacked graphene film, which can be synthesized in large scale by CVD technique. The fabricated device shows good flexibility and diverse planer geometry with custom designed integration properties. Figure 3.6a, b shows the digital images of the fabricated micropattern arrays on the PET substrate. The device exhibits excellent volumetric ED (23 mWh cm^{-3}) with PD of 1860 W cm^{-3} using PVA/H_2SO_4 hydrogel electrolyte [61].

Electrophoretic deposition of colloidal suspension on the current collector is another attractive technique for the preparation of thick layers of active materials. Hydrothermally grown graphene quantum dots (GQD) can be deposited on the flexible carbon cloth by the electrophoretic deposition technique. The symmetric FSSCs device exhibits ~24 mF cm^{-2} of areal capacitance with good capacitance retention under bent/flex conditions [62]. CBD is another popular technique to fabricate FSSCs devices. It is one of the cheapest techniques to deposit thin films/nanoflexible materials and can be employed for large area deposition through batch or continuous process. CBD grown PANI/ZnO/ZIF8/graphene/polyester textile electrode is used to fabricate FSSCs device [63]. The ED of the FSSCs device is ~235 μWh

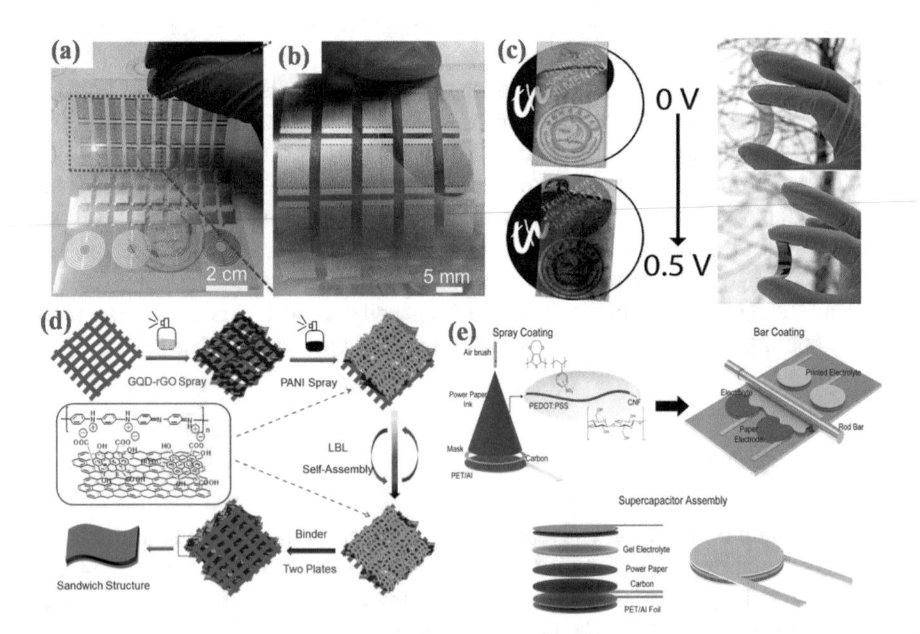

Fig. 3.6 **a**, **b** Optical images of direct laser writing of CVD grown graphene micropattern on PET substrates [61], **c** optical image of rGO–$Co_{(1-x)}$ Ni_x $(OH)_2$ materials in flat and bending state [64], **d** schematic of the synthesis of PANI/GQD rGO/CFC [65], and **e** schematic of the printing steps and components of the spray-coated printed paper-based FSSCs device [66]

cm^{-3} at a PD of 1542 μW cm^{-3} with good capacitance retention under mechanical deformations. The charged species diffuse toward the electrode surface under the electric field. This is a cost-effective deposition technique and can be used for the scaled-up manufacturing. Transition metal chalcogenides/carbonaceous material composites can be prepared by electrochemical deposition process. Electrochemical deposition is capable of reduction of carbon and deposing the metal/mixed metal chalcogenides simultaneously in one step. In addition, homogeneous deposition occurs at room temperature. Lei and co-workers deposited amorphous $Co_{(1-x)}$ Ni_x $(OH)_2$ on the rGO sheet through single-step ED technique [64]. Figure 3.6c shows the flexible electrochromic SCs device. The device has pseudo-capacitive storage performance, excellent rate capability, high Columbus efficiency, and non-diffusion limited behavior. Figure 3.6d shows the schematic of the fabrication of thin-film nanomaterial-based FSSCs by spray deposition techniques. The film prepared by spray deposition is similar to the film fabricated by ink-jet printing and electrophoretic deposition. FSSCs device has been assembled through layer-by-layer spraying of GQD/rGO and PANI solution alternatively on carbon fiber cloth. The device can be fabricated by sandwiching the PVA gel electrolyte by the flexible active electrodes. Due to the strong electrostatic interaction between the electrode material and the carbon fiber cloth, the fabricated device shows good capacitance retention of ~97.7% after 10,000 GCD cycles. The layer-by-layer spraying technique shows great advantages of easy operation, good controllability, and versatility. The fabricated FSSCs device can be used for the fabrication of foldable and wearable electronic devices [65]. Berggren et al. assembled FSSCs device by using cellulose nanofiber and PEDOT: PSS conducting polymer-based electrodes. Figure 3.6e shows the schematic of the fabrication of spray-coated printed paper-based FSSCs device. In this technique, spray deposition is used to fabricate the lightweight electrode and gel electrolytes can be prepared by bar coating technology. Solid-state SCs are flexible, mechanically robust, and have low equivalent series resistance (0.22Ω) resulting high PD (~104 W kg^{-1}) energy storage systems [66].

3.3.3 Ink-Jet Printing

Printing is a popular technique in the fabrication of FSSCs on paper substrate. Inkjet printing, screen printing, and gravure printings are some commonly used printing techniques. Over the past few decades, varieties of inks are formulated for realizing the energy storage devices that survive under high flexural stress. Two most important aspects of FSSCs device fabrication are the preparation of ink and preparation of film. The development of suitable ink with proper electrical conductivity and rheological properties (viscosity and surface tension) is more critical. Ink-jet printing involves conductive fillers, binders, solvents, and additives. In a flexible device, strain is generated within the substrate and the printed component when the device is flexed. Therefore, ink synthesis is an important factor for developing a device that possesses high degree of flexibility without compromising the properties of the device [67, 68].

Adhesion property of binder plays a crucial role in detecting the flexible nature of the ink. Uniform dispersion of the binders in ink minimizes the strain during flexing the device. Printing technique is an important task. The printed FSSCs take advantage of simple, scalable, controlled structure, cost effectiveness, and rapid fabrication processes.

Owing to the intrinsic features such as fast printing, reduced wastage of materials, scalability, cost effectiveness, and large area of fabrication, ink-jet printing technique is extensively used in portable and wearable electronics. The top-down inkjet printing method is one of the most popular approaches of writing directly on the substrate in which the pattern can be directly controlled by computer program. The nozzle size in ink-jet printer is in the range of 100–400 nm and is designed for a nominal drop volume. There are only a few reports on ink-jet printing energy storage devices due to the technical challenges of ink materials [69, 70]. FSSCs device can be fabricated by printing the SWCNT/AC@Ag-NW ink onto the conventional A4-sized paper by the commercial desktop printer. Different devices with various dimensions can be fabricated by controlling the printing technology and the device exhibits ~100 mF cm^{-2} areal capacitance with good capacitance retention and negligible degradation after 2000 cycles. Ink-jet-printed SCs can be easily connected in series or in parallel without the support of metal interconnects, allowing users to customize the battery voltage and capacitance. Figure 3.7a shows the schematic of the stepwise fabrication of ink-jet-printed FSSCs device. Ink-jet printer flat FSSCs can also be used as a monolithic integrated power supply for the Internet of things (IoT) [71]. Paper-based ink-jet printing in-plane architecture transparent graphene FSSCs device displays excellent electrochemical performance than the other types of architecture. The device exhibits high areal capacitance of ~1.586 F cm^{-2} with ~89.6% retention in capacity after 9000 cycles. Based on this outstanding performance, the ink-jet printed in-plane FSSCs device holds substantial promising flexible power source for electronic instruments. In addition to the carbonaceous materials, two-dimensional pseudo-capacitive materials like δ-MnO_2 can be used for the fabrication of FSSCs electrode. The FSSCs device has been fabricated by printing the δ-MnO_2 ink on the oxygen plasma-treated glass and polyimide film substrates by ink-jet printing technique [72]. The device shows highest volumetric capacitance of ~2.4 F cm^{-3} with 1.8×10^{-4} Wh cm^{-3} ED at a PD of 0.018 W cm^{-3}. It shows excellent mechanical flexibility and ~88% retention in capacitance after 3600 GCD cycles.

Screen printing is another low cost, simple, scalable, and useful technique for designing FSSCs device through various pattern of printing like push-through process, deposition printing, etc. The main advantage of screen printing technology is that it can efficiently produce a large number of devices in a short time. The Ag@PPy ink is used to print on the preprinted silver interdigital current collector by screen printing method to fabricate the flat FSSCs. The device retains its electrochemical properties in high bending angles with large bending time. Liu et al. fabricated more than 40 devices through screen-printed way with a very short time (<30 min) [73]. Figure 3.7b demonstrates optical photograph of micro-SCs device on various flexible substrates like cloth, PET, and paper. The PET-based device exhibits ED of ~4.33 μWh cm^{-2} in the presence of PVA-H_3PO_4 gel electrolyte with superior mechanical

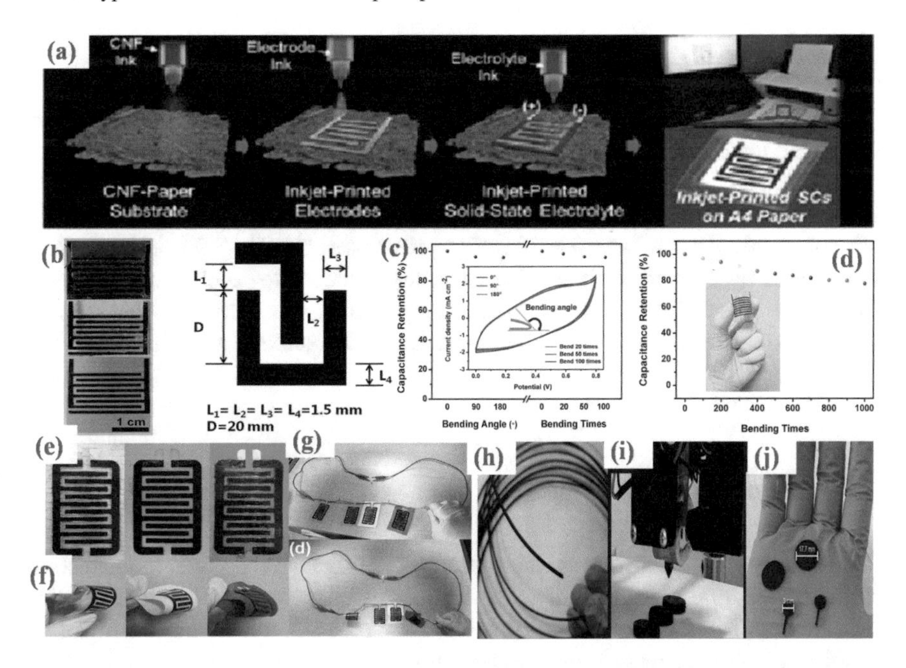

Fig. 3.7 **a** Schematic of the stepwise fabrication of ink-jet-printed FSSCs [71], **b** optical images of fabricated micro-FSSCs on various substrates (cloth, PET, and paper), **c**, **d** capacitance retention ratio of the micro-FSSCs at 100 mV s^{-1} [73], **e**, **f** digital images of the all printed solid-state flexible SCs, **g** digital images of the devices in series lighting up a yellow LED in flat and bending condition [74], **h**, **i** optical images of the 3D printable graphene/PLA and 3D printing process, and **j** different types of printed 3D electrodes [75]

flexibility and capacitance retention after many different bending cycles (up to 100 times). Figure 3.7c, d shows ~77.6% retention in specific capacitance after 1000 bending cycles [73]. Interestingly, electrochemical performance is independent on the substrate (cloth/PET/paper) used to fabricate the FSSCs device. Figure 3.7e, f shows the flexibility of different FSSCs device with different bending state. The device exhibits areal capacitance of ~5.7 mF cm^{-2} and ~80% retention after 2000 GCD cycles. Even after highly bending and stretching, the fabricated FSSCs light up a 1.9 V LED yellow light as shown in Fig. 3.7g [74].

There have been some attempts to print active and conductive electrode materials with 3D-printers for the fabrication of high performance FSSCs device. So far, there have been 3D printers, like "Voxel8®", which has a confrontational printing technology using ink and solid-state filament for printing purposes. Polymer, wax, metal, and ceramic paste are used as 3D printing materials. Control of printing thickness and structure pattern by controlling the materials flow is the main advantage of 3D printing technology. Figure 3.7h–j represents the optical image of graphene/PLA-based different 3D printing electrodes and 3D printing process [75]. 3D-printed hybrid FSSCs device is fabricated by direct writing of active graphene-based PLA filament using a FDM 3D printer. The device exhibits ~75.51 $\mu F\ g^{-1}$ specific capacitance at 200 μA applied current in the presence of 1.0 M PVA/H_2SO_4 gel-electrolyte [76]. Preparation of the printing ink and structural design are the important factors for the fabrication of 3D-printed device. In this fabrication process, current collector, active electrode material and gel electrolyte are printed on the silicon substrate. The fabricated FSSCs device exhibits good flexibility and good capacitance retention (54–58% of its initial capacitance) at 50 mV s^{-1} scan rate [77]. The 3D printing process shows good reproducibility indicating the utility of the ink to develop more sophisticated electronic devices. The crucial issue for printing a successful SCs device is the selection of electrode and electrolyte materials with excellent printability and low manufacturing cost.

3.3.4 Dip Coating

Dip coating is the facile technique to deposit active materials onto any substrate including metallic, ceramic, polymers, textiles, etc. This process is widely used in industrial applications also. In this process, at first the target materials are dissolved in solvents followed by coating onto the substrates. The morphology and the thickness of the coated substrate can be controlled by monitoring the related parameters such as immersion time, dip coating cycles, substrate surface condition, withdrawal speed, density, and viscosity of the coated substance. In addition to the normal dip coating methods, sol–gel dip coating, vacuum-assisted dip coating, photo-assisted dip coating, multilayered dip coating are also used to prepare thin-film electrodes for FSSCs fabrication. In sol–gel dip coating process, the quality of the deposited film can be controlled by adjusting the amount and preparation technique of sol and gel. Photo-assisted dip coating is used to control the evaporation process of the coating solution, because the radiation effect is conducive to film formation [78–80]. Among the above mentioned dip coating methods, the simplest and commonly used scaled-up coating technique is the direct solution dip coating. Solution dip coating is an effective method, which can be used for mechanical improvement and functional manufacturing of composite materials through interface tailoring. Self-assembled GO interlayers can be easily fabricated on the flexible substrates by this technique. Figure 3.8a demonstrates the fabrication of electrode by dip coating process. In addition to the carbonaceous materials, transition metal oxide (zinc oxide) films or nickel catalysts can be successfully decorated on the flexible cotton fibers in alkaline medium using a simple solution dip coating method [81]. Zu and co-workers developed ultrafine Fe_3O_4 nanoparticles/functionalized graphene (FG) on carbon fiber cloth by using a controllable annealing-assisted dip coating method and fabricated an FSSCs device as shown in Fig. 3.8b, c. The assembled FSSCs device shows no significant degradation in capacitive properties after folding at 45°, 90°, 135°, and 180°. The ED, PD, and cyclic stability of the FSSC device are found to be 19.2 Wh kg^{-1}, 800.2 W kg^{-1}, and 100% retention after 4000 cycles at 1 A g^{-1}, respectively [82]. Sol–gel coating technique is efficient, cost-effective, and emerging chemical

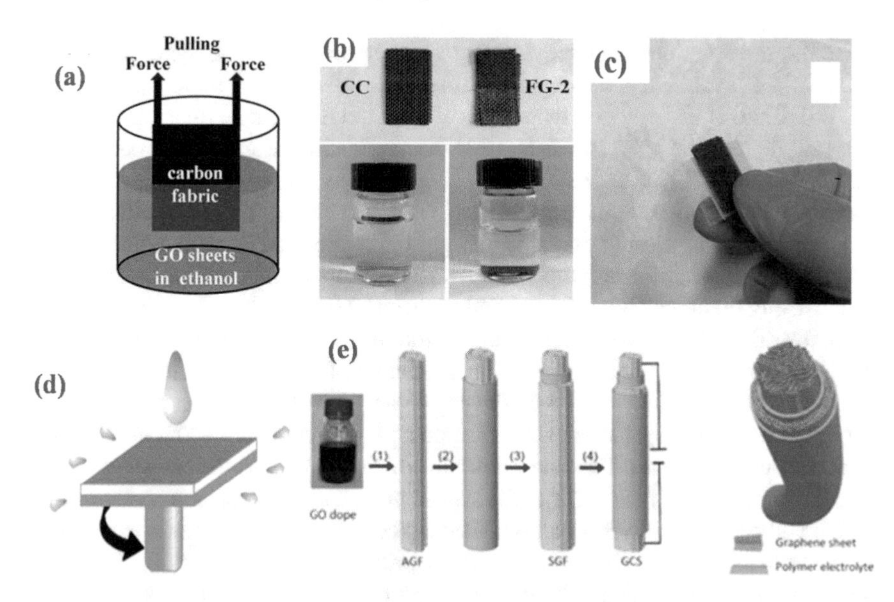

Fig. 3.8 **a** Schematic of the dip coating process [81], **b**, **c** digital images of dip coating electrodes and device [82], **d** schematic of the spin coating technique [86], **e** schematic of the FSSCs fabrication process and the cross-sectional structure of a fiber-type cylinder FSSCs device [87]

approach for synthesizing flexible electrode materials. Sol–gel coating techniques are commonly used to deposit the active materials onto the fibrous substrate and the starting precursors mainly consist of various ligands and metalloid elements [83]. Both the sol–gel dip coating and direct solution dip coating techniques are based on the solid–liquid inherent interaction. Sol–gel dip coating technique is time dependent, evaporation-induced concentration, and more complex process as compared to the direct solution dip coating technique. Spin-assisted dip coating is another useful and feasible technique for film devotion. In this process, the materials are deposited on the substrate by scattering the materials onto the center of a rotary table under centripetal force and evaporation condition. Figure 3.8d shows the schematic of the spin coating technique. In spin-assisted process, electrostatic interaction between the coating materials and substrate, centrifugal force, and viscous force are the dominating factors to adsorb and rearrange the deposit materials onto the substrate and facilitate the coating process. Solution concentration and rpm are another two important parameters in spin coating process to control the thickness and morphology of the active electrodes [84–86]. To increase the thickness and improve the uniformity of the materials, multilayer dip coating process is highly useful. The repetition of dip coating is more relevant and proficient in fabricated functional complex multilayered deposited film. Multilayered dip coating is different from layer-by-layer fabrication process. In layer-by-layer process, the substrate is alternatively dipped in the aqueous solutions of oppositely charged ions with intermediate steps of rinsing with pure water. In multilayered dip coating techniques, varieties of coating solutions are used to achieve the functional superposition by several deposited films. Synergistic effect of deposited layers plays an important role in multilayered dip coating technique [84, 86]. Zhao et al. used continuously wet spun axial graphene fiber (AGF) as the core and dip coated graphene sheath for the fabrication of graphene coaxial fiber SCs (GCS) [87]. For AGF preparation, aqueous dispersion of GO (8 mg ml^{-1}) was spun into the coagulation bath containing 375 ml of water, 125 ml of ethanol, and 25 g of $CaCl_2$. The freshly prepared GO gel fibers are washed with water–ethanol mixture and dried at room temperature for 24 h. For the fabrication of separation layer, freshly prepared PVA gel is coated onto the AGF layer by dip coating process. Then, the dried PVA-coated AGF is dipped into the 5 wt% $CaCl_2$ solution followed by dipping into the GO dispersion. The schematic of GCS fabrication is shown in Fig. 3.8e. The GCS shows high specific capacitance of ~182 F g^{-1} and ED of ~15.5 Wh kg^{-1}. Interestingly, the retention in specific capacitance is ~100% after 10,000 charge–discharge cycles. Flexible properties of the GCS were investigated by CV tests during repeated bending and the retention of specific capacitance was ~92% after 100 cycles. Recently, vacuum-assisted dip coating technique is used extensively for the entire coverage of zero defect active particles onto the substrate. This technique provides a new route for manufacturing the tubular membrane with a novel configuration [88].

3.4 Types of Flexible SCs Device

The demand of ultrathin, wearable, and foldable energy storage system is increasing day by day due to the rapid growth of smart electronic devices in IoT. Different types of stretchable, compressible, transparent, and microflexible SC devices are considered as the most suitable energy storage devices. This section emphasizes on the different types of FSSCs devices [89, 90].

3.4.1 Stretchable SCs Device

Among the different types of flexible energy storage system, stretchable SCs devices are expected to become power sources for wearable electronic devices. Fabrication of flexible SCs device is relatively simpler and can be integrated with other units to form the wearable electronic systems. The stretchable SCs device is able to deliver high PD with long-term life stability. It is able to withstand large strains without considerable degradation of capacitive performance. Light-emitting diode and display panels show >100% stretchability while retaining their function [91, 92]. To date, research efforts on flexible SCs are devoted mostly on the design and establishment of stretchable components, construction of the device, and other multifunctionalities like lightweight, thickness, transparency, etc. Fabrication of highly stretchable electrodes and design of the device are the main two important criteria for successful fabrication of highly stretchable SCs device. Stretchable SCs can be classified into three types: 1D linear, 2D planer, and 3D stereo SCs. To achieve high electrochemical performance, each type of SCs device should maintain high conductivity under large mechanical deformation [93].

Single-walled carbon nanotube films (SWCNT), graphene, conducting polymers, activated carbon, etc., are used as the electrodes in FSSCs devices [94–98]. Figure 3.9a shows the transfer of vertically aligned forest like CNT onto the elastomeric substrate by thermal annealing process. CNT-based stretchable electrode exhibits good electrochemical properties and stability up to thousands of stretching–relaxing cycles under an applied uniaxial strain (300%) or biaxial strain (300% × 300%) as shown in Fig. 3.9b. The assembled stretchable SCs of crumpled CNT forest electrode is able to sustain up to ~800% stretchability and shows ~5 mF cm^{-2} specific capacitance at 50 mV s^{-1} scan rate [96]. 2-D graphene is used to fabricate a stretchable SCs device due to its high surface area and high mechanical strength. The 2D rGO sheets form a stretchable electrode with the combination of another 2D material $Ti_3C_2T_x$ (MXene). The combination of mechanically robust rGO and excellent electrochemical activity of Ti_3C_2Tx leads to the formation of mechanically stable and flexible electrodes for SCs devices (Fig. 3.9c). It is found that the stretchable $Ti_3C_2T_x$/rGO composite electrode with 50 wt% of rGO loading is able to

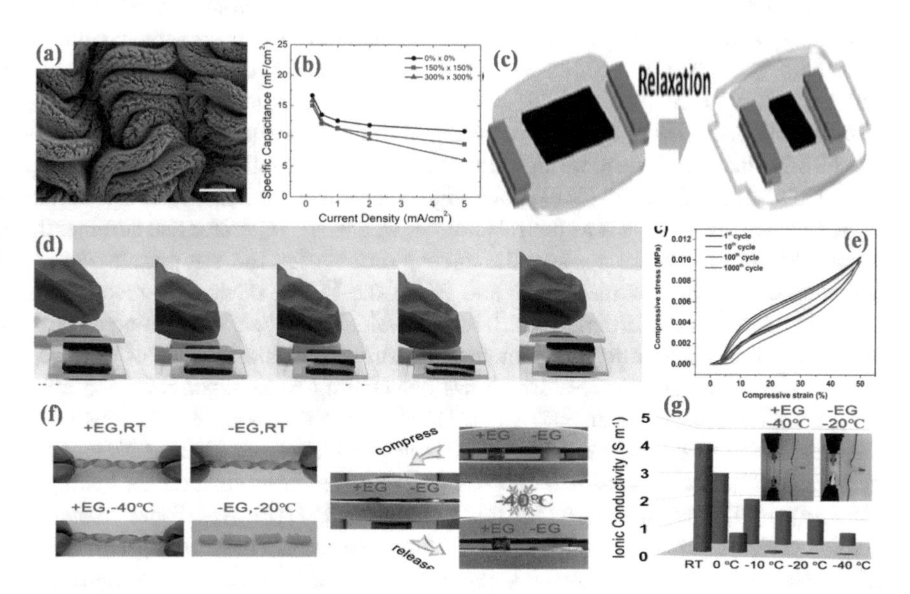

Fig. 3.9 **a** FE-SEM image of CNT-forest, **b** variation of the specific capacitance at different current densities and uniaxial strains [96], **c** schematic of a stretchable $Ti_3C_2T_X$ MXene/rGO composite thin-film electrode in different stretching condition [98], **d** compressive properties of PPy/MF-40 composite-based device at different compressed conditions, **e** corresponding cyclic stress–strain curves in different cycles at a strain of 50% [104], **f** elastic stability of the DN 2.5 + EG hydrogel at −40 °C and the DN 2.5-EG hydrogel at −20 °C, and **g** Ionic conductivity curve of the DN 2.5 + EG hydrogel and the DN 2.5-EG hydrogel from RT to −40 °C [105]

mitigate the cracks generated under large strains. The fabricated stretchable device shows ~18.6 mF cm^{-2} specific capacitance with the stretchability of up to 300% and the electrochemical stability under various strain conditions over 10,000 GCD cycles [98]. Selection of a suitable dielectric separator is another challenge in flexible SCs device since the conventional separator is not flexible in nature. Highly stretchable buckled SWCNT film is used to fabricate the FSSCs using PVA/H_2SO_4 gel, which act as both electrolyte and separator. The fabricated stretchable SCs device works well under high strain (120%) and stretching condition [99, 100].

3.4.2 Compressible SCs Device

Three-dimensional foam or sponge like material is used to fabricate the compressible SCs devices. The sponge is composed of interconnected cellulose or polyester fibers. These materials are stable against high compressible strains without any structural deformation. Polypyrole (PPy)-mediated graphene foam and CVD grown CNT sponge are used as compressible SCs electrode due to their sustainable compressible strain. However, the ED of the CNT or graphene-based FSSCs is relatively lower as compared to the conventional type of SCs. Therefore, the addition of pseudo-capacitive-type transition metal chalcogenides or PPy is very effective for the enhancement of ED [101–103]. SWCNT films show several cross-linked pathways and the SWCNTs are attached with the sponge through van der Waals force. Therefore, the SWCNT-coated sponge is considered as the good electrode for compressible FSSCs electrodes due to its high compression and the fault tolerance [103, 104]. PPy-coated melamine foam (MF) shows high mechanical flexibility in compressible FSSCs device. The MF acts as the substrate as well as the separator and sufficiently absorbs the electrolyte to confirm the interaction between the electrode materials and electrolyte. The volumetric ED and PD of the FSSCs device are recorded as 0.18 mWh cm^{-3} and 22.90 mW cm^{-3}, respectively. The device shows high electrochemical stability under different compressive stress in different strain as observed from Fig. 3.9d, e. The device also shows outstanding capacitance retention after 1000 compression cycles at 50% strain [104]. Liu and co-workers demonstrated the preparation of cross-linked PVA double-network ethylene glycol (DN + EG) hydrogel [104]. In comparison to the chemically derived cross-linked PVA hydrogel (25-fold) and DN-EG hydrogel (5.3-fold), the DN + EG hydrogel displays considerable perfection in compressive stress. In the presence of anti-freezing agent, the fabricated device shows excellent low-temperature (−40 °C) tolerance and high ionic conductivity (0.48 S m^{-1}) with 15.5 MPa of compressive stresses as illustrated in Fig. 3.9f. The FSSC shows ~100% retention in capacitance under strong compressive stress.

DN 2.5 EG hydrogel shows higher ionic conductivity as compared to that of the DN 2.5 + EG hydrogel as evidenced from Fig. 3.9g. In addition, the capacitance retention is ~86.5% after 4000 bending cycle (180°) at −30 °C confirming the outstanding compressive performance at sub-zero temperature [105, 106].

3.4.3 Transparent SCs Device

In a typical solid-state flexible transparent SCs, electrolyte and separator are transparent in nature. The electrodes are generally fabricated using ultrathin transparent materials. The use of metallic current collectors is generally avoided in transparent SCs devices. Metal nanowire is used as the current collector in transparent SCs devices due to its high optoelectronic properties, which improve the steady flow of electron between the electrode materials and circuit without any electrochemical reaction. However, the poor wire-to-wire contact in metal nanowire network is the great challenge during device fabrication and thus inhibits its practical applications [100, 107, 108]. The metal-to-metal contact area between two pitches is large, which can help improve the mechanical strength and can be used as a current collector in transparent SCs devices [109]. Generally, Ag, Ni, and Au meshes are used in transparent SC electrode due to their high electrical conductivity and mechanical flexibility. Primarily SPEs and GPEs like PVA/$LiClO_4$, PVA/LiCl, and PVA/H_2SO_4 are used in flexible transparent SCs devices. The ultrathin electrodes facilitate the electrolyte to infiltrate into the electrodes. Ultrathin-aligned SWCNTs, MWCNTs, and graphene show high electrical conductivity, large specific surface area, and good mechanical flexibility at lower mass loadings. So, these carbonaceous materials and their composites serve as flexible transparent electrode [110–117]. Aligned carbon nanotube array sheet possesses excellent optical transmittance and mechanical stretchability and is used as high-performance transparent and stretchable all-solid SCs device. It is seen that highly aligned CNTs exhibit superior performance, higher transmittance, high specific capacitance, and longer life cycle. Aligned CNT-based flexible SCs device can achieve up to 75% of transmittance at the wavelength of 550 nm and 7.3 F g^{-1} specific capacitance shown in Fig. 3.10a, b. The fabricated device can be bi-axially stretched up to 30% strain without any noticeable change in electrochemical properties even after 100 stretching cycles [113]. CVD is a useful technique to synthesize SWCNTs with thicknesses of <100 nm and exhibits ~70% transparency at 550 nm wavelength. In order to fabricate the FSSCs electrode, a substrate is needed to deposit the active materials on it. Polyethylene terephthalate (PET) is a good transparent substrate that can be used for coating of SWCNTs due to the van der Waals and electrostatic interaction between the PET and SWCNTs. SWCNTs-based flexible transparent SCs device shows ~60% transmittance at the wavelength of 550 nm [114]. Depending upon the electrode film thickness, maximum ED and PD of the SWCNT-based FSSCs reaches up to 20 W h kg^{-1} and 15 kW kg^{-1}, respectively [114]. On the contrary, MWCNTs show excellent characteristics, such as large active surface area, high electrical conductivity, and good chemical stability. In gaseous phase, MWCNT network can be deposited directly onto the polymer electrolyte membrane and results in the formation of an aerogel-type transparent electrode material. The optimal transmittance of the MWCNT-based transparent device is ~70% and shows ~1370 kW kg^{-1} PD. The operating potential window of the transparent device is found to be ~3.5 V and the capacitance retention is

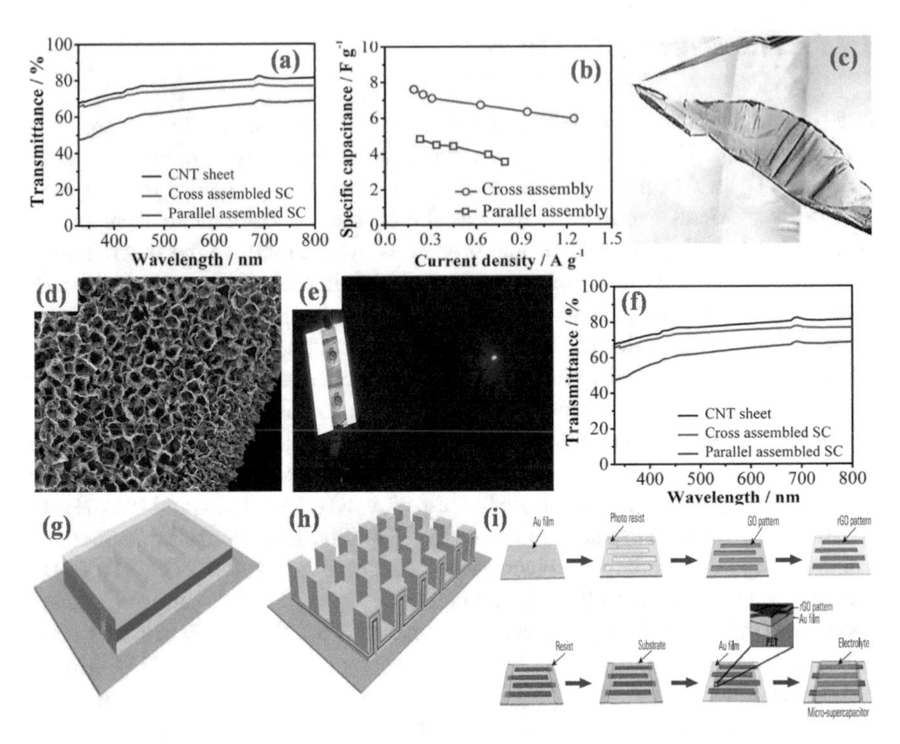

Fig. 3.10 **a** Transmittance spectra of CNT sheet on PDMS substrate, **b** Ragon plots of both types of SCs [112], **c** digital photographs of self-supporting microstructured graphene film, **d** FE-SEM image of a typical curved area of graphene film, **e** lighting of LED bulb with two SCs connected in series, **f** transmittance of a pure substrate, composite, and an integrate SC device [Insets are corresponding photographs of the electrode and device] [116], **g**, **h 2D and 3D** in-plane micro-FSSCs device [121], **i** schematic of rGO-based micro-SC fabricating process [124]

~96% after 20,000 charge-discharge cycles. The device exhibits reproducible electrochemical performance after 100 bending cycles at 180° [115]. Transfer of CVD grown single/few-layers graphene films from the metal foil to transparent substrate is another approach of fabricating transparent FSSCs electrode. The transmittance of graphene-based device is in the range of 48–72.9% and the stretchability is ~40%. Optical transmittance of graphene-based SCs device reaches to ~67% and ~2.94 W h kg^{-1} ED at 438 kW kg^{-1} PD [115, 116]. Maintenance of high transparency and surface area with high mass loading of graphene is still a challenging job due to its high specific surface area. All these factors are very important in determining the final performance of the SCs device. Figure 3.10c demonstrates the digital photograph of the synthesized self-supporting microstructured graphene film with the dimension of 63 mm × 15 mm. A wrinkled-wall self-supporting flexible and transparent graphene film is fabricated with open-hollow polyhedron like building blocks as visible from the FE-SEM image of electrode materials shown in Fig. 3.10d. The wrinkled structure enhances the surface area and the open-hollow microstructures to maintain the transmittance properties of the device (Fig. 3.10e, f). The SCs device exhibits high specific capacitance of 3.5 mF cm^{-2}, and minor change in capacitance with bending cycle is observed. It also exhibits ~552 μWh cm^{-3} ED at a PD of 561.9 mW cm^{-3} and ~94.8% retention in initial capacitance is recorded after 20,000 GCD cycles [116]. The transition metal chalcogenides are also used for the fabrication of flexible transparent SCs. 2D transition metal carbide or nitrides commonly known as MXene is considered as the promising material for flexible transparent SCs device. By spin coating the Ti_3C_2Tx nanosheet colloidal suspensions and then vacuum annealing at 200 °C, a highly transparent and conductive Ti_3C_2Tx film is prepared and used as an electrode in FSSCs device. The synthesized films with 93 and 29% transmittance demonstrate impressive volumetric capacitance (676 F cm^{-3}). Asymmetric SCs device based on $Ti_3C_2T_x$ and SWCNT exhibits ~72% transmittance, 1.6 mF cm^{-2} specific capacitance, 0.05 μWh cm^{-2} ED, and ~100 retention after 20,000 GCD cycles [92]. A few layered 2D Co-$(OH)_2$ nanosheets are vertically arranged on a modified Ag nanowire (AgNWs) by electrochemical deposition. The 2D Co-$(OH)_2$/gNWs network shows high transparency, low contact resistance, and high pseudo-capacitive properties. Judicial material design greatly improves the electrochemical properties of the hybrid network, and the device shows high areal capacitance as high as 3108 μC cm^{-2} (5180 μF cm^{-2}) and a long cycle life. The flexible transparent device exhibits 0.04 μWh cm^{-2} ED, 28.8 μW cm^{-2} PD, high mechanical flexibility, and cycle stability [118].

3.4.4 Flexible Micro-SCs Device

Rapid growth of miniaturized ultrathin wearable electronics has stimulated the research on self powered microenergy storage and conversation devices. Flexible micro-SCs (FMSCs) posses many advantages, such as high ED, PD, cyclic stability, and are useful in biomedical engineering. Furthermore, many types of flexible, bendable, twistable, compressible, and biocompatible microelectronics require micropower sources. FMSCs supply energy to the microelectronics consisting of an array of interdigitated microelectrodes in micron-scale sizes. The areal size of the FMSCs can vary from centimeter to millimeter scale, and the distance between the two neighboring electrodes is less than in micrometer range. The conventional FMSCs device adopted "sandwitched" configuration, in which two electrodes are stacked together and separated by a solid electrolyte. However, the use of thick electrolyte and electrode deteriorates electrochemical properties of the sandwiched SCs. The use of separator increases the thickness and transport resistance of the sandwiched FMSCs device. In order to overcome this lacuna, in-plane integral electrodes are fabricated [119, 120]. Based on the dimension of the device, FMSCs can be categorized mainly three types: fiber-shaped, in-plane, and 3D. Figure 3.10g–h shows the schematic diagram of 3D in-plane and stacked FMSCs, respectively. Fiber-shaped FMSCs are mainly used as a power source in fiber and textile electronics due to their parallel, twisted, and coaxial structure [121–123]. For the construction of on-chip electronic system, in-plane FMSCs are fabricated with standard microelectromechanical systems. On the other hand, 3D FMSCs with internal 3D architectures deliver much higher areal capacitance as compared to the in-plane FMSCs due to the large areal mass loading of active materials on the current collectors. However, the complicated fabrication processes and use of thick microelectrodes are the main issues in FMSCs leading to the loss of flexibility. In order to determine the flexibility of the micro-SC device, mechanical properties of the current collector, separator, electrodes, and solid-state electrolytes play a crucial role. FMSCs can be fully charged or discharged in seconds, providing an ultrahigh areal PD (>10 mW cm^{-2}) and long cycle life ($\geq$10,000 cycles). The special merit of FMSCs is the design of microelectrode fingers with interdigital configuration. Fabrication of microelectrodes and microdevice depends on the advanced microfabrication technologies such as photolithography, deposition, laser etching, and ink-jet printing. On the contrary, the current micromanufacturing technology used in most FMSCs is not compatible with the manufacturing technology of other electronic components [124–126]. In order to achieve high capacitive FMSCs, 0D GQD, 1D CNT, 2D graphene, metal carbides and nitrides (MXenes) are explored due to their large specific surface area, high electrical conductivity, and presence of redox active sites. However, easy preparation, scalability, and low cost production are still challenging for the practical applications. Development of high-performance FMSCs depends not only on the properties of the electrode materials but also on the device configuration [127, 128]. There are mainly two approaches for the fabrication of FMSCs: bottom-up and top-down. In the first technique, the active materials are deposited onto the designated position by electrophoretic deposition process. The bottom-up technique may lead to the firm adhesion of the rGO interdigital electrode on the Au film, as shown in Fig. 3.10i. The FMSCs device achieves 31.9 mW h cm^{-3} volumetric ED and 324 W cm^{-3} volumetric PD by using PVA/$H3PO_4$ as the electrolyte. In top-down process, an electrode-patterning step is performed to achieve the good electrode materials followed by the etching process. The oxygen plasma-assisted in-plane rGO FMSCs device shows ~80.7 mF cm^{-2} and 17.9 F cm^{-3} areal and volumetric capacitance, respectively [124, 129, 130].

3.5 Various Design of FSSCs Devices

High flexibility and compact design of FSSCs are suitable to incorporate in wearable electronics. Recently several innovative designs on FSSCs devices are reported for use in wearable/bendable electronics. Mainly three types of FSSCs devices are reported: sandwiched, planer, and fiber types. Fiber-type devices can be further categorized into fiber on-plane, wire, and cable types.

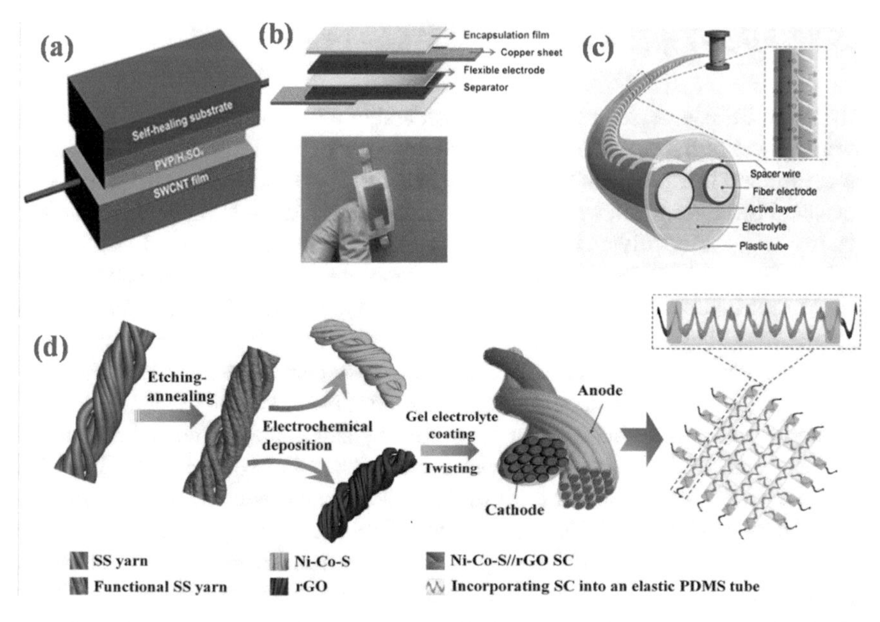

Fig. 3.11 **a** Schematic of sandwiched FSSCs device [131], **b** schematic and optical image of a $FSSC_S$ device [130], **c** fiber-type FSSCs device [94], **d** schematic of the cable-type FSSCs device fabrication [137]

3.5.1 Sandwiched-Type Device

Sandwiched-type device is the most common type of FSSCs. In this type, gel–solid-state electrolyte is sandwiched by two planer flexible electrodes. The mechanical flexibility and electrochemical properties of the sandwiched devices are investigated in twisted, bent, and folded conditions. Figure 3.11a, b shows the sandwiched-type FSSCs device. This type of device does not include separator in between the two electrodes and the probability of short circuits is high [131, 132].

3.5.2 Planer-Type Device

Planer-type device is much thinner and flexible than the conventional FSSCs device due to their unique structural design. This type of device includes three main components (active electrode material, current collector, and electrolyte) in the same horizontal plane. Planer FSSCs device offers planner channels, which facilitates fast transport of electrolyte. The ion transport properties should not be affected in different conditions like bending or folding state. Planer FSSCs device is a promising power source in various smart electronic devices. Planer FSSCs device also exhibits good volumetric capacitances compared to other FSSCs devices [94, 133, 134].

3.5.3 Fiber-Type Device

Fiber-type FSSCs device has drawn considerable attention with increasing use of wearable and portable flexible electronics. In terms of flexibility, weight, and structural variation in fiber, FSSCs have some benefits over sandwiched and planer-type devices. Figure 3.11c shows that the fiber-type FSSCs device can be fabricated by placing two gel electrolyte-coated fiber electrodes in parallel. The fiber FSSCs can be easily integrated in series or in parallel to enhance the working potential and current. Fiber-type FSSCs also meet the high energy and power requirements for practical applications. They show large PD as compared to the planar SCs. The fiber-shaped device can also be hybridized with selected pseudo-capacitive materials like metal oxide, conductive polymers, and metal hydroxides [94, 135, 136].

Single-wire-type device can be fabricated by single fiber wounded around the fiber electrode and the gel-electrolyte is sandwiched in between them. Wire-like FSSCs device is also recognized as coil-type SCs. When checking the electrochemical properties of the linear device through deformation, the gel-electrolyte may be damaged and the two electrodes may be separated from each other. Therefore, the optimized device should be designed to increase the stability of the device. Direct contact between the fiber electrodes cause serious issues of short circuits, and the leakage problem limits its practical applications. In order to mitigate this lacuna, similar type of FSSCs device can be fabricated as shown in Fig. 3.11d. Additional space is available in cable-type device, which prevents direct short-circuit problem. It is established that the spacer in cable-type device not only prevents short-circuits problem but also enhance the ions transport properties [137–139].

3.6 Evaluation of Flexible Solid-State SC Device Performance

The performance of SC devices is usually evaluated based on specific capacitance, ED, PD, and rate capability. The most important parameters of a SCs device are how much energy it can store (ED) and how fast it can be charged–discharged (PD). The ED and PD of FSSCs can be represented in terms of area, weight, and volume diagrams. Fundamental electrochemical properties of any active materials are generally tested with three-electrode configuration. In contrast, two-electrode configuration is mostly used to evaluate the electrochemical performance of a device [140].

3.6.1 Cell Capacitance Measurement

The cell capacitance of a SCs device can be measured by constant current charging–discharging method. There are several procedures for measuring the cell capacitance values of a SCs device.

3.6.1.1 USABC Test Procedure

In this process, constant current charging–discharging technique is used to characterize the capacitance performance of an energy storage device. In this process, the capacitance (C) is measured in terms of 5C rate, which corresponds to 12 min discharge time. Therefore, this procedure is more appropriate for cell capacitance calculation of batteries than SC [141].

3.6.1.2 IEC Procedure

In this procedure, capacitance (C) as well as resistance (R) of SCs can be determined. For ideal double-layer capacitor, the cell capacitance (in F) can be calculated from the GCD plot according to the equation:

$$C = I_{\text{dich}} \times \Delta t / \Delta U$$

Here, $\Delta U = 0.9U_r - 0.7U_r$

where I_{dich} is the applied constant current in A, "Δt" is the discharge time in sec, "ΔU" is the potential difference in V, and "U_r" is the rated voltage.

In case deviation from ideal behavior, the capacitance value can be calculated by the following equation;

$$C = 2E/(0.9U_r)^2 - (0.7U_r)^2$$

Here, "E" is the energy (in J) delivers at discharge time and can be calculated as follows:

$$E = I \int U(t)\mathrm{d}t$$

Whereas the discharge current is related with the "U_r" and "R" by the equation of

$$I_{\text{dich}} = U_r / 40R$$

3.6.1.3 YUNASKO Test Procedure

This testing procedure normally uses to calculate the capacitance of SCs device. Capacitance (in F) is calculated from the discharge time of the GCD curve by the following equation:

$$C = I_{\text{test}} \times \Delta t / \Delta U$$

where I_{test} is the discharge current, in A, Δt is the discharge time, in sec, ΔU is the voltage change, in V.

3.6.2 *Internal Resistance Calculation*

The internal resistance of the SC device can be expressed in terms of equivalent series resistance (ESR) andequivalent distributed resistance (EDR). ESR corresponds to all the resistive components within the cell. On the other hand, EDR includes the ESR and an additional contribution due to the charge redistribution in the pore of electrode materials.

ESR and EDR value can be calculated from the voltage drop calculation (Fig. 3.12) using the following equation, respectively [142, 143],

$$\text{ESR} = \Delta U_4 / I_{\text{dich}}$$

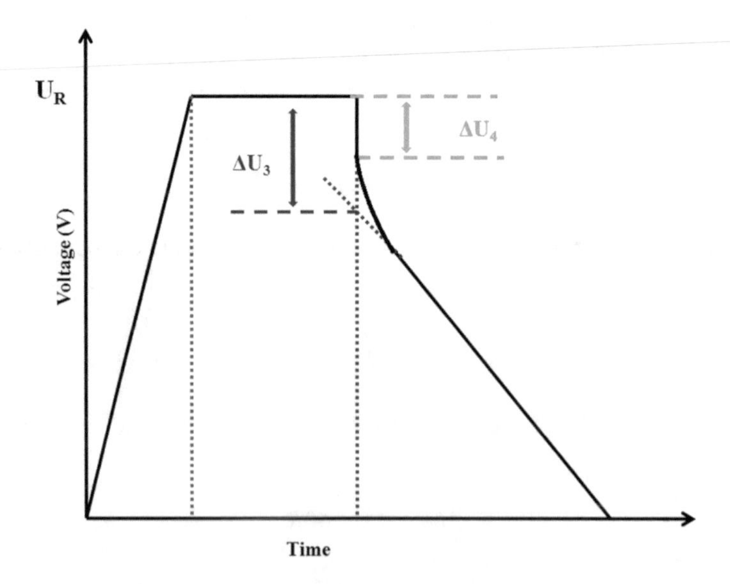

Fig. 3.12 Measurement profile for ESR (from ΔU_4) or EDR (from ΔU_3) evaluation

$$\mathrm{EDR} = \Delta U_3 / I_{\mathrm{dich}}$$

3.6.3 Energy and Power Calculation

In addition to the cell capacitance, ED and PD are two common parameters to determine the performance of a SCs. The ED and PD can be calculated according to the equations:

$$\mathrm{ED} = C \times U_r^2 / 2m \times 3600$$

And

$$\mathrm{PD} = \mathrm{ED} / \Delta t \times 1000$$

The maximum PD delivered by a SC device can also be calculated using the EDR value of the device with the help of the following equation:

$$\mathrm{PD}_{\mathrm{max}} = 0.25 U_r^2 / R \times m$$

Where the symbols refer to the inner meanings.

3.6.4 Cycle Life Test

Long life cycle is an important parameter of a SCs device. The performance of a SCs is characterized periodically from capacitance calculation to stability test. Devices can be tested at 0th, 500th, 1000th, 2000th, 5000th, and then after each 1000 GCD cycles. The temperature of the device should be under control during charging–discharging. Improvement in cycle stability favors the performance of the SCs device.

Based on the above equations, it shows that the properties of the device are mainly dependant on the potential window and electrode materials. So, improvement of device capacitance and working voltage is the direct approach to enhance the performance of the device. Nowadays, numerous techniques and procedures are employed to determine the performance of the device.

Performance of the device is mainly measured based on the amount of active materials used, which fails to reflect the actual properties of the device due to the weight/area/volume of the active material are equally important. So, the energy and PD calculation should be based on the total mass/area/volume of all the parameters including the current collector, solid-state electrolytes, separator, and packaging materials of the device. Therefore, gravimetric/areal/volumetric energy and PD calculation are the best way to represent the performance of any flexible device. Since the weight, area, or volume ED/PD are calculated based on the overall mass/area and volume measurement results of the flexible device, it is considered to be a more reliable parameter for evaluating the performance of the device [144, 145].

On the other hand, flexibility is another key parameter that determines the durability of any FSSCs equipment. However, there is no standard technique for evaluating the flexibility of FSSCs electrodes, electrolytes, or diaphragms. To date, the performance of the FSSCs can be evaluated by three most important tests: (1) compressing test, (2) bending test, and (3) stretching test. Specifically, the electrochemical performances are recorded under different bending/stretching/compressing conditions. The mechanical and reversible properties tests are exploited after repeated bending/stretching/compressing that are alternative measures for flexibility evaluation of the device. Therefore, by controlling the structural characteristics of the electrode material and modifying the configuration of the device, the best way is to enhance the flexibility of the device.

3.7 Conclusions

Development of FSSCs devices is considered as the emerging breakthrough in miniaturized energy storage systems with various lightweight, flexible, microtechnological worlds, such as wearable electronics, electronic skins, smart clothes, and implantable medical devices. Although significant progress has been achieved in the fabrication procedure of FSSCs devices by using several techniques, there is enough scope for commercial scale production and wide range of applications. A large number of flexible, porous structure and lightweight substrates including paper, textiles, metal foils, papers, and fibers are investigated to achieve the high performance of the device. Extensive efforts have been devoted toward the improvement of cell capacitance, ED, and stability. For developing a good FSSCs device, further improvement is needed in its electrodes, its fabrication techniques and also device integration aspect. The performance of any type of SCs is mainly dependent on the electrode materials. Therefore, much effort should be made to develop such electrode materials, which have the advantages of high flexibility, electrical conductivity, and large ion accessible surface area, tightly packed structure, excellent electrochemical stability, and expanded FSSCs performance. A flexible SCs fabrication technique is also an important parameter and affects the electrochemical performances. Several techniques including pencil drawing, printing, deposition, and coating are used for FSSCs device fabrication. However, ink-jet printing, 3D printing, and coating are highly promising for FSSCs device fabrication. Lastly, the integration of FSSCs device with energy harvesting (including solar cell, wind energy units, etc.) units and other micro- or flexible smart electronics devices (like sensor, medical device, photodetector, etc.) is an important developing direction. The high production cost is another disadvantage of the practical application of FSSCs technology. Therefore, future research should focus on cost-effective embryos, highly flexible and dynamic FSSCs equipment. The development of highly integrated multifunctional wearable devices (such as

SCs sensors, photoelectric SCs, and thermal SCs) in a single component will receive great attention in the near future.

Acknowledgements The authors are thankful to the Director of CSIR-CMERI. Authors are also thankful to GAP219012 project for the financial support.

References

1. Y. Liu, C. Zhao, S. Seyedin, J. Razal, J. Chen, *Flexible Energy Conversion and Storage Devices*, ed. by C, Zhi, L. Dai (Wiley-VCH Verlag GmbH & Co, 2018), pp. 01–36
2. X. Lu, M. Yu, G. Wang, Y. Tong, Y. Li, Energy Environ. Sci. **7**, 2160 (2014)
3. D. Wei, S.J. Wakeham, T.W. Ng, M.J. Thwaites, H. Brown, P. Beecher, Electrochem. Comm. **11**, 2285 (2009)
4. S. Ghosh, P. Samanta, N.C. Murmu, T. Kuila, J. Alloy. Compd. **835**, 155432 (2020)
5. B.D. Gates, Science **323**, 1566 (2009)
6. Y.Z. Zhang, Y. Wang, T. Cheng, W.Y. Lai, H. Pang, W. Huang, Chem. Soc. Rev. **44**, 5181 (2015)
7. K.D. Verma, P. Sinha, S. Banerjee K.K. Kar, in Characteristics of electrode materials for supercapacitors, in *Handbook of Nanocomposite Supercapacitor Materials I*, vol. 300, ed. by K.K. Kar (Springer Nature, Switzerland AG, 2020), p. 315
8. X. Lu, Y. Zeng, M. Yu, T. Zhai, C. Liang, S. Xie, M.S. Balogun, Y. Tong, Adv. Mater. **26**, 3148 (2014)
9. D.P. Dubal, Advances in flexible supercapacitors for portable and wearable smart gadgets, in *Emerging Materials for Energy Conversion and Storage*, ed. by K.Y. Cheong, M.A. Fraga, G. Impellizzeri (Elsevier, 209)
10. J. Deng, W. Zhuang, L. Bao, X. Wu, J. Gao, B. Wang, X. Sun, H. Peng, Carbon **149**, 63 (2019)
11. B. De, S. Banerjee, K.D. Verma, T. Pal, P.K. Manna, K.K. Kar, Transition metal oxides as electrode materials for supercapacitors, in *Handbook of Nanocomposite Supercapacitor Materials II*, vol. 302, ed. by K.K. Kar (Springer Nature, Switzerland AG, 2020), p. 89
12. F. Mokhtari, J. Foroughi, T. Zheng, Z. Cheng, G.M. Spinks, J. Mater. Chem. A **7**, 8245 (2019)
13. K. Qi, R. Hou, S. Zaman, B.Y. Xia, H. Duan, J. Mater. Chem. A **6**, 3913 (2018)
14. H. Choi, P.T. Nguyen, C.V. Tran, J.B. In, Appl. Surf. Sci. **510**, 145432 (2020)
15. M. Arvani, J. Keskinen, D. Lupo, M. Honkanen, J. Ener, Storage **29**, 101384 (2020)
16. Y. Liu, X. Miao, J. Fang, X. Zhang, S. Chen, W. Li, W. Feng, Y. Chen, W. Wang, Y. Zhang, A.C.S. Appl, Mater. Interfaces **8**, 5251 (2016)
17. M. Kumar, P. Sinha, T. Pal, K.K. Kar, Materials for supercapacitors, in *Handbook of Nanocomposite Supercapacitor Materials II*, vol 302, ed. by K.K. Kar (Springer Nature, Switzerland AG, 2020), p. 29
18. D.P. Dubal, S.H. Lee, J.G. Kim, W.B. Kim, C.D. Lokhande, J. Mater. Chem. **22**, 3044 (2012)
19. D.P. Dubal, J.G. Kim, Y. Kim, R. Holze, W.B. Kim, Energy Technol. **1**, 125 (2013)
20. C. Meng, C. Liu, L. Chen, C. Hu, S. Fan, Nano Lett. **10**, 4025 (2010)
21. W. Liu, C. Lu, H. Li, R.Y. Tay, L. Sun, X. Wang, W.L. Chow, X. Wang, B.K. Tay, Z. Chen, J. Yan, K. Feng, G. Lui, R. Tjandra, L. Rasenthiram, G. Chiu, A. Yu, Mater. Chem. A **4**, 3754 (2016)
22. G. Wang, X. Sun, F. Lu, H. Sun, M. Yu, W. Jiang, C. Liu, J. Lian, Small **8**, 452 (2012)
23. M. Sawangphruk, P. Srimuk, P. Chiochan, A. Krittayavathananon, S. Luanwuthi, J. Limtrakul, Carbon **60**, 109 (2013)
24. T.G. Yun, M. Oh, L. Hue, S. Hyun, S.M. Han, J. Power Sour. **244**, 783 (2013)
25. Q. Cheng, J. Tang, J. Ma, H. Zhang, N. Shinya, L.C. Qin, J. Phys. Chem. C **115**, 23584 (2011)
26. X. He, X. Mao, C. Zhang, W. Yang, Y. Zhou, Y. Yang, J. Xu, Mater. Electron. **31**, 2145 (2020)

27. X. Xiao, T. Li, Z. Peng, H. Jin, Q. Zhong, Q. Hu, B. Yao, Q. Luo, C. Zhang, L. Gong, J. Chen, Y. Gogotsi, J. Zhou, Nano Energy **6**, 1 (2014)
28. K. Xie, X. Qin, X. Wang, Y. Wang, H. Tao, Q. Wu, L. Yang, Z. Hu, Adv. Mater. **24**, 347 (2012)
29. H.P. Cong, X.C. Ren, P. Wang, S.H. Yu, Energy Environ. Sci. **6**, 1185 (2013)
30. Z. Taia, X. Yan, J. Lang, Q. Xue, J. Power Sour. **199**, 373 (2012)
31. Z. Zhang, T. Zhai, X. Lu, M. Yu, Y. Tong, K. Mai, J. Mater. Chem. A **1**, 505 (2013)
32. V.L. Pushparaj, M.M. Shaijumon, A. Kumar, S. Murugesan, L. Ci, R. Vajtai, R.J. Linhardt, O. Nalamasu, P.M. Ajayan, PNAS **104**, 13574 (2007)
33. S. Anwer, A.B. Ari, G. Bharath, P. Cao, S.P. Patole, S. Luo, H.T. Masood, W.J. Cantwell, K. Liao, Q. Li, L. Zheng, Adv. Mater. Interfaces **6**, 1900670 (2019)
34. Z. Zou, W. Zhou, Y. Zhang, H. Yu, C. Hu, W. Xiao, Chem. Eng. J. **357**, 45 (2019)
35. N. Li, G. Yang, Y. Sun, H. Song, H. Cui, G. yang, C. Wang, Nano Lett. **15**, 3195 (2015)
36. L. Bao, J. Zang, X. Li, Nano Lett. **11**, 1215 (2011)
37. B.E. Francisco, C.M. Jones, S.H. Lee, C.-R. Stoldt, Appl. Phys. Lett. **100**, 103902 (2012)
38. A. A. Łatoszynska, G.Z. Zukowska, I.A. Rutkowska, P.L Taberna, P. Simon, P.J. Kulesza, W. Wieczorek, J. Power Sour. **274**, 1147 (2015)
39. R. Na, P. Huo, X. Zhang, S. Zhang, Y. Du, K. Zhu, Y. Lu, M. Zhang, J. Luan, G. Wang, RSC Adv. **6**, 65186 (2016)
40. M.L. Verma, M. Minakshi, N.K. Singh, Electrochim Acta **137**, 497 (2014)
41. R.C. Agrawal, G.P. Pandey, J. Phys. D: Appl. Phys. **41**, 223001 (2008)
42. X.L. Hu, G.M. Hou, M.Q. Zhang, M.Z. Rong, W.H. Ruan, E.P. Giannelis, J. Mater. Chem. **22**, 18961 (2012)
43. L.Q. Fan, J. Zhong, J.H. Wu, J.M. Lin, Y.F. Huang, J. Mater. Chem. A **2**, 9011 (2014)
44. P. Sivaraman, A. Thakur, R.K. Kushwaha, D. Ratna, A.B. Samuiz, Electrochem. Solid State Lett. **9**, 435 (2006)
45. C.W. Huang, C.A. Wu, S.S. Hou, P.L. Kuo, C.T. Hsieh, H. Teng, Adv. Funct. Mater. **22**, 4677 (2012)
46. X. Lu, M. Yu, T. Zhai, G. Wang, S. Xie, T. Liu, C. Liang, Y. Tong, Y. Li, Nano Lett. **13**, 2628 (2013)
47. H. Yu, J. Wu, L. Fan, Y. Lin, K. Xu, Z. Tang, C. Cheng, S. Tang, J. Lin, M. Huang, Z. Lan, J. Power Sour. **198**, 402 (2012)
48. A.S. Ulihin, Y.G. Mateyshina, N.F. Uvarov, Solid State Ionics **251**, 62 (2013)
49. X. Yang, F. Zhang, L. Zhang, T. Zhang, Y. Huang, Y. Chen, Adv. Funct. Mater. **23**, 3353 (2013)
50. X. Peng, L. Peng, C. Wu, Y. Xie, Chem. Soc. Rev. **43**, 3303 (2014)
51. A.M. Hoang, G. Chen, A. Haddadi, S.A. Pour, M. Razeghi, Appl. Phys. Lett. **100**, 211101 (2012)
52. K. Gao, Z. Shao, J. Li, X. Wang, X. Peng, W. Wang, F. Wang, J. Mater. Chem. A **1**, 63 (2013)
53. G. Zheng, L. Hu, H. Wu, X. Xie, Y. Cui, Energy Environ. Sci. **4**, 3368 (2011)
54. B. Yao, L. Yuan, X. Xiao, J. Zhang, Y. Qi, J. Zhou, J. Zhou, B. Hu, W. Chen, Nano Energy **2**, 1071 (2013)
55. S. Zhu, Y. Li, H. Zhu, J. Ni, Y. Li, Small **15**, 1804037 (2019)
56. M.P. Down, C.W. Foster, X. Jib, C.E. Banks, RSC Adv. **6**, 81130 (2016)
57. D. King, J. Friend, J. Kariuki, J. Chem. Educ. **87**, 507 (2010)
58. K. Pokpas, N. Jahed, O. Tovide, P.-G. Baker, E.I. Iwuoha, Int. J. Electrochem. Sci. **9**, 5092 (2014); S. Tian, C. Zhao, P. Nie, H. Wang, X. Xue, L. Lin, L. Chang, Energy Technol. **7**, 1900680 (2019)
59. S. Tian, C. Zhao, P. Nie, H. Wang, X. Xue, L. Lin, L. Chang, Energy Technol. **7**, 1900680 (2019)
60. K. Robert, D.S. Venard, D. Deresmes, C. Douard, A. Iadecola, D. Troadec, P. Simon, N. Nuns, M. Marinova, M. Huvé, P. Roussel, T. Brousse, C. Lethien, Energy Environ. Sci. **13**, 949 (2020)
61. J. Ye, H. Tan, S. Wu, K. Ni, F. Pan, J. Liu, Z. Tao, Y. Qu, H. Ji, P. Simon, Y. Zhu, Adv. Mater. **30**, 1801384 (2018)

62. R. Tjandra, W. Liu, M. Zhang, A. Yu, J. Power Sour. **438**, 227009 (2019)
63. Y.N. Liu, L.N. Jin, H.T. Wang, X.H. Kang, S.W. Bian, J. Colloid Interf. Sci. **530**, 29 (2018)
64. F. Grote, Z.Y. Yu, J.L. Wang, S.H. Yu, Y. Lei, Small **36**, 4666 (2015)
65. S. Wang, J. Shen, Q. Wang, Y. Fan, L. Li, K. Zhang, L. Yang, W. Zhang, X. Wang, A.C.S. Appl, Energy Mater. **2**, 1077 (2019)
66. M.G. Say, R. Brooke, J. Edberg, A. Grimoldi, D. Belaineh, I. Engquist, M. Berggren, Flexible Electron. **4**, 14 (2020)
67. B.D. Gates, Q. Xu, M. Stewart, D. Ryan, C.G. Willson, G.M. Whitesides, Chem. Rev. **105**, 1171 (2005)
68. N. Matsuhisa, M. Kaltenbrunner, T. Yokota, H. Jinno, K. Kuribara, T. Sekitani, T. Someya, Nat. Commun. **6**, 7461 (2015)
69. A. Chiolerio, S. Bocchini, S. Porro, Adv. Funct. Mater. **24**, 3375 (2014)
70. H. Tao, B. Marelli, M. Yang, B. An, M.S. Onses, J.A. Rogers, D.L. Kaplan, F.G. Omenetto, Adv. Mater. **27**, 4273 (2015)
71. K.H. Choi, J.T. Yoo, C.K. Lee, S.Y. Lee, Energy Environ. Sci. **9**, 1 (2016)
72. Y. Wang, Y.Z. Zhang, D. Dubbink, J.E. ten Elshof, Nano Energy **49**, 481 (2018)
73. L. Liu, Q. Lu, S. Yang, J. Guo, Q. Tian, W. Yao, Z. Guo, V.A.L. Roy, W. Wu, Adv. Mater. Technol. **3**, 1700206 (2018)
74. Q. Lu, L. Liu, S. Yang, J. Liu, Q. Tian, W. Yao, Q. Xue, M. Li, W. Wu, J. Power Sour. **361**, 31 (2017)
75. C.W. Foster, M.P. Down, Y. Zhang, X. Ji, S.J. Rowley-Neale, G.C. Smith, P.J. Kelly, C.E. Banks, Sci. Rep. **7**, 42233 (2017)
76. A.D. Valentine, T.A. Busbee, J.W. Boley, J.R. Raney, A. Chortos, A. Kotikian, J.D. Berrigan, M.F. Durstock, J.A. Lewis, Adv. Mater. **29**, 1703817 (2017)
77. M. Areir, Y. Xu, D. Harrison, J. Fyson, Mater. Sci. Eng. B **226**, 29 (2017)
78. E.A. Gaulding, B.T. Diroll, E.D. Goodwin, Z.J. Vrtis, C.R. Kagan, C.B. Murray, Adv. Mater. **27**, 2846 (2015)
79. Y. Lu, R. Ganguli, C.A. Drewien, M.T. Anderson, C.J. Brinker, W. Gong, Y. Guo, H. Soyez, B. Dunn, M.H. Huang, J.I. Zink, Nature **389**, 364 (1997)
80. R.M. Almeida, M.C. Goncalves, S. Portal, J. Non-Cryst, Solids **345 & 346**, 562 (2004)
81. Y. Niu, X. Zhang, W. Pan, J. Zhaob, Y. Li, RSC Adv. **4**, 7511 (2014)
82. S. Su, L. Lai, R. Li, Y. Lin, H. Dai, X. Zhu, A.C.S. Appl, Energy Mater. **3**, 9379 (2020)
83. H.T. Nguyen, L. Miao, S.-T. Mura, M. Tanemura, S. Toh, K. Kaneko, M. Kawasaki, J. Cryst. Growth **271**, 245 (2004)
84. R.J. Chen, M. Huang, W.Z. Huang, Y. Shen, Y.H. Lin, C.W. Nan, J. Mater. Chem. A **2**, 13277 (2014)
85. T. Lv, M. Liu, D. Zhu, L. Gan, T. Chen, Adv. Mater. **30**, 1705489 (2018)
86. Y. Li, X. Wang, J. Sun, Chem. Soc. Rev. **41**, 5998 (2012)
87. X. Zhao, B. Zheng, T. Huanga, C. Gao, Nanoscale **7**, 9399 (2015)
88. T. Sekitani, T. Someya, Adv. Mater. **22**, 2228 (2010)
89. T. An, W. Cheng, J. Mater. Chem. A **6**, 15478 (2018)
90. M. Yousaf, H.T. H. Shi, Y. Wang, Y. Chen, Z. Ma, A. Cao, H.E. Nagui, R.P.S. Han, Adv. Energy Mater. **6**, 1600490 (2016)
91. Q. Xue, J. Sun, Y. Huang, M. Zhu, Z. Pei, H. Li, Y. Wang, N. Li, H. Zhang, C. Zhi, Small **13**, 1701827 (2017)
92. H. Sheng, X. Zhang, Y. Ma, P. Wang, J. Zhou, Q. Su, W. Lan, E. Xie, C.-J. Zhang, A.C.S. Appl, Mater. Interfaces **11**, 8992 (2019)
93. G. Wang, L. Zhang, J. Zhang, Chem. Soc. Rev. **41**, 797 (2012)
94. Z. Wu, L. Li, J.M. Yan, X.B. Zhang, Adv. Sci. **4**, 1600382 (2017)
95. M. Karnan, A.G.K. Raj, K. Subramani, S. Santhoshkumar, M. Sathish, Sustain. Energy Fuels **4**, 3029 (2020)
96. C. Cao, Y. Zhou, S. Ubnoske, J. Zang, Y. Cao, P. Henry, C.B. Parker, J.T. Glass, Adv. Energy Mater. **9**, 1900618 (2019)

97. S. Banerjee, K.K. Kar, Conducting polymers as electrode materials for supercapacitors, in *Handbook of Nanocomposite Supercapacitor Materials II*, vol. 302, ed. by K.K. Kar (Springer Nature, Switzerland AG, 2020), p. 333
98. Y. Zhou, K. Maleski, B. Anasori, J.O. Thostenson, Y. Pang, Y. Feng, K. Zeng, C.B. Parker, S. Zauscher, Y. Gogotsi, J.T. Glass, C. Cao, ACS Nano **14**, 3576 (2020)
99. Z. Niu, H. Dong, B. Zhu, J. Li, H.H. Hng, W. Zhou, X. Chen, S. Xie, Adv. Mater. **25**, 1058 (2013)
100. Y. Zhao, J. Liu, Y. Hu, H. Cheng, C. Hu, C.C. Jiang, L. Jiang, A. Cao, L. Qu, Adv. Mater. **25**, 591 (2013)
101. C. Xu, Z. Li, C. Yang, P. Zou, B. Xie, Z. Lin, Z. Zhang, B. Li, F. Kang, C.P. Wong, Adv. Mater. **28**, 4105 (2016)
102. Y. Song, H. Chen, Z. Su, X. Chen, L. Miao, J. Zhang, X. Cheng, H. Zhang, Small **13**, 1702091 (2017)
103. L. Sheng, J. Chang, L. Jiang, Z. Jiang, Z. Liu, T. Wei, Z. Fan, Adv. Funct. Mater. **28**, 1800597 (2018)
104. Q. Liu, L. Zang, X. Qiao, J. Qiu, X. Wang, L. Hu, J. Yang, C. Yang, Adv. Electron. Mater. **5**, 1900724 (2019)
105. Z. Liu, J. Zhang, J. Liu, Y. Long, L. Fang, Q. Wang, T. Liu, J. Mater. Chem. A **8**, 6219 (2020)
106. C. Zhong, Y. Deng, W. Hu, J. Qiao, L. Zhang, J, Zhang. Chem. Soc. Rev. **44**, 7484 (2015)
107. G. Cai, P. Darmawan, M. Cui, J. Wang, J. Chen, S. Magdassi, P.S. Lee, Adv. Energy Mater. **6**, 1501882 (2016)
108. A. Aliprandi, T. Moreira, C. Anichini, M.A. Stoeckel, M. Eredi, U. Sassi, M. Bruna, C. Pinheiro, C.A.T. Laia, S. Bonacchi, P. Samorì, Adv. Mater. **29**, 1703225 (2017)
109. Y.H. Liu, J.L. Xu, S. Shen, X.L. Cai, L.S. Chena, S.D. Wang, J. Mater. Chem. A **5**, 9032 (2017)
110. J.L. Xu, Y.H. Liu, X. Gao, Y. Sun, S. Shen, X. Cai, L. Chen, S.D. Wang, A.C.S. Appl, Mater. Interfaces **9**, 27649 (2017)
111. N.F. Anglada, J. Pérez-Puigdemont, J. Figueras, M.Z. Iqbal, S. Roth, Nanoscale Res. Lett. **7**, 571 (2012)
112. T. Chen, H. Peng, M. Durstock, L. Dai, Sci. Rep. **4**, 3612 (2014)
113. R. Yuksel, Z. Sarioba, A. Cirpan, P. Hiralal, H.E. Unalan, A.C.S. Appl, Mater. Interfaces **6**, 15434 (2014)
114. E. Senokos, M. Rana, M. Vila, J. Fernandez-Cestau, R.D. Costa, R. Marcilla, J.J. Vilatela, Nanoscale **12**, 16980 (2020)
115. Y. Gao, Y.S. Zhou, W. Xiong, L.J. Jiang, M. Mahjouri-samani, P. Thirugnanam, X. Huang, M.M. Wang, L. Jiang, Y.F. Lu, APL Mater. **1**, 012101 (2013)
116. N. Li, X. Huang, H. Zhang, Y. Li, C. Wang, A.C.S. Appl, Mater. Interfaces **9**, 9763 (2017)
117. C. Zhang, B. Anasori, A. Seral-Ascaso, S.H. Park, N. McEvoy, A. Shmeliov, G.S. Duesberg, J.N. Coleman, Y. Gogotsi, V. Nicolosi, Adv. Mater. **29**, 1702678 (2017)
118. N. Liu, Y, Gao. Small **13**, 1701989 (2017)
119. W.J. Hyun, E.B. Secor, C.H. Kim, M.C. Hersam, L.F. Francis, C.D. Frisbie, Adv. Energy Mater. **7**, 1700285 (2017)
120. P. Zhang, F. Wang, M. Yu, X. Zhuang, X. Feng, Chem. Soc. Rev. **47**, 7426 (2018)
121. Y. Yang, W. Gao, Chem. Soc. Rev. **48**, 1465 (2019)
122. E. Pomerantseva, F. Bonaccorso, X. Feng, Y. Cui, Y. Gogotsi, Science **366**, 969 (2019)
123. C. Lethien, J.L. Bideau, T. Brousse, Energy Environ. Sci. **12**, 96 (2019)
124. Z.S. Wu, X. Feng, H.M. Cheng, National Sci. Review **1**, 277 (2014)
125. C.A. Milroy, S. Jang, T. Fujimori, A. Dodabalapur, A. Manthiram, Small **13**, 1603786 (2017)
126. T.M. Dinh, F. Mesnilgrente, V. Connedera, N.A. Kyeremateng, D. Pecha, J. Electrochem. Soc. **162**, 2016 (2015)
127. L. Zhang, D. DeArmond, N.T. Alvarez, R. Malik, N. Oslin, C. McConnell, P.K. Adusei, Y.Y. Hsieh, V. Shanov, Small **13**, 1603114 (2017)
128. L. Liu, D. Ye, Y. Yu, L. Liu, Y. Wu, Carbon **111**, 121 (2017)

129. Z.S. Wu, K. Parvez, A. Winter, H. Vieker, X. Liu, S. Han, A. Turchanin, X. Feng, K. Müllen, Adv. Mater. **26**, 4552 (2014)
130. Y. Li, Y. Zhang, H. Zhang, T.L. Xing, G. Qiang Chen, RSC Adv. **9**, 4180 (2019)
131. H. Wang, B. Zhu, W. Jiang, Y. Yang, W.-R. Leow, H. Wang, X. Chen, Adv. Mater. **26**, 3638 (2014)
132. M.F. El-Kady, R.B. Kaner, Nat. Commun. **4**, 1475 (2013)
133. W. Si, C. Yan, Y. Chen, S. Oswald, L. Hana, O.G. Schmidt, Energy Environ. Sci. **6**, 3218 (2013)
134. J. Liang, C. Jiang, W. Wu, Nanoscale **11**, 7041 (2019)
135. L. Gao, S.J. Utama, K. Cao, H. Zhang, P. Li, S. Xu, C. Jiang, J. Song, D. Sun, Y. Lu, A.C.S. Appl, Mater. Interfaces **9**, 5409 (2017)
136. Q. Wang, X. Wang, J. Xu, X. Ouyang, X. Hou, D. Chen, R. Wang, G. Shen, Nano Energy **8**, 44 (2014)
137. Y. Chen, B. Xu, J. Wen, J. Gong, T. Hua, C.W. Kan, J. Deng, Small **14**, 1704373 (2018)
138. Y. Yang, N. Zhang, B. Zhang, Y.X. Zhang, C. Tao, J. Wang, X. Fan, A.C.S. Appl, Mater. Interfaces **9**, 40207 (2017)
139. K. Chi, Z. Zhang, J. Xi, Y. Huang, F. Xiao, S. Wang, Y. Liu, A.C.S. Appl, Mater. Interfaces **6**, 16312 (2014)
140. A. Burke, M. Miller, Electrochim. Acta **55**, 7538 (2010)
141. *Fixed Electric Double Layer Capacitors for Use in Electronic Equipment. Part 2. Sectional Specification-Electric Double Layer Capacitors for Power Application.* https://standards.iteh.ai/catalog/standards/clc/83b3ed94-e699-450d-b3c2-c110d2699c6f/en-62391-2-2006
142. *Electric Double Layer Capacitors for Use in Hybrid Electric Vehicles—Test Methods for Electrical Characteristics.* https://standards.iteh.ai/catalog/standards/clc/84a4fc8b-f6ad-4ac2-8503-d1992adf7721/en-iec-62576-2018
143. K. Li, J. Zhang, Sci China Mater. **61**, 210 (2018)
144. S. Kim, H.J. Kwon, S. Lee, H. Shim, Y. Chun, W. Choi, J. Kwack, D. Han, M. Song, S. Kim, S. Mohammadi, I. Kee, S.Y. Lee, Adv. Mater. **23**, 3511 (2011)
145. G.R. Li, Z.L. Wang, F.L. Zheng, Y.N. Ou, Y.X. Tong, J. Mater. Chem. **21**, 4217 (2011)

Chapter 4
Conducting-Polymer-Based Supercapacitors

Pallab Bhattacharya

Abstract CPs are known for their astonishing electrical and electrochemical properties. Characteristic features are tunable conductivity, structural flexibility, mild synthesis and processing conditions; chemical and structural diversity makes them excellent candidate for different fields of interest. Since the first introduction of CPs, it still remains relevant to discuss and grow rapidly in different fields of applications with various modern advancements. This chapter aims to revisit the journey and recent advancements of CPs in the field of energy storage systems like supercapacitors. Supercapacitors are one of the popular modern energy storage systems as they have many advantages like high power density, long cycle life, moderate to high capacitance, tunable rate capability, simple construction and low processing cost. Despite many advantages, supercapacitor is still facing many major challenges such as limited potential window, low energy density and sluggish rate kinetics. CPs have been considered as one of the excellent candidates for supercapacitor as they show miscellaneous redox nature, amazing electrical conductivity, good flexibility and many others. Therefore, substantial discussion is required to discuss the supercapacitors and its advantages and disadvantages, recent advancements, future challenges and new possibilities. This review focuses on the synthesis, processing and chemical modifications of various CPs with various interesting properties and their electrodes used for the advancements of supercapacitors which is the need of the hour.

List of Abbreviations

ASPB	Anionic spherical polyelectrolyte brushes
BFEE	Boron trifluoride diethyl etherate
CE	Counter electrode

P. Bhattacharya (✉)
Functional Materials Group, Advanced Materials & Processes Division, CSIR-National Metallurgical Laboratory (NML), Burmamines, East Singhbhum, Jamshedpur, Jharkhand 831007, India
e-mail: pallab.b@nmlindia.org

K. K. Kar (ed.), *Handbook of Nanocomposite Supercapacitor Materials III*,
Springer Series in Materials Science 313,
https://doi.org/10.1007/978-3-030-68364-1_4

CNT	Carbon nanotube
CyD	Cyclodextrin
CPs	Conducting polymers
CV	Cyclic voltammetry
C_s	Specific capacitance
CTAB	Cetrimonium bromide
DMC	Dimethyl carbonate
EDOT	3,4-Ethylenedioxythiophene
EDLC	Electrical double-layer capacitance
EIS	Electrochemical impedance spectroscopy
ESR	Equivalent series resistance
Et_4NBF_4	Tetraethylammonium tetrafluoroborate
EC	Ethylene carbonate
ET	Polyethylene terephthalate
ED	Light-emitting diode
FPT	Fluorophenylthiophene
FTIR	Fourier transform infrared spectroscopy
FTS	Tridecafluoro-(1,1,2,2-tetrahydrooctyl)-trichlorosilane
Gr	Graphene
GCD	Galvanostatic charge/discharge
HSC	Hybrid supercapacitor
LIB	Lithium-ion battery
MNT	α-MnO_2 nanotubes
MOF	Metal organic framework
MWCNT	Multi-walled carbon nanotube
NSA	2-Naphthalene sulfonic acid
PA	Polyacetylene
PANI	Polyaniline
PPy	Polypyrrole
PTh	Polythiophene
P3BT	Poly(3-butylthiophene)
P3MT	Poly(3-methylthiophene)
P3HT	Poly(3-hexylthiophene)
P3OT	Poly(3-octylthiophene)
P3PT	Poly(3-pentylthiophene)
PPV	Poly(*p*-phenylenevinylene)
PEDOT	Poly(3,4-ethylenedioxythiophene)
PC	Pseudocapacitor
PSS	Poly(styrene sulfonic acid)
PPP	Poly(*p*-phenylene)
PFPT	Poly(3-fluorophenyl)thiophene
p-TSA	*p*-Toluenesulfonic acid
PBTTT	Poly(2,5-bis(3-tetradecylthiophen-2-yl)thieno[3,2-b]thiophene)
PMeT	Poly(3-methyl thiophene)
PDTT	Poly(ditheno(3,4-b:3,4d) thiophene)

Poly(An-co-Py)	Poly(aniline-co-pyrrole)
PVdf-co-HFP	Poly(vinylidene fluoride-*co*-hexafluoropropylene)
PMeT	Poly(3-methyl thiophene)
PDTT	Poly(ditheno(3,4-b:3,4d) thiophene)
Poly(An-co-Py)	Poly(aniline-co-pyrrole)
PVdf-co-HFP	Poly(vinylidene fluoride-*co*-hexafluoropropylene)
PU	Polyurethane
RE	Reference electrode
SDS	Sodium dodecyl sulfate
SEM	Scanning electron microscopy
SSDP	Self-stabilized dispersion polymerization
SCE	Saturated calomel electrode
SWCNT	Single-walled carbon nanotube
SDS	Sodium dodecyl sulfate
$TBAPF_6$	Tetrabutylammonium hexafluorophosphate
THF	Tetrahydrofuran
TEM	Transmission electron microscopy
TR	Triton X-100
UV–Vis	Ultraviolet–visible spectroscopy
WE	Working electrode
XRD	X-ray diffraction

4.1 Introduction

The Nobel Prize in Chemistry for the year 2000 was awarded to Alan J. Heeger, Alan MacDiarmid and Hideki Shirakawa for the development of electrically conductive PA doped with iodine oxide. High conductivity, diversity and abundance of redox sites, excellent mechanical strength, high thermal stability and outstanding flexibility of CPs make them most desirable candidates for modern applications like energy storage systems, smart windows, LEDs, transistors, flexible screens, artificial muscles, smart textiles, etc. [1]. In energy storage application, characteristics of CPs, especially the charge transfer processes occurring during redox reactions, have attracted the researchers almost from every field of materials. Over the years, various conductive polymers such as PANI, PPy, PTh and its many derivatives like P3BT, P3MT, P3HT, P3OT, P3PT, PPV, PEDOT and its functionalized derivatives have been developed as energy storage electrodes. Charge once created on any given atom of the polymer chain, made up essentially of sigma bonds, is not mobile. A polymer exhibits excellent conductivity when there is conjugation in the polymer series, so that a resonant or mobile electron is present for conductivity. CPs have an intrinsic conductivity, varying from 10^{-14} to 10^{2} S cm^{-1}. All of these polymers in conjugated form have controlled conductivity depending on the dopant and doping level. Conducting polymers are also of interest due to the presence of various redox sites in it. In terms of conductivity, CPs possess electronic and ionic conductivity and the electrochemical reversibility between oxidized and reduced forms. Depending on the oxidation state, CPs can transition from conducting (salt form) and insulating (reduced) form. These possess good flexibility, mechanical integrity and non-toxic nature. Such diversity in the morphology, redox, ionic and electronic characteristics of CPs promoted them to be used in various high-performing supercapacitor applications as shown in Fig. 4.1.

Supercapacitors, due to their advantages like high power density (>10 kW kg^{-1}), long cycle life (>100,000 cycles), moderate to high capacitance, tunable rate capability, simple construction and low processing cost, are one of the most promising and environmentally safe candidates to meet the future energy crisis. According to the

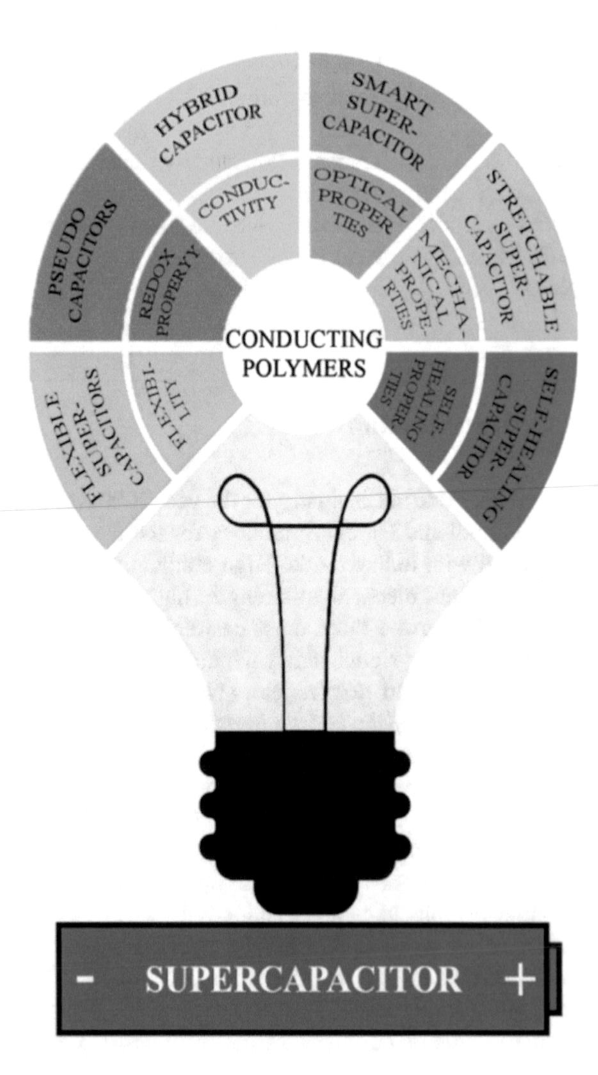

Fig. 4.1 Properties and applications of CPs

charge storage mechanisms, supercapacitor can be classified as EDLC and PC. Generally, EDLC is known for high cycle life, power density and rapid charge/discharge characteristics but fails to generate high C_s and energy density. In this situation, PCs play a major role to boost the C_s and energy density required for various modern applications. PCs can store energy via Faradaic redox reactions occurring at the electrode surface, which is unlike to the batteries where ion diffusion occurs into the bulk of the electrode with phase transformation. Till now, various inorganic materials like transition metal oxides and chalcogenides were introduced and studied extensively as PCs. Few organic materials like CPs and functionalized carbon materials such as Gr, CNT and activated porous carbons were also studied for the developments of advance PCs. Among various candidates, CPs are one of the interesting PCs that can exhibit high PC depending on the abundance of redox sites, electronic conductivity, flexibility and mechanical strength [2]. Moreover, synthesis of CP is simple and scalable which makes them a potential candidate for commercialization of energy storage systems.

In this chapter, we aim to elucidate various CPs, their synthesis methods, conductivity mechanisms, electrical and electrochemical properties in detail and the factors that significantly affect such properties. Applications of CPs in supercapacitor have been thoroughly discussed. Finally, we have demonstrated the challenges and possibilities of CP-based supercapacitors.

4.2 Conducting Polymers and Their Characteristics

4.2.1 Polyacetylene (PA)

PA is an organic polymer with a repeating unit of $(C_2H_2)_n$. This polymer is conceptually important because the discovery of PA and its high electrical conductivity contribute to the field of conductive organic polymers. Depending upon the preparatory condition, PA exists in two isomeric forms, trans and cis. When the polymerization temperature is higher than 150 °C, trans-isomer dominates whereas use of dry ice during polymerization produces more than 85% of the cis-isomer. On the other hand, polymerization at intermediate temperature results in mixture of both cis- and trans-isomers. In a synthesis of PA film, trans form appears with intense black color with a metallic luster, whereas cis form with copper-like luster. It is found in the study that trans form is suitable to produce high electrical conductivity. For example, conductivity increases from 6.02×10^{-9} S cm^{-1} (for 15% trans form) to 10^{-4} S cm^{-1} (for 80% trans polymer). However, interestingly the polymers having higher than 80% of trans-isomer show the trend of decreasing conductivity which reaches up to 4.83×10^{-6} S cm^{-1} for 100% trans polymer [3]. Not only the conductivity, but depending upon the isomeric forms PA shows different mechanical strengths too. In a study, it is observed that a polymer with 98% cis-isomer showed tensile strength of 3.8 kg mm^{-2} but a polymer with 96% trans polymer is having tensile strength of 2.4 kg mm^{-2}. At the same time, the brittleness also increases with decreasing cis content. Although the discovery of PA as a conductive organic polymer has made much effort in the development of materials science, PA has no commercial application due to its disadvantages like difficulties in synthesis and processing, air instability, low solubility, etc., which enforced the attention to be directed toward other conductive polymers.

4.2.2 Polyaniline (PANI)

PANI, the polymer of aniline monomer, is one of the impressive conductive polymers that can be chemically and electrochemically synthesized. The general structure of PANI consists of monomer units built from reduced (y) and oxidized ($1 - y$) blocks, where $0 \leq y \leq 1$ [4]. The value of y which can be continuously varied from a value of 0–1 determines the final redox state of the polymer. The different forms of PANI according to y-value are given in Table 4.1.

For PANI, a completely reduced form of repeating unit is generally made of two benzenoid rings whereas an oxidized form of the same contains one benzenoid and one quinonoid ring. The electrochemical properties of PANI are directly dependent on its chain network, the structure of the doped PANI and the growth direction. Conversely, the chemical composition controls the physical properties of PANI including its conductivity. During polymerization and post-polymerization, doping plays a vital role in the capacitive behavior of PANI. In the present context, the term "doping" is the same as in semiconductors, such as silicon or germanium, into which atoms, such as phosphorus or boron, are introduced. The polymers obtained through the doping of electron acceptor molecules (oxidation) or electron donors (reduction) are considered p-type or n-type, respectively. According to a study by Cao et al. [6], the nature of oxidizing agent significantly affects the nature of PANI synthesized. Using $(NH_4)_2S_2O_8$ and $K_2Cr_2O_7$ resulted in high polymerization yield, electrical conductivity and viscosity, whereas using KIO_3 and $FeCl_3$ gave PANI with similar conductivity, but lower viscosity. Also, it was observed that controlling over the polymerization temperature may change the viscosity as required. Therefore, it

Table 4.1 Different forms of PANI

y-value	Name	Conductivity ($S.cm^{-1}$)	Color
0 (oxidized form)	Polypernigraniline base	$<10^{-5}$	Blue
	Polyemeraldine salt	~ 15	Green
0.5	Polyemeraldine base	$<10^{-5}$	Violet
	Polynigraniline base	$<10^{-5}$	Blue
1 (reduced form)	Polyleucoemeraldine base	$<10^{-5}$	Transparent
	Polyprotoemeraldine base	$<10^{-5}$	Transparent

is important to mention that electrochemical performance of PANI-based electrodes can be tuned through conductivity, morphology and redox sites by controlling the synthesis parameters.

4.2.3 Polypyrrole (PPy)

Another organic polymer, PPy, is the polymeric form of pyrrole monomer. In pristine(undoped) state, also called as the neutral state, PPy consists of benzenoid structure resembling the aromatic or quinoid forms. The polymer conducts electricity only in oxidized state but not in neutral state. The charge remains delocalized over several pyrrole units and can form polaron (radical cation) or bipolaron (dication), in the oxidized state to conduct electricity. Further charge transfer from the polymer chain can be achieved by removing the single electron from the lower polaron level through the introduction of positive charge carriers. Liang et al. [7] reported a low-temperature synthesis (at 0 °C) of PPy by mixing pyrrole monomer with 1 M $Fe(ClO_4)_3$ which showed conductivity of 22.3 S cm^{-1} at room temperature. As the synthesis temperature is increased to 27 °C and 45 °C, the conductivity drops to 5.9 S cm^{-1} and 4.8 S cm^{-1}, respectively. In another study [8], PPy was prepared by chemical method by mixing pyrrole monomer with various proportions of $FeCl_3$ solution. The conductivity was found to increase from 8×10^{-2} S cm^{-1} to 2×10^{-1} S cm^{-1} with increasing the molar ratio ($FeCl_3$: monomer) from 0.5 to 3. This means PPy shows variable conductivities depending upon the synthesis conditions which could be beneficial for smooth charge transfer during supercapacitor applications.

4.2.4 Polythiophene (PTh)

PTh is structurally similar to PPy but replaces the nitrogen with sulfur in the heterocycle. Delocalized electrons along the conjugated backbone of PTh generate an extended π-system filled with valence electrons. The addition or removal (doping) of electrons in the π-system generates charge units called bipolaron units. Tamao et al. [9] polymerized an undoped PTh film in $LiBF_4$/benzonitrile solution and treated with gaseous ammonia. The room temperature conductivity was found to be 2×10^{-8} S cm^{-1}, but a 10^{10} times increase in conductivity was observed when the film was doped with 30 mol% of BF_4 anion per unit thiophene molecule, to form polythiophene borofluoride film. The conductivity of PTh borofluoride film was found to be 106 S cm^{-1} at 290 K. Like others, PTh is also familiar for various derivatives having desired properties for various applications including electrochemical technologies.

4.2.5 Poly(Ethylenedioxythiophene) (PEDOT)

In the late 1980s, scientists at the Bayer Research Laboratory in Germany developed a new derivative of PTh, the PEDOT with high electrical conductivity and environmental stability. Although PEDOT is insoluble in aqueous solvent, it can be made water-soluble by using PSS as the charge-balancing dopant during polymerization to produce PEDOT/PSS. The PEDOT/PSS becomes water-soluble and exerts excellent stability, good film formability, high transmittance and conductivity. In a study by Mochizuki et al. [10], the conductivity of a pristine PEDOT/PSS film was measured to be 6×10^{-2} S cm^{-1}. On addition of 5 wt% ethylene glycol, the carrier mobility was enhanced and conductivity increased to 270 S cm^{-1}. On increasing the pH from 2 to 13, the conductivity decreased and at pH > 11, it dropped rapidly. Conductivity of PEDOT/PSS system may also vary by the addition of variety of additives like ethylene glycol, dimethyl sulfoxide, etc. (detailed discussions are made in the respective sections).

4.2.6 Other Conducting Polymers

Few other conducting polymers having different characteristics are also known but not very popular in the field of energy storage due to the few disadvantages like difficulties in processing, absence of variety in redox sites and conductivity and many others. PPV is one of them, and it is an alternating copolymer of PA and polyphenylene. PPV is an insulator in pure form, but room temperature conductivity was determined up to ~10^{-10} S cm^{-1}. On doping with AsF_5 (57 wt% AsF_5), its conductivity increases by ten orders of magnitude to 3 S cm^{-1} [11]. PPP is also a CP and formed through the polymerization of *p*-phenylene units. It consists of a linear sequence of phenyl rings that forms complexes with metal-like properties. In undoped state, it is found to have a conductivity less than 10^{-12} S m^{-1}, which increases to 10^4 S m^{-1} by doping with impurities like AsF_5 [12]. It can also be doped with alkali metals to give *n*-doped materials with metallic-gold appearance. The conductivity of PPP increased from less than 10^{-10} S m^{-1} to 720 S m^{-1} when a pellet of PPP was exposed to potassium naphthalide solution in THF for 100 h [12].

The above-discussed CPs are the most important in terms of their variable conductivity and abundant redox features which makes them promising for various electrochemical applications. Detailed discussion on conductivities of various CPs and their supercapacitive performances is featured in the subsequent sections of this chapter.

4.3 Synthesis of Conducting Polymers

CPs can be prepared using various synthesis techniques as briefly discussed in the following subsections of Sect. 4.3. Various chemical routes and their characteristics used in CPs synthesis have been discussed till date. One of the most common ways to synthesize CPs is to deprotonate the monomers using a specific oxidizing agent. For example, Fe^{3+} is a popular oxidant which can initiate the polymerization of monomer M in the presence of doping anion X {as denoted in (4.1)} [13].

$$Fe^{3+}X_3^- + M \rightarrow Fe^{2+}X_2^- + M^* + X^- \tag{4.1}$$

4.3.1 Chemical Polymerization

Various CPs can be synthesized from the chemically modified (oxidized or reduced) monomers subjected through various chemical methods. For example, PA can be prepared through chemical method by the catalytic polymerization of acetylene using different kinds of Ziegler–Natta catalysts, such as $TiCl_4$-$Al(C_2H_5)_3$, $TiCl_3$-$Al(C_2H_5)_3$ and $Ti(OC_4H_9)_4$-$Al(C_2H_5)_3$ [14]. It is possible to prepare PA in the form of lustrous silvery films by using the last combination which itself has been used extensively because it gives a crystalline polymer with thickness varying from 10^{-5} mm to several mm. The films should be stored and treated under highly purified inert gas or under high volume to avoid oxidation [14].

Chemical synthesis of PANI via chemical oxidation requires three reagents primarily: an acidic medium (aqueous or organic), aniline and an oxidizing agent. Generally, hydrochloric acid (having pH value between 0 and 2) of 1 mol L^{-1} and ammonium persulfate (oxidizing agent) with an oxidant/aniline molar ratio ≤ 1.15 are used in most of the synthesis procedures to achieve high conductivity and efficiency. To restrain side reactions, the temperature of the solution is maintained between 0 and 2 °C. The duration of the reaction usually varies between 1 and 2 h. The material procured by this process is green-colored polyemeraldine salt. Polyemeraldine base can be obtained by treating polyemeraldine hydrochloride in an aqueous ammonium hydroxide solution for about 15 h. There are other oxidants which can also be used for the purpose including $K_2Cr_2O_7$, $(NH_4)_2S_2O_8$, $NaVO_3$, $Ce(SO_4)_2$, KIO_3 and $K_3(Fe(CN)_6)$ [14].

Chemical synthesis of PPy occurs via oxidation of pyrrole with an oxidant like ferric chloride. In the resulting polymer, charge compensation is afforded by $FeCl^{4-}$, which makes it conducting in the oxidized form [15]. The effect of dopant ion (due to different ferric salts) is related to the Fe^{2+}/Fe^{3+} redox potential with strong acid anions providing the most oxidizing ferric species, which affects the conductivity of PPy formed from different ferric salts. Fe^{3+} ions are more strongly coordinated by weaker acid anions, reducing their oxidizing potential.

Corradiet et al. [16] have synthesized PEDOT using $FeCl_3$, $Ce(SO_4)_2$ and $(NH_4)_2Ce(NO_3)_6$ as oxidants. EDOT has been polymerized chemically to produce a "sky-blue" CP. PEDOT nanofibers can be synthesized from an aqueous anionic surfactant solution, using a self-assembled micellar soft template methodology resulting in a high yield of ultrathin (having diameter 10 nm) and long (>5 mm) nanofibers as reported by Han et al. [17]. It was also noted that these fibers showed enhanced levels of electrical conductivity. Primarily, an aqueous micellar solution was produced by combining the anionic surfactant SDS and water. Eventually, by increasing the SDS concentration, transition to a rodlike shape was observed so as to hold more surfactant molecules and to reduce the free energy of the system [17]. An increase in the ionic strength and aggregation number of the solution can be observed when $FeCl_3$ was added to the SDS solution to assist this aggregation to the rodlike structure. Lastly, when the EDOT monomer was introduced into the solution, it moved into the rodlike micelles owing to its hydrophobicity and ultimately the monomer was polymerized into PEDOT by the oxidant $FeCl_3$. The purification was done in order to remove the surfactant and any excess $FeCl_3$, and then it was sonicated to disperse the PEDOT fibers. This nanofiber (having diameter of 30 nm) exhibited a tubelike structure [17].

Polycondensation of 2,5-dibromothiophene in THF was used for the synthesis of PTh in the presence of magnesium, and NiCl was utilized as a catalyst [18]. The THF was kept in a round bottom flask, where magnesium was added to it and then the solution was stirred under an argon atmosphere. An Mg insertion reaction occurred at each halogen bond when the dibromothiophene was added slowly in the absence of air. The reaction commenced instantaneously, and in one hour it concluded. A red powder was obtained after filtering and washing several times, using methanol to eliminate various oligomers, unreacted monomers and catalyst. By a coupling reaction of 3-bromothiophene and 4-fluorophenylmagnesium bromide in THF where $NiCl_2$ (diphenylphosphino propane) was used as a catalyst, PFPT was synthesized. A round bottom flask containing the bromothiophene and the catalyst was placed in a glove box filled with argon. This solution of bromothiophene and the catalyst was stirred and then kept at −10 °C. In the absence of air, 1 M fluorophenyl magnesium bromide in THF was added to the solution with the help of a syringe. This mixture was then stirred for 12 h, and finally to cease the reaction, 1 M HCl was added which neutralized the excess of fluorophenylmagnesium present. After filtration of the organic phase, the monomer procured was a yellow-colored powder. For recrystallization, the monomer was dissolved in methanol and eventually it was precipitated with water. Polymerization of FPT was carried out by direct oxidation in chloroform and $FeCl_3$ as an oxidant [18].

Chemical polymerization process is very vast in terms of different chemicals and techniques. The controllability of the process through various synthetic parameters is one of the major reasons for their use in large-scale production. Such chemical methods offer various opportunities to alter the backbone of CP either covalently or non-covalently. However, sometime uses of such processes become restricted due to the use of toxic chemicals and complexity in the experimental setup.

4.3.2 Electrochemical Polymerization

Electrochemical method is another widely used synthesis method because of the uniformity obtained in the film of CPs and also due to its cost-effectiveness. Three electrochemical methods can be used to synthesize CPs: (i) galvanostatic method (constant current), (ii) potentiostatic method (constant potential) and (iii) potentiodynamic method (current and potential vary with time). The electrochemical method consists of a three-electrode assembly of a CE, a RE and a WE on which the polymer is deposited. Fabrication of thin film through this process is simpler and has better control over synthesis, like entrapment of molecules or creating porosity in CPs. Separation of the film from the substrate and post-covalent modification of bulk CP is difficult. In the adjoining paragraphs, the synthesis procedure of few CPs by electrochemical method has been mentioned.

PA can be synthesized through electrochemical route using platinum foil as the cathode, nickel foil as anode and nickel bromide in acetonitrile as the electrolyte. PA was precipitated in powdered form in the cell at room temperature. A thin layer of PA may get deposited on the surface of the platinum electrode when a potential of 4–40 V was applied across the cell for 50 min [19]. PANI can also be synthesized by electrochemical process in which the electro-polymerization of aniline was carried out in a single compartment cell. The single compartment cell comprised graphite rod (WE), stainless steel (CE) and a SCE (RE). 3.1 mL of aniline (0.15 M) and 20 mL of concentrated HCl (1 M) were added to 200 ml of distilled water and kept in the cell. A potential difference of 0.7 V applied for 2 h is sufficient to get PANI deposited on graphite electrode [20]. The remaining HCl was washed by treating PANI with excess ammonia until the solution became basic. PANI can also be prepared potentiostatically using an electrolyte solution of 1 M H_2SO_4 and 0.05 M PANI (instead of aniline) at 0.75 V versus SCE.

Electrodeposition of PPy can be done from 0.1 M freshly distilled pyrrole, either in a solution of $TBAPF_6$ in propylene carbonate or in a solution of 0.1 M lithium perchlorate in acetonitrile. The PPy films were grown over 30 potentiodynamic cycles at room temperature. Three-electrode systems with platinum as a WE, platinum coil as a CE and silver as a pseudo-RE were used for the cyclic voltammogram measurement. The starting potential of 0 V and a scan rate of 100 mV s^{-1} were used. Using ionic liquids, large films can be grown potentiostatically at 2 h, 1 V. Using molecular solvents, they can be grown at 2 h, 0.85 V. The films were grown at lowest possible potentials, to minimize the risk of over-oxidation [21].

For synthesis of PEDOT, electrochemical polymerization is considered a convenient method, where a film yields on the surface of the anode. PEDOT can be electro-synthesized in the presence of both organic and aqueous solutions. But the electro-polymerization is usually carried out in organic solution (e.g., acetonitrile and propylene carbonate), due to low solubility of EDOT in water at room temperature. Polymerization of EDOT was performed electrochemically in a three-electrode cell, with propylene carbonate solution containing 0.1 M $LiClO_4$ and 0.01 M solution of monomers. Electrodepositions were carried out at a scan rate of 100 mV s^{-1} for 10 cycles, by CV between −1.4 V and +1.55 V versus SCE [22].

PTh was electrochemically synthesized in a traditional one-compartment cell with the three-electrode system. Stainless steel electrodes were used as WE and CE, whereas SCE was used as RE. PTh films were synthesized from BFEE solution containing 0.3 mol L^{-1} thiophene monomer, and a potential of 1.3 V versus SCE was applied [23]. All solutions used for this procedure were deaerated by a dry nitrogen stream and maintained at a light overpressure. The analysis of the total charges consumed for the electro-polymerization is used to determine the thickness of the films. After the completion of the polymerization reaction, a green-colored, homogeneous PTh film was obtained which was then scraped from the electrode and washed repeatedly with acetonitrile to remove impurities. Chemical method and electrochemical method are the most extensively used procedures for synthesis of CPs. However, few other methods are also available which can be employed to prepare CPs.

4.3.3 Metathesis Process of Polymerization

A chemical reaction between two compounds, which exchange part of each compound to form two different compounds, is called metathesis. It can be operated in different ways like ring-opening metathesis of cyclic olefins, metathesis of cyclic or acyclic acetylene and metathesis of diolefins. For example, Evans et al. [24] studied the metathesis of aniline derivatives and 1,2-dihydroquinoline and Masuda et al. [25] studied the synthesis of a typical conjugated acetylene-based polymer. Metathesis may be adopted for the development of different CPs with lot of advantages such as permitting additional aliquots of monomer to be polymerized, laser ablation in liquid termed as ("direct matrix-coupling") hydrogel capsule fabrication and cleavage does not deactivate the catalyst, but the process has a few disadvantages like low tolerance and degree of polymerization toward functional groups of (macro)monomers.

4.3.4 Emulsion Polymerization

The emulsion polymerization is a heterophase polymerization technique where three different phases such as water, latex and monomer can be treated through radical polymerization to obtain the desired polymers. For example, PTh can be synthesized through emulsion polymerization where chemical oxidation of thiophene was conducted by ozone-generated oxygen radical [26]. Similarly, other CPs like PANI, PPy, PEDOT, etc., may also be produced through modified emulsion polymerization using various radicals produced in the system. There are many advantages of emulsion polymerization such as use of aqueous polymerization media, applicable to broad range of monomers, excellent monodispersity, good control and high yields. Nevertheless, the major disadvantage of this process is the presence of impurities generated from the unreacted emulsifiers and radical initiators.

4.3.5 *Inclusion Polymerization*

Inclusion polymerization is a host–guest polymerization technique, where guests (monomers) are trapped in inclusion spaces of hosts (large molecules/molecular assemblies) and under appropriate conditions get polymerized. The inclusion spaces constrain the motion of monomers and accelerated the polymerization. For instance, Yuan et al. [27] encapsulated PANI into the cavities of CyDs, to form inclusion complexes. The complex was formed by in situ polymerization of *N*-phenyl-1,4-phenylenediamine encapsulated in β-CyD in advance. Another route for encapsulation was developed, wherein post-encapsulation of PANI–emeraldine base into β-CyD in aqueous solution was performed at room temperature. Monomers thus polymerize into low-dimensional and anisotropic assemblies. This technique can be used to create low-dimensional composite materials at atomic/molecular level. Though such method is known for the production of controlled structure, it is mainly limited to few composites which leave enormous scope for further development for the synthesis of different CPs structures with rich desirable properties.

4.3.6 *Solid-State Polymerization*

Synthesis of CPs like PTh and PEDOT can be carried out through a process called solid-state polymerization. This is the process where a polymer is heated in the absence of oxygen and water. It can be controlled through the variation in process temperature, pressure and/or the rate of diffusion of by-products. Such processes are mostly familiar for the industrial production of PET bottles, films, etc., due to the simplicity of the reaction setup.

4.3.7 *Plasma Polymerization*

Organic and organometallic preliminary materials are generally used in a process called plasma polymerization to develop various polymeric thin films. Plasma polymerization typically uses a plasma source to develop high-quality polymers with high chemical, thermal and mechanical stability. Such processes also allow developing desired polymeric thin films on various substrates like glassy metal surfaces, hard surfaces and on other polymeric surfaces which enables its wide acceptability in different fields of applications. In recent studies, plasma polymerization technique further extended to develop various CPs for different energy applications such as solar cell, LIBs and micro-plasma jet for polymeric batteries [28].

4.3.8 *Matrix Polymerization*

Matrix (template) polymerization is another process in which the monomer units are structured by a preformed macromolecule (template) and refers to one-phase systems in which the monomer and template are soluble in the same solvent. Such templates could be of two types, i.e., hard and soft. A hard matrix is an inorganic matrix that can interact with monomers and can change the parameters of the polymer process and the resultant polymer. Polymerization occurs at the interface between the rigid substrate, monomer and initiator solution. Soft fixtures include surfactants and polymers. In the presence of a soft template (a surfactant), the polymerization of aniline occurs around micelles of the surfactant once critical micelle concentration is attained [4]. The polymerization of aniline with the participation of a surfactant as a soft template includes the interaction of an aniline monomer or a cation radical with surfactant micelles and the formation of PANI.

The above-discussed methods have been widely used for making CPs of different structures and morphologies with different desirable properties for various applications. Doping is a vital process to modulate the different properties of CPs. There are different doping techniques available, with a lot of inherent advantages and disadvantages. A few widely known doping procedures are chemical doping, electrochemical doping, photodoping, non-redox doping and charge injection doping. Among the mentioned methods, chemical doping and electrochemical doping are widely known for their low cost and convenience. Thanh-Hai Le et al. [29] have discussed various advantages and disadvantages of different doping methods as given in Table 4.2. One of the major focuses of this chapter is energy storage and, more specifically, the supercapacitor. Conductivity is one of the important factors that control the performance of supercapacitors. From the above discussions, it is clear that electrical conductivity can be controlled through various synthetic parameters. So, we have discussed the electrical properties of different CPs with brief mechanism and various factors that affect their conductivities in Sect. 4.4.

4.4 Electrical Properties of Conducting Polymers

Even though CPs are known for their various properties such as flexibility, strength, toughness, elasticity, optical properties and microwave absorption, we are mainly interested (in this chapter) in discussing their electrical conductivities, as they are important for any supercapacitor electrodes. Generally, electrical conductivities of CPs depend on the delocalization of the available π-electrons. If delocalization of π-electrons is high, then it produces high conductivity whereas the opposite results

Table 4.2 Advantages and disadvantages of various doping methods

Doping method	Controlled variables	Advantages	Disadvantages
Chemical doping	Vapor pressure, exposure time to dopant	Simple way to obtain doping upon exposure of the sample to a vapor of the dopant or immersion into a solution with the dopant	Performed as slowly as possible to avoid inhomogeneous doping
			The doping levels obtained are not stable with respect to time
			Unexpected structural distortion may cause electrical conductivity decay
			Doping/de-doping shows low reversibility
Electrochemical doping	Amount of current passed	Doping level can be easily controlled by using an electrochemical cell with a controlled amount of current passed	Unexpected structural distortion may cause electrical conductivity decay
		Doping/de-doping is highly reversible, and clean polymer can be retrieved	
		Can be achieved with many dopant species	
Photodoping	Radiation energy of light beam	Charge carrier is formed without chemical compound (dopants)	The electrical conductivity disappears rapidly when irradiation is discontinued due to recombination of electrons and holes
		No distortion of the material structure	
Charge injection doping	Applying an appropriate potential on the polymer structure	Does not generate counterions. Minimized distortion	Coulombic interaction between charge and dopant ion is very strong and can lead to change in the energetics of the system

at the localization of the π-electrons. Distribution of π-electrons in CPs can be tuned by adjusting the parameters of doping, structural arrangements of polymeric chains, length of conjugation and the purity of the samples. It is necessary to maintain the structural and morphological order of CPs to achieve a high amount of π-electron delocalization, thus increasing electrical conductivity, which helps in smooth charge transfer during electrochemical operation. Therefore, it is possible to obtain the optimum electrical conductivity of CPs by tuning their synthesis techniques and can be used in various electrical applications as discussed below. Doping is one of the techniques which reduces the structural and morphological disorder in pristine CPs to enhance their electrical conductivity from insulating to metallic regime. The conductivity of undoped CPs, which is generally in the range of 10^{-6}–10^{-10} S cm^{-1}, can be increased by >10-fold in magnitude by doping. Since the electrical conductivity controls the charge transfer process, it seems that doped CPs are better electrodes for supercapacitor while compared to the pristine CPs. In one report, Tsukamoto et al. [30] claimed that the electrical conductivity of PA can be reached up to 10^4 S cm^{-1} by doping with iodine. Generally, doping creates self-localized excitons such as solitons, polarons and bipolarons in the polymers. These excitons behave as the charge carriers depending on the structural arrangement of CPs. For instance, in trans-PA with degenerate ground state, charged solitons become the charge carriers whereas, in cis-PA with non-degenerate ground state, bipolarons act as the charge carriers. Doping on cis-PA first creates polarons, which join together to form spinless bipolarons to act as the charge carriers. In case of other CPs (such as PPy, PTh or PPV) also, bipolarons carry the charges. The first step corresponds to the formation of free radicals by electron transfer of the aniline nitrogen atom from the 2 s energy level. The reaction between the radical cation and the resonance form of aniline radical cation occurs, in acidic medium forming dimer. The dimer is then oxidized to form a new radical cation. The formed radical cation reacts further either with the radical cation monomer or with the radical cation dimer to form respective polymer. The conduction of electrons through the π-systems is shown in Fig. 4.2 for a single PANI unit. Electronic delocalization is important throughout the chain for any CPs to attain high electronic conductivity. As shown for PANI in Fig. 4.2, electronic delocalization occurs in other CPs as well and hence they are also conductive. Depending upon the electronic structure (in the presence or absence of dopants) and extent of electronic delocalization, different CPs result in a different conductivity value which affects their charge transfer process during electrochemical measurements. Therefore, it is important to discuss the changes in conductivity values of CPs with different structures synthesized in the presence of different dopants (we have already discussed various doping processes in Table 4.2) and synthesis conditions.

Lee et al. [31] doped PANI–emeraldine salt nanoparticles with iodine and found that the conductivity increased from 3.1 S cm^{-1} to 9.35 S cm^{-1} after 10 min. Similarly, the conductivity of PANI–emeraldine base nanoparticles increased four orders from 2×10^{-5} S cm^{-1} to 0.5 S cm^{-1} on doping with iodine. Emeraldine base form of PANI was doped using *p*-TSA, and the obtained PANI-*p*-TSA polymer showed a conductivity of 16 S cm^{-1} when the ratio of dopant, emeraldine base, was kept

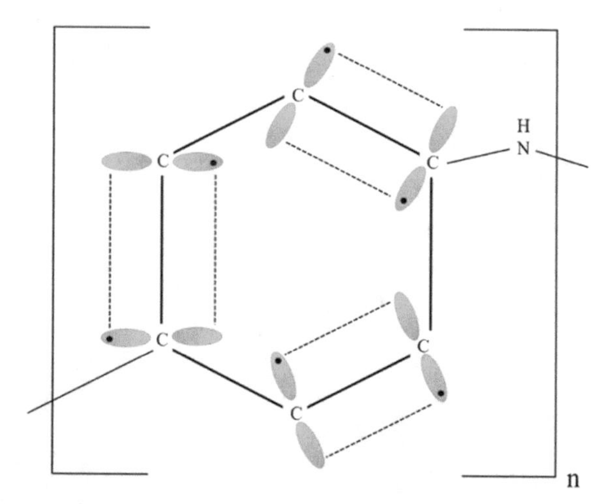

Fig. 4.2 Electronic delocalization in PANI

at 1 [32]. Dilligham et al. [33] doped polythiophene using vapor deposition. Co-evaporation with $FeCl_3$ and exposure to iodine vapor were used for doping the PTh films. The $FeCl_3$ codeposited films showed a conductivity of 10-25 S cm^{-1}, whereas iodine exposed films showed only 0.01 S cm^{-1}. They also concluded that $FeCl_3$ codeposited films were more stable on exposure to atmosphere. Gupta et al. [34] have doped PPy with NSA using electrochemical polymerization. They synthesized doped PPy films with microcup and microbowl morphology and showed conductivity of 1–50 S cm^{-1}. Dimitriev et al. [35] spin coated/cast thin films of PEDOT/PSS onto soda lime glass substrates and observed the conductivity of a 100 nm film to be 2×10^{-2} S cm^{-1}. The effect of organic additives (ethylene glycol and dimethyl sulfoxide) on conductivity was studied, and it was found that conductivity increased as the function of additive added. Substantial reduction in conductivity was observed on increasing the annealing temperature.

As evident from the discussion above, the conductivity of doped CPs depends on the factors like nature and concentration of the dopant and doping time. Size of the dopant is important to select to modulate the electrical conductivity of particular CPs. Small-sized dopants such as Na^+, Cl^- and ClO_4^- or large-sized dopants like polystyrene sulfonate and polyvinyl sulfonate have different advantages and disadvantages. Insertion of small dopants in the CP chain is easy but leaching of dopants in certain condition is also possible, whereas large-sized dopants may strongly bind with the parent CPs which inhibits the leaching of the dopants [36]. Use of common dopants in various CPs has been represented in Table 4.3 with the resultant conductivity values.

Not only the nature of the dopant and doping procedure, but the duration of doping is also crucial for determining the electrical conductivity of CPs. In this respect, Tsukamoto et al. [30] examined the effects of iodine doping in PA with a different quantity of doping agents and a different duration of doping. It is been found that initially, the conductivity of PA increases with the increase in iodine concentration

Table 4.3 Standard examples of dopants for CPs and their corresponding conductivity values

CP type	Dopant	Chemical source	Doping method	Conductivity (S cm^{-1})
Trans-PA	Na^+	$(C_{10}H_8)Na$	Solution doping	80
PPP	AsF_5	AsF_5	Vapor phase doping	1.5×10^4
PPV	CH_3SO_3H	CH_3SO_3H	Non-redox doping	10.7
	AsF_5	AsF_5	Vapor phase doping	57
Poly(3-vinylperylene)	Cl_{4-}	$(C_4H_9)_4N(ClO_4)$	Electrochemical doping	10^{-5}
PPy	AsF_6^-, PF_6^-, BF_4^-	$C_{16}H_{36}AsF_6N$, $(CH_3)_4N(PF_6)$, $(C_2H_5)_4N(BF_4)$	Electrochemical doping	30–100
	NSA	NSA	Electrochemical doping	1–50
	Cl_4^-	$LiClO_4$	Electrochemical doping	65
	Cl^-	NaCl	Electrochemical doping	10
	PSS/Cl^-	PSS/$FeCl_3$	Solution doping	4
	MeOH	MeOH	Vapor phase doping	0.74
	HSO_4^-	$(C_4H_9)_4N(HSO_4)$	Electrochemical doping	0.3
	$C_{20}H_{37}O_4SO_3^-$	$C_{20}H_{37}O_4SO_3Na$	Solution doping	4.5
PANI	$C_{10}H_{15}OSO_3^-$	$C_{10}H_{16}O_4S$	Solution doping	300
	HC1	HC1	Non-redox doping	10
	I_3^-	I_2	Vapor phase doping	9.3
	BF_4^-	HBF_4	Solution doping	(2.3×10^{-1})
PBTTT	FTS	$C_8H_4F_{13}SiCl_3$	Vapor phase doping	$604–1.1 \times 10^3$

(continued)

Table 4.3 (continued)

CP type	Dopant	Chemical source	Doping method	Conductivity (S cm^{-1})
Poly(2-(3-thienyloxy)ethanesulfonate)	Na_2SO_3	Na_2SO_3	Solution doping	5
PTh	Cl^-	$FeCl_3$	Vapor phase doping	10–25
PANI-PPy	ASPB	ASPB	Electrochemical doping	8.3

Reprinted from reference [29], copyright (2017), with permission from MDPI

but after a certain level of doping concentration it remains constant. Thus, it can be suggested that after reaching the saturation level of doping (CP attains maximum density of dopants), doping time does not change the conductivity of CPs (because diffusion of dopants becomes very slow at its saturation level in CP) anymore. From the above discussion, it is evident that the conductivity of CPs can be tuned through different processes. And according to the synthesis process and structure, CPs can generate different conductivities as presented in Table 4.3. Conductivity of CPs is an important parameter for applications in different fields.

4.5 Electrochemical Supercapacitor

High conductivity, excellent mechanical strength, high thermal stability and high flexibility placed CPs in a promising position in various modern applications including the green electrical energy storage system supercapacitor. Low manufacturing cost of CPs is an important reason that these are used in various applications like energy storage electrodes, electromagnetic shielding agents, chemical sensors, antistatic coatings, corrosion protection, etc. Nowadays, manufacturing of CP-based products like smart windows (which controls the amount of light passing through it), light-emitting diodes and transistors is of great interest in terms of commercial viability. Not only that, but CPs might also be used for flexible smart screens in near future. Due to the high flexibility and ease of fabrication, some of the CPs are very suitable candidates to be used as artificial muscles and smart textiles [1]. Anyway, this chapter will focus only on the synthesis, fabrication and performances of various electrodes used for the kinds of supercapacitors made of different CPs.

In energy storage systems like batteries and supercapacitors, CP is a topic of interest to be used as new custom electrode material with excellent performance. The CP-based supercapacitors show promising energy storage performances due to their extraordinarily good electrical conductivity (as discussed earlier) and suitable Faradaic sites (as originates from various heteroatom centers present in CPs) to produce high PC. Before the thorough discussions on the CP-based various electrochemical systems and their performances, we believe a brief discussion on the fundamentals of various electrochemical systems, their charge storing mechanisms and the important parameters required for the performance measures will be in good justice with the motive of this chapter. The electrochemical supercapacitor is a capacitor, capable of storing a large amount of energy per unit mass or volume, typically 10–100 times the energy of the conventional electrolytic capacitor. Generally, likely to batteries, a supercapacitor also consists of positive and negative electrodes, a separator and a suitable electrolyte. Contrast to batteries, electrochemical capacitors have many advantages such as high power density (>10 kW kg^{-1}), long life (>100,000 cycles), low toxicity, operation over a wide temperature range, low maintenance, acceptable performance in extreme weather and many others. Recently, electrochemical supercapacitors are used in emergency and safety systems, diesel engine starting systems, wireless power tools and hybrid electric vehicles.

Depending on the storage mechanism, electrochemical capacitors can be classified as (i) EDLC, (ii) PC and (iii) HSC. As already known, EDLCs function through electrostatic charge separation and formation of the double layer at the electrode–electrolyte interface whereas PC undergoes highly reversible Faradaic reactions to store energies. EDLCs store less energy (energy storage capacity is ~10 Wh kg^{-1}) through a fast-charging–discharging process and generally originate from large area materials (like Gr, CNTs and other nanoscale carbons) via an electric double layer at the electrode–electrolyte interface. However, PC stores a good amount of energy (energy storage capacity is 50 Wh kg^{-1}) on or near the surface of the electrode via kinetically slow (as compared to the kinetics of EDLC) Faradaic oxidation–reduction reaction. Though the electrochemical supercapacitors are known for high power density (>10 kW kg^{-1}) and excellent cycle life (>10^6 cycles), they suffer from the limited energy density (10 Wh kg^{-1}) as compared to the LIBs (~150 Wh kg^{-1}). To overcome the existing bottlenecks and achieve high energy density, recent studies are dealing with the hybridization of PC and EDLC electrodes to produce HSC. In a typical HSC, the anode is made of PC electrode material (which stores energy through Faradaic reaction) and the cathode is made of high surface area carbon material (which stores energy through EDLC formation). Many recent studies [37, 38] have revealed the promising performances of such hybrid systems made of various cathodes and anodes attached in a different fashion. However, it also appeared from the studies that fabrication of supercapacitor systems (both symmetric and asymmetric) always needs extra attention in terms of mass and charge balance, contacts between the components to control the ESR, placing of electrolytes for rapid ion and electron transfer, mechanical integrity to lengthen the cycle life and few others. Therefore, it is very important to discuss the fabrication procedures of various supercapacitor systems to achieve high capacitance, high energy density and power density, low ESR and long cycle life.

4.5.1 Fabrication Procedures of Supercapacitors

The electrodes are important components of a supercapacitor. The electrode active materials may be of different forms such as powder, semisolid like gels, thin films, etc. Electrodes from the powdery active materials are generally made by mixing active materials with suitable polymeric binders and conductive additives in a specific proportion and then casting on required current collectors. After that, the electrodes are generally fried and roll pressed to increase the contact among all the mixing constituents and finally used for electrochemical tests in both the two- and three-electrode systems. The main disadvantage of such widely accepted process is the low surface area of the final electrode (induces by the agglomeration of various constitutive additives) which generates low capacitance and prevents smooth ion movement toward active material. Compared to active materials, the additives contribute much lower capacitance, significantly decreasing both, the volumetric and gravimetric capacitances of the electrodes. Therefore, studies were also conducted to find the proper alternatives such as direct growing of active materials on current collectors like stainless steel, carbon cloth, fluorine doped tin oxide, PET, etc., and can be directly used for measurements. Freestanding electrode preparation is another possible technique where the electrode materials are grown without the presence of any external current collector and can be directly used for supercapacitor measurements. After the fabrication, discussions on various measuring parameters of supercapacitor system are also very important. Therefore, in the next section, we briefly discuss the various performance characteristics and parameters of supercapacitor system.

4.5.2 Performance Characteristics and Parameters

Researchers often use three-electrode configuration, consisting of WE, RE and CE to measure electrochemical behaviors of the electrode active materials. Generally, active materials are used as a coating on a suitable current collector or maybe directly grown on the same to produce the WE. SCEs and platinum electrodes are often used as RE and CE, respectively. GCD, EIS and CV techniques are used to characterize the electrochemical performances of supercapacitors. Parameters obtained from the said methods can be used to evaluate the performance of a supercapacitor in terms of C_s, cycle life, charge–discharge rate, etc.

The amount of charge stored by an ideal capacitor is proportional to the voltage and is given by 4.2 [13].

$$Q = C.V \tag{4.2}$$

where Q is the capacitor's charge in coulombs (C), C is the capacitance in farads (F) and V is the voltage between the devices' terminals in volts (V).

The capacitance of an electrochemical supercapacitor (specific capacitance is denoted as C_s in this chapter) can be calculated from the constant current charge–discharge (GCD) curve with 4.3 [39].

$$C_s = \frac{I\Delta t}{m\Delta V} \tag{4.3}$$

where C_s *is the specific capacitance*, I is the charge current, Δt is the discharge time, m is the total mass of active materials and ΔV is the potential window of the discharging process. The C_s can also be calculated from CV curves with 4.4 [39].

$$C_s = \frac{A}{v \times f \times m} \tag{4.4}$$

where A is the integral area of the cyclic voltammogram loop, f is the scan rate, V is the voltage window and m is the mass of the active material. The energy (E) and power (P) density are two important parameters to decide the energy storage performance of any supercapacitor. Amount of energy density (E) and power density (P) can be derived from 4.5 [13] and 4.6, respectively.

$$E = \frac{1}{2} C_s . V^2 \quad (4.5)$$

$$P = \frac{E}{t} \quad (4.6)$$

Generally, ideal capacitor does not show any power or energy loss but practically the case is opposite. Real capacitors can only operate in a "voltage window" with an upper and lower voltage limit. Any voltages outside this window can damage the device by causing electrolytic decomposition. The electrolytes can be aqueous or non-aqueous, and affect the range of voltage window. Capacitors with non-aqueous electrolytes have much wider voltage window, whereas aqueous electrolytes are safer and easier to use. During charging and discharging, power loss (P_{loss}) occurs in real capacitors, because resistance arises from electrodes, electrolyte and other electrical contacts. The sum of these resistances is called as ESR. Power loss (P_{loss}) or ESR can be measured through the given 4.6 [13].

$$P_{\text{loss}} = I^2 . \text{ESR} \quad (4.7)$$

Another important factor is leakage current. Constant voltage can be maintained without current flow in ideal capacitors, whereas real capacitors show leakage current through self-discharging process. In self-discharging process, a charged capacitor slowly gets discharged even without the external connections to its terminals due to the cause of leakage current.

4.6 Conducting-Polymer-Based Supercapacitor

CPs are widely known for their applications as electrode materials for supercapacitors. Generally, CPs undergo fast and reversible Faradaic reaction throughout the bulk volume of the electrode materials and offer high C_s as compared to various other carbonaceous electrodes. π-electrons available in the CP chain are extracted during electro-oxidation (*p*-doped state) and move along the chain; however, at the same time, the anions of the electrolyte will travel to the CP chain to balance the total electronic charge. In the electro-reduction process (*n*-doped state), opposite occurs, i.e., electrons will move toward the CP chain, and then cations from electrolyte will travel toward the same direction to neutralize the charge. In recent years, CP-based electrodes have been inspected with great attention for supercapacitors because they have some unique properties, such as ease of processability, good elasticity, facile thin-film fabrication and lightweight density. Though CPs are well known for their Faradaic properties, they may also store electrical energy through EDLC mechanism. Sometimes, both the EDLC and Faradaic mechanisms may operate together in a single CP-based electrode. PANI-based electrodes are one such candidate among many, which can operate through both EDLC and Faradaic pathways in different conditions. In the section below, CP-based supercapacitor electrodes are discussed thoroughly.

4.6.1 PANI-Based Supercapacitors

PANI is one of the extensively studied CPs in the field of development of electrode materials for supercapacitors. PANI in the form of emeraldine salt has good electrical conductivity and stores energy through EDLC mechanism, and it can also store energy through PC mechanism as it can undergo oxidation and reduction under certain electrochemical condition. In a circumstance, PANI may induce the PC mechanism due to its chemical transformation under a suitably applied potential. Till now, a lot of studies have been conducted to verify the mechanism of charge storing capacity and it is been confirmed that PANI can store energy through both ways. In this regard, one symmetric supercapacitor based on PANI electrodes was assembled and investigated under 0.5 M H_2SO_4 electrolyte to reveal the charge storage mechanism. They synthesized PANI through chemical oxidation under different pressures to produce granular and nano-network of PANI. Electrochemical analysis for both confirmed the presence of both EDLC and PC contributions. The PC contribution comes from the redox peaks appeared at ~0.33 and ~0.44 V corresponding to the leucoemeraldine–emeraldine transformation and/or weak redox couple near 0.70 V corresponding to the intermediates of hydroquinone/benzoquinone states [40]. Contribution of EDLC has been calculated by Trasatti method and reached on a decision that at the potential window of 0.25–0.55 V PC dominates whereas at the potential window of 0.60–0.80 V EDLC dominates. Since the nano-network PANI has a better surface area and porosity (as porosity helps better infiltration of electrolyte), so it showed higher EDLC (as it depends on surface area) contribution as compared to granular PANI.

As the granular and nano-network PANI showed distinct electrochemical properties, it is easy to believe that electrochemical properties of PANI-based electrodes surely depend on their characteristics of morphology, surface area, porosity, etc. Therefore, researchers have shown a lot of interest to investigate the effects of a different morphology, surface area and porosity on the electrochemical properties. To create a different morphology of any material, it is a common practice to tune the synthesis parameters suitably. Synthesis parameters such as temperature, pressure, duration, source and concentration of reactants and many others can modulate the kinetics and thermodynamics of a reaction which affects the growth of a particular material. In one such work [41], chemical oxidation of aniline was done in different reaction temperatures and produced PANI with different morphologies like granules, tubular and spherical. Electrochemical investigation of the various PANI-based electrodes developed indicates that tubular PANI is superior candidate among all three (granules, tubular and spherical PANI). In 1 M H_2SO_4 electrolyte, the obtained C_s of tubular, spherical and granules PANI is about 300, 300 and 290 F g^{-1} at a constant current of 5 mA, respectively. Though the initial capacitance is almost equal for all three different PANI, with the increase in cycles (after 500 charge–discharge cycles) retention of capacitance appears in the order of tubular (75%) >granules (35%) >spherical (57%) PANI. Though the exact reason behind the superior electrochemical properties of tubular PANI was not studied, CV plots indicate that the dominant charge storage mechanism is PC. The enhanced performance of tubular PANI may also be the outcome of its structural arrangements which makes the charge transfer process easy and smooth.

To explore the morphology-dependent electrochemical performances of PANI, it is been realized that the dimensions and shape of PANI should lie in the nanoscale region. Therefore, Park et al. [42] have developed three different morphologies of PANI, i.e., nanospheres, nanorods and nanofibers by using a polymeric stabilizer, poly(*N*-vinylpyrrolidone). They kinetically controlled the anisotropic growth of PANI at the nanometer scale in the presence of the said stabilizer in various concentrations to develop PANI of different shapes. The electrochemical investigations revealed that the shape and size of active materials are important to achieve high electrochemical performance of PANI-based electrodes. The C_s for nanospheres (71 F g^{-1}), nanorods (133 F g^{-1}) and nanofibers (192 F g^{-1}) was determined by GCD curves at current density of 0.1 A g^{-1} [42]. According to the obtained results, nanofiber-based electrode had faster electrode kinetics and better capacitance as compared to nanorods and nanospheres. Moreover, the nanofibers had structural ordering and the most outstanding oxidation/protonation level. The order of the electrochemical performances of the three morphologies of PANI developed is nanofiber > nanorods > nanospheres. All three samples showed the same shape of CV plots with distinguishable redox peaks, and they bear a linear relationship between peak current and scan rate which suggests that they stored energy through surface-redox reactions. However, calculation of electron transfer coefficient and electron transfer rate constant using Laviron theory suggests that nanofiber morphology is more suitable for easy and smooth charge transfer which supports the order of their electrochemical performances.

In later, another work has been reported by Chen and his group [43]. They developed PANI nanotubes in addition to its nanofiber and nanosphere analogues (Fig. 4.3) through a template-assisted chemical synthesis. As SEM images confirm the formation of various shapes of PANI, CV and charge–discharge analysis confirm the Faradaic reaction in aqueous H_2SO_4 electrolyte. The integral area of CV curve and charge–discharge time varies with the changes in morphology, and the obtained C_s indicates the superior electrochemical performance of PANI nanotubes (502 F g^{-1} at 1 A g^{-1}) while compared to nanofibers (404 F g^{-1} at 1 A g^{-1}) and nanospheres (345 F g^{-1} at 1 A g^{-1}) of PANI. Depending upon the morphology, porosity, pore size and its distribution, it is been observed that access of electrolyte ions is much easier and faster for the PANI nanotubes over others.

Lots of other attempts have also been made to interplay with the PANI structures to evolve the best electrochemical performances out of it. In a report, Sharma et al. [44] have developed a nano-porous hyper-cross-linked PANI, having a specific surface area of 1059 m^2g^{-1} and produced the C_s of 410 F g^{-1} at a scanning speed of 3 mV s^{-1}

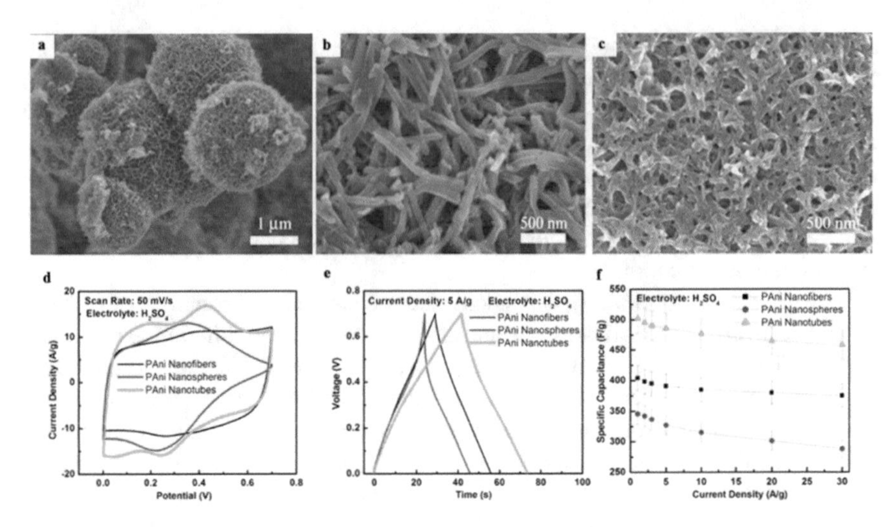

Fig. 4.3 **a** SEM images of PANI nanospheres, **b** SEM images of PANI nanotubes, **c** SEM images of PANI nanofibers, **d** CV at scan rate of 50 mV s^{-1}, **e** C_s versus current density, **f** C_s of all three PANI-based electrodes. Reprinted with permission from [43], copyright (2013), American Chemical Society

up to 1000 cycles. The high surface area of nano-porous hyper-cross-linked PANI helps to store energy through EDLC mechanism, whereas the redox sites available in the same electrode can store energy through PC mechanism simultaneously. Not only the new morphology but tuning physical properties are also important. For an instance, the length of nanofiber is also important to develop high-performance electrode materials for supercapacitors. During the polymerization of aniline, the addition of a small quantity of para-phenylenediamine produces long PANI fibers with fewer entanglements which help to achieve a C_s of 548 F g^{-1}, a power density of 127 W kg^{-1} and an energy density of 36 Wh kg^{-1} [45].

In addition to the structures and shape of CP-based active materials, the intrinsic characteristics such as doping and oxidation states can also influence their electrochemical properties. Bian et al. [46] have employed a doping and de-doping process to develop doped and de-doped PANI nanofiber and characterized their electrochemical performance. Obtained C_s is 29% higher for de-doped PANI as compared to the doped PANI. Process of de-doping helps to generate micropores in the whole polymer which assists the infiltration of electrolytes throughout the available electroactive surface to result in enhanced C_s. In another report on de-doped PANI, Anderson et al. [47] have proposed the formation of channels in the polymer matrix which helps the electrolytic ions to access the whole electroactive sites and as a result C_s and rate performance increase.

To achieve a long cycle life and faster charge transportation, adhesion between active materials and the current collector is very important. Not only the adhesion between these two but the inter-particle connection within active material is also very crucial. Therefore, direct growth/development of active materials on a conductive substrate has already attracted a lot of attention. Cho et al. [48] proposed a method known as SSDP to fabricate a PANI-based thin film with high porosity for the application as supercapacitor electrode. The modified highly porous PANI-based active material was deposited on a stainless steel substrate to form a film-type electrode material with a thickness of 10 μm and then tested as electrochemical supercapacitor in 0.5 M H_2SO_4 electrolyte. At a scan rate of 20 mV s^{-1}, it showed distinct oxidation–reduction peaks and produced a C_s of 361 F g^{-1} at a current density of 0.25 A g^{-1}, which is much higher than that of an equivalent electrode made with pristine PANI. A film made of PANI fibers through electrospinning has also shown great advancement in the performance while compared to that of PANI powders. The interconnected network of the fibers in a film helps to achieve high C_s (230 F g^{-1} at 1000th cycle) through a fast and stable Faradaic reaction [49]. In a different work, Wang et al. [50] proposed a galvanostatic deposition method to develop vertically aligned PANI nanowires on different substrates such as Au, Pt, stainless steel and graphite. The microstructural analysis confirmed the formation of PANI nanowire. CV plot depicted the clear redox peaks for PANI nanowire in various electrolytes and showed improved energy and power density. A detailed process of electrode fabrication is discussed below.

Though direct growing of active material is a fruitful technique in terms of electrochemical storage, sometimes direct growing of active material on the desired substrate is difficult. Usually, as an alternative method, a slurry of active materials with suitable additives can be made and cast on the substrate for electrochemical measurements. Generally, in this process, the mechanical cohesion between the active materials and the substrate depends on many factors like quantity and quality of the binders or conductive agents, deposition process, kind of substrates, surface nature of the substrates and many more. Therefore, in this process cycle life of the electrode becomes very short, and also the use of extra binder badly affects the storage performance in many cases.

A full cell was constructed by inserting a separator (Celgard 3501) wetted by electrolyte, between two identical electrodes and sealed into a button-like cell [43]. Though the electrochemical test under three-electrode configurations is necessary to understand the nature and initial performance of the electrode materials, for the practical application it is important to study the supercapacitor electrodes under full cell configuration. Supercapacitance performance of PANI-based full cell may produce low (e.g., 4–55 F g^{-1}) [42] to high (e.g., 300–500 F g^{-1}) [43] capacitance depending on many factors such as loading of the active material, functionalization, electrical conductivity, surface area, porosity, circuit connection and so on.

A smart supercapacitor was developed by depositing PANI on aligned CNT sheet electrodes [51]. CNT sheets were dry-drawn from CNT arrays synthesized by chemical vapor deposition technique. The CNT sheets showed conductivity of 10^2–10^3 S cm^{-1} and optical transmittance of over 90%. CNTs retained the aligned structure even at 70 wt% PANI deposition. C_s of 308.4 F g^{-1} was observed, and the capacitors changed color rapidly and reversibly among, yellow, green and blue, with variation of voltage during charge–discharge process. The smart capacitors are also stretchable and maintained C_s after stretching up to 100% for 200 cycles or bending for 1000 cycles [51].

PANI hydrogel was used as electrode material for a solid-state flexible supercapacitor [52]. Hydrogel has a 3D network and large amount of water dispersion medium. PANI has advantages like high conductivity, ease of synthesis, intrinsic flexibility and low cost. CP hydrogel offers intrinsic 3D conducting frameworks. In an aqueous solution, the swelling of porous nanostructure occurs, improving the contact between electrode and electrolyte. This causes large capacitance in the solid-state capacitor [52]. A capacitance of 430 F g^{-1} was obtained for the PANI hydrogel. Even after 100 bending cycles (bending angle of 180°), the capacitance of the device showed almost no decay. The ESR was found to decrease after 100 bending cycles, showing that contact between layers enhances with bending cycle [52].

Despite the high theoretical capacitance of PANI (2000 F g^{-1}), it is difficult to achieve high capacitance for the same in real, due to the limited availability of reactive sites in the material for the access of electrolytes [43]. Other disadvantages of PANI include its brittle nature, short cycle life, restricted rate capability, low power density at high current density, excess volume expansion on repeated charge–discharge process, etc. Therefore, a lot of attention has been given to develop new CPs with high electrochemical performance. PPy is one such candidate which attracts a lot of attention for electrochemical studies.

4.6.2 PPy-Based Supercapacitor

Like PANI, PPy also remains in semiconducting form but may transform easily from insulator to conductor under protonation. Therefore, the salt form of PPy is a promising candidate for electronic applications where high conductivity of the material is desirable. Among the various uses of PPy in electronic applications, development as electrodes of a supercapacitor is of great significance. PPy possesses advantages like great flexibility in synthesis and relatively high capacitance property with long cycling life. In a trend, it is been found that PPy possesses C_s of 480 F g^{-1} which is lower than 1284 F g^{-1} of PANI but higher than 210 F g^{-1} of PEDOT [53]. Flexibility of PPy-based electrodes is also significantly high for the fabrication of flexible energy storage devices.

A nanostructured conductive PPy hydrogel is synthesized by interfacial polymerization [54]. Polymer nanospheres are interconnected to form a unique 3D porous nanostructure. The transport of electrons and ions occurs due to large open channels between branches and nanoscale porosities in the 3D hierarchical structures. This makes the material suitable to be used as supercapacitor electrodes, having a C_s of ~380 F g^{-1} with an excellent rate capability and areal capacitance as high as ~6.4 F cm^{-2} [54]. PPy has spherical shell morphology with a shell thickness of 50–100 nm, which helps in overcoming its inherent brittleness and contributing to its flexibility. CV data in a potential window of 0.0–1.0 V versus Ag/AgCl at a scan rate of 100 mV s^{-1} under different bending conditions (with radii of curvature of ∞ (flat), 8 mm and 3 mm) remains almost the same (~3% decrease) as the curvature of the supercapacitor increases [54]. There is negligible change in the capacitance in comparison with a flat one. Analysis of the flexibility-related rate performance of a PPy hydrogel-based supercapacitor was carried out through GCD tests, and C_s was calculated and plotted against current density. The C_s retains about 81% and 61% as the current density is increased by ~10 × and ~25 × , respectively. The cycling performance of the flexible supercapacitor in 3000 charge–discharge electrochemical cycles in the potential range of 0.0–1.0 V was measured. Capacitance retention of ~90% indicated good electrochemical stability and cyclability of the assembled supercapacitor device [54].

Compared to other CPs, PPy shows a better degree of flexibility other than its greater mass density; i.e., PPy has the aptitude to provide a higher performance even with a smaller volume and is capable to adapt to different forms. Due to these advantages, PPy is considered as a promising material for the fabrication of high-performing, flexible and lightweight supercapacitors, which boosts their application in an important area of energy storage field, the portable and flexible electronic power sources. However, the main bottlenecks which still hinder the smooth functioning of PPy-based supercapacitors are (i) the differences in expected and actual capacitance and (ii) short life during charging and discharging processes. Various strategies have been proposed to eradicate such shortcomings through designing and development of specific hierarchical architecture of PPy having unique morphology with aligned multiscale porosity development of different PPy and metal oxide-based synergistic composites and/or suitable design of PPy-based supercapacitor.

In one strategy, Sahoo et al. [55] have synthesized a PPy nanofiber-based electrode through a facile chemical route using sodium alginate as template. The SEM image reveals that the synthesized PPy has a fiber like morphology, i.e., a random network of nanostructured PPy with the diameter ~100 nm containing hierarchical multiscale porosity. This porous network is exceptional for easy and fast electrolyte transport, which is the essential requirement for enhanced electrochemical properties of the electrode materials. The non-rectangular shape of CV curves of PPy nanofiber in 1 M KCl refers to its redox behavior due to broad pore size distribution. The charge–discharge curves of PPy nanofiber in KCl electrolyte performed at 1 A g^{-1} are linear as well as symmetrical, which is a typical characteristic of ideal capacitor. PPy nanofiber shows maximum capacitance of 284 F g^{-1} which is higher than that of pure PPy having conventional agglomerated particle-like morphology.

In addition to electrical and electrochemical performances, PPy is also known for its good stretchability and high mechanical strength which helps to develop high-performing stretchable supercapacitors. Wang et al. [56] have fabricated a highly reversible stretchable all-polymer supercapacitor using PPy as electrode and double network hydrogel as electrolyte. The capacitance was determined to be 79.7 mF cm^{-2}, and the performance remains unaltered even after the 1000 stretches.

Such a commendable performances results from the combination of PPy-based electrode and Agar/hydrophobically associated polyacrylamide-based double network gel electrolyte.

Accidental mechanical damage limits the life span and reliability of supercapacitors for practical applications. Hence, a supercapacitor having mechanically and electrically self-healing properties has been considered as an excellent solution. Recently, yarn-based supercapacitors have received extensive attention, offering exceptional possibilities for future wearable electronic devices. A yarn-based supercapacitor (as shown in Fig. 4.4a) having exceptional electrical and mechanical self-healing capacity has been fabricated by enveloping the magnetic electrodes with self-healing carboxylated PU shell [57]. The magnetic force caused the broken electrodes to reconnect through magnetic alignment, which promoted the recovery of electrochemical performance through self-healing (as shown in Fig. 4.4b) with the assistance of the carboxylated PU shell. In fabrication, two individual yarn electrodes are assembled with suitable solid electrolyte and self-healing shell to develop the self-healing supercapacitor [57]. An investigation of its electrochemical performance is carried out to prove the synergistic effect of the self-healing shell and magnetic electrodes. The CV curves in the voltage window from 0 V to 1 V are measured. The overall CV curves of the device appear almost rectangular as it is a combination of non-rectangular curves originated from iron oxides and rectangular curves of PPy. The GCD test of the device was conducted in between 0 and 1 V at different current densities. GCD curves deviate from the linear voltage–time relation which indicates its pseudocapacitive behavior due to the surface-confined Faradaic reaction of the electrodes which agrees with the non-rectangular shape of the CVs. The C_s of self-healing device estimated from the CV curve is 61.4 mF cm^{-2} at a scan rate of 10 mV s^{-1}. Even after numerous cycles of cutting and healing, the capacitance is maintained at a remarkably high level. The C_s obtained after the second and fourth healings (calculated from CV curves) is 50.7 and 44.1 mF cm^{-2}, respectively, with capacitance retention of 71.8% after the fourth healing cycle. Therefore, PPy has been regarded as a promising flexible electrode material for high-performance supercapacitor, owing to its intrinsic advantages, such as high electrical conductivity, interesting redox properties, high flexibility, good mechanical strength, easy processing with tunable morphological characteristics and low cost. However, PPy-based supercapacitors have also few disadvantages like others. They undergo serious structural pulverization and counterion drain effect during charging and discharging, affecting its cycling stability. Therefore, hunting for new high-performing supercapacitor is still continuing, and we believe suitable CPs with unique structural features will advance the field further.

4.6.3 PEDOT-Based Supercapacitor

After the PANI and PPy, the most electrochemically studied candidate is PEDOT which shows high conductivity (300–500 S cm^{-1}), higher potential range (1.4 V,

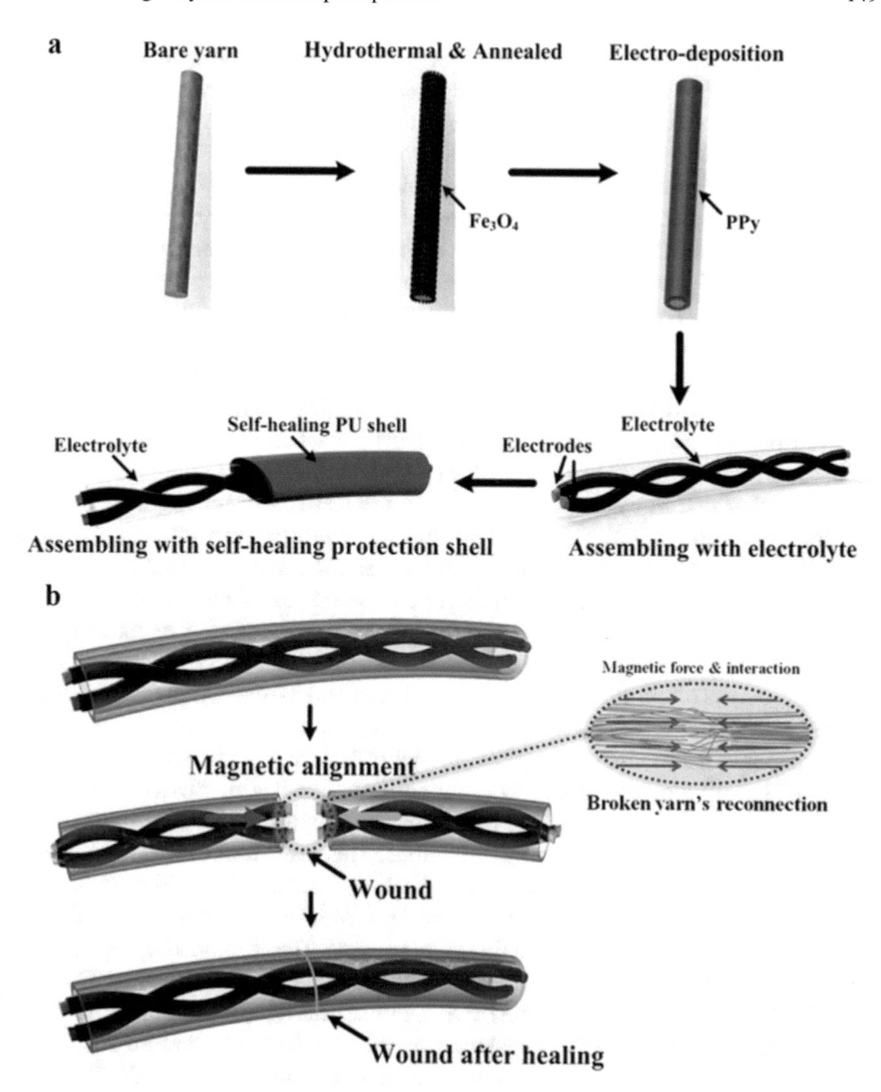

Fig. 4.4 **a** Design and manufacturing process of the magnetic-assisted self-healable supercapacitor, **b** schematic illustration of the self-healing process of supercapacitor. Reprinted with permission from [57], copyright (2015) American Chemical Society

depending on electrolyte), low oxidation potential, great mechanical strength and flexibility. Unlike PANI and PPy which can only be *p*-doped, PEDOT on the other hand can be of both categories, i.e., *n*- and *p*-doped which means more diversity in electronic arrangements and suitable for different electrochemical studies. Since this chapter is devoted toward the supercapacitor applications, we are only interested to discuss more about the various PEDOT-based energy storage systems. PEDOT is not only popular for its diversity in doping and electronic conductivity but also known for its amazing thermal and chemical stability, good film-forming properties, excellent flexibility and mechanical strength in combination with high surface area and high charge mobility suitable for fast electrochemical kinetics [58].

In a work [58], PEDOT-based thin-film electrode was developed on a non-conducting plastic substrate by chemical polymerization for electrochemical supercapacitor. The film electrode was then immersed in a liquid organic electrolyte (0.1 M $LiClO_4$ in acetonitrile) and measured in a three-electrode system where RE was Ag/AgCl and CE was a high surface platinum foil [58] to measure the supercapacitive performance. The result signified that CV response is the result of both EDLC and Faradaic redox reactions and retains even at high scan rates of 200 mV s^{-1}. At full charging, the cell voltage reaches up to 0.8 V and the energy density is measured from 1 Wh kg^{-1} to 4 Wh kg^{-1}, depending on the applied discharge current densities used during discharge. However, the power density reaches up to 2.5 kW kg^{-1} at the highest current density of 1 mA cm^{-1} [58].

Many other studies were also conducted on PEDOT-based supercapacitors in the investigation of high-performing and affordable supercapacitors [59]. In one such work [59], a PEDOT-based symmetric supercapacitor and an asymmetric supercapacitor of PEDOT (as cathode) and activated carbon (as anode) were developed and electrochemically studied. The electrochemical performance was measured in Et_4NBF_4 in propylene carbonate and $LiPF_6$ in EC/DMC electrolyte. The charge and discharge curves of the 1000th cycle for both redox and hybrid-type supercapacitors in the range 0–1 V were measured. The charge and discharge curves of redox type are parabola in nature, which is similar to that of PC and the charge and discharge curves of hybrid type are similar to a straight line, like that of EDLC. In the case of PEDOT, oxidation and reduction processes are accompanied with electrochemically reversible doping and undoping of anions. This leads to the charge storage in PEDOT, resulting in PC. The charge and discharge curves of hybrid type suggest that the characteristics of active carbon powders (MSP-20), used mainly as EDLC electrode materials, are prevalent in this voltage range. The specific discharge capacitance of redox type with Et_4NBF_4 in propylene carbonate is measured to be 22 F g^{-1} and with $LiPF_6$ in EC/DMC, to be 27 F g^{-1}. For hybrid type, the specific discharge capacitance with Et_4NBF_4 in propylene carbonate is 19 F g^{-1} and 50 F g^{-1} for $LiPF_6$ in EC/DMC. The specific discharge capacitance measured after 1000 cycles in range 0–3 V for redox type is 20 F g^{-1} for Et_4NBF_4 in propylene carbonate and 13 F g^{-1} for $LiPF_6$ in EC/DMC. For hybrid-type supercapacitor, these were found to be 30 F g^{-1} for Et_4NBF_4 in propylene carbonate and 56 F g^{-1} for $LiPF_6$ in EC/DMC. The authors concluded that good electrochemical performance with both electrolyte solutions in 0–1 V range can be obtained for hybrid-type supercapacitor with MSP-20. The hybrid-type supercapacitor with PEDOT also showed good performance with $LiPF_6$ in EC/DMC in 0–3 V range. The electrochemical performance of hybrid supercapacitor was found to be superior to that of redox supercapacitor [59].

PEDOT-based flexible supercapacitor has also been extensively studied by various groups. Cheng et al. [60] have developed a freestanding PEDOT:PSS electrodes having high flexibility (180° bending at 50% strain) with outstanding electrochemical performance. Excellent electrochemical performance of the developed flexible electrode was witnessed through tuning the doping process and modulating the number of the PEDOT:PSS layers [60]. The sheet resistance of pristine PEDOT:PSS layer (0.4 MΩ sq^{-1}) dramatically reduced to ~140 Ω sq^{-1} on addition of 6 vol.% ethylene glycol. Further on doping with surfactant (TR) (2%), the sheet resistance reduced to 110 Ω sq^{-1}. The sheet resistance of PEDOT:PSS films was also found to decrease from 145 to 27 Ω sq^{-1} when the number of layers was increased from one to five layers. They formed one symmetric, flexible and transparent all-solid-state PEDOT:PSS-based supercapacitor in sandwiched fashion and observed that the nature of CVs and GCDs alters with the change in number of PEDOT:PSS layers present in the electrode. Not only the electrochemical nature alters with the number of layers but also the performances like specific capacitance and discharging time have increased, with the increase of number of layers, as evidenced from the electrochemical results. From the GCD curves, it was concluded that supercapacitor with 2 vol.% surfactant-doped PEDOT:PSS electrodes exhibited the longest discharge time. The current density was also found to increase, with the increase in the number of PEDOT:PSS layers. The obtained areal and gravimetric C_s of the supercapacitors comprised of PEDOT:PSS electrodes with one, two, three and four layers at current density of 0.025 mA cm^{-2} was 0.23 mF cm^{-2} (6.05 F g^{-1}), 0.41 mF cm^{-2} (7.19 F g^{-1}), 0.66 mF cm^{-2} (5.50 F g^{-1}) and 1.18 mF cm^{-2} (6.94 F g^{-1}), respectively. The corresponding areal/gravimetric C_s of the electrode calculated from the GCD curve of supercapacitor at current density of 0.025 mA cm^{-2} was 0.92 mF cm^{-2} (24.20 F g^{-1}), 1.64 mF cm^{-2} (28.76 F g^{-1}), 2.64 mF cm^{-2} (22.0 F g^{-1}) and 4.72 mF cm^{-2} (27.76 F g^{-1}). The results indicate that areal C_s has improved linearly with the number of layers but the gravimetric C_s is not. Such differences are hidden in the basic fundamentals of areal and gravimetric C_s calculation. As the number of layers increases, the active material mass also increases even at a constant electrode area and as a result gravimetric C_s (unit is F g^{-1}) decreases.

Though PEDOT is widely known for its flexibility, conductivity and excellent electrochemical statistics, PEDOT-based electrodes have many limitations too. Likely, PEDOT has low to moderate theoretical C_s (210 F g^{-1}) because large molecular weight limited doping level of only 0.33, high hydrophobicity and many others like shrinkages, breaking and cracks on electrodes arises from subsequent cycling. Therefore, despite drawing attention from the every corner of energy storage field PEDOT-based electrodes have distinctive disadvantages which are limiting its commercial applications yet.

4.6.4 PTh-Based Supercapacitor

Another important CP is PTh which left a considerable footprint in the energy fields. The C_s of the PTh-based electrode is generally lower as compared with PANI or PPy, but the major benefit is that it can work comparatively at higher potential window (~1.2 V) [2]. PTh and its derivatives can be both *p*- and *n*-type CPs and can be applied for supercapacitor applications. Capacitance values of few thiophene-based CPs like

Table 4.4 Electrode-specific capacitance and capacity from CVs at 20 mV s^{-1} of polymers electrosynthesized on current collectors

Polymer	*p*-Doping			*n*-Doping		
	Capacitance (F g^{-1})	CV range (V vs. SCE)	Capacity (mAh.g^{-1})	Capacitance (F g^{-1})	CV range (V vs. SCE)	Capacity (mAh g^{-1})
PDTT1	110	1.0/−0.2	19	75	−1.5/−0.2	17
PMeT	220	1.15/−0.2	62	165	−2.0/−1.0	26
PFPT	95	1.0/−0.2	19	80	−1.7/−1.0	9

PFPT, PMeT and PDTT except PEDOT (as we already discussed above) in both *n*- and *p*-doping state are reproduced with permission in Table 4.4.

Laforgue and co-workers [18] produced a PTh-based supercapacitor and stated a specific storage level of 260 F g^{-1}. In a different work, PTh was synthesized by chemical oxidative polymerization method, in the presence and absence of three various surfactants: cationic CTAB, anionic SDS and nonionic TR, employing $FeCl_3$ as oxidant. The C_s of PTh prepared using TR surfactant was found to be 117 F g^{-1}, which was higher (up to 33%) compared to the C_s of PTh prepared without surfactant (7 F g^{-1}). A rectangular form of voltammogram was observed, suggesting capacitive behavior of PTh [23]. Another method for synthesis of unsubstituted PTh through oxidative chemical vapor deposition was reported by Nejati et al. [62]. The synthesis was based on the direct vapor-to-solid oxidative polymerization of thiophene monomer by using antimony pentachloride as an oxidant, stabilized up to the 5000 cycles with only a 10% decrease in capacitance.

PTh nanoparticles were synthesized by cationic surfactant-assisted dilute polymerization method where $FeCl_3$ was used as an oxidant and the C_s of the capacitor was 134 F g^{-1}, as reported by Gnanakanet et al. [63]. The calculated energy and power densities were 8 Wh kg^{-1} and 396 W kg^{-1}, respectively. The redox supercapacitor was fabricated by PTh nanoparticles-based symmetric electrodes using PVdF-co-HFP-based microporous polymer electrolyte containing 1 M $LiPF_6$ in ethylene carbonate/propylene carbonate (1:1 v/v). To study the cyclic stability of the system, 1000 cycles were carried out at the scan rate of 50 mV s^{-1}. A slight decaying of current was observed between 5th and 1000th cycles, thereby ensuring the small capacitance fading of the electrode material, and consequently, the stability of the electrode material was found to be moderate [63]. PTh nanofibers were also produced in a study by surfactant-assisted dilute polymerization method using $FeCl_3$ as oxidant [64]. XRD characterization was used to confirm their composition. SEM image shows fibrous structure with small globular structure. The PTh nanofibers were calcined in an inert atmosphere at 1400 °C for 2 h to get carbonaceous PTh nanofibers. The asymmetric supercapacitor was fabricated using PTh nanofibers as the cathode and carbonaceous PTh nanofibers as the anode in 6 M KOH electrolyte [64].

PTh thin films show high environmental stability, high electrical conductivity and electrochemical properties. High conductivity, stability in oxidized form and attractive electrochemical performance make it an attractive candidate to be used as electrode material in supercapacitor. PTh has been produced in bulk and thin-film forms by several physical and chemical methods. Fu et al. [65] synthesized PTh thin film by electrodeposition and reported the C_s to be 110 F g^{-1}. Chemical bath deposition method at room temperature (300 K) through oxidative polymerization of thiophene using ammonium peroxodisulfate as an oxidizing agent was used to prepare PTh thin films [66]. Globular particulates of PTh were placed on the stainless steel and glass substrates, and morphology and chain structure were examined by Raman spectroscopy and SEM techniques. SEM images of PTh thin film at two different magnifications, X5000 and X15,000, were taken, and globular particulate of PTh distributed over the entire surface was observed in SEM images. The electrochemical performance of PTh electrode was examined using CV and GCD analyses, and a C_s of 300 F g^{-1} at 5 mV s^{-1} in 0.1 M LiCO $_4$/polycarbonate electrolyte was reported. Excellent cyclability and rate capability of PTh cathode are pointed out by CV curves of PTh films for the first cycle and after 1000 cycles. PTh film shows 87% cyclic stability in $LiClO_4$/polycarbonate electrolyte. This is because of the porous morphology of PTh, which increases the ability of the electrolyte to soak into PTh electrode and decreases the electrode polarization. The decrease in capacitance observed after 1000 cycles is attributed to the doping and undoing of CP which undergoes volume change and hence is bound to undergo mechanical degradation which results in capacitance decay [66].

4.6.5 Conducting-Polymer-Based Composites

Though the discussion on the composites of CPs is not in the scope of this chapter, it will be relevant to mention few recent studies published on various CP-based composites applied in the fields of supercapacitors. Composites are known for their synergistic effects resulted from various components of composites. Generally, CPs-based composites were developed to optimize the electronic, mechanical and electrochemical storage performances of the composite supercapacitor electrodes. For example, a hybrid nanocomposite of PANI and MNT was prepared to avail the good electrical conductivity from PANI and variable redox sites from MnO_2 as the epicenter of PC. Inclusion of MNTs in PANI matrix enhances their electrochemical properties and restrains the dissolution of MNT in acidic electrolyte. MNTs provide rigid support and interlink the PANI chains, thus enhancing conductivity. In addition to conductivity, PANI also provides high surface area for MNT dispersion and redox processing. The hybrid nanocomposite shows high C_s of 626 F g^{-1} and energy density of 17 Wh kg^{-1}. Likely, Hou et al. [67] grew a Cu-MOF nanowire array on PPy membrane and used hybrid electrode material for flexible supercapacitor. The flexibility, huge surface/mass ratio and high conductivity of PPy membrane make the hybrid electrode suitable for high-performing flexible supercapacitor having areal capacitance of 252.1 mF cm^{-2}, energy density of 22.4 μWh cm^{-2} and power density of 1.1 mW cm^{-2}. Many different attempts have also been made to develop such composites with improved electrochemical properties. Antiohos et al. [68] have reported a freestanding PEDOT-PSS/SWCNTs having C_s of 104 F g^{-1} at 0.2 A g^{-1}. Poly(An-co-Py)/Cu/CNT nanocomposite was synthesized by Dhibar et al. [69] using in situ chemical oxidative polymerization where the uniform coating of $CuCl_2$-doped poly(An-co-Py) was deposited over MWCNT surface to achieve C_s of 383 F g^{-1} at 0.5 A g^{-1} [70]. Mao et al. [71] have prepared Gr/PANI nanofiber composites by in situ polymerization, and C_s of 526 F g^{-1} was obtained at a current density of 0.2 A g^{-1} with good cycling stability [71]. Sahoo et al. [72] synthesized fiber like poly(An-co-Py)/Gr nanocomposite by simple chemical polymerization and obtained C_s of 351 F g^{-1} at 10 mV s^{-1}. Sahoo et al. [70] have also synthesized Gr and PPy nanofiber using a biopolymer, sodium alginate. Fiber-like morphology, with the presence of Gr, was obtained. The composite achieved high electrical conductivity of 1.45 S cm^{-1} at room temperature and a maximum capacitance value of 466 F g^{-1} at 10 mV s^{-1} in 1 M KCl solution. Mn-PANI/SWCNT nanocomposite was synthesized depositing Mn-doped PANI on SWCNT surfaces using in situ oxidative polymerization [73]. Detailed investigation showed that transition metal doping has improved the overall electrical conductivity which helps to attain C_s of 546 F g^{-1} measured at 0.5 A g^{-1}. Similarly, PTh-based composites like Gr-SWCNTs-P3MT ternary nanocomposite were also developed which generates C_s of 561 F g^{-1} at 5 mV s^{-1} [74]. This list may continue from pages to pages by describing various CPs based on composites and their electrochemical performances, but this is under the limited scope of the present chapter. However, tons of scopes can be produced for materials, chemicals and energy researchers by conducting critical postmortem of CPs-based composite electrodes for energy storages.

4.7 Conclusion with Challenges and Possibilities

CPs have been successfully investigated in many fields like energy, materials, engineering, biomedical, electromagnetic interference, optical and many others, because they are known for simple synthesis and processing, lightweight, low cost, abundance of redox sites, optimized electronic and ionic conductivity and super-flexibility. Though success of CPs in various fields including energy is evident from previous studies, but it is still relevant to investigate further because of the new challenges appeared in recent days. The current era is of modern technologies like electric vehicles, smart electronic gadgets, smart cloths, flexible devices and so on. Modern technologies have brought new challenges with it. Such modern technologies demand high-end energy storage systems with high energy and power density and long cycle life and fast-charging characteristics. Not only the electrochemical performances are necessary but for advanced technologies such as wearable devices and smart cloths energy storage systems should also come with high flexibility and robust mechanical strength. Studies are also indicative that enhancement in potential window up to 2.7–3 V and energy density up to ~150 Wh kg^{-1} is necessary for various modern devices. Till today, Gr, CNTs, chalcogenides, transition metal oxides, MXenes, different composites and even CPs have been tried to mitigate the risen challenges, but most importantly, desired targets are yet far to be achieved and the efforts are still continuing. Though few candidates have shown promising performances, lack of cost-effectiveness and difficulties in processing becomes one of the main bottlenecks in the path of their commercial success. Therefore, we believe that the low cost, simplicity and unique electrical and redox properties of CPs are more advantages over its competitors and possibly they can mitigate the energy challenges of future. From the above discussion, we found that PANI is producing different electrochemical performances depending on the variation in morphology from sphere to tube to fiber. We also found that CPs are having different electrical properties depending on different doping conditions. A different functionalization can modulate the redox sites in CP-based electrodes and hence their PC performance. Flexibility and mechanical properties may also be tuned through various additives. So, all such findings confirmed that CPs are the suitable candidate for energy storage applications and can be tuned further by altering its structure–property relationship. The challenges are not only in the field of selection and development of suitable materials but also exist in device engineering. Proper loading, contact of active materials with current collector, fabrication fashion is also very important to control the ESR for producing high-end supercapacitors. Therefore, we believe development of chemically modified CPs and its composites with controlled morphology and hierarchical porosity including proper electrode engineering could create enormous opportunity in the future to counter the existing challenges in the energy storage systems. Recently, electrochromic supercapacitor devices were made from various CPs as their electronic and optical properties can be controlled through applied voltages. Electrochromic devices are capable to indicate the energy level by changing its color depending on the changes in voltage level. Such new developments have open up many possible avenues for the researchers, and we believe CP-based energy storage systems will remain a relevant topic to discuss beyond this chapter.

Acknowledgements I am thankful to the DST-INSPIRE-FACULTY program (Faculty Registration No.: IFA17-MS135) for the support given through a stable funding program. I am also thankful to Ms. N. Gupta and Mr. Y.S.S. Sarma for helping in manuscript preparation.

References

1. J. Kim, J. Lee, J. You, M.S. Park, M.S. Al Hossain, Y. Yamauchi, J.H. Kim, Mater. Horizons. **3**(6), 517–535 (2016). https://doi.org/10.1039/c6mh00165c
2. I. Shown, A. Ganguly, L.C. Chen, K.H. Chen, Energy Sci. Eng. **3**(1), 2–26 (2015). https://doi.org/10.1002/ese3.50
3. H. Shirakawa, T. Ito, S. Ikeda, Macromol. Chem. Phys. **179**(6), (1978). https://doi.org/10.1002/macp.1978.021790615
4. Z.A. Boeva, V.G. Sergeyev, Polym. Sci. Ser. C. **56**(1), 144–153 (2014). https://doi.org/10.1134/S1811238214010032
5. D. Nicolas-Debarnot, F. Poncin-Epaillard, Anal. Chim. Acta **475**(1), 1–15 (2003). https://doi.org/10.1016/S0003-2670(02)01229-1

6. Y. Cao, A. Andreatta, A.J. Heeger, P. Smith, Polymer. **30**(12), (1989). https://doi.org/10.1016/0032-3861(89)90266-8
7. W. Liang, J. Lei, C.R. Martin, Synth. Met. **52**(2), 227–239 (1992). https://doi.org/10.1016/0379-6779(92)90310-F
8. N. Othman, Z.A. Talib, A. Kassim, A.H. Shaari, J.Y.C. Liew, Mal. J. Fund. Appl. Sci. **5**(1), (2009). https://doi.org/10.11113/mjfas.v5n1.284
9. K. Tamao, S. Kodama, I. Nakajima, M. Kumada, A. Minato, A.K. Suzuki, Tetrahedron. **38**(22), (1982). https://doi.org/10.1016/0040-4020(82)80117-8
10. Y. Mochizuki, T. Horii, H. Okuzaki, Trans. Mater. Res. Soc. Japan. **37**(2), (2012). https://doi.org/10.14723/tmrsj.37.307
11. G.E. Wnek, J.C. Chien, F.E. Karasz, C.P. Lillya, Polymer. **20**(12), (1979). https://doi.org/10.1016/0032-3861(79)90002-8
12. D.M. Ivory, G.G. Miller, J.M. Sowa, L.W. Shacklette, R.R. Chance, R.H. Baughman, J. Chem. Phys. **71**(3), (1979). https://doi.org/10.1063/1.438420
13. R. Brooke, P. Cottis, P. Talemi, M. Fabretto, P. Murphy, D. Evans, Prog. Mater Sci. **86**, 127–146 (2017). https://doi.org/10.1016/j.pmatsci.2017.01.004
14. N. Sakmeche, S. Aeiyach, J.J. Aaron, M. Jouini, J.C. Lacroix, P.C. Lacaze, Langmuir **15**(7), 2566–2574 (1999). https://doi.org/10.1021/la980909j
15. T. Pal, S. Banerjee, P.K. Manna, K.K. Kar, in *Handbook of Nanocomposite Supercapacitor Materials I*, ed. by K.K. Kar (Springer, 2020), p. 247
16. R. Corradi, S.P. Armes, Synth. Met. **84**(1–3), 453 (1997)
17. M.G. Han, S.H. Foulger, Small. **2**(10), (2006). https://doi.org/10.1002/smll.200600135
18. A. Laforgue, P. Simon, C. Sarrazin, J.F. Fauvarque, J. Power Sources **80**(1), 42–148 (1999). https://doi.org/10.1016/S0378-7753(98)00258-4
19. S.A. Chen, H.J. Shy, J Polym Sci A Polym Chem. **23**(9), 2441–2446 (1985). https://doi.org/10.1002/pol.1985.170230909
20. S. Bhadra, N.K. Singha, D. Khastgir, J. Appl. Polym. Sci. **104**(3), 1900–1904 (2007). https://doi.org/10.1002/app.25867
21. J.M. Pringle, J. Efthimiadis, P.C. Howlett, J. Efthimiadis, D.R. MacFarlane, A.B. Chaplin, S.B. Hall, D.L. Officer, G.G. Wallace, M. Forsyth, Polym. **45**(5), 1447–1453 (2004). https://doi.org/10.1016/j.polymer.2004.01.006
22. A.S. Saraç, G. Sönmez, F.Ç. Cebeci, J. Appl. Electrochem. **33**(3), 295–301 (2003). https://doi.org/10.1023/A:1024139303585
23. B. Senthilkumar, P. Thenamirtham, R. Kalai Selvan, Appl. Surf. Sci. **257**(21), 9063–9067 (2011). https://doi.org/10.1016/j.apsusc.2011.05.100
24. P. Evans, R. Grigg, M. Monteith, Tetrahedron Lett. **40**(28), 5247–5250 (1999). https://doi.org/10.1016/S0040-4039(99)00993-4
25. T. Masuda, S.M. Abdul Karim, R. Nomura, J. Mol. Catal. A. Chem. **160**(1), 125–131 (2000). https://doi.org/10.1016/s1381-1169(00)00239-9
26. J. Nowaczyk, K. Kadac, E. Olewnik-Kruszkowska, Adv. Sci. Technol. **9**(27), 118–122 (2015). https://doi.org/10.12913/22998624/59093
27. G.L. Yuan, N. Kuramoto, M. Takeishi, Polym. Adv. Technol. **14**(6), 428–432 (2003). https://doi.org/10.1002/pat.352
28. J.A. Tembhe, S.A. Waghuley, Int J Curr Eng Sci Res. **5**, 182–186 (2018)
29. T.H. Le, Y. Kim, H. Yoon, Polymers. **9**(4), 150 (2017). https://doi.org/10.3390/polym9040150
30. J. Tsukamoto, Adv. Phy. **41**(6), 509–546 (1992). https://doi.org/10.1080/00018739200101543
31. Y.W. Lee, K. Do, T.H. Lee, S.S. Jeon, W.J. Yoon, C. Kim, J. Ko, S.S. Im, Synth. Met. **174**, 6–13 (2013). https://doi.org/10.1016/j.synthmet.2013.04.009
32. D. Poussin, H. Morgan, P.J.S. Foot, Polym. Int. **52**(3), (2003). https://doi.org/10.1002/pi.1107
33. T.R. Dillingham, D.M. Cornelison, S.W. Townsend, J. Vac. Sci. Technol. **14**(3), (1996). https://doi.org/10.1116/1.5799%5b75
34. S. Gupta, Appl. Phys. Lett. **88**(6), (2006). https://doi.org/10.1063/1.2168688
35. O.P. Dimitriev, D.A. Grinko, Y.V. Noskov, N.A. Ogurtsov, A.A. Pud, Synth. Met. **159**(21–22). https://doi.org/10.1016/j.synthmet.2009.08.022

36. N.K. Guimard, N. Gomez, C.E. Schmidt, Prog. Polym. Sci. **32**(8), 876–921 (2007). https://doi.org/10.1016/j.progpolymsci.2007.05.012
37. S. Banerjee, B. De, P. Sinha, J. Cherusseri, K.K. Kar, in *Handbook of Nanocomposite Supercapacitor Materials I*, ed. by K.K. Kar (Springer, 2020), p. 341
38. K. Machida, S. Suematsu, S. Ishimoto, K. Tamamitsu, J. Electrochem. Soc. **155**(12), A970 (2008). https://doi.org/10.1149/1.2994627
39. C. Zequine, C.K. Ranaweera, Z. Wang, S. Singh, P. Tripathi, O.N. Srivastava, B.K. Gupta, K. Ramasamy, P.K. Kahol, P.R. Dvornic, R.K. Gupta, Sci Rep. **6**, 31704 (2016). https://doi.org/10.1038/srep31704
40. C.C. Hu, J.Y. Lin, Electrochim. Acta **47**, 4055–4067 (2002). https://doi.org/10.1016/S0013-4686(02)00411-5
41. H. Kuang, Q. Cao, X. Wang, B. Jing, Q. Wang, L. Zhou, J. Appl. Polym. Sci. **130**(5), 3753–3758 (2013). https://doi.org/10.1002/app.39650
42. H.W. Park, T. Kim, J. Huh, M. Kang, J.E. Lee, H. Yoon, ACS Nano **6**(9), 7624–7633 (2012). https://doi.org/10.1021/nn3033425
43. W. Chen, R.B. Rakhi, H.N. Alshareef, J. Phys. Chem. C **117**(29), 15009–15019 (2013). https://doi.org/10.1021/jp405300p
44. V. Sharma, A. Sahoo, Y. Sharma, P. Mohanty, RSC Adv. **5**(57), 45749–45754 (2015). https://doi.org/10.1039/C5RA03016A
45. H. Guan, L.Z. Fan, H. Zhang, X. Qu, Electrochim. Acta **56**(2), 964–968 (2010). https://doi.org/10.1016/j.electacta.2010.09.078
46. C. Bian, A. Yu, Synth. Met. **160**(13), 1579–1583 (2010). https://doi.org/10.1016/j.synthmet.2010.04.019
47. M.R. Anderson, B.R. Mattes, H. Reiss, R.B. Kaner, Science **252**(5011), 1412–1415 (1991). https://doi.org/10.1126/science.252.5011.1412%JScience
48. S. Cho, K.H. Shin, J. Jang, A.C.S. Appl, Mater. Inter. **5**(18), 9186–9193 (2013). https://doi.org/10.1021/am402702y
49. S. Chaudhari, Y. Sharma, P.S. Archana, R. Jose, S. Ramakrishna, S. Mhaisalkar, M. Srinivasan, J. Appl. Polym. Sci. **129**(4), 1660–1668 (2013). https://doi.org/10.1002/app.38859
50. K. Wang, J. Huang, Z. Wei, J. Phys. Chemi. C **114**(17), 8062–8067 (2010). https://doi.org/10.1021/jp9113255
51. X. Chen, H. Lin, P. Chen, G. Guan, J. Deng, H. Peng, Adv. Mater. **26**(26), (2014). https://doi.org/10.1002/adma.201400842
52. K. Wang, X. Zhang, C. Li, H. Zhang, X. Sun, N. Xu, Y. Ma, J. Mater. Chem. A. **2**(46), (2014). https://doi.org/10.1039/c4ta04924a
53. Y. Huang, H. Li, Z. Wang, M. Zhu, Z. Pei, Q. Xue, Y. Huang, C. Zhi, Nano Energy. **22**, 422–438 (2016). https://doi.org/10.1016/j.nanoen.2016.02.047
54. Y. Shi, L. Pan, B. Liu, Y. Wang, Y. Cui, Z. Bao, G. Yu, J. Mater. Chem. A **2**(17), 6086–6091 (2014). https://doi.org/10.1039/c4ta00484a
55. S. Sahoo, S. Dhibar, C.K. Das, Polymer Lett. **6**, 965–974 (2012). https://doi.org/10.3144/expresspolymlett.2012.102
56. Y. Wang, F. Chen, Z. Liu, Z. Tang, Q. Yang, Y. Zhao, S. Du, Q. Chen, C. Zhi. Angew. Chem. **58**(44), 15707–15711 (2019). https://doi.org/10.1002/anie.201908985
57. Y. Huang, M. Zhu, W. Meng, Z. Pei, C. Liu, H. Hu, C. Zhi, ACS Nano **9**(6), 6242–6251 (2015). https://doi.org/10.1021/acsnano.5b01602
58. J.C. Carlberg, O. Inganäs, J. Electrochem. Society. **144**(4), L61–L64 (1997). https://doi.org/10.1149/1.1837553
59. K.S. Ryu, Y.G. Lee, Y.S. Hong, Y.J. Park, X. Wu, K.M. Kim, M.G. Kang, N.G Park, S.H. Chang, Electrochimica. Acta. **50**(2), 843–847 (2004). https://doi.org/10.1016/j.electacta.2004.02.055
60. T. Cheng, Y.Z. Zhang, J.D. Zhang, W.Y. Lai, W. Huang, J. Mater. Chem. A. **4**(27), 10493–10499 (2016). https://doi.org/10.1039/C6TA03537J
61. M. Mastragostino, C. Arbizzani, F. Soavi, Solid State Ion. **148**(3), 493–498 (2002). https://doi.org/10.1016/S0167-2738(02)00093-0

62. S. Nejati, T.E. Minford, Y.Y. Smolin, K.K.S. Lau, ACS Nano **8**(6), 5413–5422 (2014). https://doi.org/10.1021/nn500007c
63. S.R.P. Gnanakan, M. Rajasekhar, A. Subramania, JIJES. **4**(9), 1289–1301 (2009)
64. K. Balakrishnan, M. Kumar, S. Angaiah, Adv. Mat. Res. 938, (2014). https://doi.org/10.4028/www.scientific.net/AMR.938.151
65. C. Fu, H. Zhou, R. Liu, Z. Huang, J. Chen, Y. Kuang, Mater. Chem. Phy. **132**(2), 596–600 (2012). https://doi.org/10.1016/j.matchemphys.2011.11.074
66. B.H. Patil, S.J. Patil, C.D. Lokhande, **26**(9), 2023–2032 (2014). https://doi.org/10.1002/elan.201400284
67. R. Hou, M. Miao, Q. Wang, T. Yue, H. Liu, H.S. Park, K. Qi, B.Y. Xia, Adv. Energy Mater. **10**(1), (2019). https://doi.org/10.1002/aenm.201901892
68. D. Antiohos, G. Folkes, P. Sherrell, S. Ashraf, G.G. Wallace, P. Aitchison, A.T. Harris, J. Chen, A.I. Minett, J. Mater. Chem. **21**(40), 15987–15994 (2011). https://doi.org/10.1039/C1JM12986D
69. S. Dhibar, P. Bhattacharya, G. Hatui, C.K. Das, J. Alloys. Compounds. **625**, 64–75 (2015). https://doi.org/10.1016/j.jallcom.2014.11.108
70. S. Sahoo, S. Dhibar, G. Hatui, P. Bhattacharya, C.K. Das, Polymer **54**(3), 1033–1042 (2013). https://doi.org/10.1016/j.polymer.2012.12.042
71. L. Mao, K. Zhang, H.S. On Chan, J. Wu, J. Mater. Chem. **22**(1), 80–85 (2012). https://doi.org/10.1039/c1jm12869h
72. S. Sahoo, P. Bhattacharya, S. Dhibar, G. Hatui, T. Das, C.K. Das, J. Nanosci. Nanotechnol. **15**(9), 6931–6941 (2015). https://doi.org/10.1166/jnn.2015.10540
73. S. Dhibar, P. Bhattacharya, G. Hatui, S. Sahoo, C.K. Das, A.C.S. Sustain, Chem. Eng. **2**(5), 1114–1127 (2014). https://doi.org/10.1021/sc5000072
74. S. Dhibar, P. Bhattacharya, D. Ghosh, G. Hatui, C.K. Das, Indus. Eng. Chem. Res. **53**(33), 13030–13045 (2014). https://doi.org/10.1021/ie501407k

Chapter 5
Electrode Material Selection for Supercapacitors

Alka Jangid, Kapil Dev Verma, Prerna Sinha, and Kamal K. Kar

Abstract Supercapacitors and conventional capacitors follow a similar charge storage mechanism. However, they differ from each other, as supercapacitors store a large amount of charge due to the extremely high surface area of the conducting electrodes. The charge storage mechanism of supercapacitors is based on either formation of an electric double layer, where adsorption and desorption of ions take place, or the Faradaic process, where reversible redox reaction occurs. The efficiency of a supercapacitor depends upon its components like electrodes, electrolyte, separator and current collectors. Among all the components, the electrode plays a major role to store a large amount of charge at its surface. So, characteristics of the electrode such as porosity, surface morphology, surface area, electrical conductivity are taken into account for selecting suitable electrode material for supercapacitor. Activated carbon, CNT, graphene, carbon aerogel, metal compounds, conducting polymers, and their composites are among various materials, which have been commonly used as electrodes and discussed in detail to select the best material with the concept of various material indices using Ashby's chart in this review.

Alka Jangid, Kapil Dev Verma, Prerna Sinha—Equal Contribution.

A. Jangid · K. D. Verma · P. Sinha · K. K. Kar (✉)
Advanced Nanoengineering Materials Laboratory, Materials Science Programme, Indian Institute of Technology Kanpur, Kanpur 208016, India
e-mail: kamalkk@iitk.ac.in

A. Jangid
e-mail: alkajangid33@gmail.com

K. D. Verma
e-mail: deo.kapildev@gmail.com

P. Sinha
e-mail: findingprerna09@gmail.com

K. K. Kar
Advanced Nanoengineering Materials Laboratory, Department of Mechanical Engineering, Indian Institute of Technology Kanpur, Kanpur 208016, India

K. K. Kar (ed.), *Handbook of Nanocomposite Supercapacitor Materials III*,
Springer Series in Materials Science 313,
https://doi.org/10.1007/978-3-030-68364-1_5

5.1 Introduction

The supercapacitor is an electrochemical energy storage device. It is also known as ultracapacitor or electrochemical capacitor because of supercapacitor stores energy in form of the electric double layer at the electrode–electrolyte interface, which delivers a high capacitance value of the device [1]. The demand for energy storage devices has increased over years due to the advancement in portable electronics, the advancement of hybrid electric vehicles, and the need for large energy storage systems to store intermittent renewable energy. For this purpose, other energy devices such as batteries and fuel cells are also used [2]. Although, the high cost and poor energy density of supercapacitors are still a barrier to be utilized widely. Currently, a lot of research has been carried out to improve efficiency and reduce the cost of these devices. When compared to other energy storage devices, supercapacitors bridge the gap of energy density, and power density between conventional capacitors and batteries [3]. It shows the following advantages and disadvantages:

Advantages: Supercapacitors depict 10–100 times more life than batteries. It has a higher power density than batteries and a quick charge–discharge process [4].

Disadvantages: Limited energy density and gradual voltage loss with discharge.

The components of supercapacitors are designed similar to batteries. Main components of supercapacitors and their respective functions are listed as follows:

(a) Electrodes: Electrolyte ions accumulate over the surface of the electrode in the charge storage process. High surface area, high electrical conductivity, and porous electrode material are used for supercapacitor. The powdered electrode material is coated on the current collector using a binder. Activated carbon, metal oxides, and conducting polymers are some examples of electrode materials [5].
(b) Electrolyte: Electrolyte provides free ions for formation of electric double layer, which gets stored as charge at electrolyte-electrode interface. High ionic conductivity and its stability at the high potential window are two important characteristics of electrolyte material for supercapacitor application. Ion size of electrolyte also plays an important role in capacitance of supercapacitor [6].
(c) Separator: Separator is used between two electrodes as a sandwich to avoid the short circuit. It is in form of a thin porous membrane that allows the transfer of ions. A separator should be an electrically insulator, mechanically strong, and chemically stable [7].
(d) Current Collectors: Current collector provides mechanical support to the electrode material, which collects current from the electrode and transfers to the outer circuit. The current collector should possess high conductivity to provide a resistance-free transfer of electrons. However, the selection of the collector is based on electrolyte and electrode material. Metal foil, indium tin oxide, Ni foam, and carbon-based materials are common examples of collector materials [8].

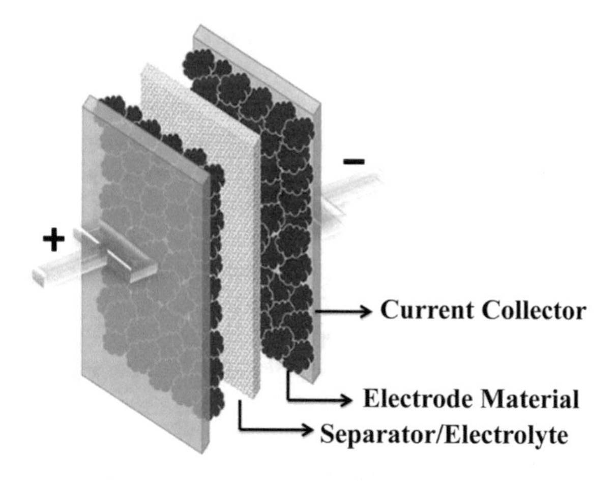

Fig. 5.1 Assembly of supercapacitor components (redrawn and reprinted with permission from [10])

(e) Binder: Electrode is coated over the current collector using active material and polymeric binder. Binder is mixed in powder form electrode material for its better adhesion with the current collector. Nafion, conductive polymers, PTFE, and natural cellulose are used as binders in supercapacitors [9].

Figure 5.1 shows the assembly of supercapacitor using current collector, electrode material, electrolyte, and separator. The electrode material is deposited over the current collector, and the electrolyte is soaked in a separator [10].

Electrolyte and electrodes are active components while binder, separator, and current collectors are passive components of the supercapacitor. Choice of electrode materials highly affects capacitance and cost of a supercapacitor. Electrodes should be mechanically stable, chemically inert, hierarchically porous, highly conductive, and stable at high temperatures. Electrode material should be compatible with electrolyte and current collector. According to the electrode material selection, supercapacitors are classified as electrochemical double layer capacitors (EDLCs), pseudocapacitors, and hybrid capacitors. EDLCs store charge by the adsorption of electrolyte ions at the electrode surface. In this vein, the internal pores of the electrode must be accessible for an electrolyte for high charge storage. So, carbon-based material such as graphene, activated carbon, and carbon nanotubes-based electrodes are desirable materials for EDLCs. Metal oxides and conductive polymers are used for making electrodes for pseudocapacitors as they undergo reversible redox reaction under potential to store charge. Hybrid capacitors use asymmetric configuration or composite electrodes, which utilizes both the charge storage mechanism to provide high-performance electrode material.

5.2 Functions of Electrodes

(1) Increase or decrease in high energy electrons in the conduction band of electrode material due to the application of voltage leads to the formation of an electric double layer. This imbalance in charge causes the movement of ions in the electrolyte and formation of an electric double layer on the electrolyte/electrode interface [5].
(2) High surface area provided by porous electrodes leads to increased capacitance compared to dielectric capacitors [5].
(3) Electrode materials also store charge by Faradaic process, where redox-active sites present at the electrode surface undergo electron transfer mechanism.

5.3 Characteristics Required for Electrodes

The electrode capacitance mainly depends upon the specific surface area. Apart from the high surface area, the electrode should possess a wide pore size distribution for the resistance-free transportation of ions. For supercapacitor application, mainly three categories of electrode materials are used: (1) carbon-based materials, (2) conductive polymers, and (3) metal oxides. All these materials show different electrical, chemical, and structural properties affecting the ultimate performance and lifetime of the supercapacitor. Figure 5.2 shows capacitance performances of carbon, conducting polymers, metal oxides, and composite materials-based electrodes. Conducting polymer and metal oxides show higher specific capacitance than carbon-based electrode material because of the Faradaic charge storage mechanism [11].

5.3.1 Conductivity

A balance between conductivity and porosity is necessary for good capacitance and rate performance. Carbon nanotubes and graphene show very high conductivities than activated carbon. Graphene is an excellent electrode material because it has high electrical conductivity, flexibility, and high specific surface area. Metal oxides show better energy storage efficacy compared to carbon materials but suffer from poor conductivity so, efforts have been made to make metal oxides more conductive. RuO_2 has variable oxidation states and high conductivity so it is widely studied for application in supercapacitors. Metal nitrides show better electrical conductivity than metal oxides [10]. Table 5.1 shows density and electrical conductivity for different types of carbon electrodes along with their specific capacitance using aqueous and organic electrolytes. CNTs and graphene show very high electrical conductivity among other electrode materials. Carbon-based electrode materials display higher specific capacitance with aqueous electrolytes in comparison with the organic electrolytes.

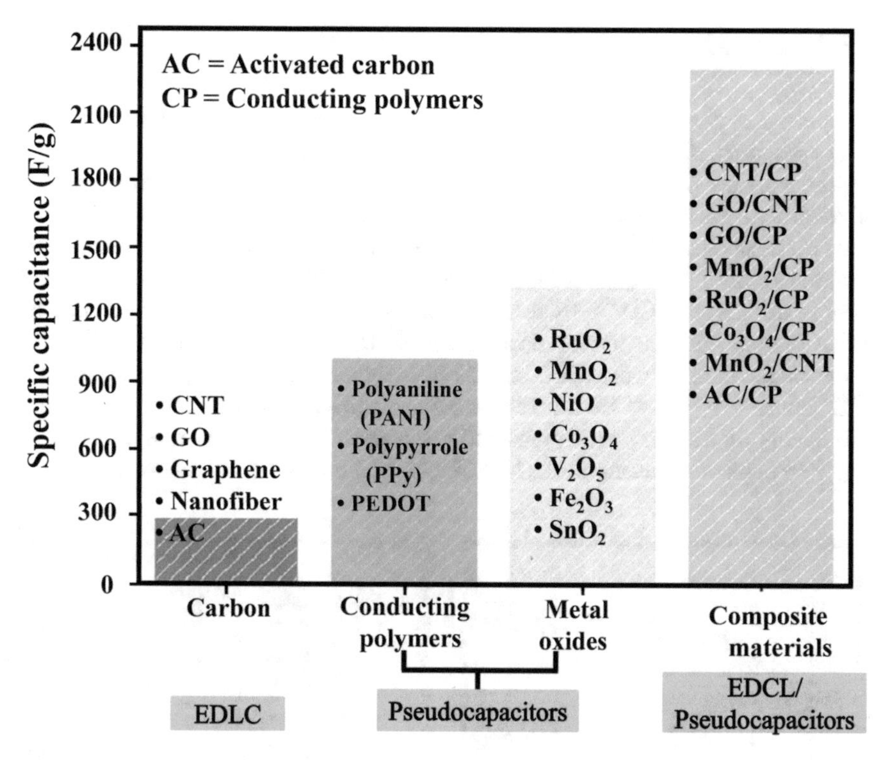

Fig. 5.2 Specific capacitance of electrode materials for different supercapacitors (redrawn and reprinted with permission from [11])

Table 5.1 Conductivity and specific capacitance of carbon-based electrodes [12]

Electrode material	Density (g cm^{-3})	Conductivity (S cm^{-1})	Specific capacitance (F g^{-1})	
			Aqueous electrolyte	Organic electrolyte
CNTs	0.6	10^4–10^5	50–100	<60
Graphene	>1	10^6	100–205	80–110
Activated carbon	0.4–0.7	0.1–1	<200	<100
Templated porous carbon	0.5–1	0.3–10	120–350	60–140
Functionalized porous carbon	0.5–0.9	>300	150–300	100–150
Activated carbon fibers	0.3–0.8	5–10	120–370	80–200
Carbon aerogels	0.5–0.7	1–10	100–125	<80

5.3.2 Porosity

The pore size close to ion size leads to maximum capacitance for electric double layer capacitors. When pore size is reduced below ion size, capacitance drops significantly. So, the optimum pore size for maximum capacitance depends on the type of electrolyte used in the supercapacitor. The size of cation and anion is comparable for most electrolytes so, the pore size required is almost same for cathode and anode [13]. In Fig. 5.3a–c, all four samples show a different type of structures and specific surface area. EDTA-1K refers to porous carbon electrodes manufactured by EDTA (ethylenediaminetetraacetic acid) and KOH, where 1K, 2K, 3K, and 4K refer to the molar ratio between KOH and EDTA. Hence, EDTA-2K has a molar ratio of 2 between KOH and EDTA. Figure 5.3a, b shows a hierarchical porous structure having micropores, mesopores and macropores. The pore size distribution in Fig. 5.3b depicts two sharp peaks at 0.58 and 1.16 nm in the micropore size range.

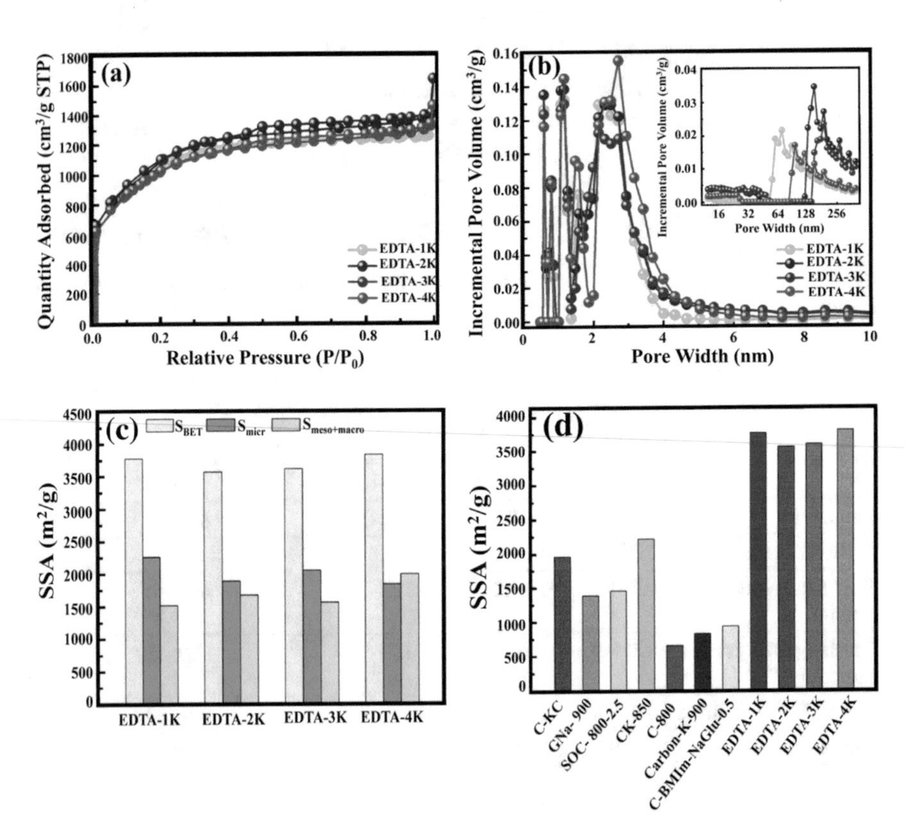

Fig. 5.3 **a** N_2 adsorption/desorption isotherms, **b** density function theory pore size distribution of samples, **c** relative specific surface area for BET, micropores, and meso + macropores, and **d** Comparison of SSA of samples with other materials (redrawn and reprinted with permission from [14])

Figure 5.3d shows a comparison of the surface of EDTA-KOH-based activated carbon with other porous carbon materials [14]. The high surface area and hierarchical pore size distribution can accommodate a higher amount of ions that enhance the charge accumulation and storage capacity.

5.3.3 Mechanical Strength

Mainly conducting polymers and metal oxide show poor mechanical strength as compared to carbon materials. This is mainly due to the structural changes that occur during the charge storage process. Carbon materials do not undergo any structural changes during the charge and discharge process due to the physical process, hence show high mechanical strength. In composite electrodes, the addition of activated carbon fibers and activated carbon particles increases bonding of reinforcement with matrix and improves mechanical strength and increase of surface area improves capacitance of electric double layer supercapacitor [5]. Figure 5.4a, b demonstrates

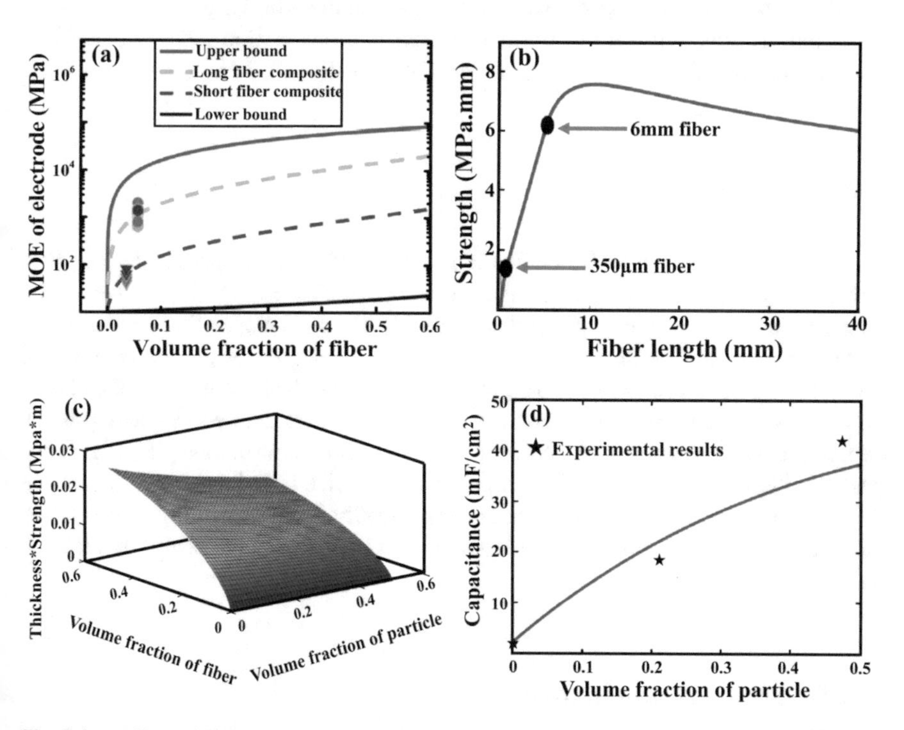

Fig. 5.4 **a** Effect of fiber volume fraction on the modulus of elasticity of electrode, **b** dependence of fiber length on strength of composite electrode, **c** effect of volume fraction of reinforcement on strength, and **d** variation in capacitance of device with particle volume fraction (redrawn and reprinted with permission from [15])

long fiber composite having a high modulus of elasticity and strength in comparison with the short fiber composite. Figure 5.4 c, d shows that activated carbon fibers (ACFs) provide better mechanical properties but less capacitance than activated carbon partials (ACPs) [15].

5.3.4 Surface Morphology

Recently, nanostructured materials are used for efficient electrodes for supercapacitors. These electrodes possess good electrochemical properties, easy charge transfer, and high storage in the charge–discharge process due to the high specific surface area. Porous carbon, graphene, conductive polymers, metal oxides, and metal hydroxides can be fabricated in nanostructured electrodes. Figure 5.5 shows SEM and TEM images of mesoporous carbon and PANI/mesoporous carbon composite materials. Figure 5.5d demonstrates the ordered growth of polyaniline on the mesoporous carbon surface. Nanostructured electrodes provide easy access to electrolyte ions leading to high specific capacitance and high rate charge–discharge ability. The high electronic conductivity of polyaniline and mesoporous carbon reduces equivalent series resistance that is the cause of high power density [16, 17].

5.3.5 Wettability

The specific capacitance of a device is not only affected by specific surface area but also by the inner surface of micropores or mesopores, which gets wetted by electrolyte. The wettability of electrodes highly affects the specific capacitance of a supercapacitor. The presence of mesopores, high degree of graphitization, and functional groups consisting of oxygen enhance the wettability and ease the charge transportation pathway. For improving wettability, pore structure must be regulated, which is done by chemical and physical methods of activating carbon [18]. Figure 5.6 shows a higher effective specific surface area of the electrode for electrolyte wetting leads to high specific capacitance and energy density. Poor wettability causes lower utilization of surface area that increases the dead volume of the electrode material [19].

5.3.6 Thermal Conductivity

High current densities can lead to the generation of a large amount of heat during charge–discharge cycles of the supercapacitor. This can cause a reduction in the capacity of the device or serious issues like an explosion in the device. Dissipation of this heat is necessary, which requires high thermal conductivity of electrode

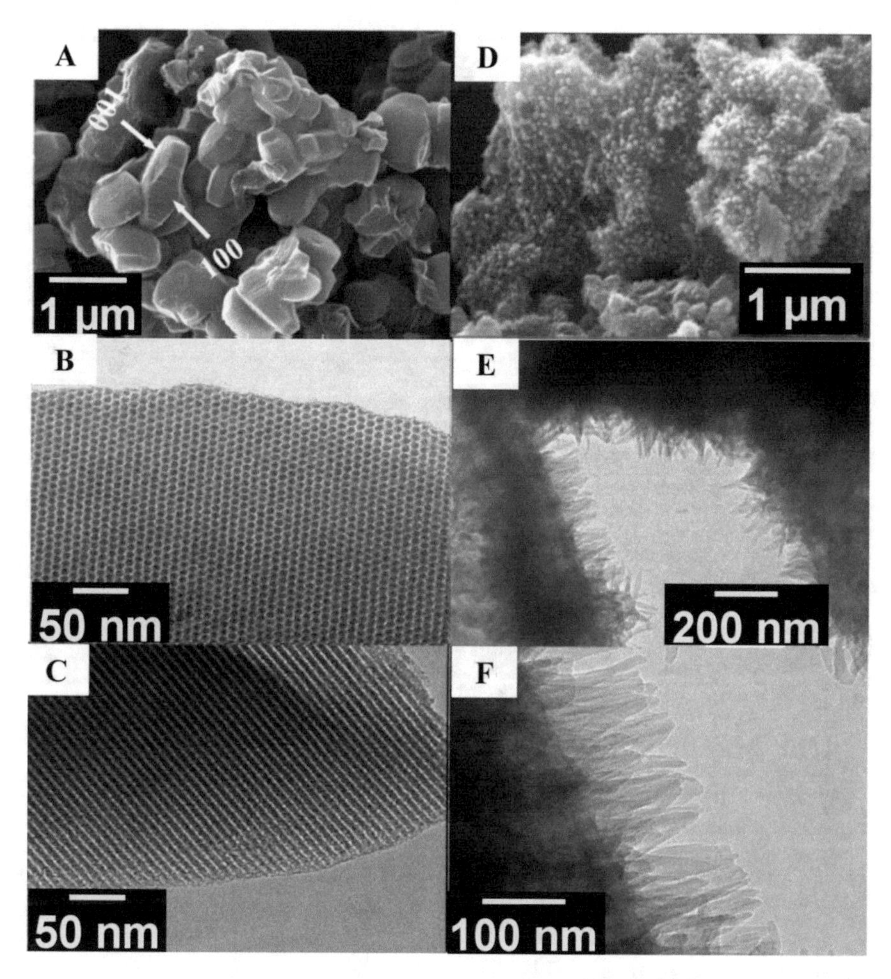

Fig. 5.5 **a** SEM micrograph of mesoporous carbon, **b**, **c** TEM image of mesoporous carbon from [001] and [100] directions, **d** SEM micrograph of PANI/mesoporous carbon composite, and **e**, **f** TEM images of PANI/mesoporous carbon at different magnifications (redrawn and reprinted with permission from [17])

materials. High thermal conductivity of electrodes reduces chances of temperature rise or temperature gradients across the supercapacitor cell. Conventional electrodes show low thermal conductivity. Graphene-MnO_2 film electrodes show high thermal conductivity up to 613.5 W m^{-1} K^{-1} while MnO_2 slurry electrode shows thermal conductivity 1.1 W m^{-1} K^{-1} [20]. Figure 5.7a, b shows the temperature profile of the cylindrical cell from the center to the outer surface with a high ohmic resistance (1 Ω cm^2) and low ohmic resistance (0.1 Ω cm^2). AC stands for electrode composed of activated carbon and OLC for electrode made of onion-like carbon. The electrode is mixed with a PTFA polymer binder [21].

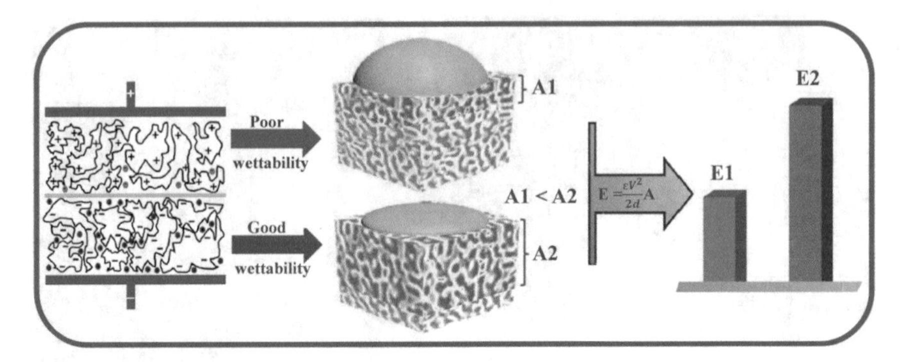

Fig. 5.6 Effect of wettability of the electrode surface on the energy density of supercapacitor. Here A_1 and A_2 denote effective specific surface area and E_1 and E_2 denote energy density (redrawn and reprinted with permission from [19])

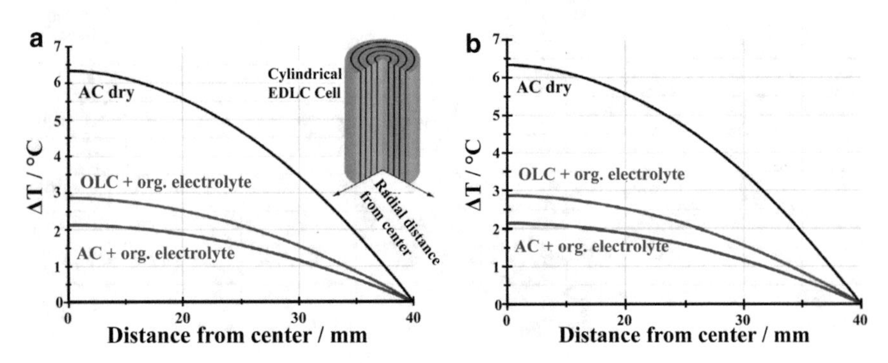

Fig. 5.7 Temperature profile in cylindrical EDLC cell **a** 1 Ω cm^2 and **b** 0.1 Ω cm^2 cell resistance (redrawn and reprinted with permission from [21])

5.3.7 Cycling Stability

Transition metal oxide and conducting polymers electrodes show good energy density but low conductivity and cycle stability. So, the development of metal oxides having good chemical and thermal conductivity is necessary, which can deliver good cycling stability. One such approach utilizes molecular cross-linking inside the tungsten oxide network that creates hybrid material and enhances chemical and cycling stability [22]. Activated carbon with high acidic group content and narrow distribution of pore size shows good cycling stability in alkaline electrolyte. Figure 5.8 shows the cycle life of activated carbon (YP-50F and YP-80F, Kuraray Europe GmbH, Vantaa, Finland) at different current densities. In Fig. 5.8a, the current density is changing from 60 to 900 mA g^{-1} after 1000 cycles. As the current density is increasing, capacitance is decreasing. At high current density (900 mA g^{-1}), YP-80F shows better performance than YP-50F). Further, long-term tests of these activated

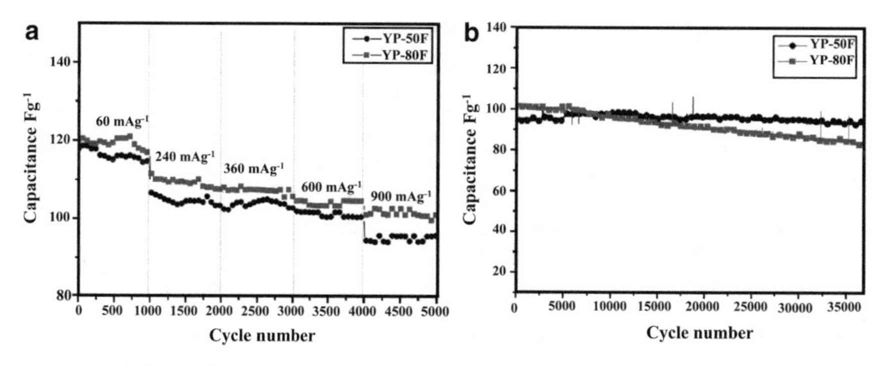

Fig. 5.8 **a** Effect of cycle number at discharge capacitance at different current loads and **b** Long-term test at 900 mA g^{-1}. YP-50F and YP-80F are two activated carbon products of Kuraray Europe GmbH (redrawn and reprinted with permission from [23])

carbons are done at 900 mA g^{-1}. 7500 cycles are the threshold after which YP-50F starts showing better performance because of the narrower pore size distribution [23].

5.3.8 *Cost*

Commonly used activated carbon is a very cheap electrode material, which costs around 1050 Rs./kg while nanostructure carbon materials such as CNTs and graphene are very costly. CNTs have high conductivity and cost around 3500 Rs./kg. Scientists showed biomass waste such as cotton waste and tamarind seeds can also be converted into cost-effective carbon electrodes, which significantly reduces the material cost. Activated carbon made from such waste shows a high surface area and high stability. Paper industry waste lignin can also be combined in manganese dioxide electrodes to improve conductivity. Except for RuO_2, metal oxide, and conducting polymers is low-cost materials, which have been explored as electrode material.

5.4 Performance of Materials Used as Electrodes

Materials for electrodes are chosen based on their properties and compatibility with other components. Currently, the following materials are commonly used for an electrode for supercapacitor application.

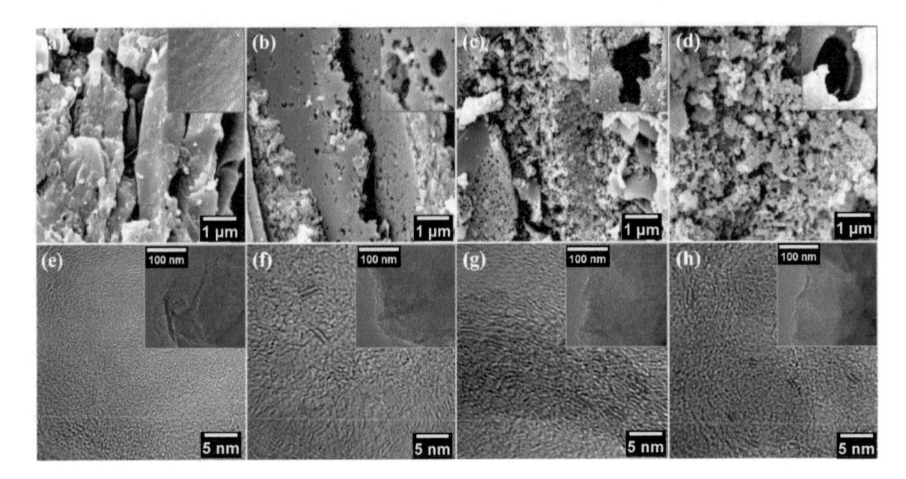

Fig. 5.9 **a**, **e** SEM and TEM image of coal-based carbon **b**, **f** SEM and TEM image of H_2O-based activated carbon **c**, **g** SEM and TEM image of KOH-based AC **d**, **h** SEM and TEM image of H_2O assisted KOH-based AC [18]

5.4.1 Activated Carbon

Activated carbon is a widely used electrode material due to the large specific surface area and good conductivity. It is cheaper compared to other carbon electrode materials (CNTs and graphene). Activated carbon is produced by carbon-rich materials such as wood, cotton waste, coconut shells, and coal by either chemical or thermal activation process [24]. In thermal activation, carbon is treated at high temperatures (>700 °C) in the presence of oxidizing gases. Chemical activation is carried out at low temperatures (<700 °C) in the presence of activating agents such as NaOH, $ZnCl_2$, and KOH [25]. Figure 5.9 shows SEM and TEM images of activated carbon using H_2O and KOH as activators. Figure 5.9a, e shows coal-based carbon, which has negligible pore size distribution. Figure 5.9b, f demonstrates H_2O-based activated carbon. H_2O-based activated carbon has more pores in comparison with coal-based carbon. Further, KOH-based activated carbon as shown in Fig. 5.9c, d, g, h shows more uniform pore size distribution.

5.4.2 CNT

CNT-based electrodes show high thermal conductivity, electrical conductivity, and mechanical strength. These have a lower specific surface area than activated carbon so have less energy density [26, 27]. When CNTs are added to activated carbon, capacitance reduces due to the less SSA but ESR reduces because of the high conductivity of CNTs. Aligned CNT structure shows better power performance compared to entangled CNT structure due to the less efficient ion transportation [28, 29]. Energy density

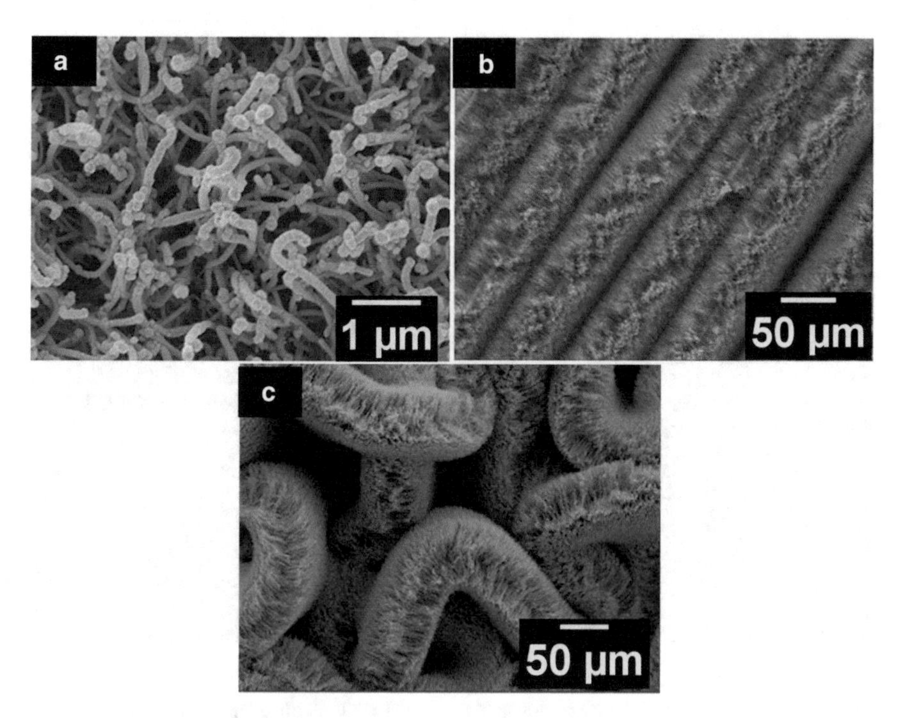

Fig. 5.10 **a** SEM micrograph of top view of Au-CNT forest, **b** SEM micrograph of Au-CNT forest with 300% uniaxial pre-strain, and C SEM micrograph of Au-CNT forest with 200% × 200% biaxial pre-strain (redrawn and reprinted with permission from [30])

can be improved by KOH activation of CNTs, which increases SSA. Figure 5.10 shows an SEM image of biaxial stretchable Au-CNT forest electrode using vacuum sputtering coater [30].

5.4.3 Graphene

Graphene is a lightweight nanostructured carbon material. It depicts a high surface area with an abundance of active sites for redox reactions, which makes graphene suitable for supercapacitor electrodes. Electrodes made of graphene show high specific capacitance. However, the purity of graphene highly depends upon the synthesis process [31, 32]. During synthesis, restacking leads to the decreased experimental surface area as compared to the theoretical value. Also, the hydrophobic nature of graphene surface limits access of aqueous electrolyte causing a decrease in capacitance. The surface area of a single graphene sheet is around 2630 m^2 g^{-1}. The high electrical conductivity of graphene allows a high thickness of electrodes without the need for conductive fillers [33]. Figure 5.11a, b shows a cross-sectional view of

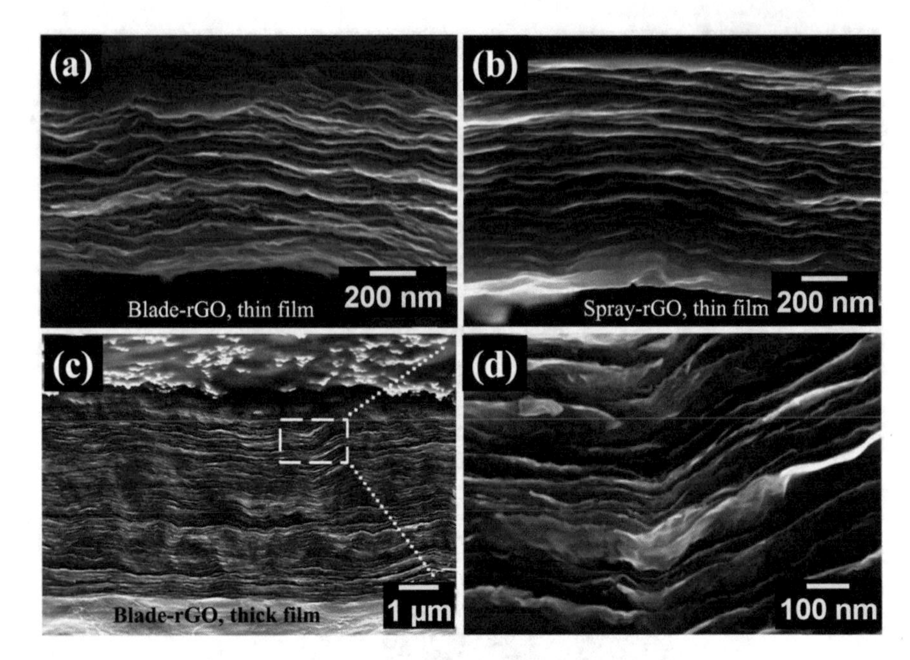

Fig. 5.11 FESEM images of **a** blade reduced graphene oxide thin film, **b** spray-rGO thin film, and **c**, **d** 4 μm thick film of blade-rGO (redrawn and reprinted with permission from [34])

FESEM image of blade-rGO, and spray-rGO thin films, respectively. Figure 5.11c, d shows the cross-sectional view of thick films, respectively [34].

5.4.4 Carbon Aerogels

Carbon aerogels are highly porous and ultra-light materials. These are formed by the sol–gel process and pyrolysis of organic aerogel afterward. In carbon aerogel electrodes, the addition of a binder is not required. Electrodes made from carbon aerogels provide very high surface area for charge storage and high conductivity resulting in supercapacitor having low equivalent series resistance, high energy density, high capacitance, and low leakage current. These devices also have a long life over a wide temperature range. Figure 5.12 shows an SEM image of BA-teta derived organic aerogel (BA-teta: bisphenol A and triethylenetetramine) and carbon aerogel [35].

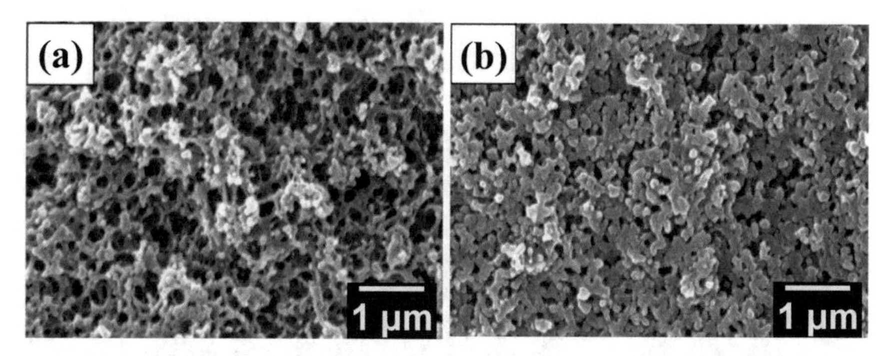

Fig. 5.12 SEM images of **a** organic aerogel and **b** carbon aerogel (redrawn and reprinted with permission from [35])

5.4.5 Carbon Quantum Dots

Quantum dots are the nanometer-sized quasi-spherical particles. Electronic properties can be tuned according to their size. These can be easily synthesized in large quantities, and their properties can be controlled by modifying their surface that makes quantum dots a good choice for graphene composites. Plasma treatment, arc discharge, hydrothermal process, and chemical oxidation are techniques used for synthesizing carbon quantum dots from carbon precursors. Figure 5.13 shows an optical image of an interconnected sheet-like assembly of carbon quantum dots (CQDs) [36].

Fig. 5.13 **a** Optical image of carbon quantum dots assembly powder (inset of sponge-like carbon quantum dots, the scale is 100 μm) and **b** SEM micrograph of carbon quantum dots ($CQDs_{-800}$) (redrawn and reprinted with permission from [36])

Fig. 5.14 **a** SEM micrograph showing coarse-grained carbide (here TiC) derived carbon powder and **b** SEM image of higher magnification showing the single grain of titanium carbide derived carbon powder (redrawn and reprinted with permission from [38])

5.4.6 Carbide Derived Carbon

Carbide-derived carbon (CDC) are carbon materials made from carbide precursors. These precursors are TiC, Mo_2C, B_4C, VC, and SiC, which are transformed into pure carbon by either chemical process or physical process. Among various CDCs, TiC derived carbon shows the highest surface area. After processing, the final product consists of a nanoporous structure that increases the surface area [37]. CDC shows better performance than activated carbon. Coarse-grained CDC shows good electrochemical properties, high capacitance, and power density. Manufacturing cost is also lower for coarse-grained CDC than micropowder and nanopowder CDC. However, a high-temperature condition is required during synthesis. Figure 5.14a shows SEM micrograph image of coarse-grained carbide (here TiC) derived carbon powder. SEM image of higher magnification showing the single grain of titanium carbide-derived carbon powder is shown in Fig. 5.14b [38].

5.4.7 Anodized Steel

Anodization of steel develops nanostructure having high surface area, which can be used as an electrode in supercapacitor. As anodization time is increased cathodic current, anodic current, and capacitance increase with an increase in surface area. However, with further increase in anodization time, anodic and cathodic currents decrease with the surface area. This suggests that optimum temperature is required to obtain a high surface area of anodized steel. Researchers have anodized cold-rolled strip steel in 10M NaOH that results in a highly porous sponge-like structure. Figure 5.15 shows SEM micrographs of anodized (for 240 min) nanostructured steel [39].

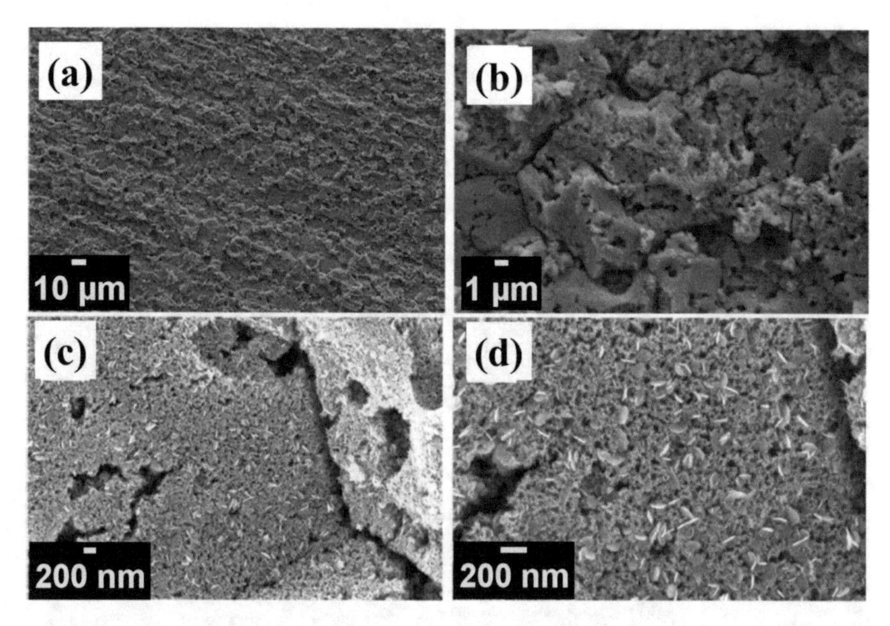

Fig. 5.15 SEM micrographs of anodized (for 240 min) nanostructured steel at **a** 1000×, **b** 10,000×, **c** 50,000×, and **d** 100,000× magnification (redrawn and reprinted with permission from [39])

5.4.8 Metal Oxides

Metal oxide electrodes have high energy storage performance as compared to carbon-based materials. These electrodes show good electrochemical performance. MnO_2, RuO_2, NiO, V_2O_5, SnO_2, and CoO are some transition metal oxides used for electrode materials [40, 41]. The multiple valence states of metal oxide undergo a redox reaction to store charge. However, poor electrical conductivity limits its performance and increases the resistance of the device. In this regard, different metals can be added in metal oxide to alter particle size and increase electrical conductivity. This addition can reduce the consumption of precious metals such as ruthenium and nickel. RuO_2 has better electrical conductivity, thermodynamic stability, chemical stability, and cyclic stability than other transition metal oxides. However, high cost and toxicity issues render its wide usage. Particles of RuO_2 also tend to agglomerate leading to the degradation in electrochemical properties [42]. Figure 5.16 shows HRSEM images of coating of Co_3O_4 on carbon fiber cloth [43]. Here, carbon cloth provides necessary conductivity to Co_3O_4, thereby improving electrode performance.

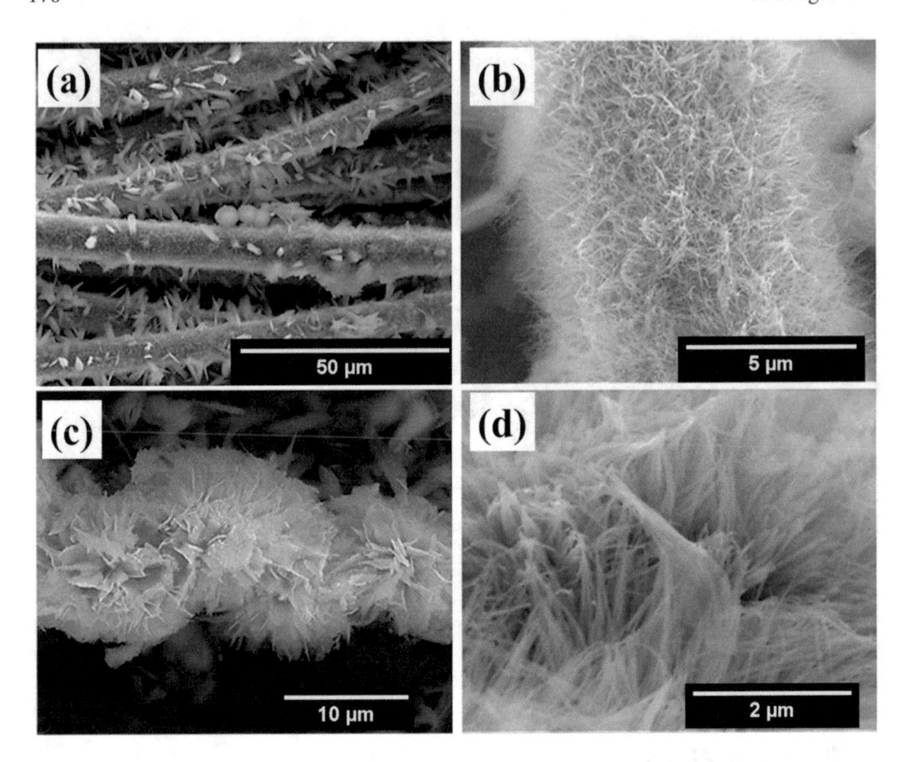

Fig. 5.16 HRSEM images showing nanowire/nanoflower surface morphology of Co_3O_4 on carbon fiber cloth at different magnifications [43]

5.4.9 Metal Nitrides

Metal nitrides show better electrical conductivity and enhanced power density than metal oxides. Among all metal nitrides, VN exhibit high electrical conductivity, large specific capacitance with variable oxidation states. These properties make VN suitable material for supercapacitor electrodes. The high surface area provides a large number of redox reaction sites. Mo_2N, Li_3N, and TiN are other metal nitrides which display good electrochemical properties. TiN shows high mechanical stability and less electrical resistivity. It has been widely studied for making anode in supercapacitors. Figure 5.17 shows SEM images of (a) chrysanthemum-like TiO_2 (b) chrysanthemum-like TiN sample. Figure 5.17c, d shows the TEM images of TiN sample [44].

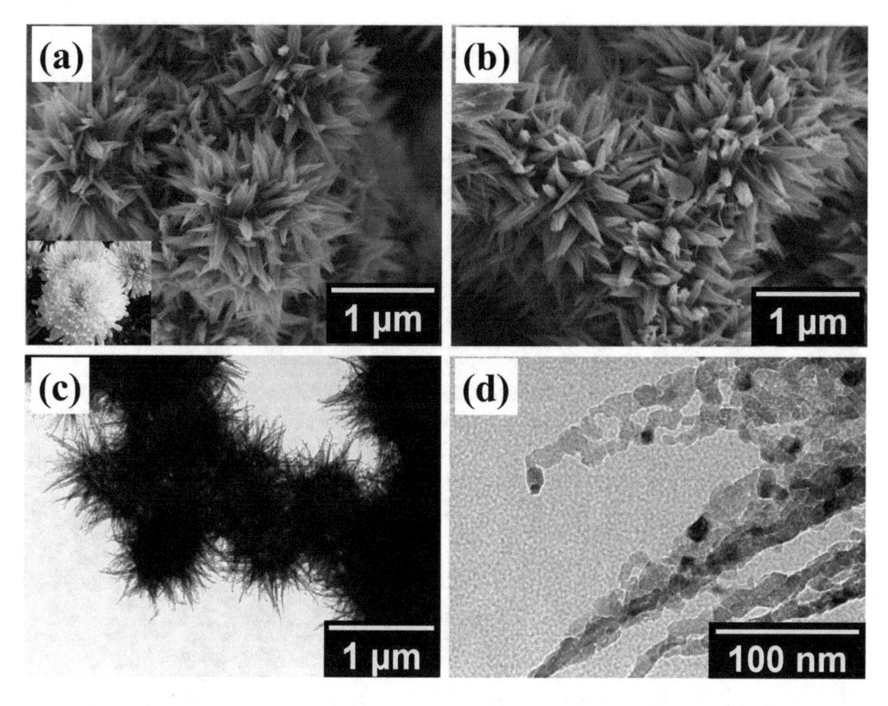

Fig. 5.17 SEM image of **a** chrysanthemum-like TiO_2, **b** chrysanthemum-like TiN, and **c, d** TEM image of chrysanthemum-like TiN at different magnifications (redrawn and reprinted with permission from [44])

5.4.10 Conducting Polymers

Conducting polymers are extensively used in making electrodes. Conducting polymers are low cost, easy fabrication, and high environmental stable material. It has high conductivity and can be operated in wide voltage window. These are of three types: (1) n-doped, e.g., polythiophene derivatives, (2) p-doped, e.g., polypyrrole (PPy), polyaniline (PANI), and (3) n-p-doped, e.g., polythiophene. PPy and PANI are widely used as pseudocapacitive electrodes due to the high capacity and electrical conductivity [53]. Conducting polymers are synthesized either by chemical or electrochemical polymerization. Chemical polymerization is relatively affordable. However, conducting polymers suffer from poor mechanical stability. During repeated charging and discharging, swelling and shrinking occur inside the polymeric chain that deteriorates the electrode performance. Since it shows extremely high electrochemical performance, conducting polymers are integrated with carbon-based material, which acts as a mechanical support and improves the conductivity. Figure 5.18 shows SEM images of different nanostructures of polypyrrole thin films [54].

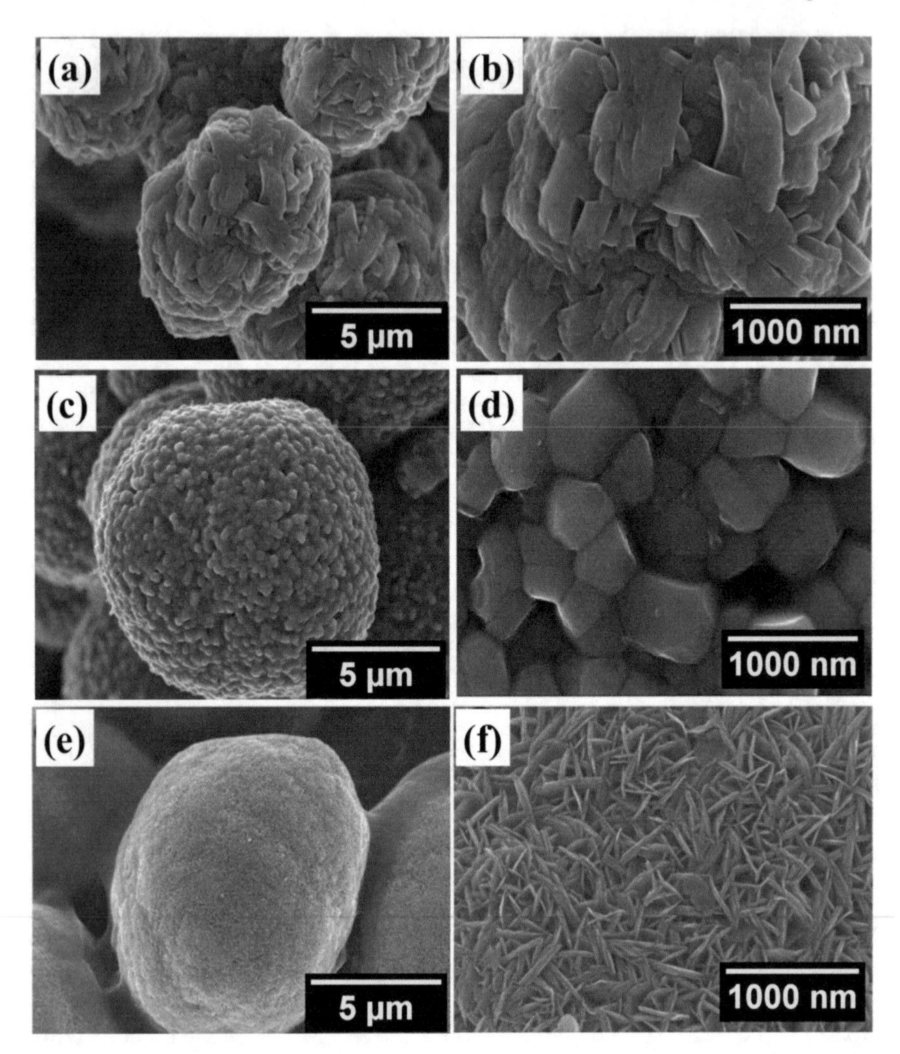

Fig. 5.18 SEM images of different nanostructures of polypyrrole thin films **a**, **b** nanobelts, **c**, **d** nanobricks, and **e**, **f** nanosheets (redrawn and reprinted with permission from [54])

5.4.11 Composite Materials

It is well known that both EDLCs and pseudocapacitive electrode materials suffer from inherent limitations. Carbon-based electrode shows high rate capability, high conductivity, and long cycle life but suffers from low capacitance, as charge storage process is limited till surface. On the other hand, pseudocapacitive electrodes such as metal oxide show high capacitance and high energy density but have poor conductivity and cycle life. Similarly, conducting polymers suffer from low mechanical strength, which starts degradation after a few cycles. In this, it is necessary to combine

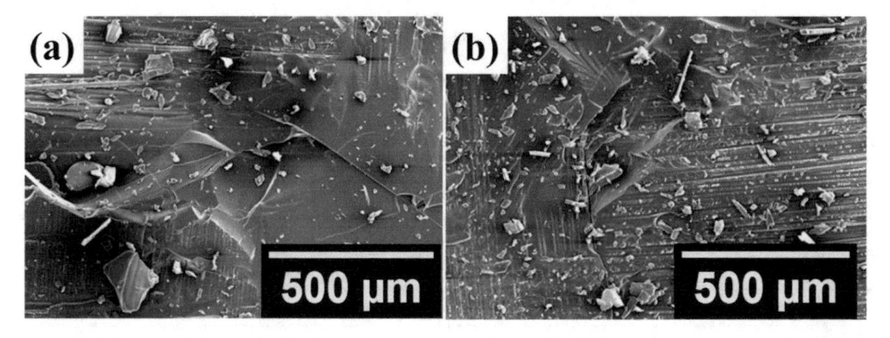

Fig. 5.19 Images showing fracture surface of **a** CF/MVR444 composite (high brittle fracture) and **b** ACF/MVR444 (enhanced adhesion in the presence of ACF). CF = carbon fiber, ACF=activated carbon fiber, and MVR444 = aerospace grade epoxy resin (redrawn and reprinted with permission from [52])

both charge storage processes, to obtain high-performance electrode material. The composite electrode aims to combine EDLC and pseudocapacitive material to provide synergistic electrochemical properties. In supercapacitors graphene is used as a composite with other carbon nanomaterials, conducting polymers and metal oxides [45, 46]. Formation of composite prevents restacking in rGO and improves capacitance by increasing surface area. In carbon nanostructure/metal oxide composite, carbon nanostructure makes charge transportation easy and provides physical support to matrix [47–50]. Metal oxide contributes to high specific capacitance and energy density in composite [51]. Figure 5.19 shows SEM images of carbon fiber and epoxy resin composite [52].

5.5 Electrode Materials Used in Commercial Supercapacitors

Carbon-based supercapacitors are currently dominating the commercial market. These materials show stability at a wide temperature range, long lifetime, high power density, and coulombic efficiency but have limited energy density (<10 Wh kg^{-1}) in organic electrolytes with a potential window of 2.5–2.8 V [55]. Table 5.2 shows different commercially available supercapacitors and their performances along with their respective electrode and electrolyte materials [57]. Activated carbon is used commercially for an EDLC device. However, its poor capacitance makes it less efficient. In order to obtain high capacitance, other carbon-based materials were introduced into EDLC electrodes such as CNTs, graphene, carbon fibers, and carbon aerogels. For further increasing capacitance, conductive polymers and metal oxides were used and pseudocapacitors came into the picture [56]. Figure 5.20 shows a Ragone plot of different types of asymmetric supercapacitor using $GrMnO_2$//$GrMoO_3$,

Table 5.2 Performance of commercial supercapacitors with different combination of electrode material and electrolytes [58]

Manufacturing company	Device name	Capacitance ($F\ g^{-1}$)	Electrode material	Electrolyte
Asahi Glass	EDLC	500–2000	Carbon	non-aqueous
AVX	Bestcap	0.022–0.56	Carbon/polymer	aqueous
Cap-XX	Supercapacitor	0.09–2.8	Carbon	non-aqueous
Cooper	PowerStor	0.47–50	Aerogel	non-aqueous
ELNA	Dynacap	0.333–100	Carbon	non-aqueous
Epcos	Ultracapacitor	5–5000	Carbon	non-aqueous
Evans	Capattery	0.01–1.5	Carbon	aqueous
Maxwell	Boostcap/PowerCache	1.8–2600	Carbon	non-aqueous
NEC	Supercapacitor	0.01–6.5	Carbon	Aqueous, organic
Nippon Chemi-Con	DLCAP	300–3000	Carbon	non-aqueous
Ness	NessCap	3–5000	Carbon	organic
Matsushita/Panasonic	Gold capacitor	0.1–2500	Carbon	organic
Tavrima/ECOND	Supercapacitor	0.13–160	Carbon	aqueous

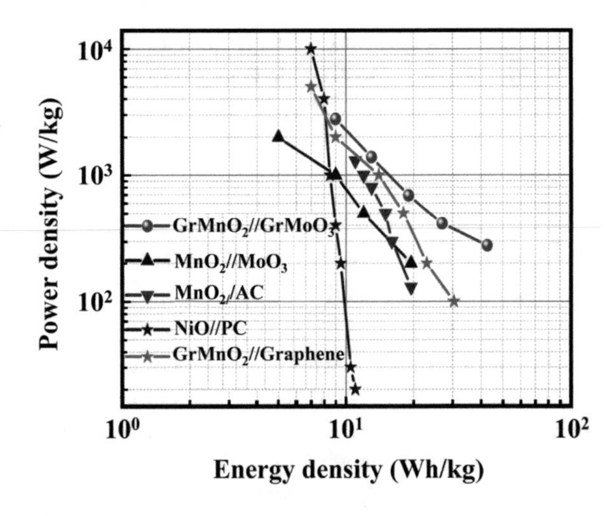

Fig. 5.20 Ragone plot showing energy and power density of asymmetric supercapacitors of different combination of electrode materials at 2.0 V (redrawn and reprinted with permission from [57])

$MnO_2//MoO_3$, MnO_2/AC, NiO//Porous carbon, and Reduced graphene oxide/MnO_2 nanowires//Reduced graphene oxide ($GrMnO_2$//Graphene) asymmetric electrodes.

5.6 Emerging Electrode Materials

- **Piezoelectric Materials**: Piezoelectric bio-based materials such as peptides have a large surface area. These can be used for making electrodes for supercapacitors. On the application of force or pressure, the material undergoes charge redistribution that induces potential. Di-phenylalanine-based peptide nanotubes are used for this purpose. This material shows high cycle stability and areal capacitance [59].
- **Flexible Materials**: Flexible materials are proposed for making stretchable and wearable electronic devices. The design of a flexible supercapacitor does not require independent current collectors and binders. Highly conductive and flexible electrodes also work as a current collector. Plastic is used for the packaging of flexible supercapacitors. The electrolyte can be liquid electrolyte or solid electrolyte [60]. Figure 5.21 shows the difference in the design of conventional and flexible supercapacitor. A flexible supercapacitor consists of a flexible electrode, leakage-proof electrolyte, and plastic package [61].

Flexible electrodes can be classified into three categories:

1. **Paper/sheet-like flexible electrode**: These are further classified based on the use of substrate into free-standing paper-like electrodes and flexible substrate supported. Examples are free-standing composite film, free-standing CNT film, free-standing graphene film, and flexible plastic substrate supported, etc.
2. **Fiber-like flexible electrode**: These are easy to form into smart clothes due to the flexibility and ease of weaving. Fiber-like flexible electrodes are also used for making microsensors for various fields such as aerospace, environment analysis, and military applications. Examples of fiber-like flexible electrodes are metal fiber supported, plastic fiber supported, CNT yarn supported, graphene fiber supported, etc.
3. **3-D porous flexible electrode**: These are porous and thick electrodes showing high energy and power density. The thickness of these electrodes is higher than 100 μm, which makes these electrodes flexible without forming an internal

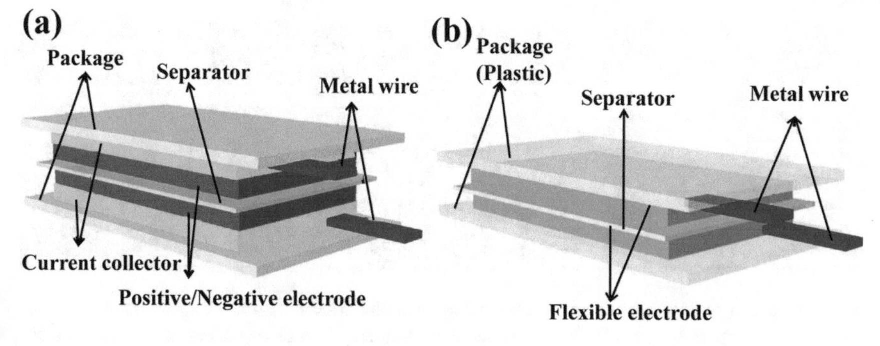

Fig. 5.21 Difference in the design of **a** conventional supercapacitor and **b** flexible supercapacitor (redrawn and reprinted with permission from 61])

crack inside the structure. Some of the examples are cotton cloth-based, carbon aerogel-based, nickel foam-based, polymeric foam-based, etc. [62].

- **Transparent Materials**: The transparent electrodes can integrate solar cells with a supercapacitor. Transparent electrodes can easily absorb solar energy that can be stored as chemical energy inside the device. The use of transparent electrodes can make supercapacitor as energy harvesting and storage devices. Au@MnO_2 nanomesh electrodes show 84.7% transparency. These electrodes are highly stretchable up to 160% strain (Fig. 5.22a, b). Gold nanowires, silver nanowires, and polypyrrole can be used for transparent electrodes but Au@MnO_2 nanomesh is suitable for making wearable electronic devices due to the stretchability. Grain boundary lithography is used for manufacturing Au nanomesh, a thin layer of MnO_2 is then formed on this mesh by electrodeposition. High thickness in the MnO_2 layer can induce stiffness and loss of conductivity upon stretching. Figure 5.22c shows a SEM image of Au nanomesh and Au@MnO_2 nanomesh. Figure 5.22d shows a TEM image of Au@MnO_2. It is clearly showing the boundary between Au mesh and Au@MnO_2 nanomesh [63].
- **MXenes**: MXenes are 2D inorganic compounds. These are carbides, carbonitrides, or nitrides of transition metals. First MXenes reported was $Ti_3C_2T_x$ in 2011. Here, T represents functional groups such as O, F, Cl, and OH. These have

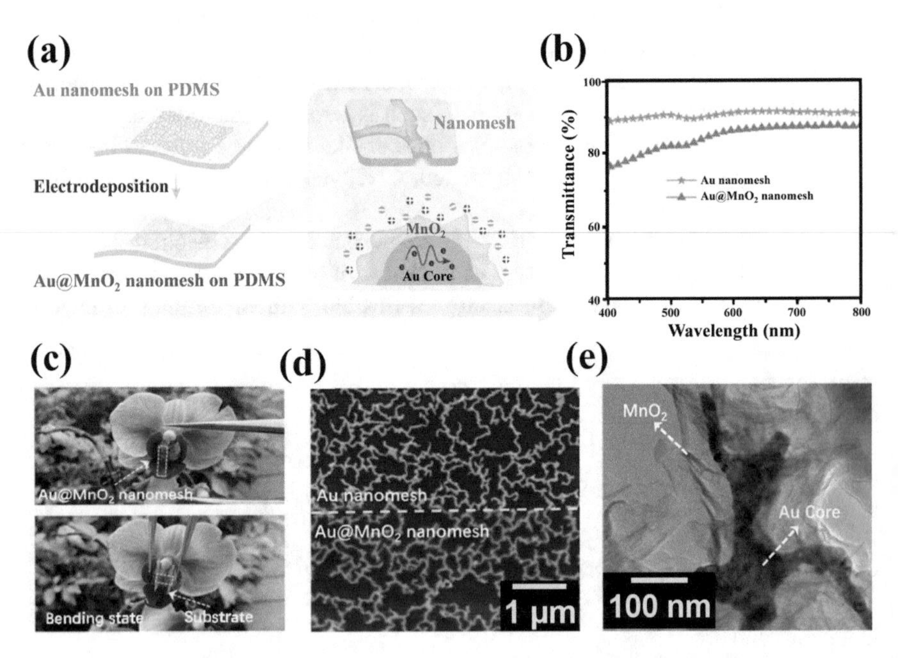

Fig. 5.22 **a** fabrication steps of Au@MnO_2 nanomesh on PDMS substrate, **b** spectrum of transmittance of Au nanomesh and Au@MnO_2 nanomesh, **c** optical images of normal and bending states of Au@MnO_2 transparent nanomesh, **d** SEM images of Au nanomesh and Au@MnO_2 nanomesh, and **e** TEM images of MnO_2 flake on Au core (redrawn and reprinted with permission from [63])

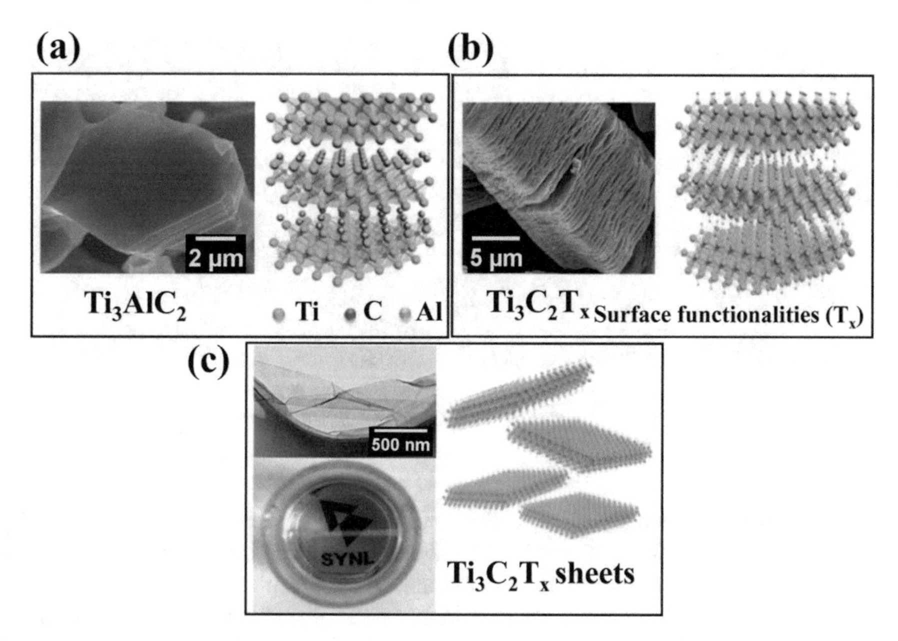

Fig. 5.23 $Ti_3C_2T_x$ (MXene) is produced by selective etching of Ti_3AlC_2 shown in fig. **a**, multilayered and single layer MXene structures shown in **b** and **c** respectively [64] (redrawn and reprinted with permission from [64, 65])

good electronic conductivity, ease of processing, good surface chemistry, and hydrophilic nature, which makes MXenes suitable for supercapacitor electrodes. Etching agents containing fluoride ions such as HF and HCl/LiF are used for producing MXenes using a selective etching technique (Fig. 5.23). But this technique is hazardous so, alternative methods are required for making these materials [64].

Black Phosphorous (BP): BP shows a high surface area and mechanical strength but due to its 2D structure, fewer active sites are present, which hinders smooth carrier transportation. So, the 3D sponge structure is formed by inhibiting the restacking of BP nanosheets. Black phosphorous sponge when used as electrode materials for supercapacitors shows higher capacitance than BP nanosheets and bulk BP crystals. Ultrathin nanosheets of BP provide a very high surface-to-volume ratio and hence deliver high charge mobility in BP sponge. Solid-state supercapacitor using BP sponge electrodes shows high stability and high specific capacitance. Figure 5.24 shows SEM images of black phosphorus sponge [66].

Metal–Organic Frameworks (MOFs): These are frameworks of compounds of metal ions and organic molecules. Research in metal–organic frameworks is growing fast, and around 20,000 MOFs have been found in just 20 years. These materials show high permeability due to the 3D porous structure and high surface area that is suitable for electrodes. However, MOFs have low intrinsic conductivity. So composites are formed to enhance conductivity. MOFs composite electrodes show flexibility, high

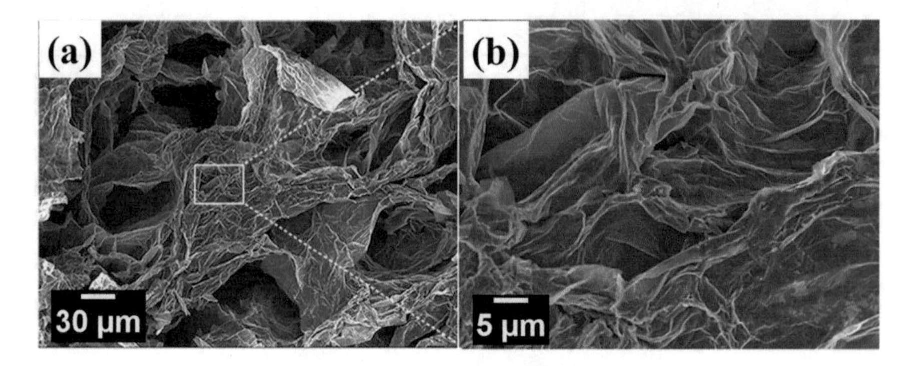

Fig. 5.24 SEM micrographs of black phosphorus sponge at different magnifications (redrawn and reprinted with permission from [66])

energy, and power density [67]. In this vein, Fig. 5.25, ZIF-67 is a cobalt-based MOF crystal. Internal resistance in ZIF-67 can be reduced by interweaving PANI with MOF on carbon cloth fibers without altering the original porous structure of MOF. This process leads to an increase in rate performance and capacitance of the supercapacitor [68].

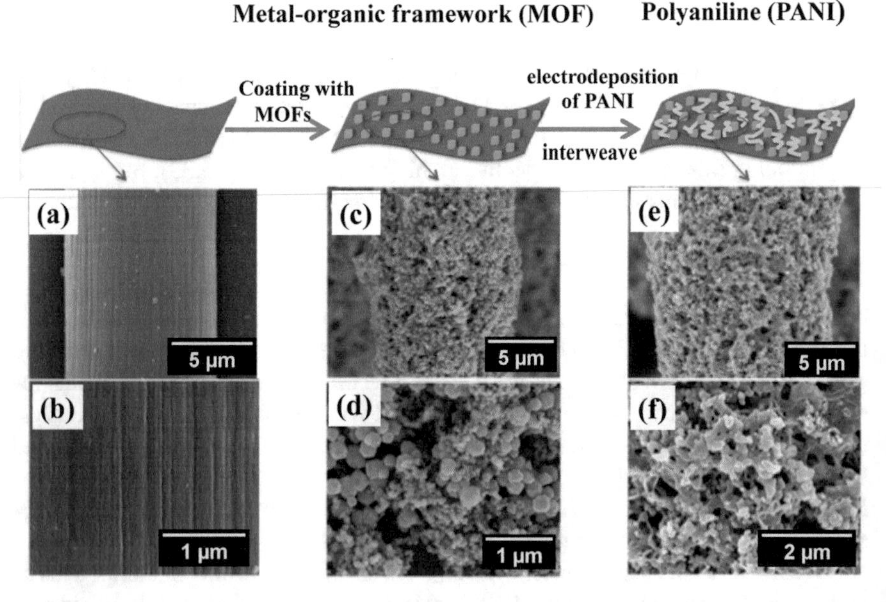

Fig. 5.25 **a** Steps for fabricating PANI-ZIF-67-CC electrode, **b** SEM images of carbon cloth fibers at different magnifications, **c**, **d** ZIF-67 coating on CCF and **e**, **f** SEM micrographs showing polymerized aniline and ZIF-67 coating (redrawn and reprinted with permission from [68])

Polyoxometalates (POM): PMOs are molecular clusters, which consist of oxygen and early transition metals such as group 5 (V and Ta) and group 6 (W and Mo) metals. These metals are in their highest oxidation state. PMOs are nanometer-sized molecular oxides containing a large number of metal atoms per molecule. The structure and size of POMs can be easily tuned in order to achieve the required properties. These materials show high ionic conductivity and are suitable for making redox-active electrodes. POMs act as electron reservoirs that makes them desirable for supercapacitor electrode materials. For making hybrid materials, nanostructured carbon materials or conducting polymers are used as a mechanical support for POMs due to their high surface area and conductivity. The hybrid electrode derived from PMOs shows excellent electrochemical properties. Figure 5.26a,b demonstrates SEM micrographs of rGO and rGO-PMo_{12} hybrid electrode materials (phosphomolybdic acid ($H_3PMo_{12}O_{40}$) (PMo_{12} in short). It can be seen in Fig. 5.26b that even after PMo_{12} cluster deposition over the rGO, rGO has an open 3D porous structure. EDS mapping in Fig. 5.26c clearly depicts the uniform distribution of PMo_{12} onto rGO nanosheet. For more insight into PMo_{12} distribution onto the rGO surface, TEM analysis is shown in Fig. 5.26d, e, f. Small black dots are representing the PMo_{12} particles [69].

5.7 Methods of Fabricating Electrodes

For obtaining high capacitance and cycle life, advanced fabrication methods should be used for making composite electrodes. Some of the fabrication methods used for supercapacitor electrodes are discussed as follows:

- **Chemical Bath Deposition**: In this process, thin films are formed on the substrate when it is put inside the chemical bath at low temperature. It is an inexpensive and simple method. Electrical conductivity is not required for substrate so, polymers can also be used. The process is used to form oxides and conductive polymer electrodes. Composite electrodes of low ESR and tunable morphology can be made using chemical bath deposition. However, this process is time-consuming [70].
- **Chemical Precipitation:** In this technique, the formation of solid precipitate in solution occurs due to the precipitation reaction. The precipitate can be formed by either converting substance into insoluble precipitate or by modifying solvent composition to decrease solubility. Various composites such as nickel oxide/carbon and MnO_2/CNTs can be fabricated by this method [70].
- **Chemical Vapor Deposition**: In this method, the substrate is exposed to suitable precursors in a reaction chamber at a high temperature. The precursor in reaction with the substrate surface forms a thin film. Carbon nanofibers and CNTs are formed by CVD using methane, acetylene, or other carbon precursors. Plasma enhanced CVD can also be used as an energy source at a lower temperature as compared to CVD to obtain dense and aligned CNTs [70].

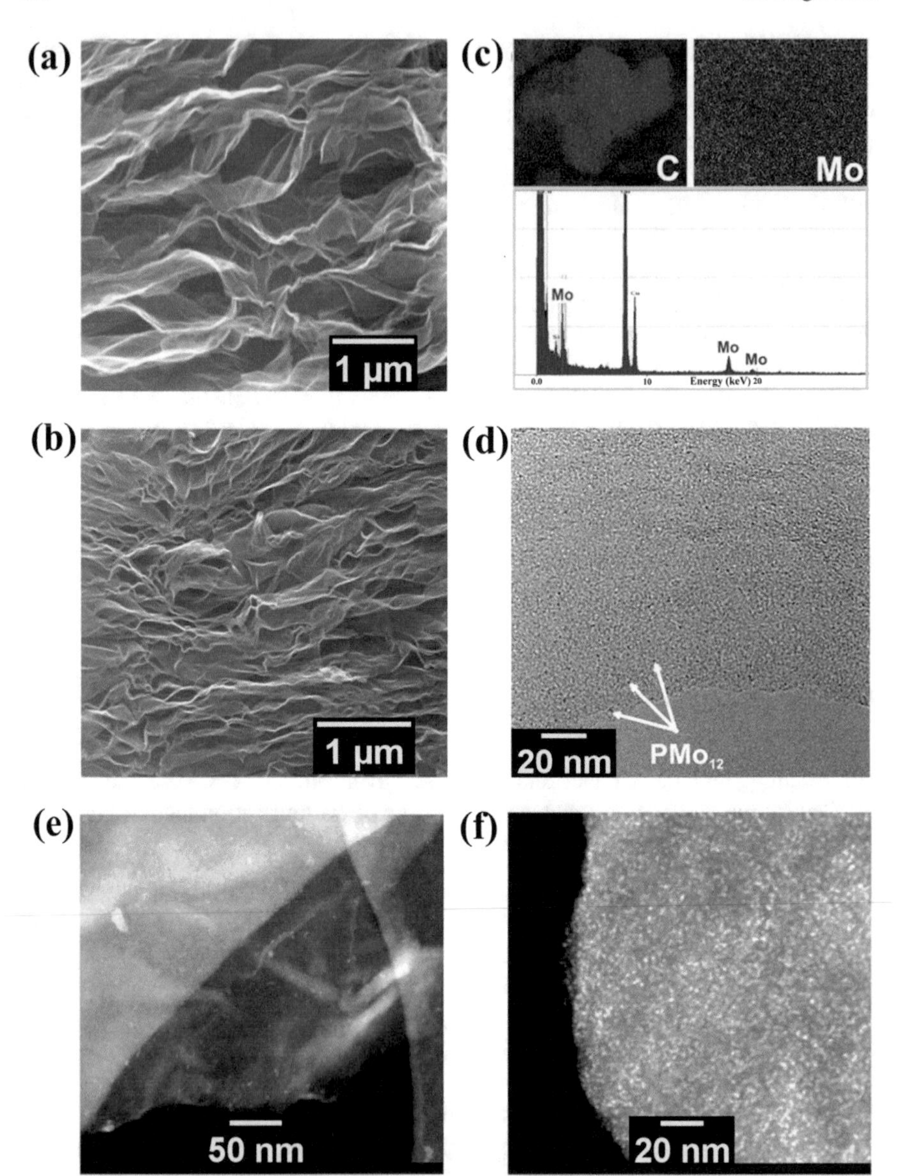

Fig. 5.26 SEM image of **a** rGO, **b** rGO-PMo_{12} hybrid, **c** EDS for rGO-PMo_{12} hybrid, **d** HRTEM image of rGO-PMo_{12} hybrid, **e** STEM image of rGO, and **f** STEM image of rGO-PMo_{12} hybrid (here PMo_{12} = phosphomolybdate) (redrawn and reprinted with permission from [69])

- **Electrochemical Deposition**: It is of various forms such as electrodeposition, electroless deposition, or electrophoretic deposition. In electrodeposition, the electric current is used to form a coating on the substrate, which is used as a cathode. Electroless deposition is used for the non-conducting substrate, where a reducing agent is used instead of an electric current [70].
- **Printing Technique**: Spray painting directly or through a template, casting and inkjet printing techniques fall under this category. The material can be deposited on any surface such as polymer or paper and very large surfaces. The solvent is mixed with material to be printed and sprayed using an inkjet printer. Then, the solvent is vaporized that leaves behind the desired material, for example, the CNT network in fabricating supercapacitor electrode [70].
- **Sol–gel Method**: Sol–gel is derived from solution gelling. The composition and structural properties of thin coating generated can be easily controlled in this process, and equipment required is also less making it an inexpensive and simple method. Carbon aerogels, carbon-ruthenium xerogels, manganese oxide, tin oxide among other materials used for electrodes are fabricated using the sol–gel method. Table 5.3 shows different electrode material fabrication techniques with their advantages and disadvantages [70].

Table 5.3 Methods of manufacturing different materials for electrodes and their advantages and disadvantages [70]

Methods	Materials	Advantages	**Disadvantages**
Electrochemical deposition method	Metal oxides, conducting polymers, ruthenium oxide-carbon composites (by electroless deposition)	Mass production, low costs, precise control on film thickness and uniformity	Process set up, current or voltage required
Chemical bath deposition (CBD)	PANI, Ruthenium oxide	Simplicity, low temperature, inexpensive, large-area substrates	Limited flexibility, low material yield
Chemical vapour deposition (CVD)	Carbon materials (CNTs, nanofibers, and graphenes)	High material yield than CBD, good film uniformity	Expensive equipment and relatively high cost
Sol-gel	Carbon aerogels, carbon-ruthenium xerogels, SnO_2, MnO_2	Low costs, controllable film texture, homogeneity, and structural properties	Complicated process
Chemical precipitation	Nickel oxide/carbon, MnO_2/nanofibers, MnO_2/CNTs	Allows synthesis of composite electrode materials, efficient, easily implemented	Waste product generated

Table 5.4 Primary, secondary and tertiary objectives for the selection of suitable electrode material for supercapacitors [5]

Primary objective	Secondary objective	Tertiary objective
High specific surface area	High chemical stability	Low cost
High electrical conductivity	Wide working temperature range	Non-toxicity
High charge storage capacity	High thermal conductivity	Easy to fabricate
Light weight	High corrosion resistance	Easy availability
	Controllable pore size	

5.8 Electrode Material Selection

The electrode is a key component related to the performance of the supercapacitor. Electrode surface area, conductivity, pore size, and compatibility between the electrolyte and current collector affect the performance of the supercapacitor. Capacitance depends on the amount of charge accumulated on the electrode surface. High contact resistance between the electrode material and current collector reduces the power density of the supercapacitor. So, for high performance of supercapacitor requires a suitable electrode whose material selection is a necessary procedure.

5.8.1 Objectives for Electrode Material Selection

Objectives for electrode material selection are to select a material, which fulfills the requirements of high performance, compatibility, economical, environment friendly, and easily available. Table 5.4 shows the primary, secondary, and tertiary objectives of the electrode material that is responsible for efficient performance. Primary objectives are the most required properties of the electrode material. Secondary objectives are also essential but these properties do not directly affect the performance of supercapacitor cells. Tertiary objectives are concern about economic and environmental aspects. Table 5.5 shows different types of electrode materials used for supercapacitor with their properties and performances.

5.8.2 Screening Using Constrains

Primary objectives for the selection of electrode material are high surface area, high electrical conductivity, high capacitance, and lightweight. Figure 5.27 demonstrates

Table 5.5 Properties of different electrode materials for supercapacitor performance [5]

Electrode material			Specific surface area ($m^2\ g^{-1}$)	Density ($kg\ m^{-3}$)	Specific capacitance ($F\ g^{-1}$)	Electrolyte	References
Carbon materials	Activated carbon	Shrimp shell	156.4	2000–2100	206 F/g at 0.1 A/g	6 M KOH	[71]
		Rice straw	2646		242 F/g at 0.5 A/g	6 M KOH	[72]
		Oily sludge	2561		248.1 F/g at 0.5 A/g	6 M KOH	[73]
		Coconut shell	2000		250 F/g	1 M H_2SO_4	[74]
		Citrus peel	1167		110 F/g at 0.1 A/g	1 M $NaClO_4$	[75]
		Sunflower seed	2585		244 F/g at 0.25 A/g	6 M KOH	[76]
		Rotten carrot	1154.9		135.5 F/g at 2.2 A/g	6 M KOH	[77]
	Graphene		2630	2266	138 F/g at 10 A/g	Aqueous electrolyte	[78, 79]
	CNT		380–2005	1600	227 F/g at 5 A/g	H_2SO_4	[80]
Metal oxide	Ruthenium oxide (RuO_2)		68.6	6970	800 F/g at 1A/g	1 M H_2SO_4	[81]
	Manganese dioxide (MnO_2)		80–120	7200	72–201 F/g	Aqueous electrolyte	[81]
	Nickel oxide (NiO)		153.2–265	6810	138.6–982 F/g at 1 A/g	6 M KOH	[82]
	Iron oxides		75	5250	170 F/g at 2 mV/s	1 M Na_2SO_4	[83]
Composite	RuO_2/CNT		750	–	1340 F/g at 25 mV/s	0.1 M H_2SO_4	[84]
	$V_2O_5XH_2O$/CNT		–	–	910 F/g at 10 mV/s	1 M $LiClO_4$/PC	[85]
	NiO-Co_3O_4		23.3	–	801 F/g at 1A/g	3 M KOH	[86, 87]
Conducting polymer	Poly-aniline (PANI)/α MnO_2		–	–	626 F/g at 2A/g	1 M H_2SO_4	[88]
	Poly-aniline (PANI)		30.9–77.1	1245	532 F/g at 1.5 A/g	1 M H_2SO_4	[89, 90]
	Poly-pyrrole (PPY)		26.5	1480	480 F/g at 10 mV/s	1 M KCl	[91, 92]
	Polythiophenes (PTH)		40.7	1400	300 F/g at 5 mV/s	0.1 M $LiClO_4$/PC	[93, 94]

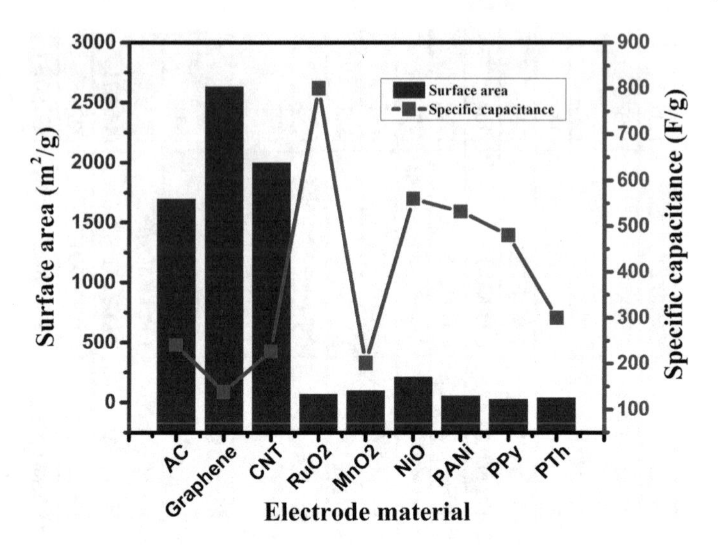

Fig. 5.27 Specific surface area and specific capacitance of commonly used electrode materials

specific surface area and specific capacitance of widely known electrode materials. Figure 5.28 demonstrates the relationship between surface area and density, as the surface area increases the density of the material decreases in carbon material and metal oxides while polymers exhibit less weight and low surface area.

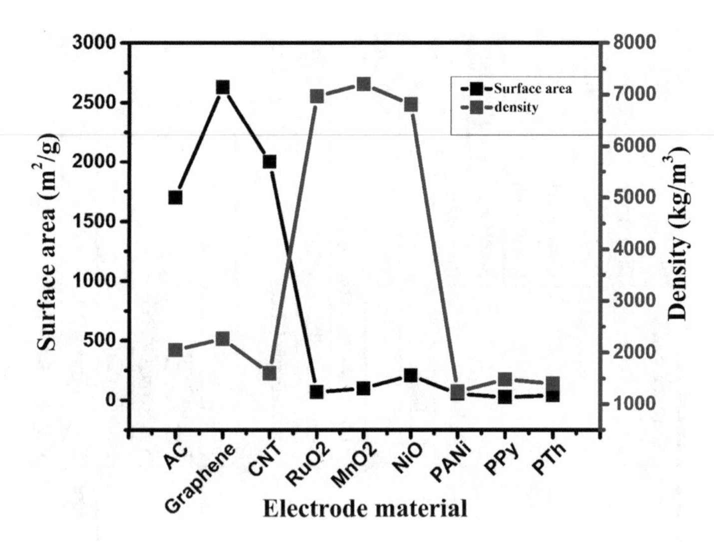

Fig. 5.28 Specific surface area and density of different electrode materials

5.8.3 Governing Equations

1. **Capacitance**: Capacitance is the ability of electrode material to store charge over its surface. The capacitance of the supercapacitor should be high. Capacitance is calculated from the following equation-

$$C = \frac{\in A}{d} \tag{5.1}$$

where $\in$ is the permittivity of dielectric material, A is the surface area of electrode, d is the effective thickness of double layer

2. **Electrical conductivity**: Electrical conductivity of an electrode is the ability to conduct electricity to the current collector. For high power density, the electrical conductivity of the electrode material should be high. The electrical conductivity of the electrode can be calculated from the following equation-

$$R = \frac{\rho l}{A_0} \tag{5.2}$$

$$\sigma = \frac{1}{\rho} \tag{5.3}$$

where σ is the electrical conductivity, ρ is the resistivity, A_0 is the cross-sectional area, and l is the length of the conducting element

3. **Mass**: Electrode material should be lightweight for high energy density. Mass of electrode material can be calculated from the following equation-

$$m = d \times L \times A_0 \tag{5.4}$$

where d is the density of material, L is the length of electrode, A_0 is the cross-sectional area of electrode.

5.8.4 Material Index

As the requirement is high electrical conductivity. So, Eqs. 5.2 and 5.3 lead to

$$R = \frac{l}{\sigma A_0} \tag{5.5}$$

Also, the requirement is low density, Eqs. 5.4 and 5.5 lead to

$$\frac{m}{(d \times l)} = \frac{l}{R\sigma} \tag{5.6}$$

Material index based on electrical conductivity and density (5.6) is represented as

$$M_1 = \frac{\sigma}{d} \tag{5.7}$$

The requirement is high capacitance, another material index based on capacitance and surface area (5.1) is represented as

$$M_2 = C \times A \tag{5.8}$$

Another material index based on surface area and density is written as

$$M_3 = \frac{A}{d} \tag{5.9}$$

Final material index, M_4 becomes (as per understanding), (or it can be in any other form like $M_3 = M_1 \times\pm\div M_2 \times\pm\div M_3$ or $a1M_1 \times\pm\div a_2M_2 \times\pm\div a_3M_3$, where a_1, a_2, and a_3 are any integers)

$$M_4 = M_1 \times M_2 \times M_3 \tag{5.10}$$

5.8.4.1 Electrical Conductivity and Density of Different Materials

The conductivity and density of some common electrode materials are given in Table 5.6.

Table 5.6 List of electrode materials with electrical conductivity and density

Sl. No.	Material	Conductivity (S m^{-1})	Density (kg m^{-3})	References
1	Activated carbon	0.142×10^5	2000–2100	–
2	Carbon nanotubes	10^6 –10^7	1600	[95]
3	Graphene	10^7 –10^8	2666	[95]
4	Ruthenium oxide	1.72×10^6	6970	[84]
5	Manganese dioxide	10^3	7210	–
6	Nickel oxide	5×10^6	6810	[96]
7	PANi	$0.1–10^{-10}$	1245	–
8	Polypyrrole	10^4	1480	–

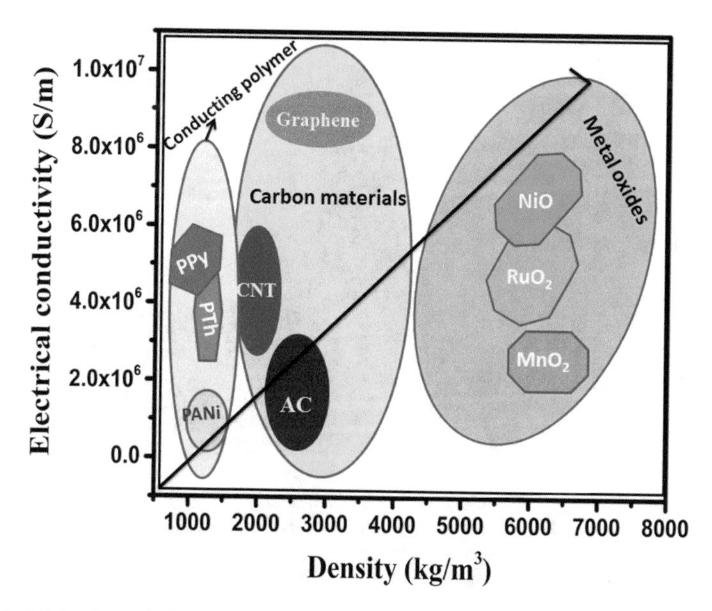

Fig. 5.29 Ashby chart of electrical conductivity and density

For supercapacitor application, the need for high electrical conductivity and low density is unavoidable. Hence, the materials just above the line drawn in the Ashby chart shown in Fig. 5.29 are the most suitable. Activated carbon, carbon nanotubes, graphene, nickel oxide, polyaniline, polypyrrole, and polythiophenes are suitable materials, which emerges in this screening.

5.8.4.2 Specific Capacitance and Surface Area of Different Electrode Materials

Material index of specific capacitance and surface area is given by 5.8

Taking log both sides of (5.8)

$$\log M_2 = \log C + \log A \quad (5.11)$$

or

$$\log C = -\log A + \log M_2 \quad (5.12)$$

In the Ashby graph (Fig. 5.30), specific capacitance and specific surface area have a negative slope.

For the high performance of the supercapacitor, there is a need for a highly effective surface area, which corresponds to high specific capacitance. For high material index, materials above the index line of Fig. 5.30 are suitable. Activated carbon,

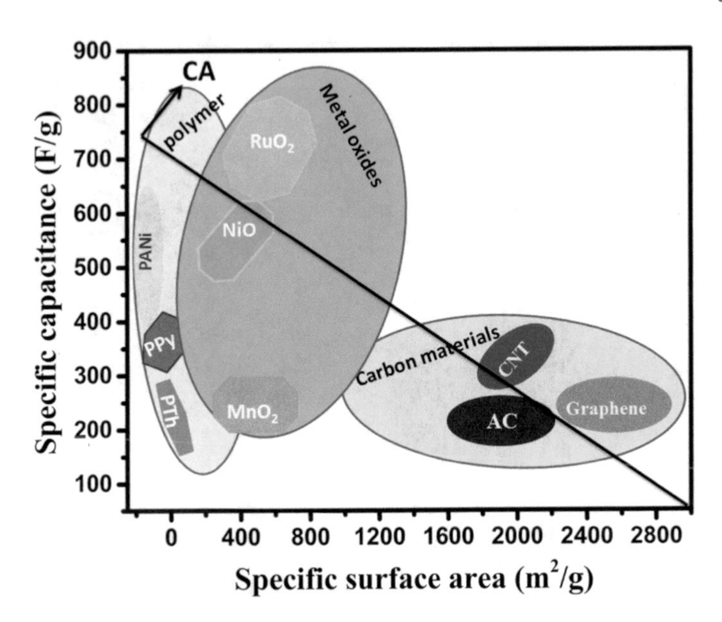

Fig. 5.30 Ashby chart of specific capacitance and specific surface area

carbon nanotubes, graphene, nickel oxide, ruthenium oxide, and polyaniline are appeared to be the suitable materials in this screening.

5.8.4.3 Surface Area and Density Relation of Different Electrode Materials

The material index of surface area and density for the supercapacitor is represented by (5.9)

Taking log in both sides of (5.9)

$$\log M_3 = \log A - \log d$$

or

$$\log A = \log d + \log M_3 \tag{5.13}$$

In the Ashby chart, Fig. 5.31 of specific surface area and density has a positive slope.

For the high performance of the supercapacitor, highly effective surface area and lightweight are desired. To achieve high material index materials above the index line as shown in Fig. 5.31 has been chosen. Activated carbon, carbon nanotubes, graphene, polyaniline (PANI), polypyrrole (PPY), and polythiophenes (PTH) are suitable materials as per the screening.

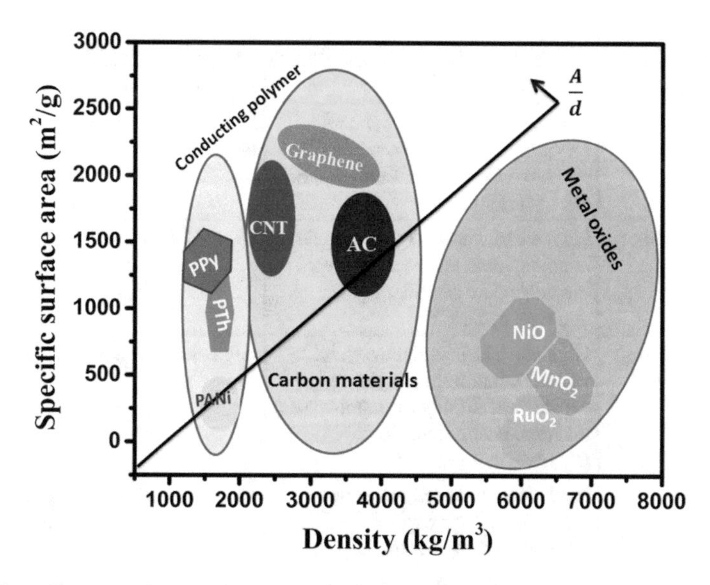

Fig. 5.31 Ashby chart of specific surface area and density

5.8.5 List of Material Index

Material indices (M_1, M_2, and M_3 and product of M_1, M_2, and M_3) are calculated and given in Table 5.7.

According to the high material index (M_4), suitable electrode materials along with merits and demerits are listed in Table 5.8.

Table 5.7 List of materials with material index value

Material	Material index M_1	Material index M_2	Material index M_3	$M_4 = M_1 \times M_2 \times M_3$
Graphene	3750.9	362.94×10^3	1.16	1579.1×10^6
Carbon nanotubes	625	454×10^3	1.25	354.68×10^6
Activated carbon	6.92	409.36×10^3	0.82	2.32×10^6
Nickel oxide	734.2	117.09×10^3	0.0307	2.63×10^6
Ruthenium oxide	246.7	54.88×10^3	9.84×10^{-3}	133.32×10^6
Manganese dioxide	0.138	21×10^3	0.0138	39.9
Polyaniline	8.03×10^{-5}	28×10^3	0.0433	9.73×10^{-2}
Polypyrrole	35.2	12.72×10^3	0.0179	80.14×10^3
Polythiophenes (PTh)	–	12.21×10^3	0.029	–

Table 5.8 Suitable electrode materials with their advantages and disadvantages

Electrode material	Properties and advantages	Disadvantages
Graphene	High surface area, high electrical conductivity, high thermal conductivity (3000 W/mK), High strength (130 GPa)	Complex fabrication mechanism, high cost
Carbon nanotubes	High surface area, high mechanical strength, good electrical property, high chemical stability, lightweight, and high cycle life	High fabrication cost
Ruthenium oxide	High electrical conductivity, wide voltage window, high specific capacitance, highly reversible redox reaction, pseudocapacitive mechanism	High density, small cycle life
Nickel oxide	Pseudocapacitive charge storage mechanism, high capacitance, low fabrication cost, low toxicity	Low potential window (0.5 V), small cycle life, high cost
Activated carbon	High surface area, low fabrication cost, high cycle life, non-toxic, chemical stability	High surface area but low electrolyte accessibility so limited capacitance, low energy and power density, low electrical conductivity
Polypyrrole	Easy synthesis, high capacitance, high cycle stability, pseudocapacitive and flexible film formation ability, high cycle stability, relatively high capacitance	Not appropriate to be utilized as electrodes active materials alone, poor doping and de-doping characteristics
Manganese dioxide	Pseudocapacitive, low fabrication cost	Low electrical conductivity, high density
Polyaniline	Easy fabrication, flexibility, simple acid/base doping/de-doping chemistry and chemical stability, controllable conductivity	Not appropriate to be utilized as electrode active materials alone, low electrical conductivity, only applicable for proton type electrolytes

5.9 Concluding Remarks

Supercapacitors bridge the gap between batteries and capacitors, which is clearly depicted in the Ragone plot. In the automotive industry applications, high energy density, high capacitance, and low cost are desirable for energy storage devices. For high specific capacitance, electrode/electrolyte interface area plays a vital role that can be increased by maximizing surface area, wide pore size distribution porosity, surface morphology, and wettability of electrode material. Electrical conductivity, thermal conductivity, mechanical strength, and cyclic stability are among other properties, which should be taken into account for electrode material selection. Electrode serves a function of providing an interface for the formation of an electric double layer of opposite charges. Carbon-based materials, transition metal oxides, conductive polymers, and their corresponding composites are commonly used electrode materials. Sol–gel, electrochemical deposition, CVD, CBD, and chemical precipitation are some of the commonly employed electrode fabricating methods. According to the electrode material selection screening, graphene, carbon nanotube, ruthenium oxide, activated carbon, nickel oxide, polypyrrole, manganese dioxide, polyaniline, etc., are some most suitable electrodes for supercapacitor application.

Acknowledgements The authors acknowledge the financial support provided by the Science and Engineering Research Board, Department of Science and Technology, India (SR/WOS-A/ET-48/2018), for carrying out this research work.

References

1. S. Banerjee, P. Sinha, K.D. Verma, T. Pal, B. De, J. Cherusseri, P.K. Manna, K.K. Kar, Capacitor to supercapacitor, in *Handbook of Nanocomposite Supercapacitor Materials I Characteristics*, ed. by K.K. Kar (Springer, Berlin, Heidelberg, 2020). https://doi.org/10.1007/978-3-030-43009-2_2
2. J. Tahalyani, J. Akhtar, J. Cherusseri, K.K. Kar, Characteristics of capacitor: fundamental aspects, in *Handbook of Nanocomposite Supercapacitor Materials I Characteristics*, ed. by K.K. Kar (Springer, Berlin, Heidelberg, 2020). https://doi.org/10.1007/978-3-030-43009-2_1
3. P. Sinha, K. Kar, Introduction to supercapacitors, in *Handbook of Nanocomposite Supercapacitor Materials II Performance*, ed. by K.K. Kar (Springer, Berlin, Heidelberg, 2020). https://doi.org/10.1007/978-3-030-52359-6_1
4. P. Sinha, K. Kar, Characteristics of supercapacitors, in *Handbook of Nanocomposite Supercapacitor Materials II Performance*, ed. by K.K. Kar (Springer, Berlin, Heidelberg, 2020). https://doi.org/10.1007/978-3-030-52359-6_3
5. K.D. Verma, P. Sinha, S. Banerjee, K.K. Kar, Characteristics of electrode materials for supercapacitors, in *Handbook of Nanocomposite Supercapacitor Materials I Characteristics*, ed. by K.K. Kar (Springer, Berlin, Heidelberg, 2020). https://doi.org/10.1007/978-3-030-43009-2_9
6. K.D. Verma, S. Banerjee, K.K. Kar, Characteristics of electrolytes, in *Handbook of Nanocomposite Supercapacitor Materials I Characteristics*, ed. by K.K. Kar (Springer, Berlin, Heidelberg, 2020). https://doi.org/10.1007/978-3-030-43009-2_10
7. K.D. Verma, P. Sinha, S. Banerjee, K.K. Kar, M.K. Ghorai, Characteristics of separator materials for supercapacitors, in *Handbook of Nanocomposite SupercapacitorMaterials I Characteristics*, ed. by K.K. Kar (Springer, Berlin, Heidelberg, 2020). https://doi.org/10.1007/978-3-030-43009-2_11
8. K.D. Verma, P. Sinha, S. Banerjee, K.K. Kar, Characteristics of current collector materials for supercapacitors, in *Handbook of Nanocomposite Supercapacitor Materials I Characteristics*, ed. by K.K. Kar (Springer, Berlin, Heidelberg, 2020). https://doi.org/10.1007/978-3-030-43009-2_12
9. T. Pal, S. Banerjee, P.K. Manna, K.K. Kar, Characteristics of conducting polymers, in *Handbook of Nanocomposite Supercapacitor Materials I Characteristics*, ed. by K.K. Kar (Springer, Berlin, Heidelberg, 2020). https://doi.org/10.1007/978-3-030-43009-2_8
10. A.K. Samantara, S. Ratha, *Materials Development for Active/Passive Components of a Supercapacitor* (Springer, Singapore, 2018)
11. I. Shown, A. Ganguly, L.C. Chen, K.H. Chen, Energy Sci. Eng. **3**, 2–26 (2014)
12. L.L. Zhang, R. Zhou, X.S. Zhao, J. Mater. Chem. **20**, 5983–5992 (2010)

13. C. Largeot, J. Portet, P.L. Chmiola, Y. Taberna, P.Simon Gogotsi, J. Am. Chem. Soc. **130**(9), 2730–2731 (2008)
14. H. Liang, T. Sun, L. Xu, C. Sun, D. Wang, J. Mater. Sci.: Mater. Electron. **30**, 13636–13646 (2019)
15. N. Anjuma, M. Grotaa, D. Li, C. Shen, J. Energy Storage **29**, 101460 (2020)
16. S. Chen, W. Xing, J. Duan, X. Hu, S.Z. Qiao, J. Mater. Chem. A **1**(9), 2941–2954 (2013)
17. Y.G. Wang, H.Q. Li, Y.Y. Xia, Adv. Mater. **18**(19), 2619–2623 (2006)
18. D. Dong, Y. Zhang, T. Wang, J. Wang, C.E. Romero, W. Pan, Mater. Chem. Phys. **252**, 123381 (2020)
19. T. Liu, K. Wang, Y. Chen, S. Zhao, Y. Han, Green. Energy Environ. **4**(2), 171–179 (2019)
20. B. Zhao, X.Z. Fu, R. Sun, C.P. Wong, Sustain. Energy Fuels **1**(10), 2145–2154 (2017)
21. O.S. Burheim, M. Aslan, J.S. Atchison, V. Presser, J. Power Sources **246**, 160–166 (2014)
22. A. Spokoyny, D. Jung, M. Muni, G. Marin, R. Ramachandran, M. El-Kady, R.B. Kaner, J. Mater. Chem. A **8**, 18015–18023 (2020)
23. B. Karamanova, A. Stoyanova, M. Schipochka, C. Girginov, R. Stoyanova, J. Alloys Compd. **803**, 882–890 (2019)
24. P. Sinha, S. Banerjee, K.K. Kar, Activated carbon as electrode materials for supercapacitors, in *Handbook of Nanocomposite Supercapacitor Materials II Performance*, ed. by K.K. Kar (Springer, Berlin, Heidelberg, 2020). https://doi.org/10.1007/978-3-030-52359-6_5
25. L.L. Zhang, X.S. Zhao, Chem. Soc. Rev. **38**(9), 2520 (2009)
26. B. De, S. Banerjee, K.D. Verma, T. Pal, P.K. Manna, K.K. Kar, Carbon nanotube as electrode materials for supercapacitors, in *Handbook of Nanocomposite Supercapacitor Materials II Performance*, ed. by K.K. Kar (Springer, Berlin, Heidelberg, 2020). https://doi.org/10.1007/978-3-030-52359-6_9
27. S. Banerjee, K.K. Kar, Characteristics of carbon nanotubes, in *Handbook of Nanocomposite Supercapacitor Materials I Characteristics*, ed. by K.K. Kar (Springer, Berlin, Heidelberg, 2020). https://doi.org/10.1007/978-3-030-43009-2_6
28. B. De, S. Banerjee, K.D. Verma, T. Pal, P.K. Manna, K.K. Kar, Carbon nanofiber as electrode materials for supercapacitors, in *Handbook of Nanocomposite Supercapacitor Materials II Performance*, ed. by K.K. Kar (Springer, Berlin, Heidelberg, 2020). https://doi.org/10.1007/978-3-030-52359-6_7
29. R. Sharma, K.K. Kar, Characteristics of carbon nanofibers, in *Handbook of Nanocomposite Supercapacitor Materials I Characteristics*, ed. by K.K. Kar (Springer, Berlin, Heidelberg, 2020). https://doi.org/10.1007/978-3-030-43009-2_7
30. Y. Zhou, C. Cao, Y. Cao, Q. Han, C.B. Parker, J. T. Glass Matter **2**, 1307–1323 (2020)
31. B. De, S. Banerjee, T. Pal, K.D. Verma, P.K. Manna, K.K. Kar, Graphene/reduced graphene oxide as electrode materials for supercapacitors, in *Handbook of Nanocomposite Supercapacitor Materials II Performance*, ed. by K.K. Kar (Springer, Berlin, Heidelberg, 2020). https://doi.org/10.1007/978-3-030-52359-6_11
32. P. Chamoli, S. Banerjee, K.K. Raina, K.K. Kar, Characteristics of Graphene/Reduced Graphene Oxide, in *Handbook of Nanocomposite Supercapacitor Materials I Characteristics*, ed. by K.K. Kar (Springer, Berlin, Heidelberg, 2020). https://doi.org/10.1007/978-3-030-43009-2_5
33. M.D. Stoller, S. Park, Y. Zhu, J. An, R.S. Ruoff, Nano Lett. **8**(10), 3498–3502 (2008)
34. B. Wang, J. Liu, Y. Zhao, Y. Li, W. Xian, M. Amjadipour, J. Macleod, N. Motta, A.C.S. Appl, Mater. Interfaces **8**(34), 22316–22323 (2016)
35. P. Katanyoota, T. Chaisuwan, A. Wongchaisuwat, S. Wongkasemjit, MAT SCI ENG B **167**(1), 36–42 (2010)
36. G. Chen, S. Wu, L. Hui, Y. Zhao, J. Ye, Z. Tan, W. Zeng, Z. Tao, L. Yang, Y. Zhu, Sci. Rep. **6**(1), 19028 (2016)
37. P. Sinha, S. Banerjee, K.K. Kar, Characteristics of activated carbon, in *Handbook of Nanocomposite Supercapacitor Materials I Characteristics*, ed. by K.K. Kar (Springer, Berlin, Heidelberg, 2020). https://doi.org/10.1007/978-3-030-43009-2_4
38. B. Dyatkin, O. Gogotsi, B. Malinovskiy, Y. Zozulya, P. Simon, Y. Gogotsi, J. Power Sources **306**, 32–41 (2016)

39. J.S. Sagu, K.G.U. Wijayantha, M. Bohm, S. Bohm, T.K. Rout, A.C.S. Appl, Mater. Interfaces **8**(9), 6277–6285 (2016)
40. B. De, S. Banerjee, K.D. Verma, T. Pal, P.K. Manna, K.K. Kar, Transition metal oxides as electrode materials for supercapacitors, in *Handbook of Nanocomposite Supercapacitor Materials II Performance*, ed. by K.K. Kar (Springer, Berlin, Heidelberg, 2020). https://doi.org/10.1007/978-3-030-52359-6_4
41. A. Tyagi, S. Banerjee, J. Cherusseri, K.K. Kar, Characteristics of transition metal oxides, in *Handbook of Nanocomposite Supercapacitor Materials I Characteristics*, ed. by K.K. Kar (Springer, Berlin, Heidelberg, 2020). https://doi.org/10.1007/978-3-030-43009-2_3
42. Z. Wu, L. Li, J. Yan, X. Zhang. Adv. Sci. **4**(6), 1600382 (2017)
43. N. Padmanathan, S. Selladurai, K.M. Razeeb, RSC Adv. **5**(17), 12700–12709 (2015)
44. X. Hou, Q. Li, L. Zhang, T. Yang, J. Chen, L. Su, J. Power Sources **396**, 319–326 (2018)
45. B. De, S. Banerjee, T. Pal, K.D. Verma, A. Tyagi, P.K. Manna, K.K. Kar, Transition metal oxide/electronically conducting polymer composites as electrode materials for supercapacitors, in *Handbook of Nanocomposite Supercapacitor Materials II Performance*, ed. by K.K. Kar (Springer, Berlin, Heidelberg, 2020). https://doi.org/10.1007/978-3-030-52359-6_14
46. B. De, S. Banerjee, T. Pal, K.D. Verma, A. Tyagi, P.K. Manna, K.K. Kar, Transition metal oxide-/carbon-/electronically conducting polymer-based ternary composites as electrode materials for supercapacitors, in *Handbook of Nanocomposite Supercapacitor Materials II Performance*, ed. by K.K. Kar (Springer, Berlin, Heidelberg, 2020). https://doi.org/10.1007/978-3-030-52359-6_15
47. P. Sinha, S. Banerjee, K.K. Kar, Transition metal oxide/activated carbon-based composites as electrode materials for supercapacitors, in *Handbook of Nanocomposite Supercapacitor Materials II Performance*, ed. by K.K. Kar (Springer, Berlin, Heidelberg, 2020). https://doi.org/10.1007/978-3-030-52359-6_6
48. B. De, S. Banerjee, K.D. Verma, T. Pal, P.K. Manna, K.K. Kar, Transition Metal Oxide/Carbon Nanofiber Composites as Electrode Materials for Supercapacitors, in *Handbook of Nanocomposite Supercapacitor Materials II Performance*, ed. by K.K. Kar (Springer, Berlin, Heidelberg, 2020). https://doi.org/10.1007/978-3-030-52359-6_8
49. B. De, S. Banerjee, T. Pal, K.D. Verma, A. Tyagi, P.K. Manna, K.K. Kar, Transition Metal Oxide/Carbon Nanotube Composites as Electrode Materials for Supercapacitors, in *Handbook of Nanocomposite Supercapacitor Materials II Performance*, ed. by K.K. Kar (Springer, Berlin, Heidelberg, 2020). https://doi.org/10.1007/978-3-030-52359-6_10
50. B. De, P. Sinha, S. Banerjee, T. Pal, K.D. Verma, A. Tyagi, P.K. Manna, K.K. Kar, Transition metal oxide/graphene/reduced graphene oxide composites as electrode materials for supercapacitors, in *Handbook of Nanocomposite Supercapacitor Materials II Performance*, ed. by K.K. Kar (Springer, Berlin, Heidelberg, 2020). https://doi.org/10.1007/978-3-030-52359-6_12
51. M. Zhi, C. Xiang, J. Li, M. Li, N. Wu, Nanoscale **5**(1), 72–88 (2013)
52. N. Shirshova, H. Qian, M.S.P. Shaffer, J.H.G. Steinke, E.S. Greenhalgh, P.T. Curtis, A. Kucernak, A. Bismarck, Compos. Part A Appl. Sci. Manuf. **46**, 96–107 (2013)
53. S. Banerjee, K.K. Kar, Conducting polymers as electrode materials for supercapacitors, in *Handbook of Nanocomposite Supercapacitor Materials II Performance*, ed. by K.K. Kar (Springer, Berlin, Heidelberg, 2020). https://doi.org/10.1007/978-3-030-52359-6_13
54. D.P. Dubal, S.H. Lee, J.G. Kim, W.B. Kim, C.D. Lokhande, J. Mater. Chem. **22**(7), 3044 (2012)
55. L. Miao, Z. Song, D. Zhu, L. Li, L. Gan, M. Liu, Mater. Adv. **1**(5), 945–966 (2020)
56. I. Dincer, *Comprehensive Energy Systems* (Elsevier, 2018)
57. J. Chang, M. Jin, F. Yao, T.H. Kim, V.T. Le, H. Yue, F. Gunes, B. Li, A. Ghosh, S. Xie, Y.H. Lee, Adv. Funct. Mater. **23**(40), 5074–5083 (2013)
58. A.G. Pandolfo, A.F. Hollenkamp, J. Power Sources **157**(1), 11–27 (2006)
59. H. Yuan, T. Lei, Y. Qin, J.H. He, R. Yang, J. Phys. D Appl. Phys. **52**(19), 194002 (2019)
60. S. Banerjee, B. De, P. Sinha, J. Cherusseri, K.K. Kar, Applications of supercapacitors, in *Handbook of Nanocomposite Supercapacitor Materials I Characteristics*, ed. by K.K. Kar (Springer, Berlin, Heidelberg, 2020). https://doi.org/10.1007/978-3-030-43009-2_13
61. S. Shi, C. Xu, C. Yang, J. Li, H. Du, B. Li, F. Kang, Particuology **11**(4), 371–377 (2013)

62. L. Dong, C. Xu, Y. Li, Z.H. Huang, F. Kang, Q.H. Yang, X. Zhao, J. Mater. Chem. A **4**(13), 4659–4685 (2016)
63. J. Yang, T. Hong, J. Deng, Y. Wang, F. Lei, J. Zhang, B. Yu, Z. Wu, X. Zhang, C.F. Guo, Chem. Commun. **55**, 13737–13740 (2019)
64. M. Hu, H. Zhang, T. Hu, B. Fan, X. Wang, Z. Li. Chem. Soc. Rev. **49**(18), 6666–6693 (2020)
65. X. Yu, X. Cai, H. Cui, S.W. Lee, X.F. Yu, B. Liu, Nanoscale **9**, 17859–17864 (2017)
66. M. Wen, D. Liu, Y. Kang, J. Wang, H. Huang, J. Li, P.K. Chu, X.F. Yu, Mater. Horiz. **6**, 176–181 (2018)
67. J. Cherusseri, D. Pandey, K.S. Kumar, J. Thomas, L. Zhai, Nanoscale **12**(34), 17649–17662 (2020)
68. L. Wang, X. Feng, L. Ren, Q. Piao, J. Zhong, Y. Wang, H. Li, Y. Chen, B. Wang, J. Am. Chem. Soc. **137**(15), 4920–4923 (2015)
69. D.P. Dubal, J.S. Guevara, D. Tonti, E. Enciso, P.G. Romero, J. Mater. Chem. A **3**(46), 23483–23492 (2015)
70. M. Vangari, T. Pryor, L. Jiang, J. Energy Eng. **139**(2), 72–79 (2013)
71. J. Qu, C. Geng, S. Lv, G. Shao, S. Ma, M. Wu, Electrochim. Acta **176**, 982 (2015)
72. L. Zhu, F. Shen, R.L. Smith, L. Yan, L. Li, X. Qi, Chem. Eng. J. **316**, 770 (2017)
73. X. Li, K. Liu, Z. Liu, Z. Wang, B. Li, D. Zhang, Electrochim. Acta **240**, 43 (2017)
74. K. Yang, J. Peng, C. Srinivasakannan, L. Zhang, H. Xia, X. Duan, Bioresour. Technol. **101**, 6163 (2010)
75. N.R. Kim, Y.S. Yun, M.Y. Song, S.J. Hong, M. Kang, C. Leal, Y.W. Park, H.J. Jin, A.C.S. Appl, Mater. Interfaces **8**, 3175 (2016)
76. X. Li, W. Xing, S. Zhuo, J. Zhou, F. Li, S.Z. Qiao, G.Q. Lu, Bioresour. Technol. **102**, 1118 (2011)
77. S. Ahmed, A. Ahmed, M. Rafat, J. Saudi Chem. Soc. **22**, 993 (2018)
78. A.S. Lemine, M.M. Zagho, T.M. Altahtamouni, N. Bensalah, Int. J. Energy Res. **42**, 4284 (2018)
79. S. Banerjee, P. Benjwal, M. Singh, K.K. Kar, Appl. Surf. Sci. **439**, 560 (2018)
80. R. Sharma, K.K. Kar, RSC Adv. **5**, 66518 (2015)
81. X. Wu, Y. Zeng, H. Gao, J. Su, J. Liu, Z. Zhu, J. Mater. Chem. A **1**, 469 (2013)
82. C.-C. Hu, W.-C. Chen, K.-H. Chang, J. Electrochem. Soc. **151**, A281 (2004)
83. S.-Y. Wang, K.-C. Ho, S.-L. Kuo, N.-L. Wu, J. Electrochem. Soc. **153**, A75 (2006)
84. R.A. Fisher, M. . Watt, W. Jud Ready, ECS J. Solid State Sci. Technol. **2**, M3170 (2013)
85. I.-H. Kim, J.-H. Kim, B.-W. Cho, Y.-H. Lee, K.-B. Kim, J. Electrochem. Soc. **153**, A989 (2006)
86. X.W. Wang, D.L. Zheng, P.Z. Yang, X.E. Wang, Q.Q. Zhu, P.F. Ma, L.Y. Sun, Chem. Phys. Lett. **667**, 260 (2017)
87. R. Sharma, A.K. Yadav, V. Panwar, K.K. Kar, J. Reinf. Plast. Compos. **34**, 941 (2015)
88. R.I. Jaidev, A.K. Jafri, S.Ramaprabhu Mishra, J. Mater. Chem. **21**, 17601 (2011)
89. N.H. Khdary, M.E. Abdesalam, G. EL Enany, J. Electrochem. Soc. **161**, G63 (2014)
90. K.K. Kar, J.K. Pandey, S. Rana, *Handbook of Polymer Nanocomposites. Processing, Performance and Application* (Springer, Berlin Heidelberg, Berlin, Heidelberg, 2015)
91. K. Nishio, M. Fujimoto, O. Ando, H. Ono, T. Murayama, J. Appl. Electrochem. **26**, 425 (1996)
92. L.-Z. Fan, J. Maier, Electrochem. Commun. **8**, 937 (2006)
93. M.M. Mulunda, Z. Zhang, E. Nies, C. van Goethem, I.F.J. Vankelecom, G. Koeckelberghs, Macromol. Chem. Phys. **219**, 1800024 (2018)
94. Q. Meng, K. Cai, Y. Chen, L. Chen, Nano Energy **36**, 268 (2017)
95. S.A. Meguid, G.J. Weng, *Micromechanics and Nanomechanics of Composite Solids* (Springer International Publishing, Cham, 2018)
96. K.K. Kar, *Composite Materials*, 1st edn. (Springer, Berlin Heidelberg, Berlin, Heidelberg, 2017)

Chapter 6
Separator Material Selection for Supercapacitors

Alka Jangid, Kapil Dev Verma, Prerna Sinha, and Kamal K. Kar

Abstract Supercapacitors have gained crucial advantages among various energy storage devices such as batteries, capacitors, and fuel cells. The efficiency of supercapacitors depends on various aspects that depend on its components. These components include electrodes, electrolyte, current collectors, and separator. Electrode store charges, electrolyte provide necessary ions, current collector transfers the charge from the electrode to external circuit, and separator acts as a membrane, which prevents the device from short circuit. The choice of separator material plays a vital role in the design of a supercapacitor. Its main function is to separate cathode and anode electrode material in supercapacitors to prevent short circuit. It is mainly present in the form of a porous membrane in order to provide easy ion transfer. The common material used as separator includes glass fiber, cellulose, ceramic fibers, or polymeric film materials. This chapter mainly describes functions served and characteristics required for separators and its materials, respectively, which are chosen according to those functions. Finally, the selection of separator material is justified with the help of various material indices using Ashby's chart.

Alka Jangid and Kapil Dev Verma are Equal contribution.

A. Jangid · K. D. Verma · P. Sinha · K. K. Kar (✉)
Advanced Nanoengineering Materials Laboratory, Materials Science Programme, Indian Institute of Technology Kanpur, Kanpur 208016, India
e-mail: kamalkk@iitk.ac.in

A. Jangid
e-mail: alkajangid33@gmail.com

K. D. Verma
e-mail: deo.kapildev@gmail.com

P. Sinha
e-mail: findingprerna09@gmail.com

K. K. Kar
Advanced Nanoengineering Materials Laboratory, Department of Mechanical Engineering, Indian Institute of Technology Kanpur, Kanpur 208016, India

K. K. Kar (ed.), *Handbook of Nanocomposite Supercapacitor Materials III*,
Springer Series in Materials Science 313,
https://doi.org/10.1007/978-3-030-68364-1_6

6.1 Introduction

Supercapacitors are electrochemical capacitors, which display a very high energy density than conventional capacitors. It exhibits characteristics of both capacitor [1] and battery [2]. The special features of capacitors and capacitors to supercapacitors are reported elsewhere [1, 3–5]. These are also known as ultracapacitors and electric double-layer capacitor (EDLC). Supercapacitors are used as an energy storage device in the field of applications like renewable energy storage, electric vehicles, medical applications, wearable electronics, etc. [6, 7]. It delivers fast charge and discharge time and longer cycle life than batteries along with higher capacitance and energy density than conventional capacitors. Hence, supercapacitors provide advantages over batteries and conventional capacitors. However, supercapacitors have low specific energy density than batteries, which can be improved by widening the working potential window of the cell. Also, linear discharge voltage and high cost are other major disadvantages of using supercapacitors in many applications [2]. Supercapacitor performance depends upon the materials of electrode [8], electrolyte [9], separator [10], and current collector [11].

For understanding the functions of a separator, one should first understand the design of the supercapacitor. The construction of a supercapacitor is similar to that of the electrolytic capacitor. It consists of two current collector metallic foils, which are coated with the electrode material. These materials can be carbon materials, transition metal oxide, and conducting polymers with a large specific surface area. Collector foil connects these electrode materials to external electric terminals [1]. A porous membrane is sandwiched between two electrodes, which is known as a separator. This membrane acts as an electrical insulator that prevents short circuits between different polarity electrodes allowing only electrolyte ions to pass though it. The housing is then sealed airtight.

The overall performance of a supercapacitor depends upon the following major parts of a supercapacitor [12].

- Current collectors:—These are made up of high conductivity metals such as aluminum, stainless steel, and copper foil. The main function of the current collector is collecting current from the electrode and passing it to the external terminal [11].
- Electrodes—These are in form of coatings applied on collector foils and should have high electrical conductivity, corrosion resistance, and surface area per unit volume. Electrodes store charge electrostatically in supercapacitor [8]. Transition metal oxides [13], activated carbon [14], carbon nanofibers [15], carbon nanotubes [16], reduced graphene oxide/graphene [17, 18], conducting polymers [19], other emerging materials and their composites are used electrode materials [20–26]. The characteristics of these materials, i.e., transition metal oxides [27], activated carbon [28], carbon nanofibers [29], carbon nanotubes [30], reduced graphene oxide/graphene [31], conducting polymers [32], etc., are reported elsewhere.

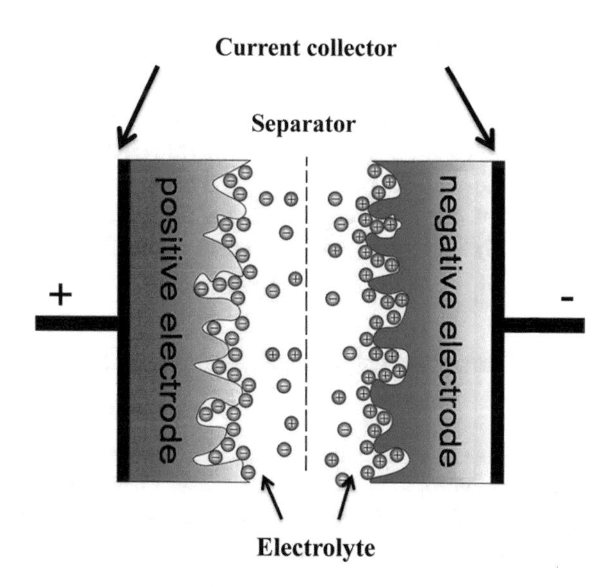

Fig. 6.1 Working of separator in a supercapacitor (redrawn and reprinted with permission from [34])

- Electrolyte—It connects two electrodes ionically. When electrodes are polarized, ions in electrolyte form the double layer of ions of different polarity near electrodes. It can be organic, aqueous, or ionic depending on chemical composition [9].
- Separator—This is in the form of a thin porous membrane having negligible electric conductivity. These membranes are permeable for ion transfer from the electrolyte solution. It works as a reservoir for electrolyte and also prevents the electrical short circuit by keeping electrodes apart (Fig. 6.1). These can be made of cellulose, glass fiber, polymer membranes, etc. [10, 33].
- Casing—It can be of different shapes such as rectangular or cylindrical and is made of metal compatible with current collectors.
- External terminals:—These connect the supercapacitor to the outer circuit.

The efficiency of the supercapacitors depends on various parts such as electrodes, electrolyte, separator, and current collectors. By altering or modifying the properties and materials used for these parts, the efficiency of the supercapacitor can be effectively improved. A separator is a porous membrane, which mechanically separates the two electrodes and works as an electrical insulator layer between electrodes to prevent short circuits, and it also works as the electrolyte reservoir. Hence, separator is a physical barrier between electrodes made of dielectric porous material allowing only electrolyte ions to transfer. It separates electrodes so it should have high chemical stability, mechanical strength, electrical resistance, and high porosity [6]. The material of separator is selected based on the type of electrodes, electrolyte, service temperature, and potential range. It is made of woven glass fibers, ceramic fibers, or nonwoven polymer films. Research is going on for developing separators of less cost, which include bio-based eggshell membranes or other cellulose-based material.

6.2 Functions of Separators

If we summarize the function of separators, they serve the main purpose of physically separating electrodes of different polarities. Along with it, separators also (1) act as an insulator layer between anode and cathode to prevent short circuit and discharge by electron current of supercapacitor (2) act as a permeable membrane for transfer of ions of electrolyte, and (3) it works as the electrolyte reservoir [10].

6.3 Commercial Manufacturers of Separators

Celgard—Conventional supercapacitors are made using porous polymer separators such as polypropylene (PP) in the case of the Celgard membrane. These membranes are 20 μm thicknesses. Polypropylene (PP) is inert chemically and inexpensive. It can withstand high acidic and basic conditions. Though it is insoluble for most solvents, it has a tendency to swell on interaction with polar solvents. Celgard membranes are prepared by the extrusion of PP films. The alignment of polymer chains takes place during cooling and crystallites form. Around 50–300% stretching at a temperature just below the melting temperature leads to a porous structure [35].

Exxon Mobil—It uses polyethylene (PE) for making separators. The commercial name of membrane for this supplier is Tonen. It is 25 μm thick separators. A wet process is used for manufacturing a single layer of PE.

Nafion—Nafion is the most common membrane used in commercial supercapacitors. It is made of Teflon backbone and side chains terminated by sulfonic acid groups. It has high chemical inertness and efficiency. But it is expensive to manufacture and has limited availability of raw materials, which limit large-scale production of separator.

Separion (Degussa)—Composite separators show high thermal stability and wettability, but they show poor mechanical strength and flexibility for folding in assembly [36]. So to overcome this disadvantage, a Separion separator is developed. Degussa (manufacturer) developed Separion (commercial membrane), which is composed of ceramic-PET-ceramic. It is manufactured by a wet process. It combines properties of the nonwoven polymer and ceramic materials both. The nonwoven mat is coated with a ceramic layer on both sides of this separator.

Parameters for commercial separators are mentioned in Table 6.1, and its effect on the supercapacitor performance due to different commercial separators is demonstrated in Fig. 6.2.

Table 6.1 Parameters for commercial separators by various manufacturers [37]

Manufacturer	Celgard	Exxon mobil	Celgard	Separion
Composition	PP	PE	PP-PE-PP	Al_2O_3/SiO_2
Thickness (μm)	25	25	25	25
Porosity (%)	55	36	41	>45
Melting temperature	163	135	134/166	210
Thermal shrinkage (%)	3	6	2.5	<1

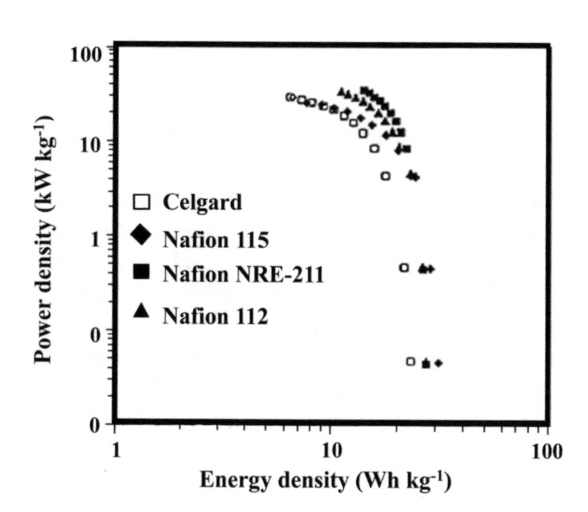

Fig. 6.2 Ragone plots for supercapacitors using different commercial separators. Current-discharge was kept constant at 1.0 V (redrawn and reprinted with permission from [38])

6.4 Characteristics Required for Separators

To improve the efficiency of the supercapacitor, researches are being done by modifying materials of various parts of the supercapacitor and properties of those materials such as surface morphology, porosity, durability, contact resistance, and chemical inertness. For good performance of supercapacitor, the separator should have a thin-film structure, high electronic resistance, low ionic resistance, high surface area, high porosity (around 50% for organic electrolyte), high mechanical strength, and high electrolyte uptake capacity. Performance parameters of organic electrolytes based on the supercapacitor are demonstrated in Table 6.2.

6.4.1 Mechanical Strength

The separator membrane should be able to withstand vibrations and folding in the fabrication process and should maintain physical separation between electrodes. Mechanical strength of separator is measured in terms of tear resistance (ASTM

Table 6.2 Performance parameters for supercapacitor using organic electrolyte [39]

Parameters	Target
Active layer thickness	100 μm
Volumetric capacitance	100 F/cm^3
Electrolyte	0.05 S/cm
Cell voltage	2.5 V
Current collector thickness	25 μm
Separator porosity	50%
Separator thickness	25 μm
Distributed resistance in pores	0.5 S/cm

D-1004), puncture strength (ASTM D-822), and tensile strength. Young's modulus can describe all these terms. Tear resistance is defined as resistance against tearing. Puncture strength is the maximum force required to penetrate a material. The separator membrane should have high mechanical strength in machine direction [5]. For separator, the minimum required mechanical strength is 1000 kg cm^{-2} and puncture strength 300 g for standard thickness. The tensile strength of the separator depends on the ratio of the machine direction and transverse direction stretches. Pore shape of stretched film is machine direction oriented. If the difference between machine direction and transverse direction is more, then splitting may occur, which causes a problem in assembling, and deteriorate the device life span. Figure 6.3 depicts the microstructural deformation in the melt-extruded HDPE film after applying the uniaxial loading.

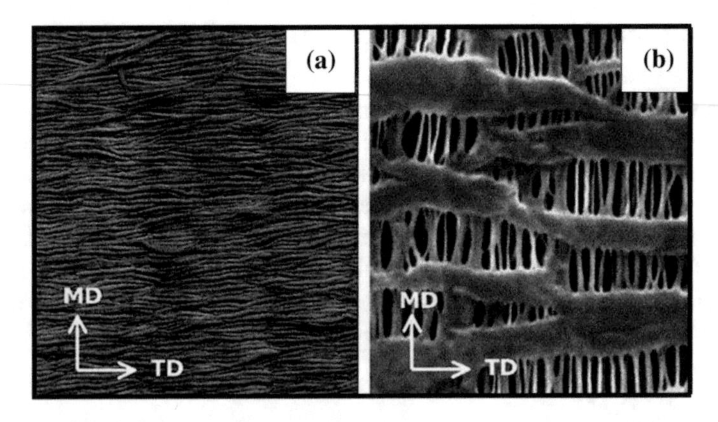

Fig. 6.3 Uniaxially oriented HDPE before and after uniaxial stretching (redrawn and reprinted with permission from [37])

6.4.2 Permeability

Uniform permeability in separator results in longer cycle life of supercapacitor. Variation in permeability causes uneven current density distribution. The addition of such separator to the device assembly adds to the ionic resistance of the supercapacitor that reduces the overall efficiency of the supercapacitor cell. It causes an increase in resistance up to 6 times. MacMullin number is defined as the ratio of the resistance of separator filled with electrolyte and resistance of electrolyte kept separately. MacMullin number is an important factor for cycle life and can be estimated by air permeability. Ion permeability allows ion transfer and ionic conduction in the separator. High ionic permeability of separator increases the performance of supercapacitors. So, the thickness of the separator should be small with high porosity [10].

6.4.3 Chemical Stability

To improve device efficiency, separator material should possess high chemical inertness. Separator material should not react with electrolyte and must be chosen according to electrolyte chemical composition [9]. If not properly taken care of, this reaction between separator material and electrolyte can change the conductivity of electrolyte, and separator porosity will reduce that overall decreases the device life span. The separator should withstand extreme oxidative or reductive conditions in case of a fully charged battery/supercapacitor. It should retain mechanical strength even in a corrosion environment. Nafion membrane is one such separator that shows higher chemical stability in oxidative conditions than other hydrocarbon membranes.

6.4.4 Dimensional Stability

The separator should retain its original dimensions under the range of service temperature and pressure. These dimensions should not get affected by electrolyte uptake or temperature change. On electrolyte uptake, various separator materials show volume expansion, while some materials on high temperatures show thermal shrinkage. When the temperature of the supercapacitor assembly rises above the softening temperature of the polymer membrane, shrinkage takes place due to an increase in density. It occurs due to different densities in crystalline and amorphous membranes. In the polyethylene membrane, thermal shrinkage is 10% at 120 °C for 10 min of time [32]. This reduces the device stability. So, a separator material should maintain the dimension stability over a wide temperature range.

6.4.5 Wettability

It depends on the number of pores or pore density, size of pores, affinity of separator material towards electrolyte, and nature of separator material. The separator should wet easily in the electrolyte and retain it permanently for better performance and high cycle life. Wetting agents can be used for improving the wettability of microporous membranes made of polymers. Surfactants improve the wettability of film but not electrolyte retention capacity. So, to improve electrolyte retention and wettability both, Taskier showed the use of surfactant and hydrophilic polymer on polyolefin membranes [40]. Wettability of separator can be determined by placing a droplet of electrolyte on separator material, and time taken to absorb that droplet and contact angle are observed. Separion separators (ceramic) show excellent wettability [40–42].

6.4.6 Porosity

The charge transfer rate of the electrolytic ion depends upon the porosity of the separator. The efficiency and power density of the supercapacitor mainly depend upon the porosity of the separator [38]. The size of the pores, relative to the size of ions present in electrolyte, also matters in deciding the capacity of electrolyte storage. These pores play important role in storing these ions. This porosity should be uniform throughout the thickness and length for better performance. However, a high amount of porosity leads to higher shrinkage rates, reducing the efficiency of a supercapacitor. Cell resistance reduces with increasing porosity [12]. The effect of porosity on equivalent series resistance (ESR) is shown in Fig. 6.4b.

6.4.7 Thickness

Separator thickness can be in the range of a few tens of micrometers. This thickness should be uniform throughout for good performance of the cell. A thicker separator provides higher ionic resistance, which directly affects the performance of the supercapacitor. Thickness is kept less for high energy and power density of the cell. It is also in form of a porous membrane for reducing equivalent series resistance. Figure 6.4 demonstrates the variation in the equivalent series resistance (ESR) of the cell using different separator parameters. Figure 6.4a represents that ESR linearly depends upon the thickness of the separator. Figure 6.4b,c shows a clear trend that, as the porosity and pore size increase, the ESR values decrease [43].

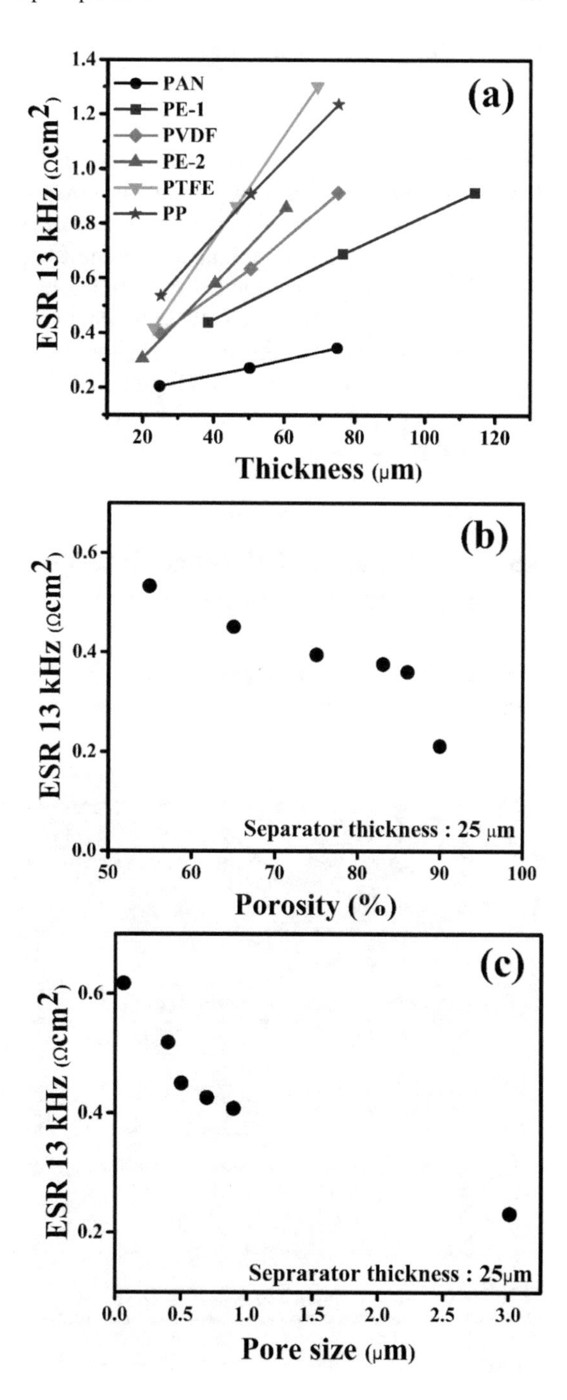

Fig. 6.4 a variation in ESR with a thickness of separator membrane, b variation in ESR with the percentage change in the porosity of separator membrane, c variation in ESR with a change in pore size for separator material (keeping thickness of separator 25 μm) (redrawn and reprinted with permission from [43])

6.4.8 Surface Morphology

A separator should have ever surface morphology with high porosity. The size of pores should be comparable to the size of ions in electrolyte for effective ion transfer. The smaller size of pores results in low electrolyte storage capacity. In this vein, a microporous polymer membrane is coated on separator film to enhance electrolyte filling. Polymer coating on porous membrane changes surface morphology of separator. This coating can be applied by solution method or phase inversion method. The polymeric coating enhances ionic conductivity, wettability, and overall electrochemical performance of the device. Figure 6.5 shows a microscopic image of the separator before and after coating. Figure 6.5a-1, a-2 shows the morphology of the bare separator, and Fig. 6.5b-1, b-2 shows the tuned morphology of the separator after polymeric coating.

6.5 Performance of Various Materials Used for Making Separators

Material selection for separator depends upon the functions served by separator material and characteristics required. Commonly used separator materials are as follows.

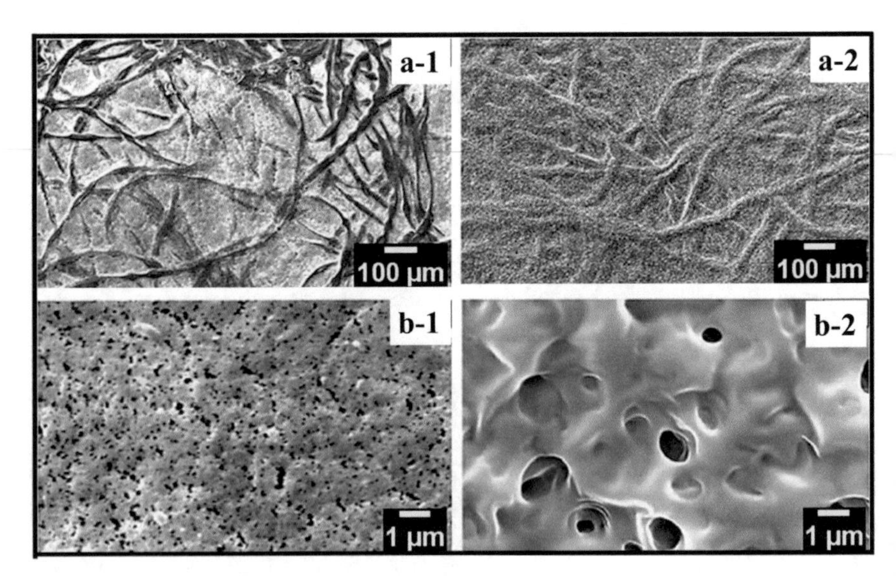

Fig. 6.5 SEM images showing surface morphology of separators a-1 bare separator, a-2 coated separator (×100), b-1 bare separator, b-2 coated separator (×10,000) (redrawn and reprinted with permission from [44])

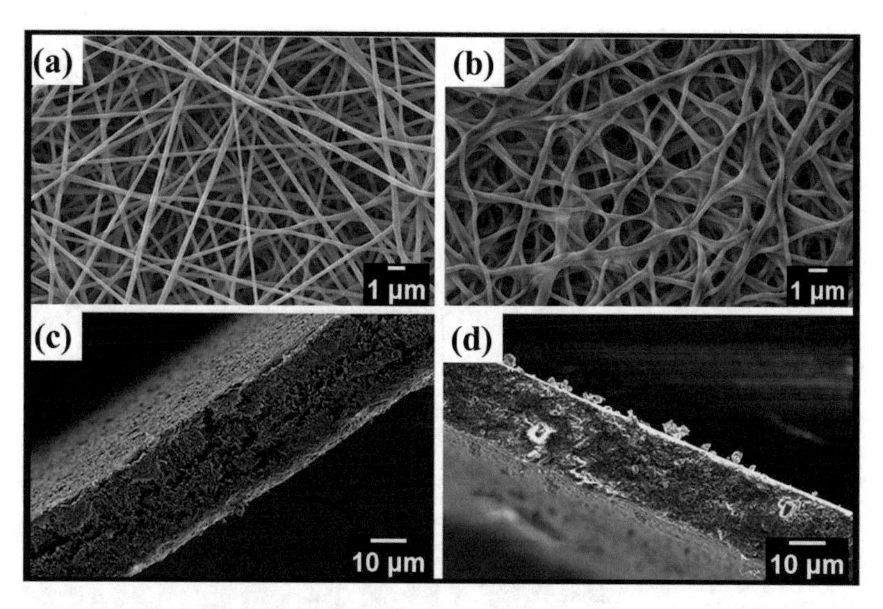

Fig. 6.6 FESEM images of surface and the cross sections of a, c electrospun polyvinyl alcohol/polyacrylic acid nanofiber membrane. b, d PVA/PAA nanofiber membrane-reinforced gel separator. (Scale: a, b 1 μm and c, d 10 μm) (redrawn and reprinted with permission from [46])

6.5.1 *Polymer Membrane*

Separators are made of polypropylene (PP) or polyethylene (PE). Polymer separators show the problem of wettability with various electrolytes and difficulty in the removal of moisture. But the nonwoven fabric surface of polyamide (PA) and polypropylene (PP) can be modified by low energy plasma technology. By this technique, wettability and mechanical strength are improved, and internal resistance is effectively decreased [19, 45]. For making the high-performance polymer composite, Fig. 6.6 displays the morphology of electrospun PVA/PAA membrane and dried PVA/PAA membrane-reinforced PVA/H_3PO_4 gel separator.

6.5.2 *Woven Ceramic Fiber*

With aqueous electrolyte, the woven ceramic separator is also used. These provide high mechanical strength and corrosion resistance. Ceramic separators are an important part of high temperature-based supercapacitors due to the low shrinkage and high thermal stability of ceramic materials. These separators work efficiently up to 120 °C and can withstand the temperature up to 350 °C without shrinkage. These high temperature-based supercapacitors are used in military and drilling applications [47]. Figure 6.7 shows the top and side view of the microscopic image of

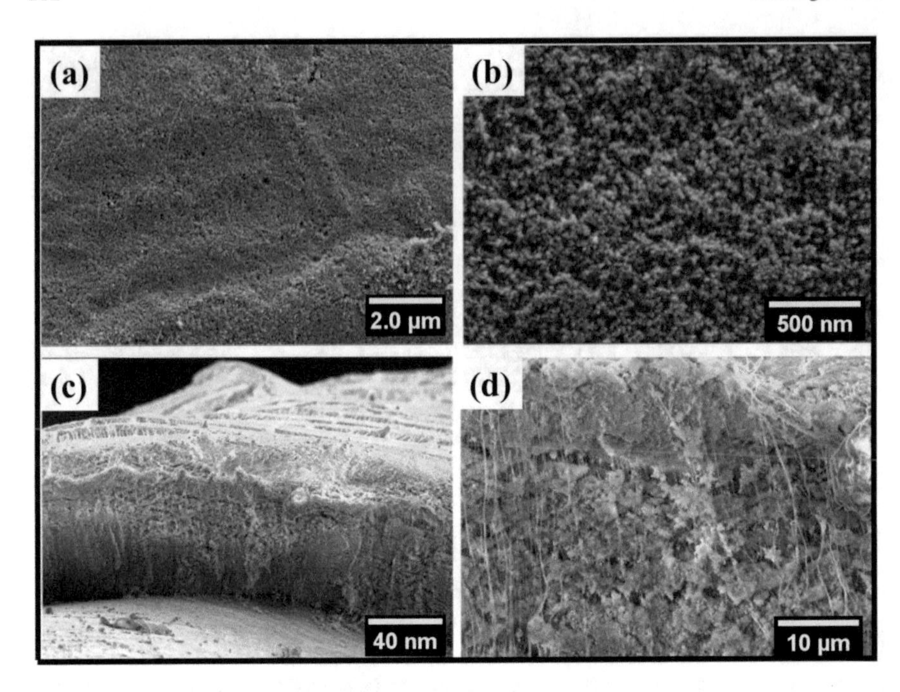

Fig. 6.7 SEM images of fumed SiO_2/polytetrafluoroethylene (80:20 wt%) separator from top and cross-sectional view (redrawn and reprinted with permission from [47])

the SiO_2/PTFE separator. Figure 6.7a, b represents the top view of the material, depicting high porosity of the separator that offer low equivalent series resistance (ESR). Also, Fig. 6.7c, d shows the cross-sectional view, depicting spider-webs like entanglement between SiO_2 and polytetrafluoroethylene (PTFE) that deliver high mechanical strength.

6.5.3 *Woven Glass Fiber*

These are used in case of aqueous electrolyte-based supercapacitors. Separators made from glass fibers have high wettability, good electrochemical performance, high thermal stability, and low shrinkage. The glass fiber shows a highly porous structure, which leads to high electrolyte uptake. However, low mechanical strength and limited flexibility restrict its wide usage as separator material for supercapacitor devices [48].

6.5.4 Composite Separators

High dielectric constant materials are used as a separator to obtain high energy and power density of the device [24, 49]. Various dielectric materials such as polymers or ceramics are efficiently used as composite separators. However, both materials individually have major merits and demerits. Ceramics have high dielectric constant, high thermal stability, and ionic conductivity, but suffer from low mechanical stability when manufactured as separators. On the other hand, polymers as separators show good chemical resistance and mechanical strength but have lower dielectric constant and poor thermal stability. Polymer/ceramic composite as separators tends to overcome limitations and display enhanced dielectric constant and tensile strength than polymer separators. These composites are porous and provide excellent wettability with high-temperature stability but are less stable mechanically. Figure 6.8 displays the top and cross-sectional view of different composite separators. The top view demonstrates the porosity of the separator, and the cross-sectional view gives an idea about the thickness and entanglement of the material.

6.5.5 GO Films

Single graphene oxide (GO) layers are insulators having a resistance greater than 10^{12} Ω at room temperature. Graphene oxide paper (GOP) is thick graphene oxide layers separated by substrate. It is made by slow evaporation of graphene oxide solution or by using the filtration method. It has great mechanical properties. It can also be used in the high-temperature range [50, 51]. Figure 6.9 shows the cross-sectional area of the graphene oxide paper. Paper sheets are layered non-homogeneously, which is the cause of folds.

6.5.6 Cellulose

Cellulose paper is also used as a separator material. Cellulose separators are made from submicron fibers. Diameters of these fibers are in the range of 0.1–1 μm. It shows very high wettability, which lowers the internal ionic resistance. For Li-ion electrolyte, cellulose separator uptake is ~203%. Where electrolyte uptake % is equal to the change in mass of separator on wetting by electrolyte/mass of dry separator $\times 100$. However, cellulose separators show low specific capacity and high self-discharge rate. The high self-discharge rate can be reduced by modifying it with ionic polyelectrolyte [52, 53]. Figure 6.10 shows the microscopic image of the cellulose separator.

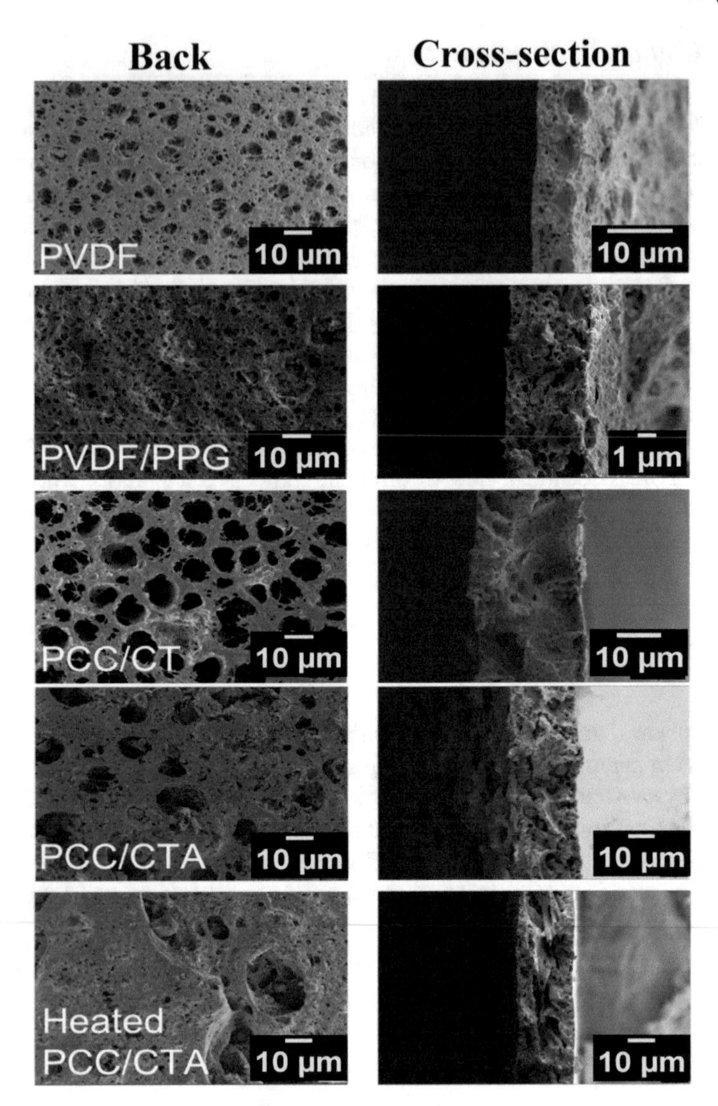

Fig. 6.8 SEM images of the composite separator surface (redrawn and reprinted with permission from [42])

6.5.7 *Eggshell Membrane*

For reducing cost purposes, some alternative separator materials are also used like natural materials as separators materials. Separator made from eggshell has higher mechanical stability, low water uptake (<10%), and high thermal stability (up to 220 °C). It shows low resistance, high specific capacitance, and short charge-discharge time. It also exhibits better mechanical properties than PP [54]. Figure 6.11

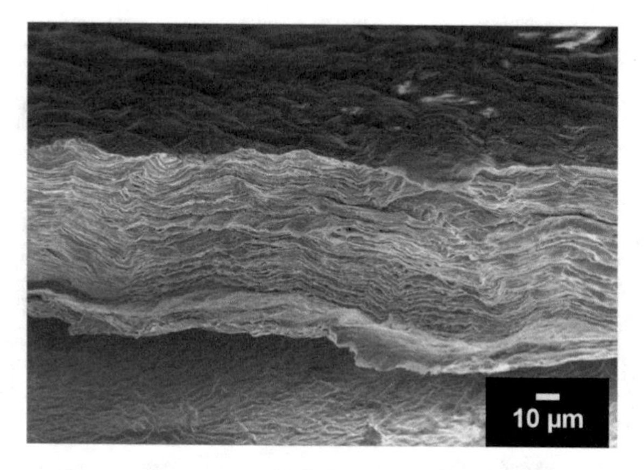

Fig. 6.9 Micrograph of a cross section of graphene oxide paper (GOP) (redrawn and reprinted with permission from [51])

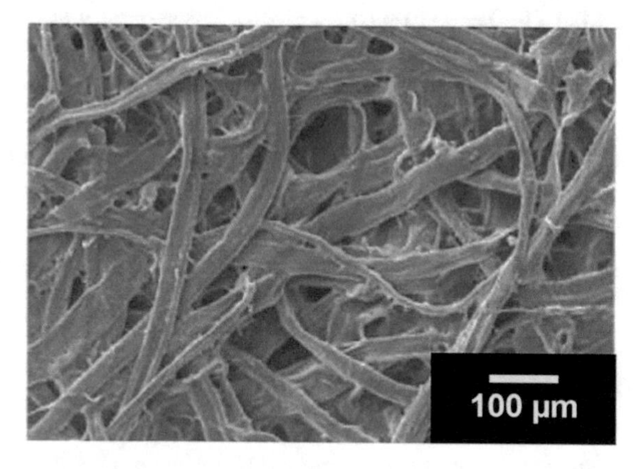

Fig. 6.10 SEM image of porous structure cellulose (redrawn and reprinted with permission from [53])

shows a microscopic image of the eggshell membrane. It has a microporous structure and interconnected eggshell membrane fiber, which helps in the fast diffusion of electrolyte ions.

6.5.8 Piezoelectric Materials

Yau Lu et al. described a self-charging supercapacitor using mechanical energy [55]. The device integrated from piezoelectric material can be used in wearable electronics due to its freedom of flexibility of various parts of supercapacitors, which so can be easily woven into wearable clothes. One such device consists of PDMS-rGO/C film

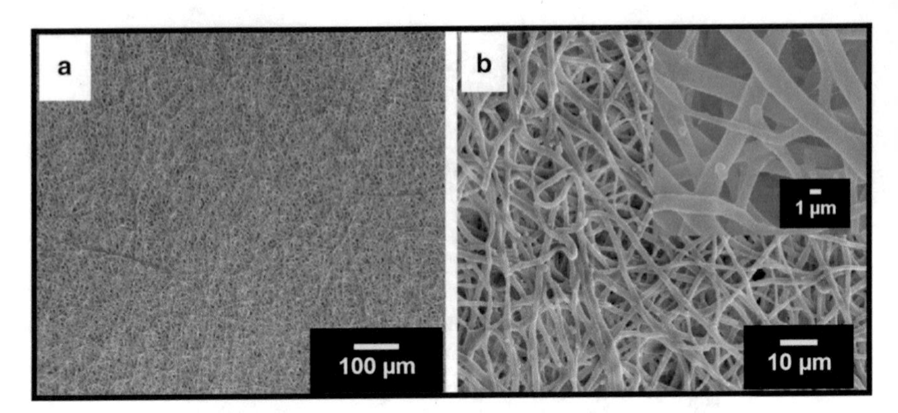

Fig. 6.11 SEM images showing the surface morphology of the eggshell membrane (redrawn and reprinted with permission from [54])

as electrodes and P(VDF-TrFE) film (piezoelectric) as separator and PVA/H_3PO_4 as electrolyte. When this supercapacitor is pressed by a finger, mechanical energy is converted into electrochemical energy. This supercapacitor could be charged up to 0.45 V in 0.17 s, and the discharging current is about 6.4 μA. Figure 6.12 shows the galvanostatic charge-discharge (GCD) curve of piezoelectric self-charging supercapacitor at different bending angles. During bending, the cell from no pressure to strong pressure cell started charging due to piezoelectric separator and after releasing the pressure cell maintained 4 s discharge time [55].

6.6 Design of Separator in Supercapacitors

Supercapacitors are manufactured in various designs such as cylindrical, coin cell, and prismatic supercapacitors depending on applications, casing material, and capacitance. With an increase in rated capacitance, the volume of the supercapacitor increases. For increasing capacitance, more charge is needed to be stored.

Construction of a supercapacitor can be either stacked or wound type. Depending on these construction types, separator material is taken of different shapes and sizes. Figure 6.13 shows commercially available different shapes of supercapacitor.

6.7 Separator Material Selection for Supercapacitors

Supercapacitor performance depends upon the material properties of its components. If we talk about the role of the separator in the performance of cell, then molar conductivity and permeability of the separator directly affect the power density of the supercapacitor cell [57]. Separators are used along with liquid electrolyte. This

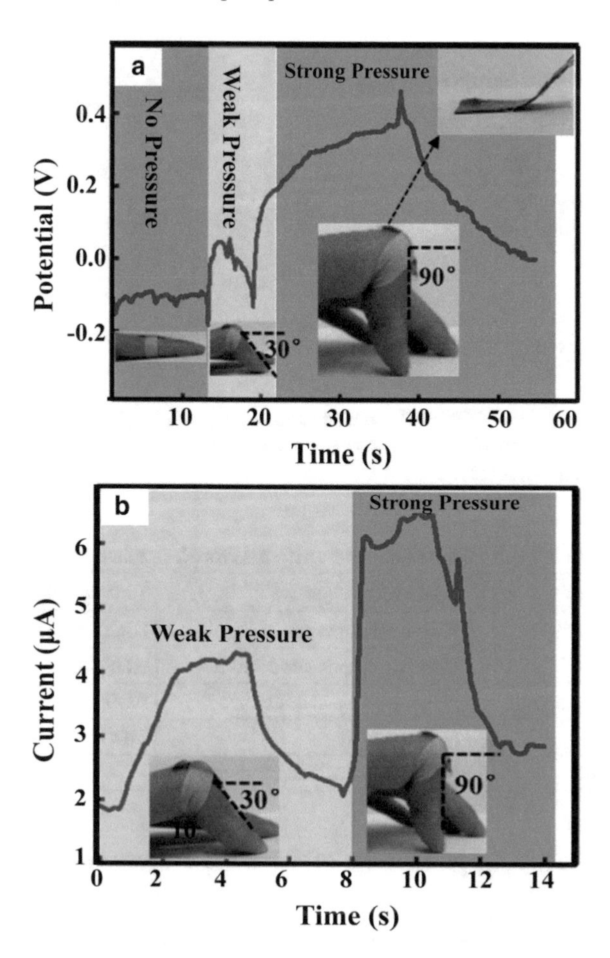

Fig. 6.12 a charging-discharging curves and b short circuit current curves for piezoelectric self-charging supercapacitor device with respect to the angle made by finger. The device is tied to the finger joint (redrawn and reprinted with permission from [55])

semi-permeable material contains electrolyte in its pores and compensates electrolyte ions to the electrodes during charging. Separator material should be strong enough to bear the external pressure or small shocks to avoid short circuit. Separator material should have high wettability so that it can carry sufficient electrolytes. It should be chemically inert and high thermal stability. Other important properties of separator material should be non-toxic and environmentally friendly. Commonly used separator materials for commercial purposes are cellulose and polypropylene, but here, we will consider different types of separator materials for further analysis [10, 58].

Fig. 6.13 Supercapacitors of different series. **a** SCC series (cylindrical supercapacitor) and **b** SCP series (prismatic supercapacitor). **a** and **b** are taken from supercapacitor manufacturer AVX Corporation product datasheet [56]

Table 6.3 Objectives for selection of separator materials for electrochemical supercapacitor

Primary objective	Secondary objective	Tertiary objective
High ionic conductivity	Minimal thickness	High corrosion resistance
High wettability	Low electronic conductivity	Non-toxic
High mechanical strength	High thermal stability	Easily available
High permeability	Lightweight	Low cost
Optimum porosity	High chemical inertness	

6.7.1 Objectives for Selection of Separator Material

For the selection of the separator material, there are some specific parameters, which should be checked according to the demand of supercapacitor cell. The objectives of the separator material selection are divided into three categories according to the priorities, which is tabulated in Table 6.3. Primary objectives are the most required properties of the separator material. Secondary objectives are also essential, but these properties do not directly affect the performance of supercapacitor cells. Tertiary objectives are concern about the economic and environmental effects. Table 6.4 shows a list of various properties of different types of separators materials.

6.7.2 Screening Using Constrains

The essential properties of the separator, which are required for the supercapacitor application, are high ionic conductivity, optimum porosity, and high mechanical strength. The porosity of the separator affects mechanical strength. Separators of high

Table 6.4 Properties of different separator materials [6]

Separator	Ionic conductivity (mS/cm)	Resistance (Ω)	Degree of electrolyte uptake (%)	Porosity (%)	Tensile strength (MPa)	Contact angle (°)	Thickness (μm)	References
Cladophora cellulose	0.82	–	–	44	137.6	–	10-40	[59–61]
Cellulose	1.74	–	340	75	–	–	25	[62, 63]
Polypropylene (PP) (Celgard)	0.503	2.15	134	41	139.16	56	20	[58, 64–66]
Polypropylene (PP)	1.05	–	120	55	–	82	25	[62]
Poly(vinylidene fluoride) (PVDF)	(0.3–3.7)	0.95	405.10	75	21.3	0	10	[57, 65, 67, 68]
PE (Tonen)	0.6	–	54	41.50	70.76	35	20	[59, 69–71]
PE	0.25	2.5	110.7	45	0.313	43.4	12	[72]
PE	0.59	1.65	100	41	–	49.2	–	[73]
Ceramic coating separator	1.10	–	71.20	41.20	–	–	24	[59]
PVA	0.52–1.27	–	225–260	–	45–70	–	100–120	[60, 74–76]
Egg shell membrane	–	19	81	–	6.59	–	14	[77]
Aquagel	72	–	–	–	–	–	125	[78]
Polyamide acid (PA)	0.80	1.58	2065	92.20	–	3.5	35	[79, 80]
Polyamide acid (PA)		1–2		38–87		17	200	[45]
polyethylene terephthalate (PET)	2.27	2.50	500	89	12	–	40	[81]
PI	2.5		2010	92.1		3.7	–	[79]
Polyimide (PI)	0.00173	–	250 ± 1.1	72.4	17.5 ± 1.4	–	39 ± 1.7	[82, 83]
Polyacrylonitrile (PAN)	2–2.14	–	395–479	–	9	–	100	[57, 84]

(continued)

Table 6.4 (continued)

Separator	Ionic conductivity (mS/cm)	Resistance (Ω)	Degree of electrolyte uptake (%)	Porosity (%)	Tensile strength (MPa)	Contact angle (°)	Thickness (μm)	References
PVA/5 wt%PVC	0.613	8–30	44.3	–	–	–	150–200	[85]
10%PVA/PVP	0.79	1.94	226	63	–	8	–	[85]
Polytetrafluoroethylene (PTFE)/polyethylene (PE)	0.96	–	171.5	65–66	0.35	–	18	[72]
poly(butylene terephthalate) (PBT)	0.27	–	–	75	–	–	55	[86]
PMMA	2.8	–	300	57	–	–	85	[87, 88]

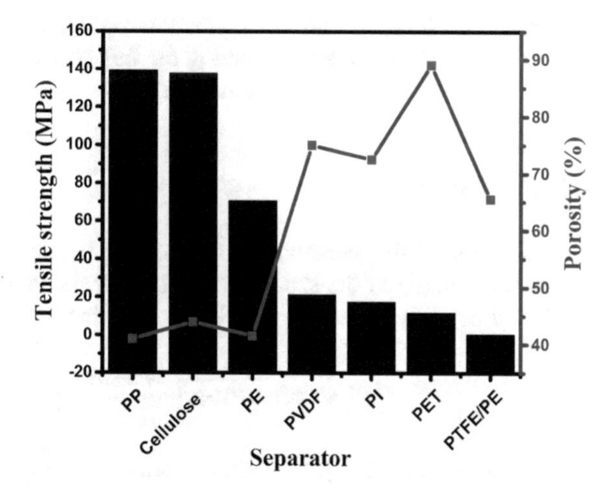

Fig. 6.14 Tensile strength and porosity of different separators

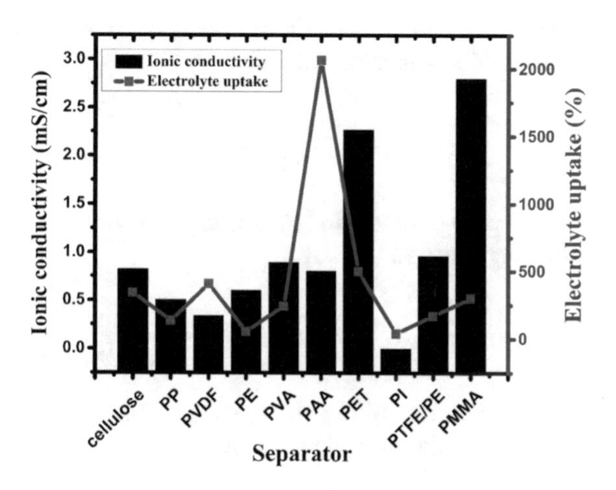

Fig. 6.15 Ionic conductivity and electrolyte uptake of different separators

porosity have low mechanical strength (Fig. 6.14). Separator materials should have high electrolyte uptake and high ionic conductivity for high performance (Fig. 6.15).

6.7.3 Governing Equations

6.7.3.1 Ionic Conductivity

Ionic conductivity is concern about the resistance-free path for the movement of ion in the separator material, which can be calculated from the following equation

$$\sigma = t/RA_0 \tag{6.1}$$

Here, σ (mS/cm) is representing the ionic conductivity, t (cm) is the thickness of the separator, R (ohm) is representing the bulk resistance of the separator, and A_0 (cm^2) is the area of the separator material.

6.7.3.2 Porosity

The porosity of the separator is represented by the void volume. The apparent geometric volume in the separator can be calculated by the total pore volume by total volume of the separator, which is written in the following equation

$$\text{Porosity}(P) = \left(\frac{W - W_0}{\rho_e \times V}\right) \times 100\% \tag{6.2}$$

Here, P indicates the porosity of the separator, W is the weight of the separator after immerging in the electrolyte, and W_0 is the weight of the separator before immerging in the electrolyte. ρ_e denotes the density of the electrolyte, and V indicates the total volume of the separator material.

6.7.3.3 Degree of Electrolyte Uptake

The percentage of electrolyte uptake in separator material is calculated from the following equation

$$\text{DoE uptake}(U) = \left(\frac{W - W_0}{W_0}\right) \times 100\% \tag{6.3}$$

Here, W_0 and W are the masses of the separator material before and after immersing in the electrolyte solution.

6.7.3.4 Tensile Strength

The tensile strength of the separator is calculated from the following equation

$$\frac{F}{t \times b} \tag{6.4}$$

Where S is the tensile strength, F is the force, t is the length of the separator, and b is the width of the separator.

6.7.3.5 Weight

The weight of the separator is calculated from the following equation

$$W = \rho \times l \times t \times b \tag{6.5}$$

where W is the weight of the separator, ρ is the density of the separator, l is the length, t is the thickness and b is the width of the separator.

6.7.3.6 Permeability

The permeability is the ability of a porous material to allow the electrolyte to pass through it. Permeability of the separator is calculated from the following equation

$$B = \frac{1\mu\upsilon}{\Delta P} \tag{6.6}$$

Here, B denotes the permeability coefficient, 1 represents the thickness of the separator, μ is the electrolyte viscosity, υ represents the fluid velocity in the pores of the separator, and ΔP represents the pressure drop across the separator.

6.7.4 Material Indexes

Material index for a high degree of electrolyte uptake and porosity:

From 6.2 and 6.3, porosity and degree of electrolyte uptake are represented as

$$\text{Porosity}(P) = \left(\frac{W - W_0}{\rho_e \times V}\right)100\%$$

and

$$\text{DoE uptake}(U) = \left(\frac{W - W_0}{W_0}\right) \times 100\%$$

Degree of electrolyte uptake will be high for the separators of high porosity, and the material index is represented as

$$M_1 = P \times U \tag{6.7}$$

Material index for high ionic conductivity and mechanical strength:

From the 6.1, the ionic conductivity is represented as

or

$$\sigma = t/\text{RA}_0 \tag{6.8}$$

and from the 6.4, strength is represented as

$$S = \frac{F}{t \times b}$$

or

$$t = \frac{F}{S \times b} \tag{6.9}$$

From the 6.8 and 6.9, another material index is represented as

$$M_2 = \sigma \times S \tag{6.10}$$

Final material index according the objective of the separators should be (it can be in any other form like $M_3 = M_1 \times \pm\div M_2 \times \pm\div M_3$ or $a_1 M_1 \times \pm\div a_2 M_2 \times \pm\div a_3 M_3$)

where a_1, a_2, and a_3 are any integers.

M_3 is another material index

$$M_3 = M_1 \times M_2 \tag{6.11}$$

6.7.5 *Material Property Chart*

Ashby chart provides the materials with specific property within a specified range. The material index line decides the appropriateness of the material for the intended application.

6.7.5.1 Ashby Chart for Material Index (M_1)

From 6.7, the material index is represented as

$$M_1 = P \times U$$

Taking log both sides

$$\log M_1 = \log P + \log U$$

or

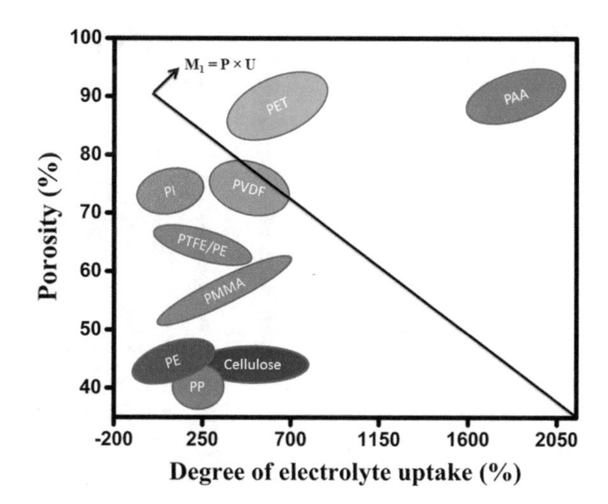

Fig. 6.16 Ashby chart of porosity and degree of electrolyte uptake

$$\log P = -\log U + \log M_1 \tag{6.12}$$

In the Ashby chart, the governing line (material index line) is related to the porosity, and the degree of electrolyte uptake is shown in Fig. 6.16. The materials present above the material index line ensure the need for the high material index is to be selected.

For high performance of the supercapacitor, a high degree of electrolytes uptake is needed such that electrolyte can provide a large number of ions to the electrode for charge accumulation and Faradaic reaction. If the degree of electrolyte uptake remains low, then the capacitance of the supercapacitor will also be low. According to the index line in Fig. 6.16, the suitable materials to be used as a separator for supercapacitor devices are polyamide acid (PA), polyethylene terephthalate (PET), poly (vinylidene fluoride) (PVDF), polyimide (PI), and poly (methyl methacrylate) (PMMA).

6.7.5.2 Ashby Chart for Material Index M_2

From the 6.10, the other material index is represented as

$$M_2 = \sigma \times S$$

Taking log both sides of the equation

$$\log M_2 = \log \sigma + \log S$$

Or

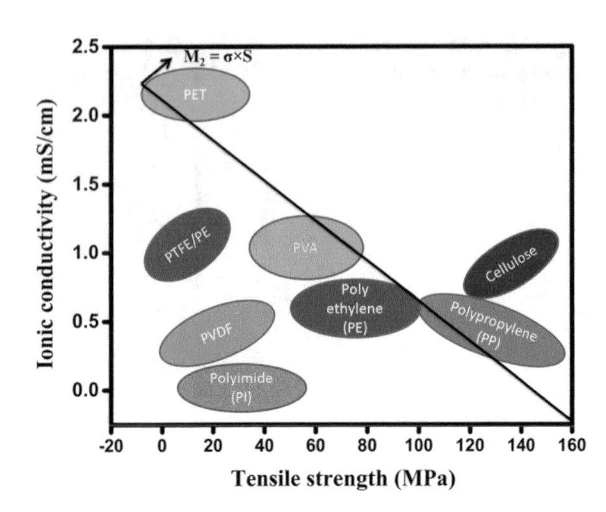

Fig. 6.17 Ashby chart of ionic conductivity and tensile strength

$$\log \sigma = \log M_2 - \log S \tag{6.13}$$

In the Ashby chart, the governing line related to high ionic conductivity and tensile strength is shown in Fig. 6.17. The materials above the material index line to ensure the need for the high material index are to be chosen.

For high performance and flexible supercapacitors, high ionic conductivity, and high tensile strength of separator material are needed. High ionic conductivity means low diffusion resistance, and hence, the equivalent series resistance of the supercapacitor device will be low. During the assembly of the supercapacitor device, the applied pressure can damage the separator. Again for flexible electronics, a lightweight and flexible separator is required, and hence, high tensile strength of the separator remains an essential property. According to the material index line in Fig. 6.17, the suitable materials for the supercapacitor separator are polypropylene, cellulose, PET, PVA, and polyethylene.

6.7.6 *List of Material Indexes*

Material indexes M_1 and M_2 are calculated using the material properties given in Table 6.4. For the high performance of the supercapacitor, the high material index is 'what is needed the most'. Separator material with the highest M_3 (material index) is the most favorable material according to the constrain put during the analyses (Table 6.5).

According to the high material index (M_3), suitable materials are listed in Table 6.6 with their properties, advantages, and disadvantages.

Table 6.5 Material index of different separator materials

Separator	M_1	M_2	$M_3 = M_1 \times M_2$
Cellulose	14,960	112.832	1.687×10^6
PP	5494	69.99748	0.3845×10^6
PE	2241	42.456	0.09514×10^6
PET	44,500	27.24	1.21218×10^6
PI	2823.6	0.02975	84.002
PTFE/PE	11,233.25	0.336	3774.37
PVDF	30,382.5	7.1355	0.216×10^6

6.8 Concluding Remarks

The performance of the supercapacitor can be improved by improving the properties of electrodes, current collectors, electrolytes, and separator material. This chapter focuses on characteristics, functions, and materials selection of separator membranes for supercapacitor applications. The separator should have high ionic conductivity, high electric resistivity, high mechanical strength, high wettability, and high chemical and dimensional stability. Optimum porosity and thickness of the separator are required for high energy and power densities of the device. Polymeric membrane, ceramic fiber, glass fibers, graphene oxide films, composite, and cellulose are common materials that are currently used as separators for supercapacitors. Eggshell and piezoelectric materials are being researched for making low cost and self-charging supercapacitors, respectively. As per the comparative analysis of separator material selection for supercapacitor cells, the most appropriate materials for the separator are cellulose, polyethylene terephthalate (PET), polypropylene (PP), PVDF, polyethylene (PE), and polyimide (PI).

Table 6.6 Different separator material with their advantages and disadvantages

Separator material	Properties and advantages	Disadvantages
Cellulose	Renewable natural polymer, excellent thermal stability, low fabrication cost, high ionic conductivity, high electrochemical stability, and good wettability	Large pore sizes, high flammability, low mechanical stability
PET	High electrolyte uptake, high wettability, good thermal stability, high mechanical strength, good thermal shrinkage, and excellent electronic insulator, high ionic conductivity	High cost of fabrication
PP	High mechanical strength, high ionic conductivity, low fabrication cost, high compatibility, good chemical stability, thermal shutdown, and electronic insulator	Low wettability (high contact angle), low thermal stability, low cyclic stability, low electrolyte uptake, large pore size
PVDF	Good electrochemical stability, high cycle stability, high degree of electrolyte uptake, porosity, highly porous structure, high permittivity, high dimension stability	Low ionic conductivity, low mechanical strength
PE	High electrochemical stability, high mechanical strength, thermal shutdown, stable cycle performance, less electrolyte leakage, and good electrical insulator	Low thermal stability, low wettability, low electrolyte uptake, low ionic conductivity, large pore size
PTFE/PE	PTFE is common binder, high thermal stability, chemical stability, biological compatibility, high porosity, good wettability, good electronic insulator	Low mechanical strength, high cost of fabrication
PI	High chemical stability, thermal stability (>200 °C), excellent electrical insulator, inherent self-extinguishing ability, high electrolyte uptake, high wettability, high porosity	Low ionic conductivity, high cost of fabrication, low tensile strength

Acknowledgements The authors acknowledge the financial support provided by the Science and Engineering Research Board, Department of Science and Technology, India (SR/WOS-A/ET-48/2018) for carrying out this research work.

References

1. J. Tahalyani, J. Akhtar, J. Cherusseri, K.K. Kar, Characteristics of capacitor: fundamental aspects, in *Handbook of Nanocomposite Supercapacitor Materials I Characteristics*, ed. by K.K. Kar (Springer, Berlin, Heidelberg, 2020). https://doi.org/10.1007/978-3-030-43009-2_1
2. R. Kumar, S. Sahoo, E. Joanni, R.K. Singh, K. Maegawa, W.K. Tan, G. Kawamura, K.K. Kar, A. Matsuda, Mater. Today (2020). https://doi.org/10.1016/j.mattod.2020.04.010
3. S. Banerjee, P. Sinha, K.D. Verma, T. Pal, B. De, J. Cherusseri, P.K. Manna, K.K. Kar, Capacitor to supercapacitor, in *Handbook of Nanocomposite Supercapacitor Materials I Characteristics*, ed. by K.K. Kar (Springer, Berlin, Heidelberg, 2020). https://doi.org/10.1007/978-3-030-43009-2_2
4. P. Sinha, K.K. Kar, Introduction to supercapacitors, in *Handbook of Nanocomposite Supercapacitor Materials II Performance*, ed. by K.K. Kar (Springer, Berlin, Heidelberg, 2020). https://doi.org/10.1007/978-3-030-52359-6_1
5. P. Sinha, K.K. Kar, Characteristics of supercapacitors, in *Handbook of Nanocomposite Supercapacitor Materials II Performance*, ed. by K.K. Kar (Springer, Berlin, Heidelberg, 2020). https://doi.org/10.1007/978-3-030-52359-6_3
6. S. Banerjee, B. De, P. Sinha, Jayesh Cherusseri, K.K. Kar, Applications of supercapacitors, in *Handbook of Nanocomposite Supercapacitor Materials I Characteristics*, ed. by K.K. Kar (Springer, Berlin, Heidelberg, 2020). https://doi.org/10.1007/978-3-030-43009-2_13
7. R. Nigam, K. D. Verma, T. Pal, K.K. Kar, Applications of supercapacitors, in *Handbook of Nanocomposite Supercapacitor Materials II Performance*, ed. by K.K. Kar (Springer, Berlin, Heidelberg, 2020). https://doi.org/10.1007/978-3-030-52359-6_17
8. K.D. Verma, P. Sinha, S. Banerjee, K.K. Kar, Characteristics of electrode materials for supercapacitors, in *Handbook of Nanocomposite Supercapacitor Materials I Characteristics*, ed. by K.K. Kar (Springer, Berlin Heidelberg, 2020). https://doi.org/10.1007/978-3-030-43009-2_9
9. K.D. Verma, S. Banerjee, K.K. Kar (2020) Characteristics of electrolytes, in *Handbook of Nanocomposite Supercapacitor Materials I Characteristics*, ed. by K.K. Kar (Springer, Berlin, Heidelberg, 2020). https://doi.org/10.1007/978-3-030-43009-2_10
10. K.D. Verma, P. Sinha, S. Banerjee, K.K. Kar, M.K. Ghorai, Characteristics of separator materials for supercapacitors, in *Handbook of Nanocomposite SupercapacitorMaterials I Characteristics*, ed. by K.K. Kar (Springer, Berlin, Heidelberg, 2020). https://doi.org/10.1007/978-3-030-43009-2_11
11. K.D. Verma, P. Sinha, S. Banerjee, K.K. Kar, Characteristics of current collector materials for supercapacitors, in *Handbook of Nanocomposite Supercapacitor Materials I Characteristics*, ed. by K.K. Kar (Springer, Berlin, 2020). https://doi.org/10.1007/978-3-030-43009-2_12
12. M. Kumar, P. Sinha, T. Pal, K. K. Kar, Materials for supercapacitors, in *Handbook of Nanocomposite Supercapacitor Materials II Performance*, ed. by K.K. Kar (Springer, Berlin, Heidelberg, 2020). https://doi.org/10.1007/978-3-030-52359-6_2
13. B. De, S. Banerjee, K. D. Verma, T. Pal, P. K. Manna, K.K. Kar, Transition metal oxides as electrode materials for supercapacitors, in *Handbook of Nanocomposite Supercapacitor Materials II Performance*, ed. by K.K. Kar (Springer, Berlin, Heidelberg, 2020). https://doi.org/10.1007/978-3-030-52359-6_4
14. P. Sinha, S. Banerjee, K.K. Kar, Activated carbon as electrode materials for supercapacitors, in *Handbook of Nanocomposite Supercapacitor Materials II Performance*, ed. by K.K. Kar (Springer, Berlin, Heidelberg, 2020). https://doi.org/10.1007/978-3-030-52359-6_5
15. B. De, S. Banerjee, K.D. Verma, T. Pal, P.K. Manna, K.K. Kar, Carbon nanofiber as electrode materials for supercapacitors, in *Handbook of Nanocomposite Supercapacitor Materials II Performance*, ed. by K.K. Kar (Springer, Berlin, Heidelberg, 2020). https://doi.org/10.1007/978-3-030-52359-6_7
16. B. De, S. Banerjee, K. D. Verma, T. Pal, P. K. Manna, K.K. Kar, Carbon nanotube as electrode materials for supercapacitors, in *Handbook of Nanocomposite Supercapacitor Materials II Performance*, ed. by K.K. Kar (Springer, Berlin, Heidelberg, 2020). https://doi.org/10.1007/978-3-030-52359-6_9

17. B. De, S. Banerjee, T. Pal, K. D. Verma, P. K. Manna, K.K. Kar, Graphene/reduced graphene oxide as electrode materials for supercapacitors, in *Handbook of Nanocomposite Supercapacitor Materials II Performance*, ed. by K.K. Kar (Springer, Berlin, Heidelberg, 2020). https://doi.org/10.1007/978-3-030-52359-6_11
18. R. Kumar, S. Sahoo, E. Joanni, R.K. Singh, W.K. Tan, K.K. Kar, A. Matsuda, Prog. Energy Combust. Sci. **75**, 100786 (2019)
19. S. Banerjee, K. K. Kar, Conducting polymers as electrode materials for supercapacitors, in *Handbook of Nanocomposite Supercapacitor Materials II Performance*, ed. by K.K. Kar (Springer, Berlin, Heidelberg, 2020). https://doi.org/10.1007/978-3-030-52359-6_13
20. P. Sinha, S. Banerjee, K.K. Kar, Transition metal oxide/activated carbon-based composites as electrode materials for supercapacitors, in *Handbook of Nanocomposite Supercapacitor Materials II Performance*, ed. by K.K. Kar (Springer, Berlin, Heidelberg, 2020). https://doi.org/10.1007/978-3-030-52359-6_6
21. B. De, S. Banerjee, K. D. Verma, T. Pal, P. K. Manna, K.K. Kar, Transition metal oxide/carbon nanofiber composites as electrode materials for supercapacitors, in *Handbook of Nanocomposite Supercapacitor Materials II Performance*, ed. by K.K. Kar (Springer, Berlin, Heidelberg, 2020). https://doi.org/10.1007/978-3-030-52359-6_8
22. B. De, S. Banerjee, T. Pal, K. D. Verma, A. Tyagi, P.K. Manna, K.K. Kar, Transition metal oxide/carbon nanotube composites as electrode materials for supercapacitors, in *Handbook of Nanocomposite Supercapacitor Materials II Performance*, ed. by K.K. Kar (Springer, Berlin, Heidelberg, 2020). https://doi.org/10.1007/978-3-030-52359-6_10
23. B. De, P. Sinha, S. Banerjee, T. Pal, K. D. Verma, A. Tyagi, P. K. Manna, K.K. Kar, Transition metal oxide/graphene/reduced graphene oxide composites as electrode materials for supercapacitor, in *Handbook of Nanocomposite Supercapacitor Materials II Performance*, ed. by K.K. Kar (Springer, Berlin, Heidelberg, 2020). https://doi.org/10.1007/978-3-030-52359-6_12
24. B. De, S. Banerjee, T. Pal, K. D. Verma, A. Tyagi, P. K. Manna, K.K. Kar, Transition metal oxide/electronically conducting polymer composites as electrode materials for supercapacitors, in *Handbook of Nanocomposite Supercapacitor Materials II Performance*, ed. by K.K. Kar (Springer, Berlin, Heidelberg, 2020). https://doi.org/10.1007/978-3-030-52359-6_14
25. B. De, S. Banerjee, T. Pal, K. D. Verma, A. Tyagi, P. K. Manna, K.K. Kar, Transition metal oxide-/carbon-/electronically conducting polymer-based ternary composites as electrode materials for supercapacitors, in *Handbook of Nanocomposite Supercapacitor Materials II Performance*, ed. by K.K. Kar (Springer, Berlin, Heidelberg, 2020). https://doi.org/10.1007/978-3-030-52359-6_15
26. P. Sinha, B. De, S. Banerjee, T. Pal, K. D. Verma, P. K. Manna, K.K. Kar, Recent trends in supercapacitor electrode materials and devices, in *Handbook of Nanocomposite Supercapacitor Materials II Performance*, ed. by K.K. Kar (Springer, Berlin, Heidelberg, 2020). https://doi.org/10.1007/978-3-030-52359-6_16
27. A. Tyagi, S. Banerjee, J. Cherusseri, K. K. Kar, Characteristics of transition metal oxides, in *Handbook of Nanocomposite Supercapacitor Materials I Characteristics*, ed. by K.K. Kar (Springer, Berlin, Heidelberg, 2020). https://doi.org/10.1007/978-3-030-43009-2_3
28. P. Sinha, S. Banerjee, K. K. Kar, Characteristics of activated carbon, in *Handbook of Nanocomposite Supercapacitor Materials I Characteristics*, ed. by K.K. Kar (Springer, Berlin, Heidelberg, 2020). https://doi.org/10.1007/978-3-030-43009-2_4
29. R. Sharma, K.K. Kar, Characteristics of carbon nanofibers, in *Handbook of Nanocomposite Supercapacitor Materials I Characteristics*, ed. by K.K. Kar (Springer, Berlin, Heidelberg, 2020). https://doi.org/10.1007/978-3-030-43009-2_7
30. S. Banerjee, K. K. Kar, Characteristics of carbon nanotubes, in *Handbook of Nanocomposite Supercapacitor Materials I Characteristics*, ed. by K.K. Kar (Springer, Berlin, Heidelberg, 2020). https://doi.org/10.1007/978-3-030-43009-2_6
31. P. Chamoli, S. Banerjee, K.K. Raina, K.K. Kar, Characteristics of Graphene/Reduced Graphene Oxide, in *Handbook of Nanocomposite Supercapacitor Materials I Characteristics*, ed. by K.K. Kar (Springer, Berlin, Heidelberg, 2020). https://doi.org/10.1007/978-3-030-43009-2_5

32. T. Pal, S. Banerjee, P.K. Manna, K.K. Kar, Characteristics of Conducting Polymers, in *Handbook of Nanocomposite Supercapacitor Materials I Characteristics*, ed. by K.K. Kar (Springer, Berlin, Heidelberg, 2020). https://doi.org/10.1007/978-3-030-43009-2_8
33. I. Dincer, *Comprehensive Energy Systems,* (Elsevier, 2018)
34. F. Béguin, V. Presser, A. Balducci, E. Frackowiak, Adv. Mater. **26**, 2219–2251 (2014)
35. K.I. Ozoemena, S. Chen, *Nanostructure Science and Technology Nanomaterials in Advanced Batteries and Supercapacitors* (Springer, 2016)
36. G. Xiong, A. Kundu, T.S. Fisher, Thermal management in electrochemical energy storage systems, in *Thermal Effects in Supercapacitors* (Springer, 2015)
37. S.S. Zhang, J. Power Sources **164**, 351–364 (2007)
38. X.R. Liu, P.G. Pickup, Energy Environ. Sci. **1**, 494–500 (2008)
39. A. Schneuwly, R. Gallay, *proceedin in PCIM* (PCIM, Boston, 2000)
40. H.T. Taskier, US Patent **4**, 359–510, (1982)
41. J.L. Gineste, G. Pourcelly, J. Memb. Sci. **107**, 155–164 (1995)
42. C.O. Alvarez-Sanchez, J.A. Lasalde-Ramírez, E.O. Ortiz-Quiles, R. Massó-Ferret, E. Nicolau, Energy Sci. Eng. **7**, 730–740 (2019)
43. A. Laforgue, L. Robitaille, J. Electrochem. Soc. **159**(7), 929–936 (2012)
44. Y.M. Lee, N.S. Choi, J.A. Lee, W.H. Seol, K.Y. Cho, H.Y. Jung, J.W. Kim, J.K. Park, J. Power Sources **146**, 431–435 (2005)
45. B. Szubzda, A. Szmaja, M. Ozimek, S. Mazurkiewicz, Appl. Phys. A Mater. Sci. Process. **117**, 1801–1809 (2014)
46. Y. Miao, J. Yan, Y. Huang, W. Fan, T. Liu, RSC Adv. **5**, 26189–26196 (2015)
47. B. Qin, Y. Han, Y. Ren, D. Sui, Y. Zhou, M. Zhang, Z. Sun, Y. Ma, Y. Chen, Energy Technol. **6**, 306–311 (2018)
48. K.K. Kar, S.D. Sharma, T.K. Sah, P. Kumar, J. Reinf. Plast. Compos. **26**, 269–283 (2007)
49. J. Hao, W. Si, X.X. Xi, R. Guo, A.S. Bhalla, L.E. Cross, ApplPhys Lett. **76**, 3100–3102 (2000)
50. Y.M. Shulga, S.A. Baskakov, V.A. Smirnov, N.Y. Shulga, K.G. Belay, G.L. Gutsev, J. Power Sources **245**, 33–36 (2014)
51. Y.M. Shulga, S.A. Baskakov, Y.V. Baskakova, Y.M. Volfkovich, N.Y. Shulga, E.A. Skryleva, Y.N. Parkhomenko, K.G. Belay, G.L. Gutsev, A.Y. Rychagov, V.E. Sosenkin, I.D. Kovalev, J. Power Sources **279**, 722–730 (2015)
52. X.Z. Sun, X. Zhang, B. Huang, Y.W. Ma, Acta Phys. Chim. Sin. **30**, 485–491 (2014)
53. H. Wang, Q. Zhou, B. Yao, H. Ma, M. Zhang, C. Li, G. Shi, Adv. Mater. Interfaces **5**, 1–7 (2018)
54. H. Yu, Q. Tang, J. Wu, Y. Lin, L. Fan, M. Huang, J. Lin, Y. Li, F. Yu, J. Power Sources **206**, 463–468 (2012)
55. Y. Lu, Y. Jiang, Z. Lou, R. Shi, D. Chen, G. Shen, Prog. Nat. Sci. **30**, 174–179 (2020)
56. SuperCapacitor. http://www.avx.com/products/supercapacitors/. Accessed 23 Aug 2020
57. H. Lee, M. Yanilmaz, O. Toprakci, K. Fu, X. Zhang, Energy Environ. Sci. **7**, 3857 (2014)
58. H. Cai, X. Tong, K. Chen, Y. Shen, J. Wu, Y. Xiang, Z. Wang, J. Li, Polymers **10**, (2018)
59. C. Shi, J. Dai, C. Li, X. Shen, L. Peng, P. Zhang, D. Wu, D. Sun, J. Zhao, Polymers **9**, 10 (2017)
60. Y.S. Ye, M.Y. Cheng, X.L. Xie, J. Rick, Y.J. Huang, F.C. Chang, B.J. Hwang, J. Power Sources **239**, 424 (2013)
61. B. Writer, *Lithium-Ion Batteries* (Springer International Publishing, Cham, 2019)
62. G. Ding, B. Qin, Z. Liu, J. Zhang, B. Zhang, P. Hu, C. Zhang, G. Xu, J. Yao, G. Cui, J. Electrochem. Soc. **162**, A834 (2015)
63. K.K. Kar, A. Hodzic, *Carbon Nanotube Based Nanocomposites: Recent Developments.*, 1st ed. (Research publishing, 2011)
64. K. Tõnurist, T. Thomberg, A. Jänes, E. Lust, J. Electrochem. Soc. **160**, A449 (2013)
65. D. Wu, L. Deng, Y. Sun, K.S. Teh, C. Shi, Q. Tan, J. Zhao, D. Sun, L. Lin, RSC Adv. **7**, 24410 (2017)
66. K.K. Kar, S.D. Sharma, P. Kumar, Plast. Rubber Compos. **36**, 274 (2007)
67. S.K. Rath, S. Dubey, G.S. Kumar, S. Kumar, A.K. Patra, J. Bahadur, A.K. Singh, G. Harikrishnan, T.U. Patro, J. Mater. Sci. **49**, 103 (2014)

68. S. Banerjee, K.K. Kar, M. Das, Recent patents. Mater. Sci. **7**, 173 (2014)
69. R.S. Baldwin, W. R. Bennett, E.K. Wong, M.R. Lewton, M.K. Harris, *Battery Separator Characterization and Evaluation Procedures for NASA's Advanced Lithium-Ion Batteries.* (2010)
70. K.K. Kar, *Carbon Nanotubes: Synthesis, Characterization and Applications* (Research publishing Services, Singapore, 2011)
71. K.K. Kar, S.D. Sharma, P. Kumar, J. Ramkumar, R.K. Appaji, K.R.N. Reddy, J. Appl. Polym. Sci. **105**, 3333 (2007)
72. K. Zhang, W. Xiao, J. Liu, C. Yan, Polymers **10**, 1409 (2018)
73. W. Xiao, K. Zhang, J. Liu, C. Yan, J. Mater. Sci.: Mater. Electron. **28**, 17516 (2017)
74. A.A. Mohamad, N.S. Mohamed, M.Z.A. Yahya, R. Othman, S. Ramesh, Y. Alias, A.K. Arof, Solid State Ionics **156**, 171 (2003)
75. K.K. Kar, A. Hodzic, *Developments in Nanocomposites*, 1st ed. (Research publishing, 2014)
76. A. Kumar, R. Sharma, M. Suresh, M.K. Das, K.K. Kar, J. Elastomers Plast. **49**, 513 (2017)
77. N.S.M. Nor, M. Deraman, R. Omar, E. Taer, Awitdrus, R. Farma, N. H. Basri, B.N.M. Dolah, AIP Conf. Proc. **1586**, 68 (2014)
78. S.T. Mayer, J.L. Kaschmitter, R.W. Pekala, 5402306 (1995)
79. X. Luo, X. Lu, G. Zhou, X. Zhao, Y. Ouyang, X. Zhu, Y.E. Miao, T. Liu, A.C.S. Appl, Mater. Interfaces **10**, 42198 (2018)
80. K.K. Kar, J.K. Pandey, S. Rana, *Handbook of Polymer Nanocomposites. Processing Performance and Application* (Springer, Berlin, Heidelberg, 2015)
81. J. Hao, G. Lei, Z. Li, L. Wu, Q. Xiao, L. Wang, J. Memb. Sci. **428**, 11 (2013)
82. C.E. Lin, H. Zhang, Y.Z. Song, Y. Zhang, J.J. Yuan, B.K. Zhu, J. Mater. Chem. A **6**, 991 (2018)
83. K.K. Kar, *Composite Materials* (Springer, Berlin Heidelberg, 2017)
84. N. Scharnagl, H. Buschatz, Desalination **139**, 191 (2001)
85. C.C. Yang, G.M. Wu, Mater. Chem. Phys. **114**, 948 (2009)
86. C.J. Orendorff, T.N. Lambert, C.A. Chavez, M. Bencomo, K.R. Fenton, Adv. Energy Mater. **3**, 314 (2013)
87. M. Yanilmaz, X. Zhang, Polymers **7**, 629 (2015)
88. K.K. Kar, S.D. Sharma, P. Kumar, J. Ramkumar, R.K. Appaji, K.R.N. Reddy, Polym. Compos. **28**, 637 (2007)

Chapter 7
Electrolyte Material Selection for Supercapacitors

Kapil Dev Verma, Alka Jangid, Prerna Sinha, and Kamal K. Kar

Abstract Non-conventional energy storage devices such as batteries, supercapacitors, and fuel cells are finding more applications due to increasing environmental concerns for energy distribution and storage. Supercapacitors charge very fast and have higher energy density than a conventional capacitor and higher power density than batteries. A lot of research is being done to improve the efficiency and performance of supercapacitors by making the right choice for electrodes, electrolytes, separators, and current collectors. Among all the components, electrolytes serve the purpose of balancing charge in supercapacitor and provide necessary ions to form an electrical connection between electrodes. The electrolyte materials used in supercapacitor can be classified as organic, aqueous, ionic liquids, solid-state, and redox-active electrolytes and are chosen according to their properties, ultimate applications, and physical state of the supercapacitor. This chapter explains the functions of electrolytes, classification of electrolytes, i.e., aqueous electrolytes, organic electrolytes, ionic electrolytes, etc., characteristics required for electrolytes, i.e., conductivity,

---

Kapil Dev Verma, Alka Jangid and Prerna Sinha—Contributed equally.

K. D. Verma · A. Jangid · P. Sinha · K. K. Kar (✉)
Advanced Nanoengineering Materials Laboratory, Materials Science Programme, Indian Institute of Technology Kanpur, Kanpur 208016, India
e-mail: kamalkk@iitk.ac.in

K. D. Verma
e-mail: deo.kapildev@gmail.com

A. Jangid
e-mail: alkajangid33@gmail.com

P. Sinha
e-mail: findingprerna09@gmail.com

K. K. Kar
Advanced Nanoengineering Materials Laboratory, Department of Mechanical Engineering, Indian Institute of Technology Kanpur, Kanpur 208016, India

K. K. Kar (ed.), *Handbook of Nanocomposite Supercapacitor Materials III*,
Springer Series in Materials Science 313,
https://doi.org/10.1007/978-3-030-68364-1_7

viscosity, ionic concentration, electrochemical stability, thermal stability, dissociation, toxicity, volatility and flammability, cost, etc., performance of various electrolytes, performance metrics and their relationships, selection of electrolyte material in detail with the support of various material indices using Ashby's chart.

7.1 Introduction

Supercapacitors are relatively new energy storage devices. Other devices are batteries, capacitors, and fuel cells [1]. Due to the depletion of fossil fuels and global warming, a lot of research is being carried out in the field of energy storage systems. The lifespan, power density, and efficiency of these devices are important factors for selecting a suitable device for a particular application. Supercapacitors are electrochemical capacitors with very high specific capacitance. It can store a high amount of charge due to the high surface area of porous electrodes and short distance between electrodes [2]. Supercapacitors deliver high energy density than conventional capacitors making it desirable for automobile applications [3]. The supercapacitor has been first developed in 1957 by H. Becker and the term supercapacitor was first used in 1978. On the basis of the storage principle, supercapacitors are classified into three categories of electric double-layer capacitor (EDLC), pseudocapacitor, and hybrid capacitor [4]. Figure 7.1 shows the charge storage mechanism by the movement of ions of electrolyte during charging and discharging of EDLC, pseudocapacitor, and hybrid capacitor [5].

Electric double-layer capacitors have carbon as electrode material. This includes nanostructured carbon such as CNT, graphene, or amorphous carbon such as activated carbon or other porous allotropes of carbon [6]. It stores charge at electrodes/electrolyte interface in the form of an electric double layer, which is commonly known as electrostatic charge storage [7]. Ions form a layer at oppositely charged electrodes. EDLC has high power density, high rate capability, and long cycle life. However, it shows low capacitance and poor energy density, as the charge storage process is limited to surface sites.

Pseudocapacitors electrode material includes conducting polymer or transition metal oxide. It stores charges by forming an electric double layer as well by reversible redox reactions, i.e., faradaic process, where electrolyte ions undergo fast intercalation/deintercalation onto electrodes during the charging/discharging process. The charge storage process is accomplished by electrosorption, redox reactions, and intercalation processes [8].

Hybrid capacitors consist of asymmetric or composite electrodes in which the EDLC electrode gives high power density and the pseudocapacitor electrode gives high energy density. Hybrid capacitors can also be symmetric, where composite electrodes are used that provide synergistic charge storage behavior of both capacitive and faradaic material. Hybrid capacitors show enhanced performance than EDLC and pseudocapacitors by addressing their individual limitations [9–11].

The assembly of supercapacitor consists of four components.

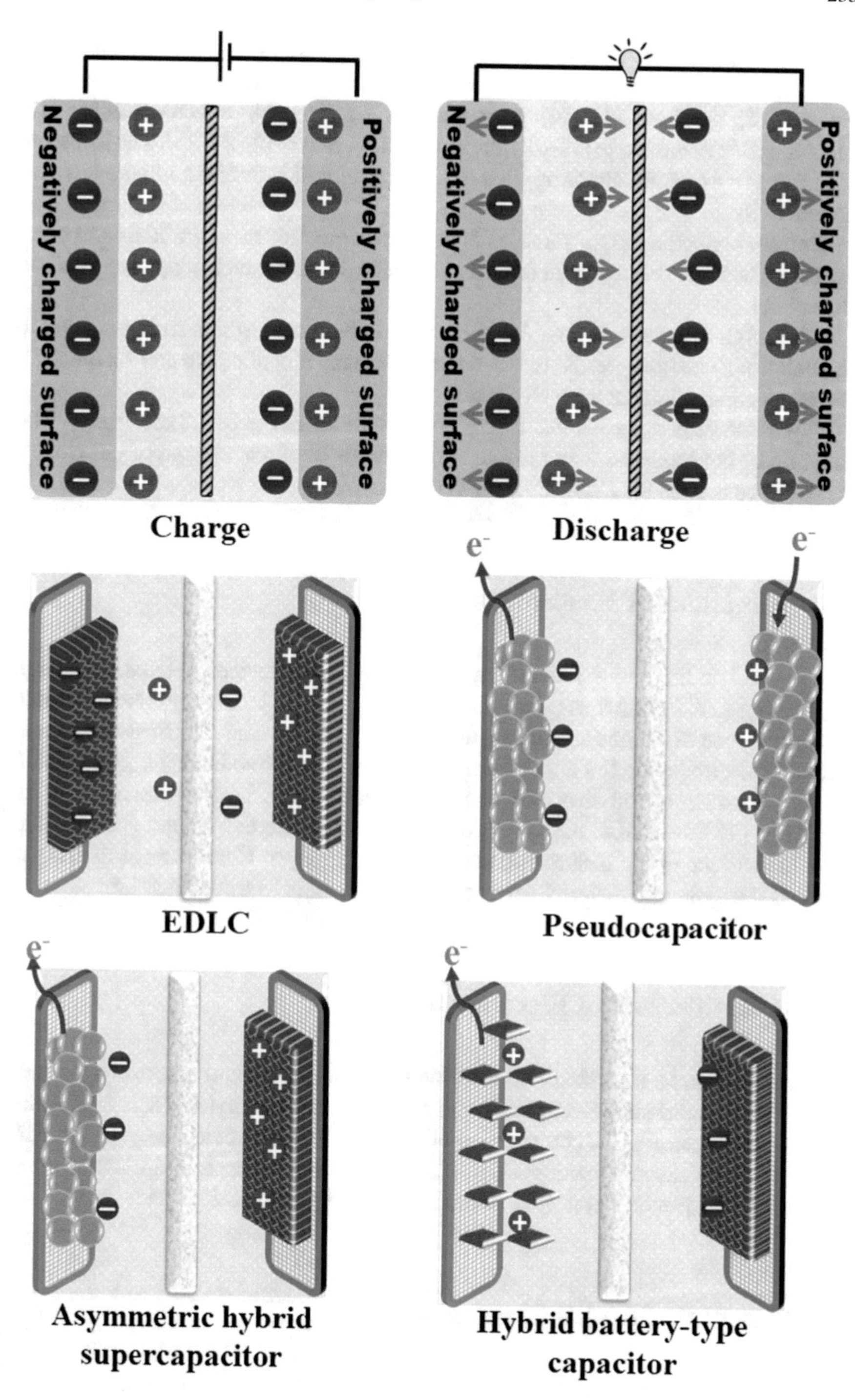

Fig. 7.1 Different types of supercapacitors (concept and permission was taken [5])

Electrodes: Electrodes are electric conductors, where charge gets stored during charging. Most common materials for electrodes in supercapacitors are activated carbon [12], graphene [13, 14], carbon nanotubes [15, 16], carbon nanofiber [17, 18], transition metal oxide [19–23], and conducting polymers [24, 25]. The electrode is in form of a coating and is applied to the current collector having high electrical conductivity.

Current collectors: Their function is to connect electrodes to terminals and outer circuits. These are mostly metal foils of aluminum, stainless steel, copper, and carbon fibers [26].

Electrolyte: Electrolyte contains solvent and chemicals, which form ions due to dissociation. The ions present in the electrolyte make it conductive and establish an electrical connection between electrodes [27].

Separator: It is a porous membrane acting as a barrier between electrodes. The function of the separator is to prevent short circuits by direct contact of electrodes. It should be porous for ionic conduction [28].

7.2 Functions of Electrolytes

The electrolyte has a vital role to play in supercapacitor devices. It balances charge by forming/transferring ions between two electrodes. It is electrically conductive due to the presence of cations and anions that dissociate to form an electrical connection between electrodes on the application of voltage. These free ions get accumulated over the surface of the electrode or intercalated into the layered structure of the electrode [27]. In EDLC based supercapacitors, electrolyte ions get accumulated over the surface of the electrode and form a double layer. Here, solvent molecules work as the dielectric material between electrolyte and electrode material.

7.3 Classification of Electrolytes

Electrolytes can be classified into three wide categories of liquid electrolyte, solid-state or quasi-solid-state electrolytes, and redox-active electrolytes. These are further classified according to the combination of solvent and ionic salt as shown in Fig. 7.2. Based on chemical composition, electrolytes used in supercapacitors are kept in three categories of aqueous electrolyte, organic electrolytes, and ionic liquids [29].

7.3.1 Aqueous Electrolytes

Aqueous electrolytes are prepared using aqueous solutions of acids, bases, and salts with high ionic concentrations. Solutes may be HCl, H_2SO_4, NaOH, and Na_2SO_4.

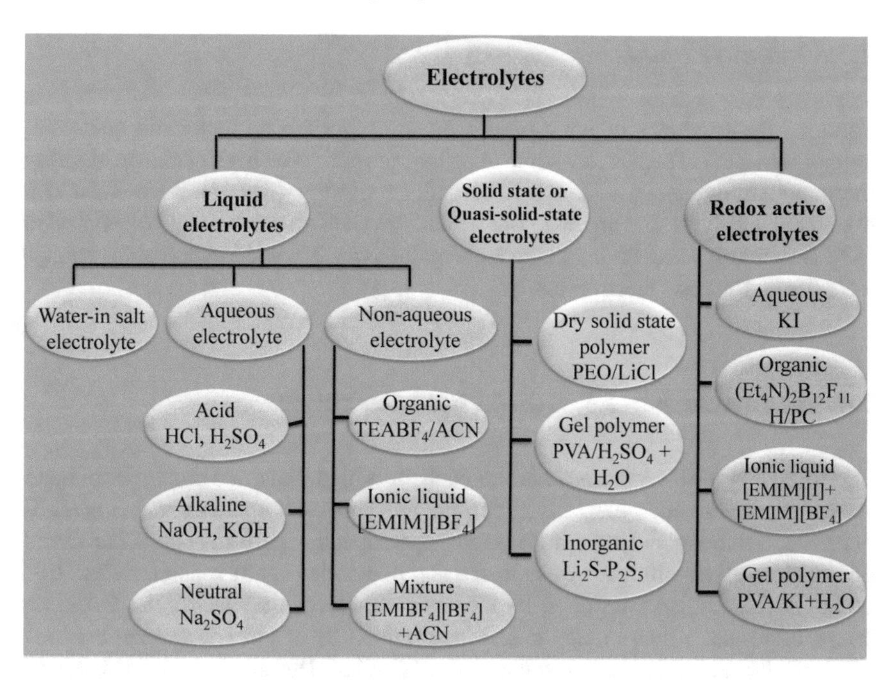

Fig. 7.2 Classification of electrolytes used in supercapacitors (redrawn and reprinted with permission from [29])

These electrolytes, when used in supercapacitor, show low resistance and hence, provide high capacitance and power density [30]. Drawbacks are low cycling stability along with leakage problems. Due to the presence of water, aqueous electrolytes show a small voltage window (1.23 V) due to electrolysis of water above 1.23 V. It decreases the operating potential window hence energy density of the device also decreases [5].

7.3.2 Organic Electrolytes

Organic electrolytes have organic solvents such as propylene carbonate and acetonitrile. Tetraethyl-ammonium tetrafluoroborate dissolved in acetonitrile is used in commercial supercapacitors [31]. These electrolytes have a wide voltage window (3.5 V). Water content in these electrolytes is kept very low because water content more than 5 ppm may cause a lowering in voltage of supercapacitor. These are abundantly used in commercial devices but it contains large size molecules making it less conductive than the aqueous electrolyte. Also, organic electrolytes are costlier and toxic chemicals, which are hazardous to the environment.

7.3.3 Ionic Liquids

Some of the examples of ionic liquids are imidazolium and aliphatic quaternary ammonium salts. These show some excellent properties as high chemical stability, thermal stability, and a wide range of voltage window (2–7 V). As compared to organic electrolytes, ILs are less volatile and have a high conductivity of 10 mS/cm [32]. But, maintaining high voltage and conductivity in a wide temperature range is a challenge for these electrolytes.

7.4 Characteristics Required for Electrolytes

For improving the performance of supercapacitors, materials, and properties of materials used as electrodes, electrolyte, current collector, and separator are being modified. Electrolytes play important role in supercapacitor performance. Characteristics of electrolytes influence capacitance and power density of supercapacitor. ESR of the device is also controlled by the conductivity of the electrolyte. Operating temperature and voltage range is also determined by electrolyte properties. Ionic conductivity and dissociation voltage of electrolyte determine power density and energy density of supercapacitor, respectively. In this regard, choosing the right electrolyte is necessary for long cycling life and safety [33]. Table 7.1 shows ionic liquid and aqueous electrolyte-based supercapacitor along with non-faradaic and faradaic capacitive storage-based electrode materials. Non-faradaic capacitive storages show higher power density and longer cycle life than faradaic capacitive storage with aqueous electrolyte and on the other hand faradaic capacitive storage shows higher energy density.

Table 7.1 Performance of supercapacitors with a different combination of electrodes and electrolytes (NFCS stands for non-faradaic capacitive storage and CFS stands for capacitive faradaic storage) [32, 34]

	EDLC	Pseudocapacitor	
Electrode material	NFCS	NFCS + CFS	CFS + CFS
Electrolyte	Ionic liquid, aqueous	Aqueous	Aqueous
Specific energy (Wh kg^{-1})	10.2 (ionic liquid) 6.7 (aqueous)	3.6	26.6
Max. specific power (kW kg^{-1})	111.6	24.7	13
Cycling life (cycles)	>10,000	>5000	>5000

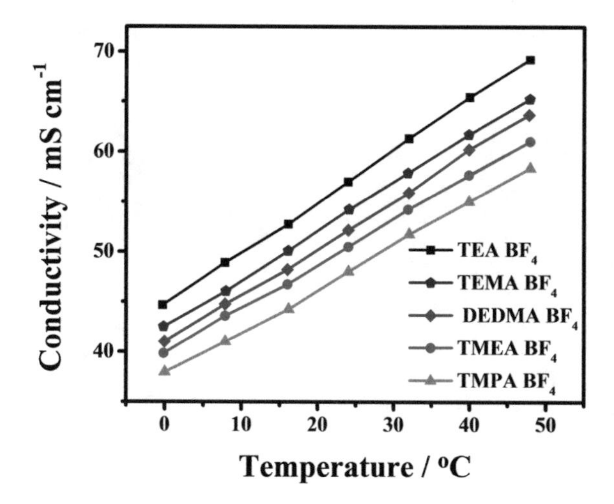

Fig. 7.3 Graph showing variation in conductivity with temperature (redrawn and reprinted with permission from [35])

7.4.1 Conductivity

Chemicals present in solvent dissociate and form cations and anions. This dissociation and the number of ions are dependent on temperature and concentration. Till a limit, as the number of ions increase, the conductivity of electrolyte also increases. Conductivity and ionic mobility are important properties of electrolytes in determining its performance. Solid electrolytes based on polymer and pure oxides show very low conductivity. Conductivity is higher for aqueous electrolytes compared to non-aqueous and solid electrolytes because it has less dynamic viscosity. The conductivity is a function of ionic mobility, charge of one electron, concentration of charge carriers, and valence of mobile ion charges [27].

A low concentration of salt in a solvent causes more free ions so the optimum concentration of salt leads to higher conductivity of the electrolyte. HCl has the smallest ions with the highest conductivity and shows the highest specific capacitance in comparison with LiCl, NaCl, KOH electrolytes [29]. Ionic conductivity also depends upon the temperature of the electrolyte. Figure 7.3 demonstrates the variation of ionic conductivity of organic electrolytes with temperature. As the temperature is increasing the ionic conductivity of organic electrolytes is also increasing [35].

7.4.2 Viscosity

The concentration of ions affects the viscosity, ionic conductivity, and charge storage mechanism of electrolyte. As shown in Fig. 7.4a that ionic liquid ($EMIMBF_4$ with acetonitrile) concentration is low in supercapacitor cell so there is a lot of unused surface of electrode material, which is not participating in charge storage. Figure 7.4b shows the optimum concentration of ionic liquid electrolyte, where all ions are

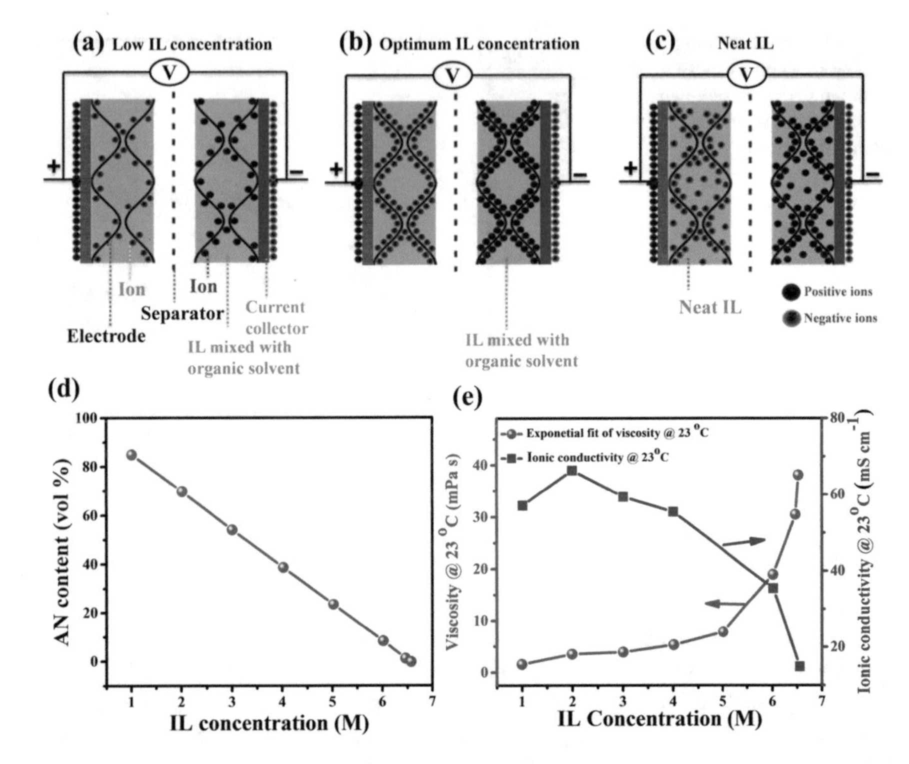

Fig. 7.4 Charge stored in EDLC in case of **a** low IL concentration, **b** optimum IL concentration, **c** neat IL, **d** graph showing relation between acetonitrile content and IL concentration, and **e** variation in viscosity and ionic conductivity with IL concentration [36]

uniformly distributed in the whole electrode material. Figure 7.4c shows neat electrolyte, which indicate a high concentration of electrolyte. A high concentration of electrolyte ions lowers ionic conductivity, where ions are not able to reach the electrode surface during the charging time. Figure 7.4d indicates that as the $EMIMBF_4$ concentration increases acetonitrile (AN) content is linearly decreasing. In Fig. 7.4e, viscosity increases exponentially with an increase in the concentration of IL [36]. Conductivity first increases then decreases with the concentration of salt in the electrolyte. With optimum concentration, the balance between viscosity and the number of free ions can be established to obtain peak ionic conductivity of the electrolytes. Viscosity controls ionic mobility, as low viscous solvents show high mobility of ions and high conductivity.

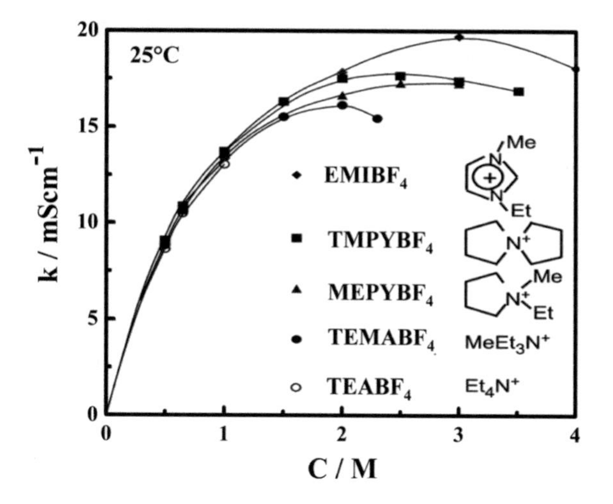

Fig. 7.5 Effect of concentration on the conductivity of quaternary ammonium salts consisting of electrolytes (redrawn and reprinted with permission from [38])

7.4.3 *Ion Concentration*

With a high concentration of salt, more cations and anions form neutral ions so the numbers of free ions decrease. This phenomenon reduces the conductivity of the electrolyte. As shown in Fig. 7.5, with increasing concentration of salt in the electrolyte, conductivity increases and attains peak conductivity and then reduces on further increase in concentration. In the case of gel polymer electrolytes, with an increase in salt concentration, conductivity increases as mobile ion concentration increases as there is more volume for the accommodation of electrolytic ions. After a certain point polymer to salt concentration conductivity decreases possibly due to agglomeration and lowering mobile ion concentration. The amorphous nature of polymer electrolytes also enhances with increasing salt concentration [37].

7.4.4 *Electrochemical Stability*

The electrochemical stability of electrolyte depends on its chemical composition and compatibility of electrode materials with a chemical composition of the electrolyte. This property controls the potential window, safety, and cycling stability of the supercapacitor. Obtaining data on the electrochemical stability of electrolytes at various temperatures is important to decide the temperature range for the supercapacitor. Liquid electrolytes show a wide potential window for electrochemical stability. The higher electrochemical stable potential window leads to higher energy density and specific capacitance so, more research is being done to enhance the properties of ionic liquids. For measuring electrochemical stability, linear sweep voltammetry and cyclic voltammetry techniques are used. Figure 7.6 shows a plot between specific charge and scan rate of HCl, KCl, NaCl, and LiCl-based supercapacitor. HCl has the

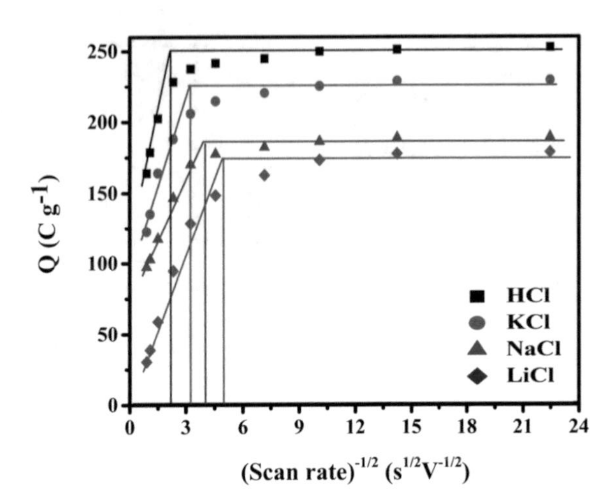

Fig. 7.6 A graph between specific charge (Q) and scan rate of cyclic voltammetry data for aqueous electrolytes (redrawn and reprinted with permission from [39])

smallest hydrated sphere (H^+) and the largest ionic conductivity among KCl, NaCl, and LiCl. Because of the small hydrated sphere radius, ions can adsorb more on the electrode–electrolyte interface that further increases the specific charge accumulation. Figure 7.6 depicts that as the hydrated ion radius decreases ($Li^+ > Na^+ > K^+ > H^+$) specific charge storage of the supercapacitor increases [39].

7.4.5 *Thermal Stability*

Electrolytes may decompose during the charge–discharge process due to the generation of heat and increased temperature. Thermal stability of electrolyte is a crucial parameter from a safety point of view for supercapacitors. It mainly depends on electrolyte and interaction between electrolyte and electrodes. In the case of dependence on electrolytes, properties such as composition, presence of additives are determining factors. The device should withstand a temperature range of –25 to 60 °C so electrolytes should be stable in this temperature range. At higher temperatures, dissociation increases and conductivity improves, which enhances the power delivering capacity for most electrolytes. Figure 7.7 shows that pure PVA has very low ionic conductivity and as KOH gets mixed with PVA, conductivity sharply increases. Ionic conductivity of these gel polymer electrolytes decreases with the increase in temperature [40].

7.4.6 *Dissociation*

Resistance in electrolytes or electrodes leads to a voltage drop in the supercapacitor. To avoid this IR drop, electrolytes should have high ionic conductivity. The

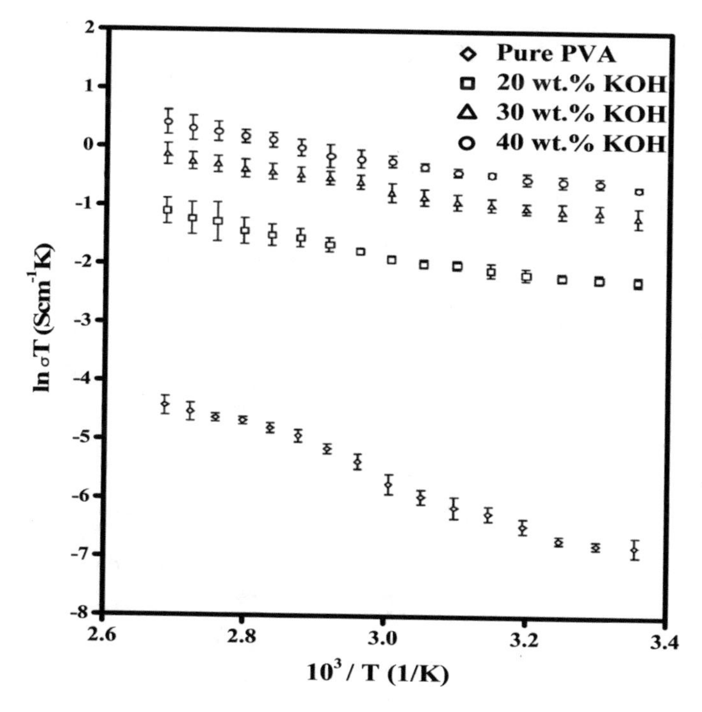

Fig. 7.7 Graphical representation of the effect of temperature on ionic conductivity of gel polymer electrolyte. Results are for the various concentration of KOH in poly (vinyl alcohol) (redrawn and reprinted with permission from [40])

maximum operation voltage allowed for a supercapacitor is dependent on the dissociation voltage of the electrolyte. So, electrolyte having wide potential stability is preferred. The concentration of free ions and the mobility of these ions depend on the ease of dissociation and stability of salt in a solvent. Dissociation of salt is determined by equilibrium. The degree of dissociation is denoted by α, and it is 1 for strong electrolytes. Dissociation voltage controls the energy density of supercapacitors. Organic electrolytes have a wide voltage window so, it is abundantly used in commercial applications. Figure 7.8a shows cyclic voltammetry of ionic gel polymer electrolytes-based supercapacitor at 50 mV/s scan rate. Ionic gel polymer electrolytes-based supercapacitor shows a high potential window, where ion gels are stable till 3.5 V of the potential window. Figure 7.8b depicts thermogravimetric analysis (TGA) of ionic gel polymer electrolytes. TGA curves clearly show that these electrolytes are thermally stable till 300 °C.

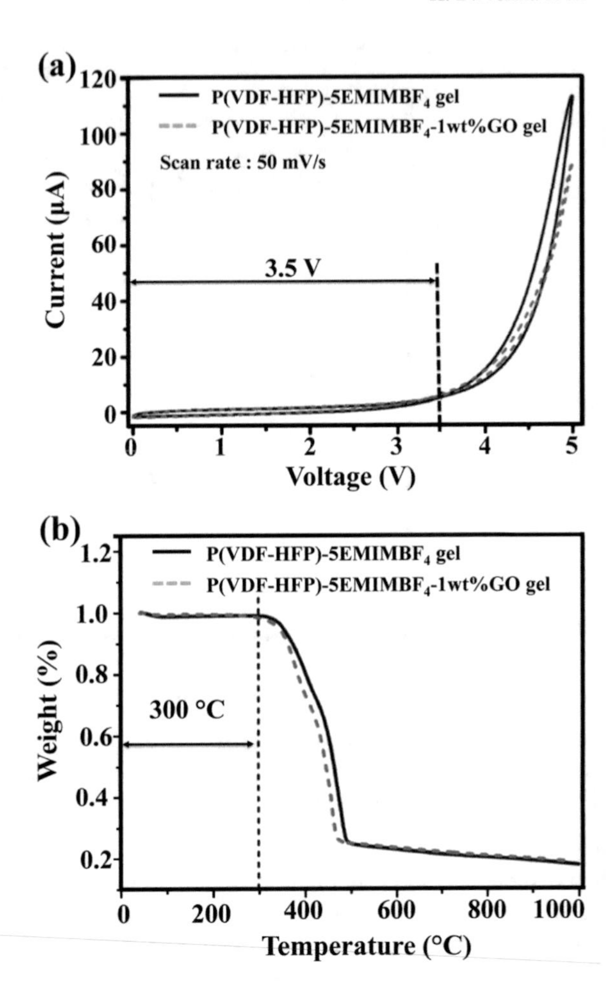

Fig. 7.8 **a** Cyclic voltammogram at scan rate 50 mV/s showing ionic gel polymer electrolytes are stable at the wide potential range and **b** TGA thermogram (carried out in nitrogen atmosphere) with heating rate 10 °C per minute shows high thermal stability of ion gels (redrawn and reprinted with permission from [41])

7.4.7 Toxicity, Volatility, and Flammability

These properties are important for safety and environmental point of view. Ionic liquid electrolytes have very low volatility as compared to organic electrolytes and are non-flammable. Organic electrolytes contain flammable and volatile solvents. So, a slow purification process is carried out for reducing such solvents. More attention is given to less toxic electrolytes in recent years. Room-temperature ionic liquids are less toxic electrolytes. Propylene carbonate and acetonitrile are commonly used as solvents in organic electrolytes. Acetonitrile is toxic and harmful to the environment so propylene carbonate is used to avoid these issues [42]. On the other hand, aqueous electrolytes are non-toxic, nonvolatile, and non-flammable, however, its narrow potential window limits its application on a commercial scale.

7.4.8 *Cost*

Organic electrolytes have one disadvantage of the high cost. Hydrogel redox-active electrolytes are preferred for their low cost compared to organic electrolytes and good energy density. These electrolytes are also easier to fabricate and safer. Carbon-based cells and non-aqueous electrolyte combination in supercapacitors are the cheapest and deliver the best performance. These devices are being used in automobile applications. International Advanced Research Center for Powder Metallurgy and New Materials (ARCI) developed a low-cost supercapacitor from the waste cotton. Natural neutral seawater is used as a cost-effective and environment-friendly electrolyte in supercapacitor. Aqueous electrolytes, mainly dilute sulfuric acid, have a low-cost advantage over non-aqueous electrolyte-based EDLCs [43].

All these aspects are considered when choosing electrolytes for supercapacitors. Choice of electrolyte also depends on the electrode material, size of the supercapacitor, and physical state of the electrolyte. The chemical and physical properties of electrolytes affect the performance of supercapacitors.

7.5 Performance of Various Electrolytes

Electrolytes are generally solutions of acid, base, and salts. In the case of aqueous electrolytes, a potential window is used around 1 V to avoid the decomposition of water. But organic electrolytes provide a voltage window up to 2.5 V. The difference in the working voltage increases the energy density of organic electrolytes-based supercapacitor by sixfold than aqueous electrolytes-based supercapacitor.

7.5.1 *Organic Electrolytes*

Organic solvents with conducting salts are organic electrolytes. The size of solvated ions is larger compared to aqueous electrolytes. So, carbon electrodes with a large number of macrospores are preferred for organic electrolytes. Figure 7.9 shows the effect of pore size of electrode material and dipole moment of solvent on the capacitance of the supercapacitor at 0–1.5 V potential window. It can be clearly seen that as the pore size of the electrode material reaches the ion size (0.5 nm), capacitance sharply increases at every dipole moment of solvent. The pattern of pore size dependence of capacitance changes with a solvent dipole moment. At 3.40 and 2.55 nm, Debye capacitance is almost constant after 1 nm pore size but below 2.55 Debye, capacitance oscillates with pore size. Organic electrolytes give good electric performance but not suitable from a safety point of view. Also, the fabrication of these electrolytes-based supercapacitors is costly [44].

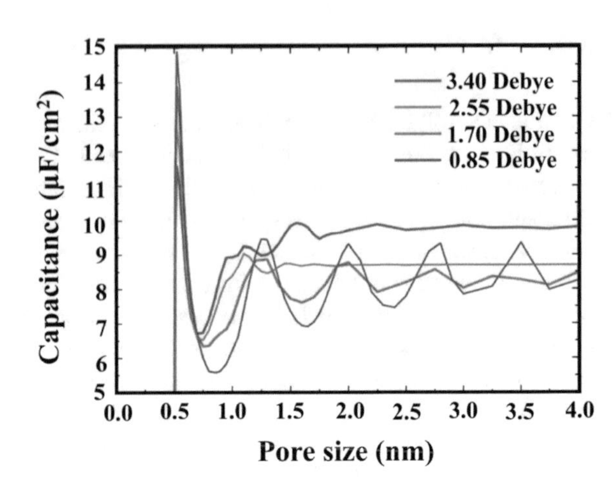

Fig. 7.9 Effect of pore size and dipole moments of solvent for organic electrolytes with different polarities on capacitance. The curve shows a peak at 0.5 nm pore size (redrawn and reprinted with permission from [44])

7.5.2 *Aqueous Electrolytes*

In aqueous electrolytes, as concentration increases transport of ions becomes easier but after a point, at higher concentrations viscosity increases because of low water hydration causing low ion mobility. Among KOH, KCl, KNO_3, and K_2SO_4 electrolytes-based supercapacitor, KOH electrolyte gives the highest specific capacitance. Hydration sphere radius is smallest for OH^- ion and highest for SO_4^{2-} ion. OH^- ions have higher mobility and conductivity, which is the cause of high power density and specific capacitance. Figure 7.10a shows the cyclic voltammetry (CV) curve of different aqueous electrolytes and Fig. 7.10b demonstrates the variation in capacitance with scan rate. Higher hydration ion sized electrolyte provides low specific capacitance [45].

7.5.3 *Ionic Liquid (IL) Electrolytes*

These are salts composed of organic or inorganic anions and organic cations. The electrochemical properties of ionic liquids mainly depend on anions. Properties of ILs can be easily controlled due to various possible combinations of cations and anions. There are three types of protic, aprotic, and zwitterionic liquid. Aprotic ILs show higher cell voltage and find more applications in supercapacitors. ILs have the disadvantage of high cost and purity requirement but these are non-flammable and have high ionic conductivity. Figure 7.11a shows the variation of specific capacitance along with current density and Fig. 7.11b shows IR drop variation with current density. Among [EMIM][NTF_2], [EMIM][DCA], and [EMIM][BF_4] IL electrolytes, radius of NTF_2^- is largest due to which [EMIM][NTF_2] shows least specific capacitance. IR drop in [EMIM][DCA] is very low due to less internal resistance [46].

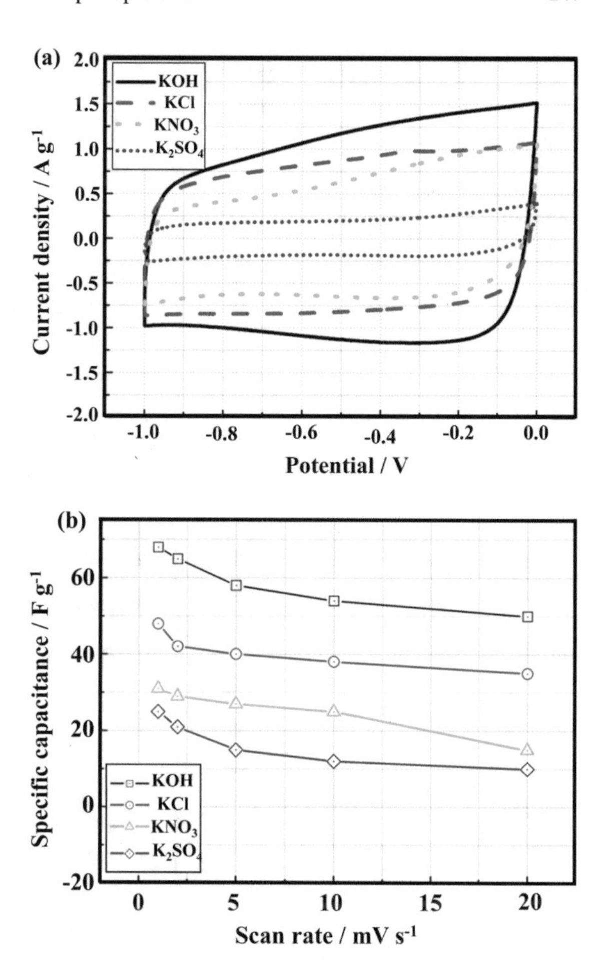

Fig. 7.10 **a** Cyclic voltammograms for various aqueous electrolytes of activated calcium carbide-derived carbon supercapacitors and **b** specific capacitance of supercapacitor for various scan rates (Here 6 M KOH, 2 M KCl, 1 M KNO_3, 0.5 M K_2SO_4) (redrawn and reprinted with permission from [45])

7.5.4 Solid and Quasi-Solid-State Electrolytes

These electrolytes are used in wearable electronics and stretchable supercapacitors to avoid electrolyte leakage problems. Solid-state electrolytes are polymer-based that bind the ions by providing a matrix. Gel electrolytes are known as quasi-solid-state electrolytes due to the presence of liquid. Liquid makes it more ionic conducting and easily stretchable. Figure 7.12a–d depicts the CV performance of different gel electrolyte based supercapacitor using PVA as host gel polymers. PVA-H_3PO_4 shows the best capacity among other gel polymers taken here due to the smallest ion size of H^+ compared to Na^+, K^+, Cl^-, and OH^- [47].

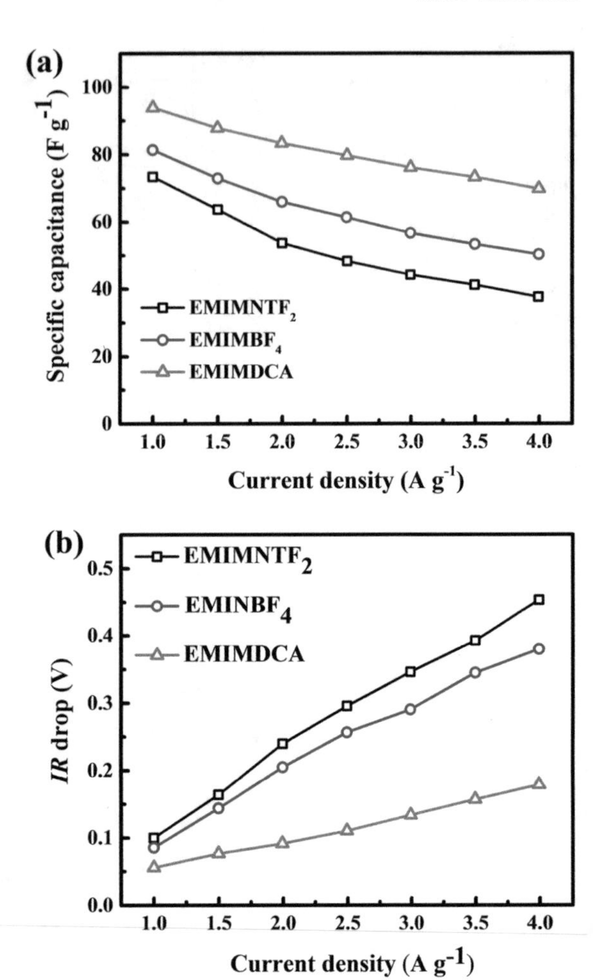

Fig. 7.11 **a** Graph showing specific capacitance of ILs at varying current density for graphene sheet electrodes and **b** IR drop of ILs with increasing current density (redrawn and reprinted with permission from [46])

7.5.5 Redox-Active Electrolytes

Redox-active electrolytes contribute to pseudocapacitance in supercapacitors along with electrode materials. Pseudocapacitance charge storage is based on faradic charge transfer and depends on the surface area [48]. Figure 7.13 shows CV curves, where redox peaks appear in the redox-active electrolyte. In Fig. 7.13, ARS is Alizarin Red S is redox reactive. ARS added gel polymer electrolyte undergoes redox activity, which provides additional capacitance to the device. On the addition of ARS in PVA-H_2SO_4, ionic conductivity increases and shows higher capacitance than PVA-H_2SO_4 electrolyte-based supercapacitor [49].

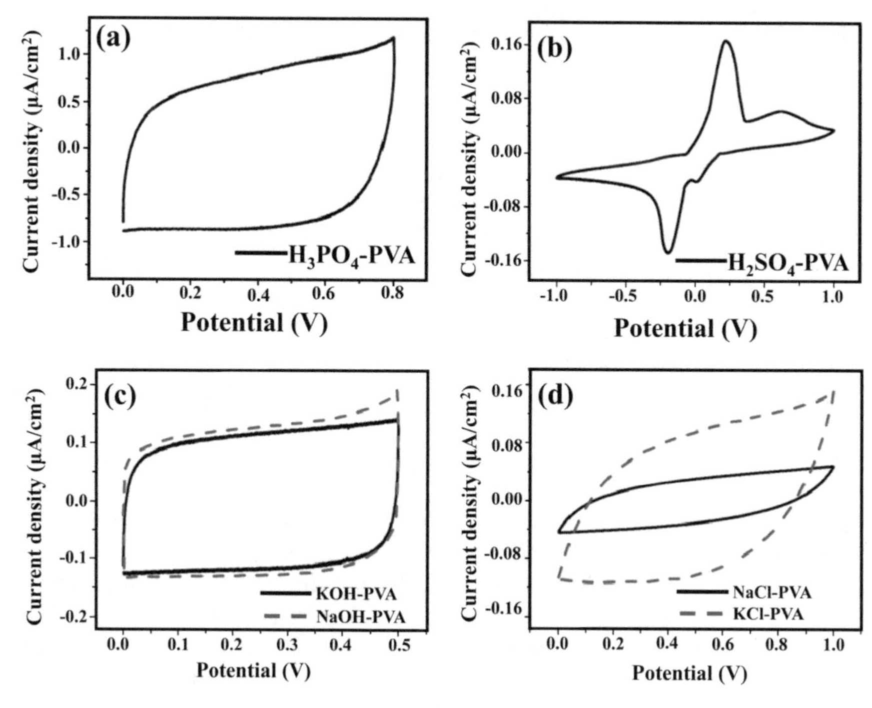

Fig. 7.12 Performance of various gel electrolytes used in solid-state supercapacitors at 10 mV per second scan rate **a** CV curve for H_3PO_4-PVA electrolyte, **b** H_2SO_4-PVA electrolyte, **c** KOH-PVA and NaOH-PVA electrolytes, and **d** NaCl-PVA and KCl-PVA electrolytes (redrawn and reprinted with permission from [47])

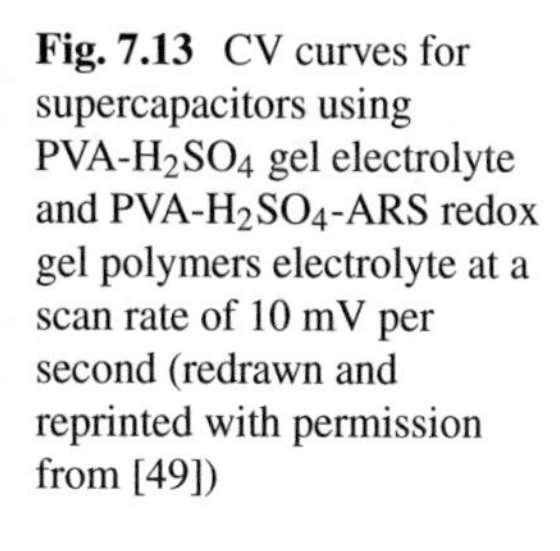

Fig. 7.13 CV curves for supercapacitors using PVA-H_2SO_4 gel electrolyte and PVA-H_2SO_4-ARS redox gel polymers electrolyte at a scan rate of 10 mV per second (redrawn and reprinted with permission from [49])

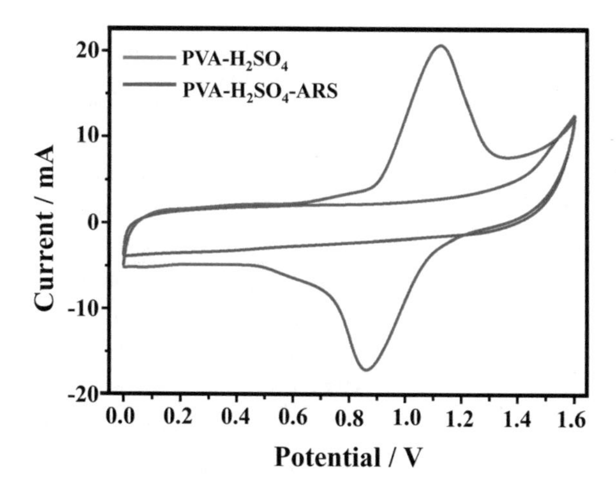

7.6 Electrolytes Used in Commercial Supercapacitors

For hybrid electric vehicles, non-aqueous electrolytes and carbon electrodes based, supercapacitors are the most popular. They show excellent performance at an affordable cost. Organic electrolytes consisting of solvents propylene carbonate or acetonitrile with 1 M tetraethyl-ammonium solute are mostly used in commercial supercapacitors. These electrolytes are compatible with the current collector made of aluminum. Supercapacitors having acetonitrile as an organic solvent for electrolyte has more capacity as compared to propylene carbonate-based supercapacitors. Saft, Panasonic UPC, Ness, and Maxwell use acetonitrile-based electrolytes. CCR, Panasonic UPA, Panasonic UPB, and EPCOS use propylene carbonate-based electrolytes. Panasonic UPA is used for low current and high capacity but UPB is used for high current and low capacity applications. Table 7.2 shows a list of electrolyte solvents used in different commercially available supercapacitors along with their capacitance.

Active carbon electrodes with organic electrolytes achieve capacitance up to 100 F/g having a cell voltage of 2.7 V. Though with aqueous electrolytes, the higher capacitance of 200 F/g can be achieved, however, the operating voltage is restricted up to 1 V. KOH, H_2SO_4, and Na_2SO_4 are commonly used aqueous electrolytes. Figure 7.14 demonstrates performance evaluation of asymmetric and symmetric supercapacitors using different types of electrolytes (Na_2SO_4 and KOH). The Nyquist plot is shown in Fig. 7.14a, which shows a small semi-circle region at high frequency initially and a linear region at low frequency afterward. Ragone plot in Fig. 7.14b shows that neutral electrolyte gives better performance than the alkaline electrolyte. Figure 7.14c shows high retention of capacity up to 90% even after 10,000 cycles [51].

Table 7.2 Capacitance and electrolytes used in various commercial supercapacitors

Manufacturer	Cell ID	Solvent in electrolytes	Capacitance (F)
Saft	SAFT ($n = 1$)	Acetonitrile	3424
Maxwell	PC2500 ($n = 2$)	Acetonitrile	3229
CCR	CCR2000 ($n = 1$)	Propylene carbonate	2439
CCR	CCR3000 ($n = 1$)	Propylene carbonate	2909
Panasonic	UPC ($n = 2$)	Acetonitrile	1737
Panasonic	UPA ($n = 2$)	Propylene carbonate	1851
Panasonic	UPB ($n = 2$)	Propylene carbonate	982
Ness	NESS ($n = 4$)	Acetonitrile	2444
EPCOS	EPI2 ($n = 2$)	Propylene carbonate	1202

Here n is the number of cells used for the experiment and get average capacitance data [50]

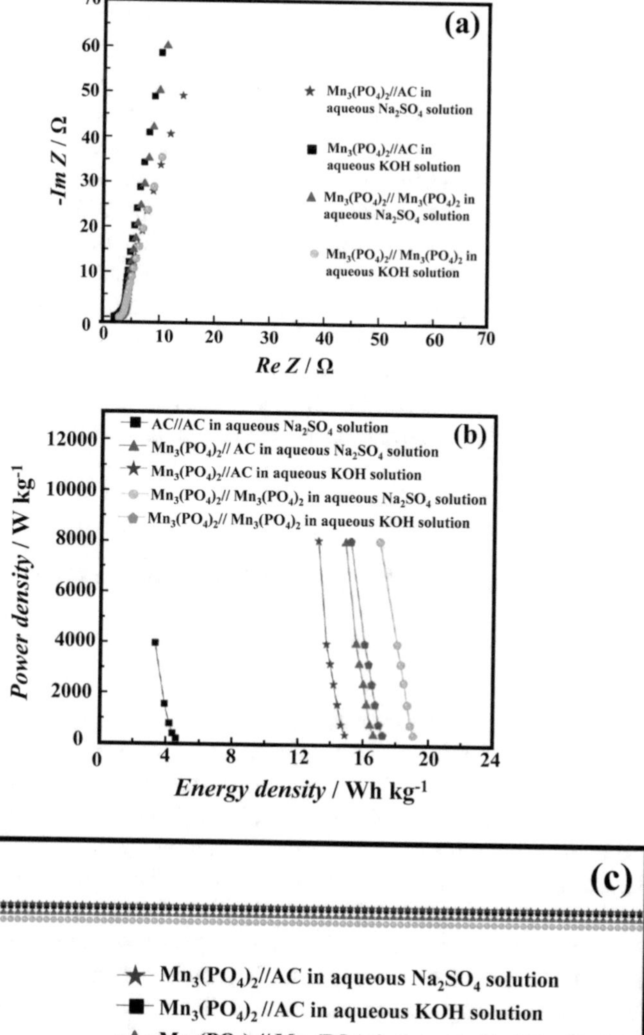

Fig. 7.14 Performance of supercapacitors with different electrode materials and electrolytes **a** Nyquist plot, **b** Ragone plot, and **c** cycle performance (redrawn and reprinted with permission from [51])

7.7 Electrolyte Material Selection for Supercapacitors

The electrolyte is another key component related to the performance of the supercapacitor. On the application of potential, the ions present in the electrolyte get adsorbed on the electrode–electrolyte interface, which induces electric double-layer capacitance. In the pseudocapacitive electrode, the electrolytic ions undergo a faradaic reversible redox reaction to provide pseudocapacitance. Electrolyte ions also intercalate on the lattice of the electrode that provides pseudocapacitance. The operating potential window, equivalent series resistance (ESR), and capacitance of the electrochemical supercapacitor device strongly depend upon the type and concentration of electrolyte. Capacitance depends on the number of charges accumulated on the electrode surface. Small size electrolyte ions can enter into the microporous and mesoporous structure of the electrode, which increases the capacitance [52, 53]. Also, the stability of electrolyte determines the electrochemical stable potential window of the overall device.

7.7.1 Performance Metrics and Relationships

7.7.1.1 Cyclic Voltammetry

Cyclic voltammetry (CV) is an electrochemical technique to investigate the reduction and oxidation process of the supercapacitor device. The curve is used for the capacitance calculation of the supercapacitor device. The CV curve represents the current generated on the application of voltage. In CV, potential varies linearly with the time, which is represented in terms of scan rate (mV/s). For EDLCs, the cyclic voltammetry curve is rectangular. In the case of the pseudocapacitance, the cyclic voltammetry curve has oxidation and reduction humps. Electric double layer and pseudocapacitor both contribute to the total capacitance of the supercapacitor, therefore, the CV curve of a supercapacitor is pseudo-rectangular in shape. In order to understand the energy storage and power delivery mechanism, the fundamentals of the capacitors have been discussed in this section. EDLC capacitance (for parallel plate model) can be calculated with the same equation that has been used for the capacitor as follows [54].

$$C = \frac{\varepsilon A}{d} \tag{7.1}$$

where C is the capacitance in farads, ε is the permittivity of dielectric, A is the area of the electrode surface, d is the distance between the outer Helmholtz plane, and inner Helmholtz plane.

When the voltage difference across the electrode is V and the charge stored in the interference is Q with capacitance C then

$$Q = C \times V \tag{7.2}$$

If the voltage increases with time, then the voltage varies linearly with rate of variation of voltage, (v). Hence,

$$V = V_o + v \times t \tag{7.3}$$

By differentiation of (7.2) with time [3]

$$\frac{\mathrm{d}Q}{\mathrm{d}t} = C\frac{\mathrm{d}V}{\mathrm{d}t} \tag{7.4}$$

From (7.4)

$$I = C \times v \tag{7.5}$$

Current (I) in the capacitor is directly proportional to the voltage variation and independent to the applied voltage.

7.7.1.2 Galvanostatic Charge–Discharge (GCD)

Although the CV curve is used for capacitance calculation of the supercapacitors, however, galvanostatic charge–discharge (GCD) is often used for capacitance calculation. GCD charge–discharge process is almost similar to the real-world application. GCD curve provides a voltage–time plot on the application of constant current until it reaches the maximum voltage. GCD shows real performance because it works with constant current rather than the constant voltage sweep rate as in the case of CV. For an ideal supercapacitor, the charge–discharge curve should be symmetric with a constant discharge curve slope. However, for practical applications supercapacitor does not always show a linear plot and symmetric curve. This happens due to the pseudocapacitance or IR drop and charge transfer resistance across the interface of the electrode and electrolyte, which is the major cause of electrolyte degradation [54].

7.7.1.3 Electrochemical Impedance Spectroscopy

In the electrochemical impedance spectroscopy (EIS), sinusoidal voltage or current is applied and the response is measured and computed to get the impedance at a given frequency. This is done at a large frequency range and the supercapacitor behavior has been analyzed using the frequency response analysis (FRA) curve.

7.7.1.4 Capacitance

Capacity is the ability of the device to store the charge. The capacitance of supercapacitor is evaluated by the following methods.

CV test

The average capacitance of a supercapacitor can be obtained by integrating the CV curve using the following (7.6) [55]

$$C_s = \int i \frac{dE}{(2vm(E_2 - E_1))} \tag{7.6}$$

where C_s (F/g) is the specific capacitance of supercapacitor cell, i (A) is the current, v (V/s) is the scan rate, m (g) is the mass of total activated material loaded in the supercapacitor, $(E_2 - E_1)$ is the total potential window.

Two electrode specific capacitance by CV test: For two electrode symmetric supercapacitor, both electrodes are in series with equal capacitance [54]

$$\frac{1}{C_s} = \frac{1}{C_e} + \frac{1}{C_e} \tag{7.7}$$

$$C_e = 4 \times \int i \frac{dE}{(2vm(E_2 - E_1))} \tag{7.8}$$

where C_s (F/g) is the specific capacitance of the supercapacitor cell and C_e represents the specific capacitance of single electrode.

EIS test

From electrochemical impedance spectroscopy data, specific capacitance has been computed from low frequency response (which is the Z'' part) using following (7.9) and (7.10) [56].

$$C_s = \frac{-1}{2\Pi f Z'' m_s} \tag{7.9}$$

$$C_e = \frac{-1}{\Pi f Z'' m_e} \tag{7.10}$$

where C_s represents the specific capacitance of supercapacitor cell, f denotes the frequency, Z'' stands for the imaginary impedance at frequency f, C_e is the specific capacitance of a single electrode, m_s and m_e are the masses of cell and electrode material, respectively.

GCD test

The specific capacitance of the supercapacitor cell is calculated from the following (7.11) [54]

$$C_s = \frac{i \Delta t}{m \Delta V} \tag{7.11}$$

where C_s is the specific capacitance of supercapacitor cell, i (A) is current, m (g) is the mass of total activated material loaded in a supercapacitor, Δt (s) is total discharge time, and ΔV is the potential window including IR drop.

Single electrode specific capacitance by GCD test

The specific capacitance of a single electrode of a supercapacitor cell is calculated from following (7.12)

$$C_e = 4 \times \frac{i \Delta t}{m \times \Delta V} \tag{7.12}$$

where C_e is the specific capacitance of a single electrode of a supercapacitor cell.

7.7.1.5 Energy and Power Density

The energy density (Wh/kg) and power density (W/kg) of the supercapacitor cell are obtained from the following (7.13) and (7.14) or (7.15), respectively [54].

$$E = C_s (\Delta V)^2 \tag{7.13}$$

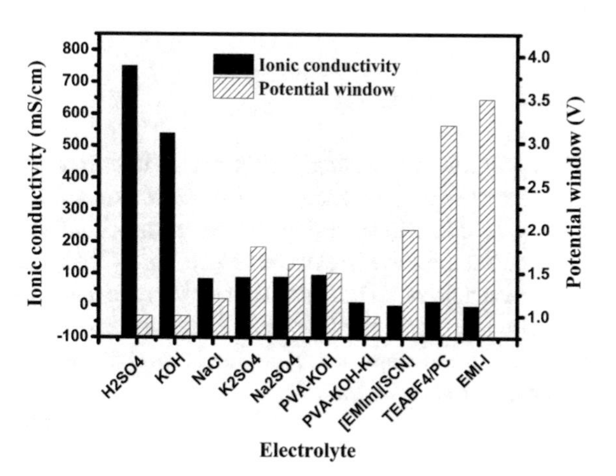

Fig. 7.15 Ionic conductivity with potential window for different electrolyte materials

$$P = \frac{E}{\Delta t} \tag{7.14}$$

or

$$P = \frac{V^2}{4R_s} \tag{7.15}$$

where E is the energy density of the supercapacitor cell, C_s is the specific capacitance of the supercapacitor cell calculated from (7.12), ΔV is the potential window, P is the power density of the supercapacitor cell, Δt is discharge time and R_s is the ESR (equivalent series resistance). Energy and power density both are dependent on the square of the potential window. Wide working potential window, high capacitance, and low ESR are the required parameters to achieve high energy density and power density.

7.7.1.6 Equivalent Series Resistance (ESR)

ESR represents the internal resistance or equivalent series resistance, which includes electrolyte resistance, electrode material internal resistance, and contact resistance between the electrode and current collector when connected to a cell [57]. For high power density, low ESR is required. Therefore, electrolyte ionic conductivity should be high to obtain high power density. Aqueous electrolytes have high ionic conductivity compared to organic and IL electrolytes. However, aqueous electrolytes have the drawback of a low potential window compared to organic and IL electrolytes. For the high performance of supercapacitor, an ideal electrolyte needs to be developed that contains high ionic conductivity and can be operated in a wide potential window [58].

7.7.1.7 Cycle Life

The cycle life of a supercapacitor makes it more reliable than batteries. Charging a device to its maximum energy density and then discharging the device fully represents one cycle life. Lifespan depends upon the chemical and physical stability of the electrode and electrolyte of the device. EDLC has a high cycle life compared to the pseudocapacitors. EDLC uses stable electrode materials like activated carbon, which can give long cycle life (10^5–10^6) because here the charge is stored electrostatically while for pseudocapacitor cycle life is low because of redox reaction on the surface of the electrode material [59].

7.7.1.8 Self-discharge Rate

Self-discharging is a process, where the potential of a charged device starts decreasing without applying any external load. A high self-discharge rate reduces the performance of the supercapacitor. Factors that increase the self-discharge rate are as follows:

- short circuit of the electrodes due to attachment of oppositely charged electrodes.
- charge redistribution over the electrode. Charge redistribution is the process of relaxation of charge because of a potential gradient.
- charging of the supercapacitor above the potential window can also be the probable reason for voltage leakage.
- corrosion of the electrode and current collector may be other reasons for current leakage.

7.7.1.9 Coulombic Efficiency

Coulombic efficiency is the ratio of discharging to charging time of the supercapacitor device and describes the charge transfer efficiency of the device facilitating the electrochemical reaction given by the following (7.16) [60]

$$\eta = \frac{t_D}{t_c} \times 100\% \tag{7.16}$$

where t_D is the discharge time and t_C is the charging time for galvanostatic charging-discharging.

7.7.2 *Objectives for Selection of Electrolyte Material*

The components should be multifunctional to function well in a device. For example, the component should support a load, hold pressure, transmit heat, and so forth. These specifications must be achieved, to assemble a high-efficiency device. While designing such component, the designer has an objective: (a) to make it economical, (b) light-weight, (c) high performance, and (d) safe, or perhaps some combination of these. Table 7.3 demonstrates primary, secondary, and tertiary objectives of the electrolyte material that are responsible for efficient capacitive performance. The electrolyte properties of different electrolyte are given in Table 7.4.

Table 7.3 Objectives of efficient electrolyte material for electrochemical supercapacitors [27]

Primary objective	Secondary objective	Tertiary objective
High potential window	Optimum viscosity	Environmental friendly
High ionic conductivity	Low cost	Easy to handle
Small size hydrated ions	High chemical stability	Easily available
High dissociation	Low volatility	Non-toxic
Large operating temperature range	Non-flammability	
	Optimum concentration	

7.7.3 Screening Using Constrains

The primary objective of an electrolyte material selection is a high potential window and high ionic conductivity. Figure 7.15 demonstrates the behavior of different electrolytes with their ionic conductivity and potential window. Till now, no electrolyte can fulfill both objectives simultaneously. The aqueous electrolyte has high ionic conductivity but low potential window while organic and ionic liquid electrolytes have a high potential window but low ionic conductivity [87].

Figure 7.16 demonstrates different electrolyte behavior with ionic conductivity and equivalent series resistance (ESR) of the supercapacitor device. For the high power density of the supercapacitor, ESR should be low. It is clearly demonstrated that the ESR of the supercapacitor device is inversely proportional to the ionic conductivity of the electrolyte. Aqueous electrolytes have higher ionic conductivity in comparison with organic and ionic electrolytes.

Governing Equations

Energy density—Energy density of the supercapacitor is energy stored in the supercapacitor per unit mass. Energy density is represented as (7.13)

$$E = C_s(\Delta V)^2$$

where E, C_s, and V stand for the energy density, specific capacitance, and potential window, respectively.

Power density—Power density is the rate of energy transfer per unit mass of the supercapacitor. Power density is represented as (7.15)

$$P = \frac{V^2}{4R_s}$$

where P, V, R_s, and m denote the power density, potential window, equivalent series resistance, and mass of the electrode material, respectively.

Table 7.4 Different electrolyte properties for supercapacitor performance

Electrolyte		Electrode material	Potential window (V)	Energy density (kWh/kg)	Power density (W/kg)	ESR (Ω)	Ionic conductivity (σ) (S/cm)	References
Aqueous electrolyte	1 M H_2SO_4	CF	−0.1 to 0.9	35	–	0.15	750×10^{-3}	[61]
	0.5 M H_2SO_4	RuO_2/Graphene	0–1.2	20.28	600	0.6	–	[62, 63]
	1 M H_2SO_4	Polyaniline	0–0.7	11.3	106.7	–	–	[64]
	6 M KOH	AC-Corn cob	0–1	15	17	0.7	–	[65]
	1 M KOH	$Ni(OH)_2$/RuO_2 /Graphene	0–1.5	21.65	14	–	–	[66, 67]
	6 M KOH	NiO	0–0.45	10.2	–	0.1	540×10^{-3}	[68]
	1.5 M Li_2SO_4	CF/CNT	0–1	28.43	3700	–	–	[69, 70]
	0.5 M K_2SO_4	V_2O_5-0.6H_2O	0–1.8	24	500		88.6×10^{-3}	[71]
	1 M Na_2SO_4	RuO_2	0–1.6	18.7	500	–	91.1×10^{-3}	[72]
Organic electrolyte	1 M $TEABF_4$/PC	AC	2.4–4	46	–	0.1	13×10^{-3}	[73]
	TEABF/HFIP	Glassy carbon	−2 to 2	–	–	0.66	15×10^{-3}	[74]
	0.7 M Et_4NBF_4/ADN	Carbon	−2.1 to 1.65	28	3000	12	4.3×10^{-3}	[75]
	SBP-BF_4/PC	AC	0–3.2	43Wh/kg	6938 W/kg	1.2	17×10^{-3}	[76]
	1.5 M $TEMABF_4$/PC	AC	0–3.2	–	–	1.25	14.6×10^{-3}	[76]
	1.5 M $TEABF_4$/AN	AC	0–3	44	9000 W/kg	0.4	4.9×10^{-3}	[77]
IL electrolyte	[BMIM][BF_4]/AN	Micro wave expanded graphite oxide	0–3.5		338,000 kW/kg	5.7	2.1×10^{-3}	[78]
	[EMIm][SCN]	AC	−1 to 1	–	–	–	2.1×10^{-3}	[79]
	EMI-FSI	AC	0–2	44	9 kW/kg	2.5	15.5×10^{-3}	[77]

(continued)

Table 7.4 (continued)

Electrolyte		Electrode material	Potential window (V)	Energy density (kWh/kg)	Power density (W/kg)	ESR (Ω)	Ionic conductivity (σ) (S/cm)	References
Solid or quasi-solid-state electrolyte	PVA-5 M KOH	NiO	0–1.5	26.1	–	1.15	9.7×10^{-2}	[80, 81]
	PVA-KOH-KI	AC	0–1	7.8	15.34 kW/kg	1.17	12.73×10^{-3}	[82, 83]
Redox-active electrolyte	4 M KOH—p-phenylenediamine	MnO_2	−0.4–0.6	–	–	1.34	–	[84]
	0.38 M hydroquinone Dihydroxybenzenes + 1 M H_2SO_4	AC	0–0.8	–	–	0.3	–	[85]
	2 M NaI	AC	0–1.2	–	–	0.3	275×10^{-3}	[85]
	EMI-I	Porous carbon	0–3.5	175.6	4994.5	2.12	0.34×10^{-3}	[86]

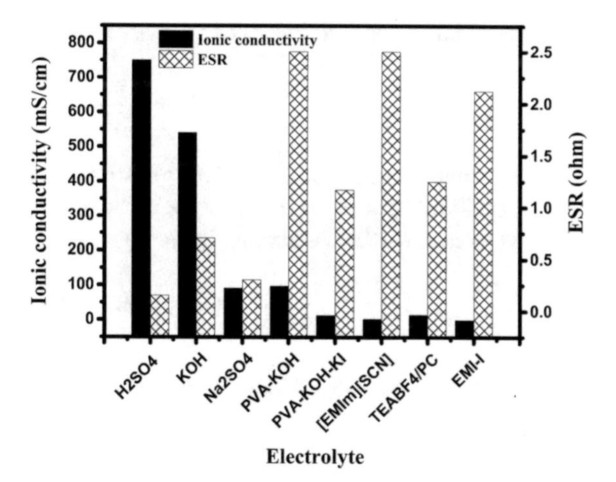

Fig. 7.16 Ionic conductivity and ESR of aqueous, organic, and ionic liquid electrolytes

Ionic conductivity—Ionic conductivity is the movement of ions from one place to another place through the medium. Ionic conductivity is given as

$$\sigma = nZe\mu \tag{7.17}$$

where σ, n, μ, and e indicate the ionic conductivity, carrier concentration, carrier mobility, and charge of an electron, respectively.

Material Indexes

From (7.13) the energy density is written as

$$E = \frac{1}{2}C_s(\Delta V)^2$$

The best materials for high-performance supercapacitors are those with high values of the material index. Material index of energy density and potential window as the requirements are high potential window and energy density

$$M_1 = \mathrm{EV}^2 \tag{7.18}$$

Similarly, the power density is written as based on (7.15)

$$P = \frac{V^2}{4Rsm}$$

Material index for high power density and low equivalent series resistance

$$M_2 = \frac{P}{\mathrm{ESR}} \tag{7.19}$$

Material index for high ionic conductivity and low equivalent series resistance

$$M_3 = \frac{\boldsymbol{\sigma}}{\mathbf{ESR}} \tag{7.20}$$

Final material index for electrolyte material is represented as (it can be in any other form like $M_4 = M_1 \times \pm \div M_2 \times \pm \div M_3$ or $a_1 M_1 \times \pm \div a_2 M_2 \times \pm \div a_3 M_3$, where a_1, a_2, and a_3 are any integers)

$$M_4 = M_1 \times M_2 \times M_3 \tag{7.21}$$

Material index of energy density and potential window

From the Eq. 7.18

$$M_1 = E \times V^2$$

Taking log both side

$$\log M_1 = \log E + 2 \log V$$

or

$$\log E = -2 \log V + \log M_1 \tag{7.22}$$

Figure 7.17 shows the Ashby chart of potential window versus energy density.

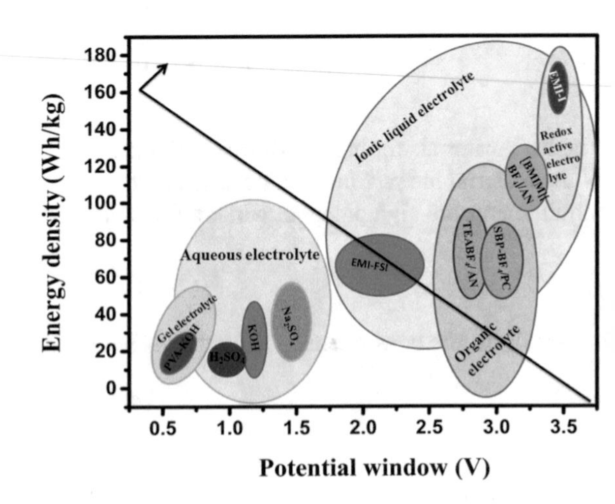

Fig. 7.17 Ashby chart of energy density versus potential window

For the high performance of the electrochemical supercapacitor, there is a need for high energy density and a high working potential window. To fulfill these requirements, electrolyte materials above the index line are suitable for supercapacitor application. Organic electrolyte like $TEABF_4$/AN and SBP-BF/PC, ionic liquid electrolyte like EMI-FSI and [BMIM][BF_4]/AN, redox-active electrolyte like EMI-I are some examples of suitable electrolyte materials.

Material index of power density and equivalent series resistance

From (7.21), another material index based on high power density and low equivalent series resistance is written as

$$M_2 = \frac{\boldsymbol{P}}{\mathbf{ESR}}$$

Taking log both side

$$\log M_2 = \log P - \log \mathrm{ESR}$$

or,

$$\log P = \log \mathrm{ESR} + \log M_2 \tag{7.23}$$

In the Ashby chart of power density and equivalent series resistance (ESR) index line is shown in Fig. 7.18.

High power density and low equivalent series resistance are the potential requirements for supercapacitor applications. Electrolytes above the index line are suitable for supercapacitors. Aqueous electrolytes, organic electrolytes, gel polymer electrolytes, and ionic liquid electrolytes are also suitable electrolytes.

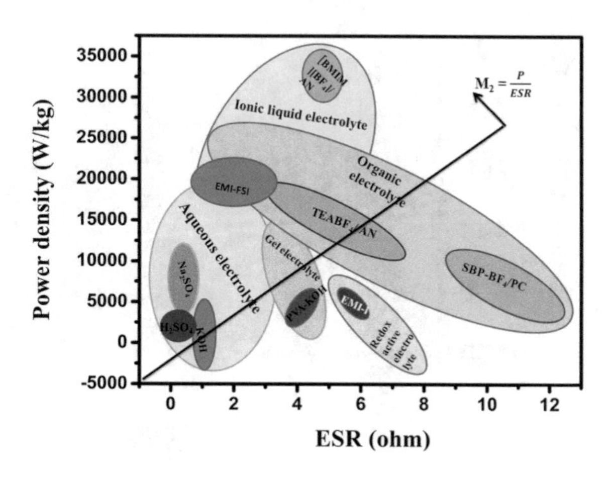

Fig. 7.18 Ashby chart of power density and ESR

Material index of ionic conductivity and equivalent series resistance

From (7.22) the third material index based on high ionic conductivity and low equivalent series resistance is represented as

$$M_3 = \frac{\sigma}{\mathrm{ESR}}$$

Taking log both side

$$\log M_3 = \log \sigma - \log \mathrm{ESR}$$

or

$$\log \sigma = \log \mathrm{ESR} + \log M_3 \tag{7.24}$$

The Ashby chart of ionic conductivity and equivalent series resistance index line is shown in Fig. 7.19.

For the high performance of the supercapacitor, we have the requirements of high ionic conductivity and low ESR. Hence, material selection should be above the index

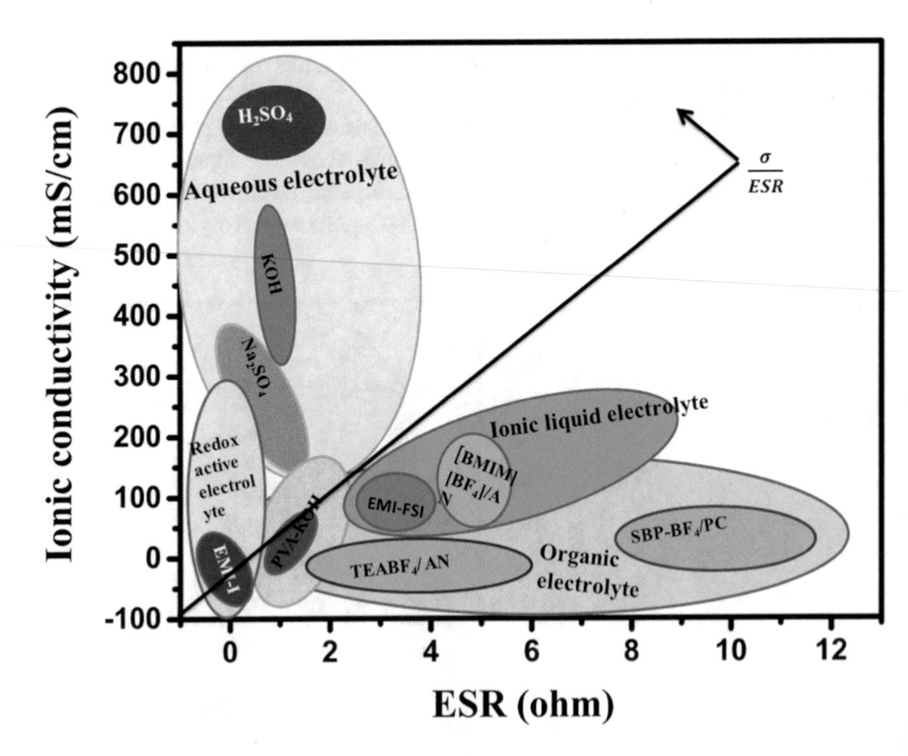

Fig. 7.19 Ashby chart of ionic conductivity and ESR

Table 7.5 List of material index

Electrolyte	Material index M_1	Material index M_2	Material index M_3	$M_4 = M_1 \times M_2 \times M_3$
H_2SO_4	5.537	1000	1250	6.92×10^6
KOH	15	24.28	771.4	2.81×10^5
Na_2SO_4	47.87	1666.6	303.6	2.42×10^7
BP-BF_4/PC	440.32	250	1.41	1.55×10^5
TEABF_4/AN	396	22,500	12.25	1.09×10^8
[BMIM][BF_4]/AN	906.5	5929.8	0.368	1.98×10^6
EMI-FSI	176	3600	6.2	3.92×10^6
PVA-KOH-KI	7.8	13,111.1	10.88	1.11×10^6
EMI-I	2151.1	2355.89	0.16	8.12×10^5

line. Aqueous electrolytes, redox-active electrolytes, and gel polymer electrolytes are suitable materials in this respect.

7.7.4 List of Material Indexes

Material indexes for selected materials are calculated and given in Table 7.5. According to material index M_4, suitable electrolytes are listed accordingly in Table 7.6.

7.8 Concluding Remarks

Energy density, power density, and lifetime are important factors among others while choosing an energy storage device for a particular application. For supercapacitors, these factors greatly depend on the characteristics of electrolytes. High ionic conductivity, optimum ion concentration, low viscosity, high thermal and electrochemical stability, and wide voltage window are preferred in electrolytes. For commercial use of electrolytes toxicity, volatility and cost are also considered. Organic electrolytes have good electrical properties but high flammability and volatility are major issues. Ionic liquids show good conductivity and non-flammability but are expensive. On the other hand, aqueous electrolytes provide low viscosity and good conductivity at low cost but possess a narrow working potential window. Hence, depending upon the type of application, these electrolytes are chosen according to their properties for commercial supercapacitors.

Table 7.6 Electrolytes for supercapacitor application according to material index

Electrolyte	Properties and advantages	Disadvantages
$TEABF_4$/AN	Organic electrolyte, high potential window, low viscosity, high chemical stability	Toxic, flammable, high cost, low ionic conductivity in comparison with aqueous electrolyte
Na_2SO_4	Neutral electrolyte, large operating temperature, chemical stable, non-flammable, low volatile, low cost, environmental friendly	Low potential window in comparison with organic and IL electrolytes
H_2SO_4	Acid electrolyte, high ionic conductivity, low cost, low viscosity, easily available	Low potential window, low chemical stability, not easy to handle, toxic
EMI-FSI	Organic electrolyte, high potential window, low viscosity, high thermal stability	Toxic, low ionic conductivity than the aqueous electrolyte, high cost, flammable
[BMIM][BF_4]/AN	Ionic liquid electrolyte, high potential window, high thermal stability	Low ionic conductivity than aqueous electrolyte and organic electrolyte, toxic
EMI-I	Redox-active electrolyte, high pseudocapacitance, high potential window, easy	Low ionic conductivity, high viscosity, flammability, high cost, toxic
KOH	Alkaline electrolyte, high ionic conductivity, low viscosity, low cost, non-flammability, environmental friendly	Low potential window, toxic
BP-BF_4 /PC	Organic electrolyte, high potential window, high thermal stability	Low ionic conductivity, toxic, high cost
PVA-KOH-KI	Gel polymer electrolyte, high mechanical stability, high potential window, high thermal stability	Low ionic conductivity, high cost, toxic

Acknowledgements The authors acknowledge the financial support provided by the Science and Engineering Research Board, Department of Science and Technology, India (SR/WOS-A/ET-48/2018) for carrying out this research work.

References

1. S. Banerjee, P. Sinha, K.D. Verma, T. Pal, B. De, J. Cherusseri, P.K. Manna, K.K. Kar, Capacitor to supercapacitor, in *Handbook of Nanocomposite Supercapacitor Materials I Characteristics*, ed. by K.K. Kar (Springer, Berlin, Heidelberg, 2020). https://doi.org/10.1007/978-3-030-43009-2_2
2. J. Tahalyani, J. Akhtar, J. Cherusseri, K.K. Kar, Characteristics of capacitor: fundamental aspects, in *Handbook of Nanocomposite Supercapacitor Materials I Characteristics*, ed. by K.K. Kar (Springer, Berlin, Heidelberg, 2020). https://doi.org/10.1007/978-3-030-43009-2_1

3. P. Sinha, K.K. Kar, Introduction to supercapacitors, in *Handbook of Nanocomposite Supercapacitor Materials II Performance*, ed. by K.K. Kar (Springer, Berlin, Heidelberg, 2020). https://doi.org/10.1007/978-3-030-52359-6_1
4. P. Sinha, K.K. Kar, Characteristics of supercapacitors, in *Handbook of Nanocomposite Supercapacitor Materials II Performance*, ed. by K.K. Kar (Springer, Berlin, Heidelberg, 2020). https://doi.org/10.1007/978-3-030-52359-6_3
5. K.K. Kar, *Handbook of Nanocomposite Supercapacitor Materials II, Performance*, in ed. by K.K. Kar (Springer Nature Switzerland AG, 2020), eBook ISBN: 978-3-030-52359-6, Hardcover ISBN: 978-3-030-52358-9. https://doi.org/10.1007/978-3-030-52359-6
6. P. Sinha, S. Banerjee, K.K. Kar, Activated carbon as electrode materials for supercapacitors, in *Handbook of Nanocomposite Supercapacitor Materials II Performance*, ed. by K.K. Kar (Springer, Berlin, Heidelberg, 2020). https://doi.org/10.1007/978-3-030-52359-6_5
7. K.D. Verma, P. Sinha, S. Banerjee, K.K. Kar, Characteristics of electrode materials for supercapacitors, in *Handbook of Nanocomposite Supercapacitor Materials I Characteristics*, ed. by K.K. Kar (Springer, Berlin Heidelberg, 2020). https://doi.org/10.1007/978-3-030-43009-2_9
8. B. De, S. Banerjee, K.D. Verma, T. Pal, P.K. Manna, K.K. Kar, Transition metal oxides as electrode materials for supercapacitors, in *Handbook of Nanocomposite Supercapacitor Materials II Performance*, ed. by K.K. Kar (Springer, Berlin, Heidelberg, 2020). https://doi.org/10.1007/978-3-030-52359-6_4
9. B. De, S. Banerjee, T. Pal, K.D. Verma, A. Tyagi, P.K. Manna, K.K. Kar, Transition metal oxide/electronically conducting polymer composites as electrode materials for supercapacitors, in *Handbook of Nanocomposite Supercapacitor Materials II Performance*, ed. by K.K. Kar (Springer, Berlin, Heidelberg, 2020). https://doi.org/10.1007/978-3-030-52359-6_14
10. B. De, S. Banerjee, T. Pal, K.D. Verma, A. Tyagi, P.K. Manna, K.K. Kar, Transition metal oxide-/carbon-/electronically conducting polymer-based ternary composites as electrode materials for supercapacitors, in *Handbook of Nanocomposite Supercapacitor Materials II Performance*, ed. by K.K. Kar (Springer, Berlin, Heidelberg, 2020). https://doi.org/10.1007/978-3-030-52359-6_15
11. A. Muzaffar, M.B. Ahamed, K. Deshmukh, J. Thirumalai, Renew. Sust. Energ. Rev. **101**, 123–145 (2019)
12. P. Sinha, S. Banerjee, K.K. Kar, Characteristics of activated carbon, in *Handbook of Nanocomposite Supercapacitor Materials I Characteristics*, ed. by K.K. Kar (Springer, Berlin, Heidelberg, 2020). https://doi.org/10.1007/978-3-030-43009-2_4
13. B. De, S. Banerjee, T. Pal, K.D. Verma , P.K. Manna, K.K. Kar, Graphene/reduced graphene oxide as electrode materials for supercapacitors, in *Handbook of Nanocomposite Supercapacitor Materials II Performance*, ed. by K.K. Kar (Springer, Berlin, Heidelberg, 2020). https://doi.org/10.1007/978-3-030-52359-6_11
14. P. Chamoli, S. Banerjee, K.K. Raina, K.K. Kar, Characteristics of graphene/reduced graphene oxide, in *Handbook of Nanocomposite Supercapacitor Materials I Characteristics*, ed. by K.K. Kar (Springer, Berlin, Heidelberg, 2020). https://doi.org/10.1007/978-3-030-43009-2_5
15. B. De, S. Banerjee, K.D. Verma, T. Pal, P.K. Manna, K.K. Kar, Carbon nanotube as electrode materials for supercapacitors, in *Handbook of Nanocomposite Supercapacitor Materials II Performance*, ed. by K.K. Kar (Springer, Berlin, Heidelberg, 2020). https://doi.org/10.1007/978-3-030-52359-6_9
16. S. Banerjee, K.K. Kar, Characteristics of carbon nanotubes, in *Handbook of Nanocomposite Supercapacitor Materials I Characteristics*, ed. by K.K. Kar (Springer, Berlin, Heidelberg, 2020). https://doi.org/10.1007/978-3-030-43009-2_6
17. B. De, S. Banerjee, K.D. Verma, T. Pal, P.K. Manna, K.K. Kar, Carbon nanofiber as electrode materials for supercapacitors, in *Handbook of Nanocomposite Supercapacitor Materials II Performance*, ed. by K.K. Kar (Springer, Berlin, Heidelberg, 2020). https://doi.org/10.1007/978-3-030-52359-6_7
18. R. Sharma, K.K. Kar, Characteristics of carbon nanofibers, in *Handbook of Nanocomposite Supercapacitor Materials I Characteristics*, ed. by K.K. Kar (Springer, Berlin, Heidelberg, 2020). https://doi.org/10.1007/978-3-030-43009-2_7

19. A. Tyagi, S. Banerjee, J. Cherusseri, K.K. Kar, Characteristics of transition metal oxides, in *Handbook of Nanocomposite Supercapacitor Materials I Characteristics*, ed. by K.K. Kar (Springer, Berlin, Heidelberg, 2020). https://doi.org/10.1007/978-3-030-43009-2_3
20. P. Sinha, S. Banerjee, K.K. Kar, Transition metal oxide/activated carbon-based composites as electrode materials for supercapacitors, in *Handbook of Nanocomposite Supercapacitor Materials II Performance*, ed. by K.K. Kar (Springer, Berlin, Heidelberg, 2020). https://doi.org/10.1007/978-3-030-52359-6_6
21. B. De, S. Banerjee, K.D. Verma, T. Pal, P.K. Manna, K.K. Kar, Transition metal oxide/carbon nanofiber composites as electrode materials for supercapacitors, in *Handbook of Nanocomposite Supercapacitor Materials II Performance*, ed. by K.K. Kar (Springer, Berlin, Heidelberg, 2020). https://doi.org/10.1007/978-3-030-52359-6_8
22. B. De, S. Banerjee, T. Pal, K.D. Verma, A. Tyagi, P.K. Manna, K.K. Kar, Transition metal oxide/carbon nanotube composites as electrode materials for supercapacitors, in *Handbook of Nanocomposite Supercapacitor Materials II Performance*, ed. by K.K. Kar (Springer, Berlin, Heidelberg, 2020). https://doi.org/10.1007/978-3-030-52359-6_10
23. B. De, P. Sinha, S. Banerjee, T. Pal, K.D. Verma, A. Tyagi, P.K. Manna, K.K. Kar, Transition metal oxide/graphene/reduced graphene oxide composites as electrode materials for supercapacitors, in *Handbook of Nanocomposite Supercapacitor Materials II Performance*, ed. by K.K. Kar (Springer, Berlin, Heidelberg, 2020). https://doi.org/10.1007/978-3-030-52359-6_12
24. S. Banerjee, K.K. Kar, Conducting polymers as electrode materials for supercapacitors, in *Handbook of Nanocomposite Supercapacitor Materials II Performance*, ed. by K.K. Kar (Springer, Berlin, Heidelberg, 2020). https://doi.org/10.1007/978-3-030-52359-6_13
25. T. Pal, S. Banerjee, P.K. Manna, K.K. Kar, Characteristics of conducting polymers, in *Handbook of Nanocomposite Supercapacitor Materials I Characteristics*, ed. by K.K. Kar (Springer, Berlin, Heidelberg, 2020). https://doi.org/10.1007/978-3-030-43009-2_8
26. K.D. Verma, P. Sinha, S. Banerjee, K.K. Kar, Characteristics of current collector materials for supercapacitors, in *Handbook of Nanocomposite Supercapacitor Materials I Characteristics*, ed. by K.K. Kar (Springer, Berlin, Heidelberg, 2020). https://doi.org/10.1007/978-3-030-43009-2_12
27. K.D. Verma, S. Banerjee, K.K. Kar, Characteristics of electrolytes, in *Handbook of Nanocomposite Supercapacitor Materials I Characteristics*, ed. by K.K. Kar (Springer, Berlin, Heidelberg, 2020). https://doi.org/10.1007/978-3-030-43009-2_10
28. K. D. Verma, P. Sinha, S. Banerjee, K.K. Kar, M.K. Ghorai, Characteristics of separator materials for supercapacitors, in *Handbook of Nanocomposite Supercapacitor Materials I Characteristics*, ed. by K.K. Kar (Springer, Berlin, Heidelberg, 2020). https://doi.org/10.1007/978-3-030-43009-2_11
29. B. Pal, S. Yang, S. Ramesh, V. Thangadurai, R. Jose, Nanoscale Adv. **1**, 3807–3835 (2019)
30. B. Viswanathan, *Energy Sources,* 1st edn. (Elsevier, 2017)
31. X.Y. Chen, C. Chen, Z.J. Zhang, D.H. Xie, J Mater Chem A **1**, 1093–1911 (2013)
32. L. Yu, G.Z. Chen, Front. Chem. **7**, 272 (2019)
33. J. Yan, Q. Wang, T. Wei, Z. Fan, Adv Energy Mater **4**, 1300816 (2014)
34. L. Yu, G.Z. Chen, J. Power Sour. **326**, 604–612 (2016)
35. A.R. Koh, B. Hwang, K.C. Roh, K. Kim, Phys. Chem. Chem. Phys. **16**, 15146–15151 (2014)
36. S.I. Wong, H. Lin, J. Sunarso, B.T. Wong, B. Jia, Appl. Mater. Today **18**, 100522 (2020)
37. A. Kumar, R. Sharma, M. Suresh, M.K. Das, K.K. Kar, J. Elastomers Plast. **49**, 513–526 (2016)
38. M. Ue, Electrochemistry **75**, 565–572 (2007)
39. J. Zhu, Y. Xu, J. Wang, J. Lin, X. Sun, S. Mao, Phys. Chem. Chem. Phys. **17**, 28666–28673 (2015)
40. A.A. Mohamad, N.S. Mohamed, M.Z. Yahya, R. Othman, S. Ramesh, Y. Alias, A.K. Arof, **156**, 171–177 (2003)
41. X. Yang, F. Zhang, L. Zhang, T. Zhang, Y. Huang, Y. Chen, Adv. Funct. Mater. **23**, 3353–3360 (2013)
42. G. Cheruvally, J.K. Kim, J.W. Choi, J.H. Ahn, Y.J. Shin, J. Manuel, P. Raghavan, K.-W. Kim, H.J. Ahn, D.S. Choi, C.E. Song. J. Power Sour. **172**, 863–869 (2007)

43. S. Somiya, *Handbook of Advanced Ceramics Materials, Applications, Processing, and Properties* 2nd edn. (Elsevier, 2013)
44. D. Jiang, J. Wu, Nanoscale **6**, 5545–5550 (2014)
45. H. Wu, X. Wang, L. Jiang, C. Wu, Q. Zhao, X. Liu, B. Hu, L. Yi, J. Power Sour. **226**, 202–209 (2013)
46. M. Shi, S. Kou, X. Yan, Chemsuschem **7**, 3053–3062 (2014)
47. Q. Chen, X. Li, X. Zang, Y. Cao, Y. He, P. Li, K. Wang, J. Wei, D. Wu, H. Zhu, RSC Adv. **4**, 36253–36256 (2014)
48. R. Narayanan, P.R. Bandaru, J. Electrochem. Soc. **162**, A86–A91 (2015)
49. K. Sun, F. Ran, G. Zhao, Y. Zhu, Y. Zheng, M. Ma, X. Zheng, G. Ma, Z. Lei, RSC Adv. **6**, 55225–55232 (2016)
50. A. Chu, P. Braatz, J. Power Sour. **112**, 236–246 (2002)
51. X.J. Ma, W.B. Zhang, L.B. Kong, Y.C. Luo, L. Kang, RSC Adv. **6**, 40077–40085 (2016)
52. M. Kumar, P. Sinha, T. Pal, K.K. Kar, Materials for supercapacitors, in *Handbook of Nanocomposite Supercapacitor Materials II Performance*, ed. by K.K. Kar (Springer, Berlin, Heidelberg, 2020). https://doi.org/10.1007/978-3-030-52359-6_2
53. A. Kumar, R. Sharma, M. Suresh, M.K. Das, K.K. Kar, J. Elastomers Plast. **49**, 513 (2017)
54. R.B. Marichi, V. Sahu, R.K. Sharma, G. Singh, in *Handbook Ecomaterials* (Springer International Publishing, Cham, 2019), p. 855–880
55. L. Zhu, F. Shen, R.L. Smith, L. Yan, L. Li, X. Qi, Chem. Eng. J. **316**, 770 (2017)
56. N.S.M. Nor, M. Deraman, R. Omar, E. Taer, Awitdrus, R. Farma, N.H. Basri, B.N.M. Dolah, AIP Conf. Proc. **1586**, 68 (2014)
57. B.-A. Mei, O. Munteshari, J. Lau, B. Dunn, L. Pilon, J. Phys. Chem. C **122**, 194 (2018)
58. A. Kumar, R. Sharma, M.K. Das, P. Gajbhiye, K.K. Kar, Electrochim. Acta **215**, 1 (2016)
59. C. Zhong, Y. Deng, W. Hu, D. Sun, X. Han, J. Qiao, J. Zhang, *Electrolytes for Electrochemical Supercapacitors*, 1st edn. (CRC Press, 2016)
60. R.I. Jaidev, A.K. Jafri, S.R. Mishra, J. Mater. Chem. **21**, 17601 (2011)
61. Z. Jin, X. Yan, Y. Yu, G. Zhao, J. Mater. Chem. A **2**, 11706 (2014)
62. L. Deng, J. Wang, G. Zhu, L. Kang, Z. Hao, Z. Lei, Z. Yang, Z.H. Liu, J. Power Sources **248**, 407 (2014)
63. S. Banerjee, P. Benjwal, M. Singh, K.K. Kar, Appl. Surf. Sci. **439**, 560 (2018)
64. Q. Wang, J. Yan, Z. Fan, T. Wei, M. Zhang, X. Jing, J. Power Sources **247**, 197 (2014)
65. W.H. Qu, Y.Y. Xu, A.H. Lu, X.Q. Zhang, W.C. Li, Bioresour. Technol. **189**, 285 (2015)
66. H. Wang, Y. Liang, T. Mirfakhrai, Z. Chen, H.S. Casalongue, H. Dai, Nano Res. **4**, 729 (2011)
67. K.K. Kar, *Composite Materials* (Springer, Berlin Heidelberg, 2017).
68. J.W. Lee, T. Ahn, J.H. Kim, J.M. Ko, J.D. Kim, Electrochim. Acta **56**, 4849 (2011)
69. A. Dang, T. Li, C. Xiong, T. Zhao, Y. Shang, H. Liu, X. Chen, H. Li, Q. Zhuang, S. Zhang, Compos. Part B Eng. **141**, 250 (2018)
70. S. Banerjee, K.K. Kar, J. Appl. Polym. Sci. **133**, 42952 (2016)
71. Q.T. Qu, Y. Shi, L.L. Li, W.L. Guo, Y.P. Wu, H.P. Zhang, S.Y. Guan, R. Holze, Electrochem. Commun. **11**, 1325 (2009)
72. H. Xia, Y. Shirley Meng, G. Yuan, C. Cui, L. Lu, Electrochem. Solid-State Lett. **15**, A60 (2012)
73. S. Ishimoto, Y. Asakawa, M. Shinya, K. Naoi, J. Electrochem. Soc. **156**, A563 (2009)
74. R. Francke, D. Cericola, R. Kötz, D. Weingarth, S.R. Waldvogel, Electrochim. Acta **62**, 372 (2012)
75. A. Brandt, P. Isken, A. Lex-Balducci, A. Balducci, J. Power Sour. **204**, 213 (2012)
76. X. Yu, D. Ruan, C. Wu, J. Wang, Z. Shi, J. Power Sour. **265**, 309 (2014)
77. F. Markoulidis, C. Lei, C. Lekakou, Electrochim. Acta **249**, 122 (2017)
78. T. Kim, G. Jung, S. Yoo, K.S. Suh, R.S. Ruoff, ACS Nano **7**, 6899 (2013)
79. G. Hua Sun, K. Xi Li, C. Gong Sun, J. Power Sour. **162**, 1444 (2006)
80. C. Yuan, X. Zhang, Q. Wu, B. Gao, Solid State Ion. **177**, 1237 (2006)
81. S. Banerjee, K.K. Kar, J. Environ. Chem. Eng. **4**, 299 (2016)
82. H. Yu, J. Wu, L. Fan, K. Xu, X. Zhong, Y. Lin, J. Lin, Electrochim. Acta **56**, 6881 (2011)

83. K.K. Kar, J.K. Pandey, S. Rana, *Handbook of Polymer Nanocomposites. Processing, Performance and Application* (Springer Berlin Heidelberg, Berlin, Heidelberg, 2015)
84. L. Su, L. Gong, H. Lü, Q. Xü, J. Power Sour. **248**, 212 (2014)
85. E. Frackowiak, M. Meller, J. Menzel, D. Gastol, K. Fic, Faraday Discuss. **172**, 179 (2014)
86. D.J. You, Z. Yin, Y.K. Ahn, S.H. Lee, J. Yoo, Y.S. Kim, RSC Adv. **7**, 55702 (2017)
87. K.K. Kar, A. Hodzic, *Developments in Nanocomposites*, 1st edn. (Research publishing, 2014)

Chapter 8
Current Collector Material Selection for Supercapacitors

Harish Trivedi, Kapil Dev Verma, Prerna Sinha, and Kamal K. Kar

Abstract The supercapacitor is a step-up device in the field of energy storage and has a lot of research and development scope in terms of design, its parts fabrication, and energy storage mechanism. The main function of the current collector is to collect and conduct electric current from electrodes to power sources. It also provides mechanical support to electrodes. To meet the required properties of the current collector materials should have minimum contact resistance, high electric conductivity, and good bonding capacity with electrodes. The bonding capacity can be increased by modifying the surface of the current collector, which is mainly carried out using industrial picoseconds laser device. Different types of materials are being used for the current collector, where the selection of materials depends upon the cost of materials and its suitability toward particular applications. Most commonly used conventional metals like copper (Cu), aluminum (Al), nickel (Ni), etc. are being replaced by advanced materials such as nanostructured or composite materials. In addition to this, the demand for flexible electronics is growing rapidly nowadays, and these devices require a material with enhanced properties. This review discusses various components of supercapacitors, i.e., electrode materials, electrolyte materials, separators, binders and current collectors, functions of current collectors, specifications of current collectors, various materials used as current collectors,

H. Trivedi · K. D. Verma · P. Sinha · K. K. Kar (✉)
Advanced Nanoengineering Materials Laboratory, Materials Science Programme, Indian Institute of Technology Kanpur, Kanpur 208016, India
e-mail: kamalkk@iitk.ac.in

H. Trivedi
e-mail: harihzl@gmail.com

K. D. Verma
e-mail: deo.kapildev@gmail.com

P. Sinha
e-mail: findingprerna09@gmail.com

K. K. Kar
Advanced Nanoengineering Materials Laboratory, Department of Mechanical Engineering, Indian Institute of Technology Kanpur, Kanpur 208016, India

K. K. Kar (ed.), *Handbook of Nanocomposite Supercapacitor Materials III*,
Springer Series in Materials Science 313,
https://doi.org/10.1007/978-3-030-68364-1_8

various parameters that affect the performance of current collectors, i.e., thickness, temperature, electrolytes, etc., dimensions of current collectors, screening of current collectors using constraints, various governing equations used for electrical conductivity, thermal conductivity, tensile strength, mass, bending strength, etc., and various material indexes used to select the best materials using Ashby charts.

8.1 Introduction

A supercapacitor is a modern energy storage device that can bridge the gap between batteries and conventional capacitors. The supercapacitor has advanced characteristics like higher capacitance and energy density compared to the traditional capacitor, which makes it capable to store a large amount of energy [1, 2]. It works on the electrostatic charge storage principle of a capacitor where adsorption takes place at electrode and electrolyte interface, also few materials undergo reversible redox reaction process to store charge [2]. The supercapacitor also has higher device stability, long cycle life, and fast charge/discharge rate. It is used in various fields like industrial equipment, electric vehicles, and energy production system [3, 4]. In a conventional capacitor, no Faradaic process occurs and energy is stored as static electricity; it contains two metal plates with dielectric insulating materials as a sandwich structure. The capacitor is a lightweight device and does not contain harmful and toxic chemicals, unlike battery. Capacitor charge–discharge cycles could be a zillion times without any significant deterioration in components. Figure 8.1 shows the comparison between the properties of battery and capacitor. The battery has a higher energy density in comparison to the conventional capacitor, whereas the capacitor holds several advantages such as lightweight, low cost, high charge–discharge rate, long lifespan, and use of non-toxic materials [5].

The poor energy density of a conventional capacitor can be increased by either selecting better dielectric material or increasing the exposed area of conducting plates. On the other hand, battery stores a large amount of energy but has poor cycle

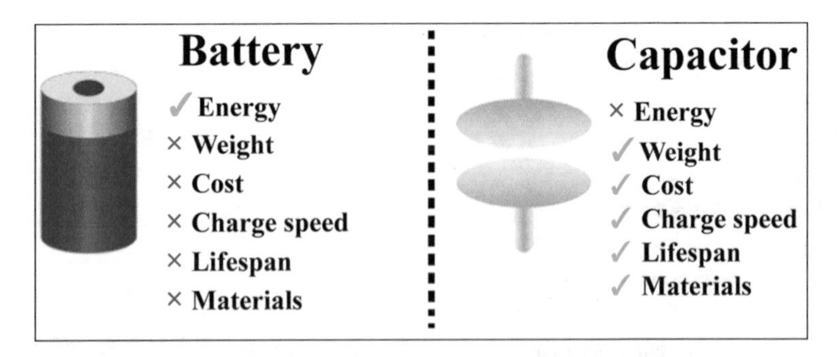

Fig. 8.1 Comparison of properties between batteries and capacitors pseudocapacitance (redrawn and reprinted with permission from [5])

life, takes a longer time to charge, contains toxic materials, and includes expensive components. To collect the advantages of capacitor and battery and minimize their limitations, the other energy storage device named supercapacitor is introduced [5]. A synonym of the supercapacitor is an ultra-capacitor or double-layer capacitor. The supercapacitor is an electrolyte capacitor that provides high capacitance and energy density than a conventional capacitor and high power density than batteries. However, supercapacitor still has some limitations like relatively low energy density and high manufacturing cost in comparison to the batteries [6]. The research in supercapacitors is a great concern, academically and commercially [7]. The present chapter introduces the components of supercapacitor along with the material used. The chapter further describes the function and properties of a material to function as a current collector.

8.2 Components of Supercapacitors

Supercapacitor consists of electrode, electrolyte, separator, binder, and current collector [8]. Supercapacitor performance depends upon the material selection for these components. During charging, electrolyte ions get accumulated over the surface of the opposite charge electrode to form a double layer. During discharging, these ions go back to the electrolyte and excess charge in the electrode gets transferred to the current collector, which is sent to the external circuit. Figure 8.2 shows different components in supercapacitor along with charge storage mechanism during charging, and electrolyte ions form an electric double layer over the surface of the carbon electrode.

8.2.1 Electrode Materials

The electrode is a solid electric conductor over which electrolyte ions are adsorbed and de-adsorbed. For high capacitance, electrode material should have a high surface area. Electric double-layer formation and pseudocapacitive Faradaic reaction are two-charge storage mechanism through which charges are stored at the electrode material [9]. Different types of materials can be used for fabrications of electrodes as follows.

8.2.1.1 Carbon Materials (Capacitive Charge Storage)

Carbon materials are widely used as electrode materials due to high surface area, low cost, wide availability, and established electrode production technology. An electrochemical double layer is formed between electrode and electrolyte at the interface and capacitance of the supercapacitor depends upon the active surface area of the electrode, which is available for electrolyte ions. Electrochemical performance

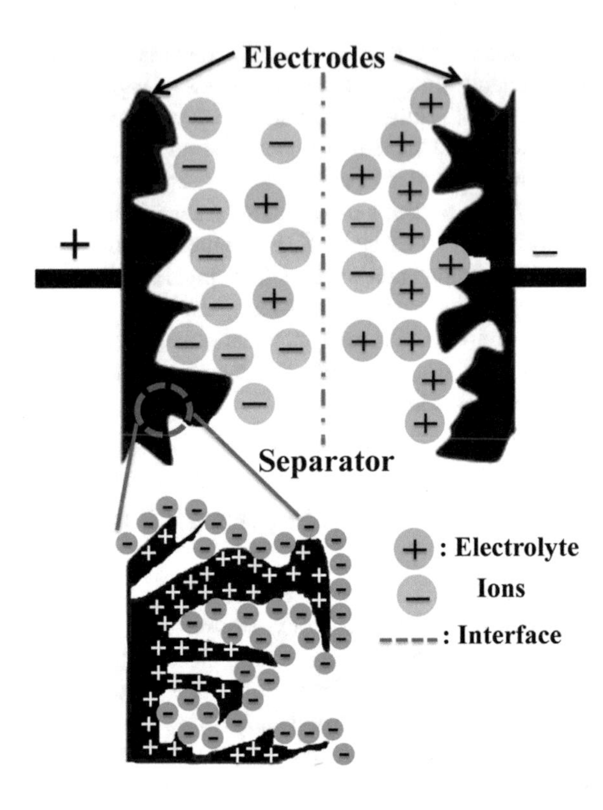

Fig. 8.2 Charge storage mechanism for EDLC supercapacitor along with different components pseudocapacitance (redrawn and reprinted with permission from [7])

is influenced by pore shape, pore size, pore size distribution, and surface area [10, 11]. Examples of carbon materials as electrodes materials are activated carbon [12], carbon nanofiber [13], carbon nanotube (CNT) [14], graphene [15], etc.

8.2.1.2 Metal Oxides and Conducting Polymer (Faradaic Charge Storage)

Metal oxide and conducting polymer deliver high specific capacitance. Due to high specific capacitance and low resistance, metal oxides and conducting polymers are also a good alternative for electrode materials [16, 17]. Unlike carbon materials, where the charge storage is limited to the availably of active surface, in transition material oxide and conducting polymer whole material undergo reversible redox reaction to store charge. Bimetallic metal oxides like $NiCo_2O_4$ are more efficient compared to single metallic oxides for overcoming the low electric conductivity of metal oxides. Examples are nickel oxide (NiO), cobalt oxide (Co_3O_4), ruthenium dioxide (RuO_2), manganese oxide (MnO_2), etc. [18–20]. Also, conducting polymer, p doped conducting polymer is generally used. Examples of such conducting polymers are polyaniline, polypyrrole, etc.

8.2.1.3 Composite-Based Electrode Materials

For improving the performance of the supercapacitor, composite electrode materials are used like carbon–carbon composites [21–24], carbon-metal oxides composites [25–28], and carbon conducting polymers composites [19, 29, 30]. These types of electrodes use both electric double layer and pseudocapacitive mechanism to store charge.

Figure 8.3 shows the list of traditional and emerging electrode materials used for supercapacitor application. This includes carbon materials, conducting polymers, metal oxides, metals nitrites, MXenes, metal organic framework, polyoxometalates, black phosphorous electrode materials used for supercapacitor application [31].

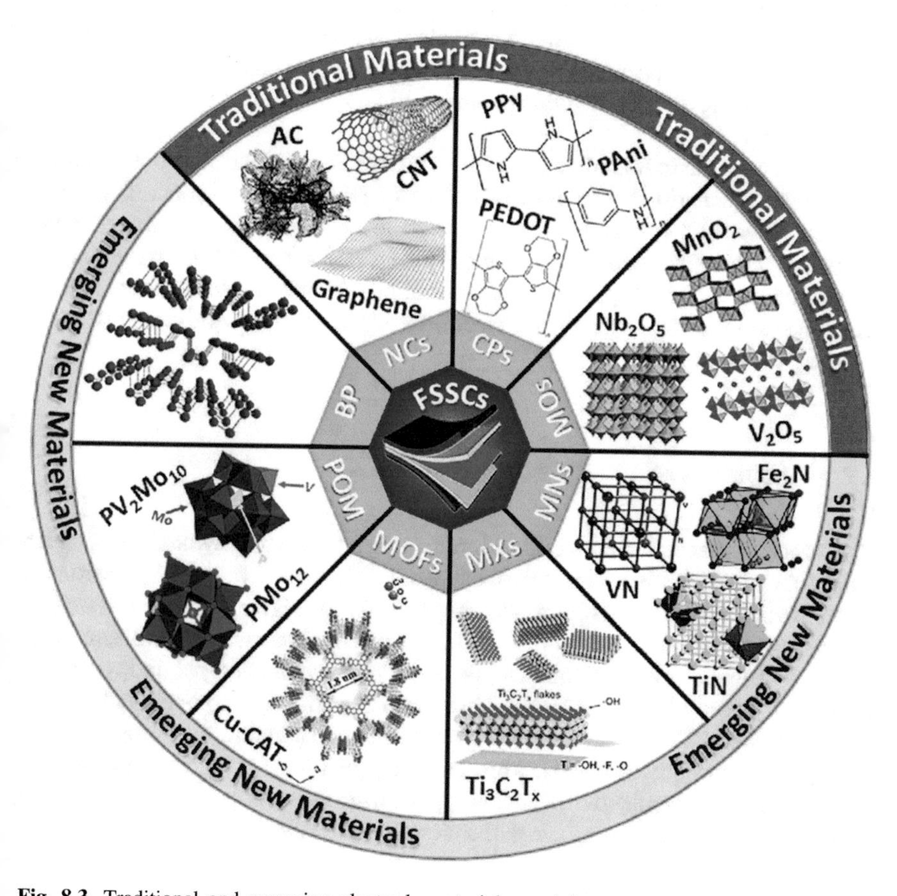

Fig. 8.3 Traditional and emerging electrode materials used for supercapacitor (reprinted with permission from [31])

8.2.2 Electrolyte Materials

Electrolytes are composed of ions, which are dissolved in solvents. On the application of potential, the electrolyte provides a particular anion and cation. The conductivity of the electrolyte is directly proportional to the number of ions present after electrolysis. Electrolyte provides conductive media between two electrodes by dissociation of ions. Electrochemical stability of electrolyte plays a measure role in determining the working potential of supercapacitors [32]. The organic electrolytes in the ionic liquid are better than the water medium due to the evolution of gases like H_2 and O_2 from water at a low voltage of 1.23 V. However, organic electrolytes and ionic liquids are expensive and display poor conductivity as compared to the aqueous-based electrolyte. To achieve high capacitance and high energy density electrolyte ions, size should be equal to or less than the size of electrodes pores [9]. Electrolyte conductivity is inversely proportional to the resistance of the supercapacitor. The thermal stability of the supercapacitor depends on the viscosity and freezing temperature of the electrolyte.

It is not possible to find all the required qualities in one electrolyte but according to the current scenario, most required qualities are low cost, broad potential window, a broad range of working temperature, high electrochemical stability, adequate viscosity, high ionic conductivity, etc [33]. Figure 8.4 shows the different types of electrolytes used for supercapacitor applications. Mainly, electrolytes are divided into the aqueous electrolyte, organic electrolyte, solid or quasi-solid electrolyte, and redox-active electrolyte.

8.2.3 Separators

Separators are made of materials like plastic, rubber, polymer, polyolefin, etc., which can act as an insulator between the two conducting electrodes. Required properties of separator material are high electrolyte ion permeability, high electrolyte wettability, low cost, high decomposition temperature, good mechanical strength, sufficient water uptake, and low swelling properties [34, 35]. Polyethylene and polypropylene are traditionally used as separators, but these materials have low porosity. Polyvinylidene difluoride (PVDF) nanofibers are made via electrospinning and porosity can be controlled accordingly, so that it has low interfacial resistance, a higher diffusion rate, and required high conductivity. Secondary characteristics of the separator include minimal thickness, low weight, and chemical inertness [36].

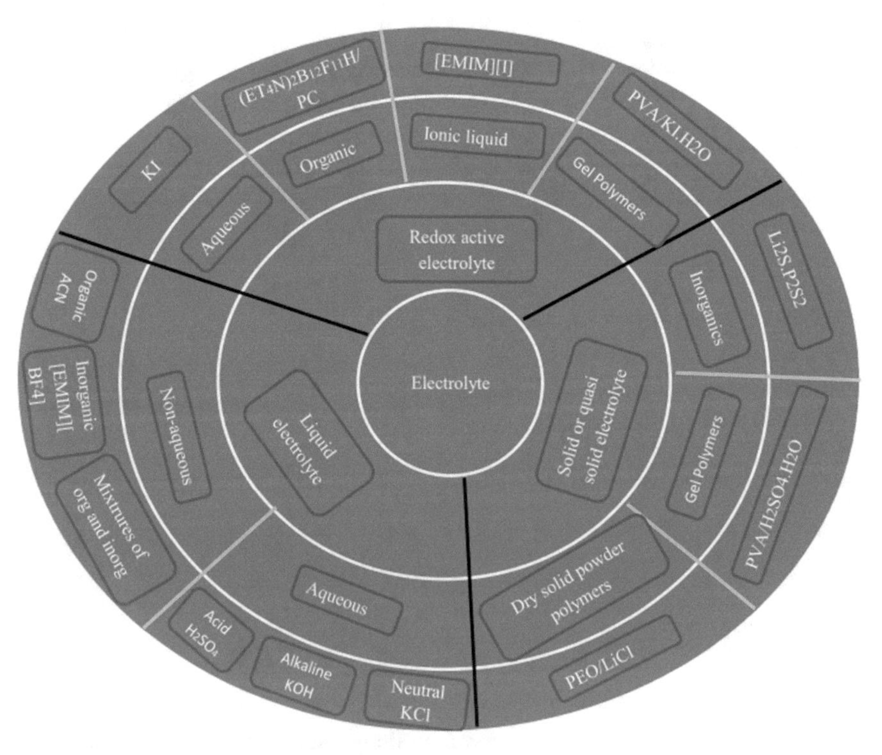

Fig. 8.4 Types of electrolytes used for supercapacitor

8.2.4 Binder

To bind active materials with conductive agents and cohering with electrode materials, a binder is used in the supercapacitor. Binder is used to providing appropriate strength and pore size to the electrode material. The selection of an appropriate binder is very important as it influences the electrochemical performance of supercapacitors by covering active pores available on the surface of electrodes [37]. Figure 8.5 shows a schematic of the fabrication of electrode for use of supercapacitor. For low contact resistance and high performance, the active material is mixed with conducting carbon and binder in the optimum ratio [38]. Some of the commonly used binders are discussed as follows:

Polytetrafluoroethylene (PTFE) PTFE provides good chemical stability due to the presence of CF_2-CF_2 units. It also displays hydrophobic and insulating nature, which decreases the wettability of electrode by electrolyte ions; hence, internal resistance of the device increases [37].

$$-(CF_2\,CF_2)_n-$$

Chemical structure of Polytetrafluoroethylene (PTFE)

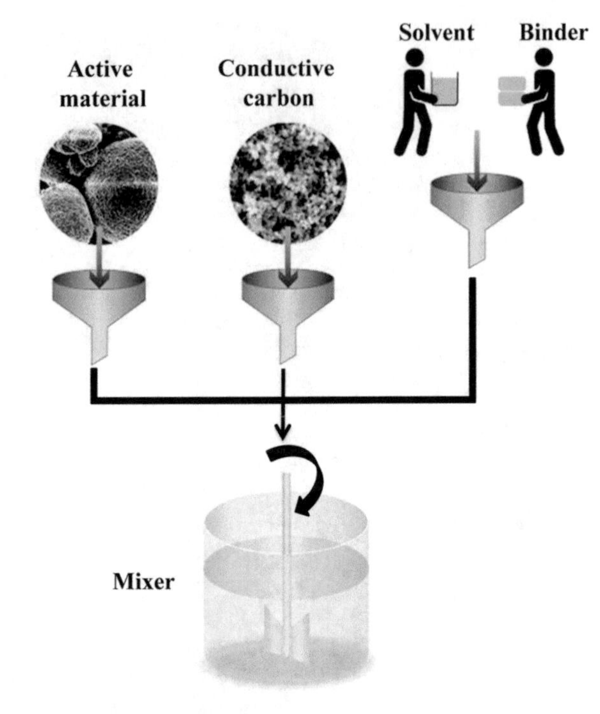

Fig. 8.5 Schematic of the mixing steps of activated material, conductive carbon, solvent, and binder for the electrode material preparation (redrawn and reprinted with permission from [38])

Nafion Nafion has very good hydrophilicity and chemical stability, and hydrophilicity is due to the presence of the sulfonic acid group as an ion-exchange group. However, nafion is costly as compared to PTFE [37].

$$-(CF_2-CF_2)_x-(\underset{|}{C}F-CF_2)_y- \quad (O-CF_2-\underset{\underset{CF_3}{|}}{C}F)_z-O-CF_2-CF_2-SO_3^-Na^+$$

Chemical structure of Nafion

Polyvinylidene difluoride (PVDF) PVDF shows poor hydrophilicity and has polarity due to its interactive groups (CH_2-CF_2). It is chemically stable and shows good chemical and corrosion resistance properties [37].

$$-(CH_2CF_2)_n-$$

Chemical structure of polyvinylidene difluoride (PVDF)

Other examples of binders are carboxymethylcellulose (CMC), polyvinylidene chloride (PVDC), graphene/PVDF composite, etc [37].

Figure 8.6 shows water-ethanol, fluorine-free, and biopolymers and their derivatives based on binder materials. Binder polymers are classified into three categories according to their processability, chemical composition, and natural availability as (a) water/ethanol processable, (b) F-free chemical composition, and (c) biopolymers and derivatives [38].

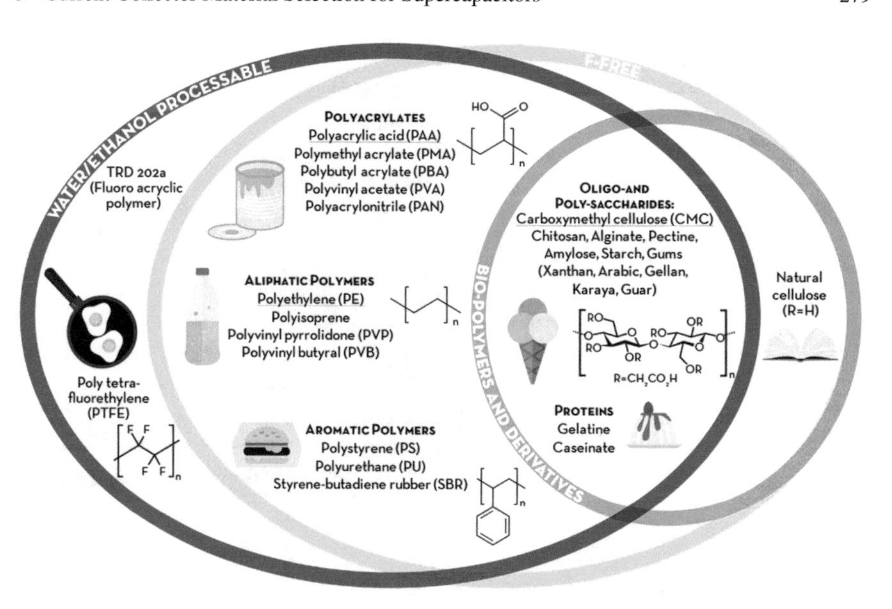

Fig. 8.6 Schematic representation of different classes of binders used for supercapacitors application (reprinted with permission from [38]

8.2.5 *Current Collector*

The current collector is made up of metal foils that are connected with electrodes to terminals of the supercapacitor [39]. It must be ensured while selecting materials for the current collector that they should not get corroded by electrolytes like sulfuric acid and other aqueous and non-aqueous materials used in the supercapacitor. Apart from electrolyte and electrode resistance, resistance arises due to current collector also plays an important role in determining the performance of supercapacitor. Metallic foam is a common example of three-dimensional current collectors, which are commonly used in supercapacitor devices [40]. In order to achieve high performance, the contact resistance of the electrode and current collector should be low. This minimizes the equivalent series resistance (ESR) and increases the power density of the device.

Operating cell voltage, reliability, and a lifetime of electrolytes are attributed to the chemical and electrochemical nature of the current collector. Active electrode materials optimization also depends on the morphology of the current collector [41]. Figure 8.7 shows the schematic of the supercapacitor cell. The current collector is attached to the electrode material to collect the electric charge and transfer it to the external circuit. Functions, specification, and various materials used for the current collector are described in the forthcoming sections.

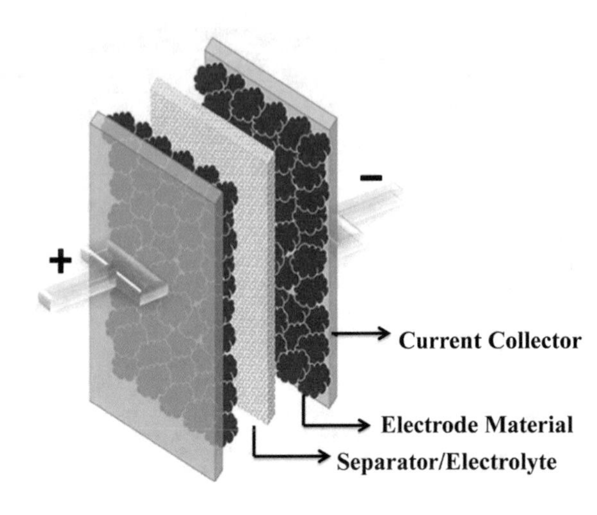

Fig. 8.7 Illustration of the current collector in the supercapacitor (redrawn and reprinted with permission from [42])

8.2.5.1 Functions

During charge-discharge cycles, transportation of charge carriers from/to electroactive materials is the main role of the current collector in supercapacitor. The interfacial resistance between the current collector and electroactive materials governs the charge transport efficiency of the current collector [40]. Surface modifications or chemical vapor deposition on the current collector surface increase binding capacity with the electrode and reduces the contact resistance between them [43]. The current collector also provides support to the electrode, and while assembling the supercapacitor, electrode slurry is deposited on current collector foil or foam to provide efficient adhesion [44].

8.2.5.2 Specifications

Low contact resistance, high electric conductivity, strong bonding with the electrode and, good mechanical strength are important specifications of a good current collector. Various materials have been proposed as a current collector; however, there is no ideal material yet. Carbon cloth is widely used as a current collector, which displays good electrical conductivity but it has low mechanical strength. Stainless steel, on the other hand, has high electrical and thermal conductivity, along with high mechanical strength but adds a lot of weight to the device. Aluminum, nickel, copper are lightweight, high electrical conductivity but they have cost issues. Also, some metal-based current collector easily gets oxidized in a normal environment that deteriorates the device performance. Binder-free and low-cost current collectors can be fabricated by combining carbon materials with conducting polymers and metal oxides. Also, the performance of the current collector can be increased by decreasing the "dead area" of the binder materials [45].

Table 8.1 Electrochemical performance parameters (resistance, capacitance, and energy density) of different current collector-based supercapacitor [44]

S. No.	Current collector	Resistance (Ω)	Capacitance (F/g)	Energy density (Wh/kg)
1	Cu-foil	150	1.66	0.23
2	Al-foil	43	5	0.73
3	PCB copper clad	0.78	166	23

Corrosion can be controlled by applying a protective coating of commonly used materials like steel and similar metals. The most commonly used protective layer for the current collectors is graphite foil and used to protect steel. The protective coating offers minimal resistance but adhesion between the protective layer, and current controller is a great deal. Also, in most cases, electrolyte enters the spaces between the protecting layer and current controller, which may corrode the current collector. In this regard, multilayer protection can be used to provide optimum protection and required adhesion. Carbon materials have p-type conductivity that is commonly used as an active mass in supercapacitor so the protective coating should also have p-type conductivity to minimize galvanic corrosion [46].

According to preferences we can classify specifications or characteristics of current collector materials into the three categories.

1. Primary specifications: High electrical conductivity, low contact resistance, high corrosion resistance, high mechanical flexibility, high thermal stability.
2. Secondary specifications: High chemical stability, high electrochemical stable, low density, high thermal conductivity, high compression strength.
3. Tertiary specifications: Environmentally friendly, easy to fabricate, low cost [39].

The variation in parameters can be observed in Table 8.1 The properties of a commonly used low-cost current collector such as Cu-foil, Al-foil and, PCB copper clad are compared in the same electrolyte.

From the Table 8.1, it can be said that PCB copper clad improves charge transfer and through channels and also reduces interfacial resistance between the current collector and electroactive materials [44].

8.2.5.3 Various Materials Used in Current Collector

The wide variety of the current collector materials like metals, carbon cloth, conducting polymers are being used to meet the required properties like electrochemical, physical, mechanical, electrical, and processing characteristics. Inexpensive metals and alloys limit its usage when corrosive electrolytes like sulfuric acid (H_2SO_4) are being used. Materials such as Ti, Al, Ni, Ag, Nb, Ta, W, and some of their alloys are a few other current collectors used in supercapacitor applications [46].

Gold (Au) is being used traditionally for corrosive electrolytes like H_2SO_4, however, it is an expensive material, which increases the overall cost of the device. To reduce the cost of supercapacitor alternative cheap materials are investigated such as indium tin oxides (ITO), electric conductive polymers, and carbon-based materials. ITO is not only stable in the acid corrosive environment but also provides transparent nature and relatively high electric conductivity. However, it gets unstable toward HCl and starts corrosion, the corrosion rate increase rapidly when HCl concentration increases from 0.4 to 1 M [41].

The performance of the supercapacitors is badly affected by corrosion or degradation of current collector metals. For example, a small quantity of iron metal present in electrolyte, due to corrosion can badly deteriorate the device performance by decomposing electrolyte voltage, which catastrophically reduces the device voltage leading to the device's failure [46].

Aluminum (Al)

Aluminum is the most commonly used material as a current collector and has a significant effect on cyclic stability and the electrochemical performance of supercapacitors [39]. The internal resistance of Al is decreased up to 0.4 Ω/cm^2 by coating with carbon nanofibers with the increase in its cyclic stability. Graphene coating on Al current collectors also protects it from corrosion.

The performance of the current collector can be improved by increasing interface bonding strength between the carbonaceous electrode and Al current collector. To improve the performance of supercapacitor interface engineering is a very important strategy, it has two important interfaces:

- electroactive material and electrolyte interface
- electroactive material and current collector interface.

Strong interface bonding prevents delamination between the components in harsh conditions. The bonding strength can be improved by increasing the surface area of the Al current collector, which can be easily carried out using industrial picoseconds laser device. During this process, Al current collector foil is treated with a laser to develop a rough surface that contains hierarchical micro and nanostructures and improves the bonding strength between the current collector and electrode material due to the enlarged surface area of Al-foil [47]. Figure 8.8 shows the top and cross-sectional views of the laser-treated aluminum current collector. SEM images of pristine and laser-treated aluminum current collector along with electroactive material are shown in Fig. 8.9. No cracks and fallen found in the SEM image of laser-Al compared to pristine-Al. Laser-treated aluminum current collector has a rough and higher surface area surface that provides better contact with electroactive material [47].

During laser treatment, chemical compositions also get changed in the Al current collector. Table 8.2 shows the variation in oxygen concentration on the Al surface along with its resistance after laser treatment of pristine Al. After laser treatment, it is

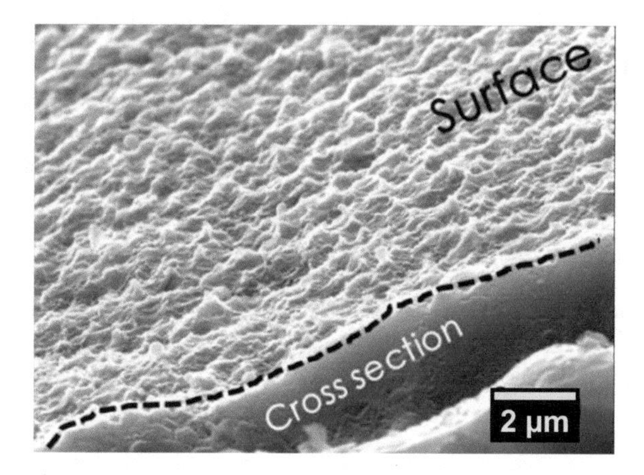

Fig. 8.8 Laser-treated surface showing hierarchical micro and nanostructures at the aluminum surface (reprinted with permission from [47])

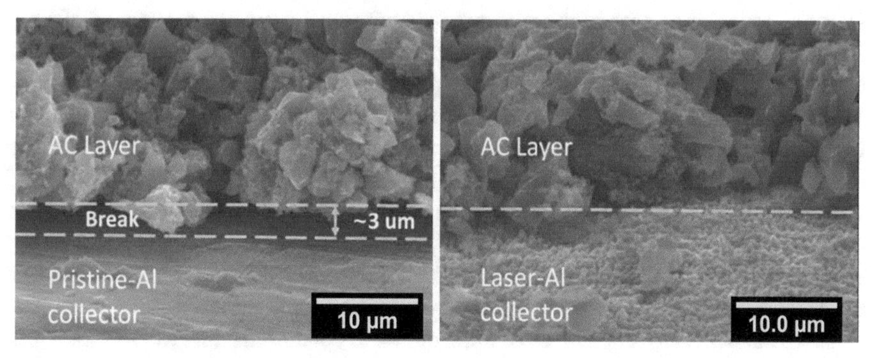

Fig. 8.9 SEM images of a pristine and laser-treated aluminum current collector (redrawn and reprinted with permission from [47])

Table 8.2 XPS analysis of pristine-Al and laser-treated aluminum [47]

Sample	Concentration of elements		Resistance (Ω/m)
	Oxygen (O)	Aluminum (Al)	
Pristine-Al	56.35	43.64	0.68
Laser-Al	55.39	44.60	0.33

observed that at the surface of Al-foil the ratio of aluminum and oxygen changed. The AL surface contains Al_2O_3 and after laser treatment, the percentage of O_2 decreased from 56.35 to 55.39%, and the percentage of Al content increased by 0.96%. The rising Al percentage also increases the electric conductivity and performance of the current collector [47].

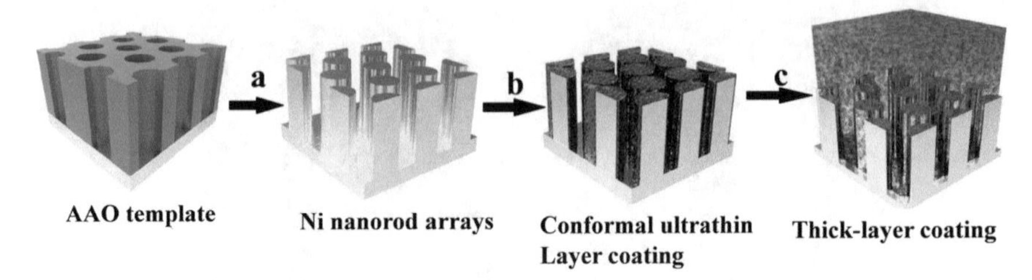

Fig. 8.10 Schematic diagram of fabrication of nanostructured supercapacitor electrode (redrawn and reprinted with permission from [40])

Nickel (Ni)

If the alkaline electrolyte is being used nickel is an excellent choice as the current collector material. It displays good chemical and electrochemical stability in alkaline media at a relatively lower cost. To improve the loading amount of active electrode material and surface area of the current collector nickel foam is used instead of film. However, the presence of nickel oxide and hydroxide as a layer on the surface provides additional pseudocapacitance, which provides error in the device performance [41].

Holm's theory suggests that constriction/spreading resistance is generated when two electrical objects come in contact and make an interface. The interface resistance inversely depends on the number of contacts spots. By increasing the number of contact spots between interfaces, interfacial resistance can be reduced. If nickel foam current collector surface is deposited by vertically oriented graphene, the contact area of the interface is increased via bridge formation between nickel foam and electroactive materials, hence interfacial resistance decreases.

In nanostructured current collectors of nickel, nanorods arrays of nickel are followed by depositing electroactive material MnO_2. This arrangement improves the surface area and reduces interface resistance between the electroactive material and current collector. Figure 8.10 shows a schematic diagram of the coating of electrode material on the nickel nanorods array.

(a) nanostructured current collector formed of Ni nanorods arrays via the AAO template,
(b) MnO_2 ultra-thin conformal layer coating via deposition on nickel nanorods arrays, and
(c) MnO_2 thick layer coating by extending deposition time [40].

Finally, it can be concluded that a nanostructured current collector with a thick layer of electroactive material can boost up the energy storage capacity of the supercapacitor [40].

In the journey of flexible electronics, the flexible and conductive current collector is required, which support paper-based technology. For a flexible current collector, the Ni layer is deposited on the filter paper via electroless deposition. This provides good mechanical strength and conductivity having a low mass density of 0.35 g/cm^3.

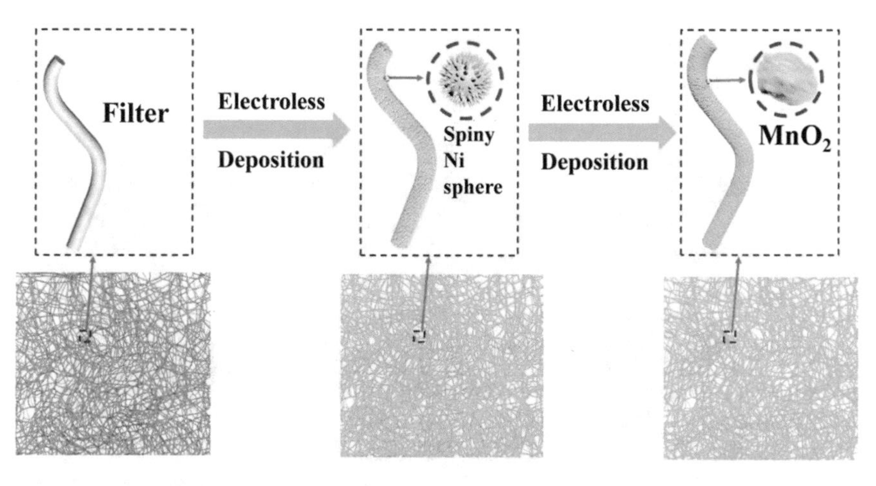

Fig. 8.11 Fabrication process of Ni + paper + MnO_2 supercapacitor electrode (redrawn and reprinted with permission from [48])

The electrochemical performance increases due to low contact resistance between the current collector and electrodes. However, after 5000 cycles, the sheet resistance increases to 2.7 Ω/cm^2 with respect to the initial resistance of 0.87 Ω/cm^2.

As illustrated in Fig. 8.11, Ni nanospheres are deposited on carbon fiber via electroless deposition, and different mass of MnO_2 (as electroactive material) on the surface of the Ni layer can be deposited via controlling soaking time. Coating of Ni can be identified by the color changing of filter paper from white to gold [48].

Figure 8.12 shows the deposition of nickel (Ni) on a drilled stainless steel sheet filled with epoxy by the electro-deposition technique. Although the "true performance" of supercapacitor that means high energy density is still unsatisfactory, the performance depends on the active mass loading and material used for the current collector. Leser drilled stainless steel sheet is used to develop pore structured nickel (Ni) current collector with appreciable flexibility [49].

Copper (Cu)

Copper (Cu) is a commonly available material with good electrical and thermal conductivity at a relatively low cost and has wide scope for modern electronics. To enhance the performance of Cu current collector, its surface modification can be done by graphene via chemical vapor deposition. This process enhances adhesion with the graphite electrode by increasing the hydrophobic nature of the copper current collector. Contact resistance, cyclic stability, and discharge capacity also increase significantly in this process. Figure 8.13 shows chemical vapor deposition (CVD) of graphene on the copper current collector. Surface modification improves the electrochemical properties due to the decrease of charge transfer resistance by the graphene layer on copper (Cu) film [43].

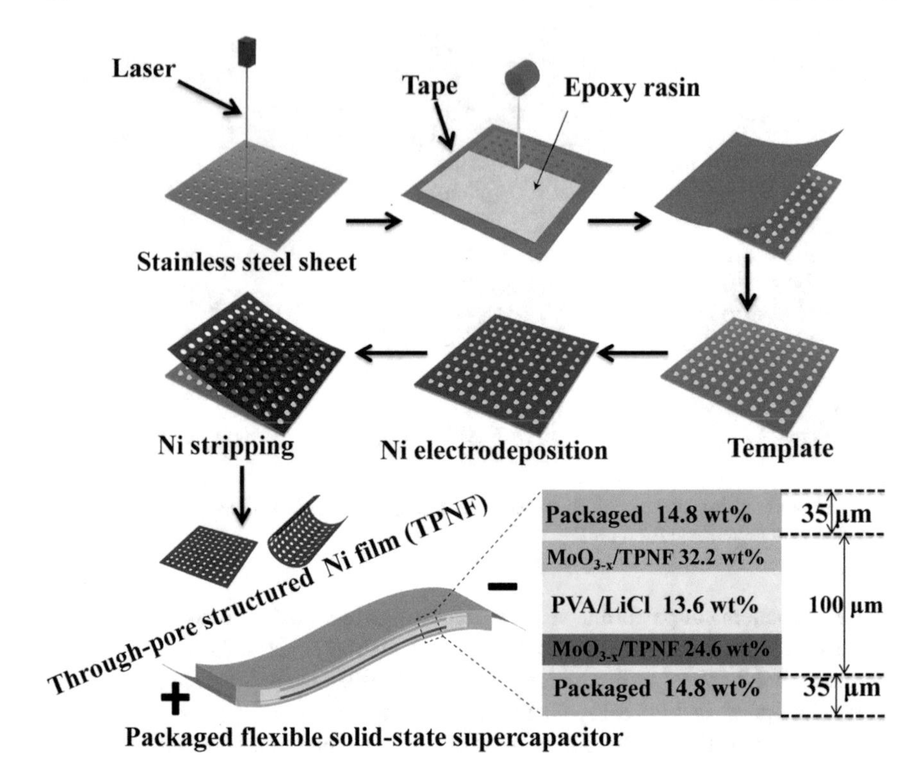

Fig. 8.12 Illustrating fabrication of through-pore structured Ni film (TPNF) (redrawn and reprinted with permission from [49])

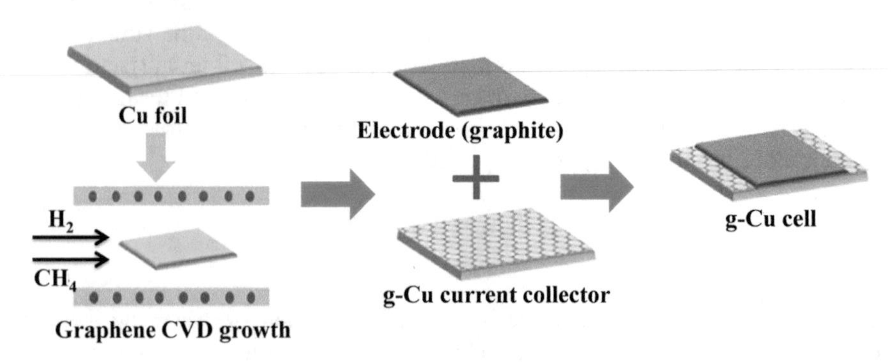

Fig. 8.13 Illustrating chemical vapor deposition (CVD) of graphene on copper (Cu) current collector (redrawn and reprinted with permission from [43])

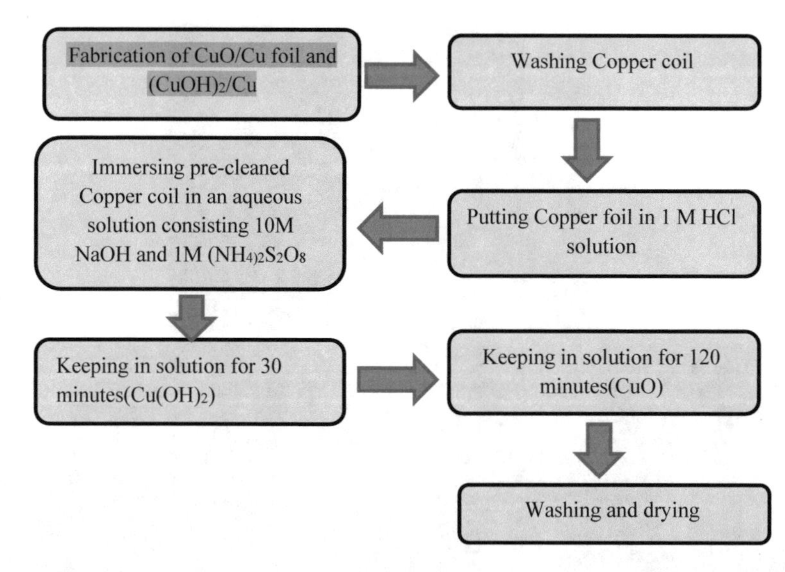

Fig. 8.14 Flowchart for the generation of $Cu(OH)_2$ and CuO on copper substrate (redrawn reprinted with permission from [45]

Copper oxide (CuO), copper hydroxide ($Cu(OH)_2$), and hydrophobic copper materials on copper substrate protect the copper from corrosion and also increase the binding of electroactive materials to the current collector. The methods of developing copper oxide (CuO) and copper hydroxide ($Cu(OH)_2$) on the copper substrate are illustrated in the following scheme (Fig. 8.14).

The liquid–solid reaction method as described above is cost-effective and synthesized $Cu(OH)_2$ arrays as nanowire and CuO microsphere as urchins. The formation of different copper compounds can be confirmed by FESEM as shown in Fig. 8.15 [45].

Lead (Pb)

Lead (Pb) and its oxide are determined as the most suitable current collector in sulfuric acid aqueous-based electrolyte because they are very stable and available at low cost. However, a thin layer of oxide is formed while repeated cyclic use on the surface of unprotected lead and its alloy; this increases the specific resistance, decreases the stability, and reduces the overall performance of supercapacitor. To minimize the oxidation process, the protecting layer is developed on lead and its alloys current collectors that control corrosion in aqueous solution electrolyte like sulfuric acid. However, the protective layer may reduce the quality parameters of the current collector or supercapacitor by increasing the resistance toward the flow of charges. Some of the literature suggests the introduction of tin in 1-3% to take the lead for the fabrication of current collector. Also, the paste of carbon powder,

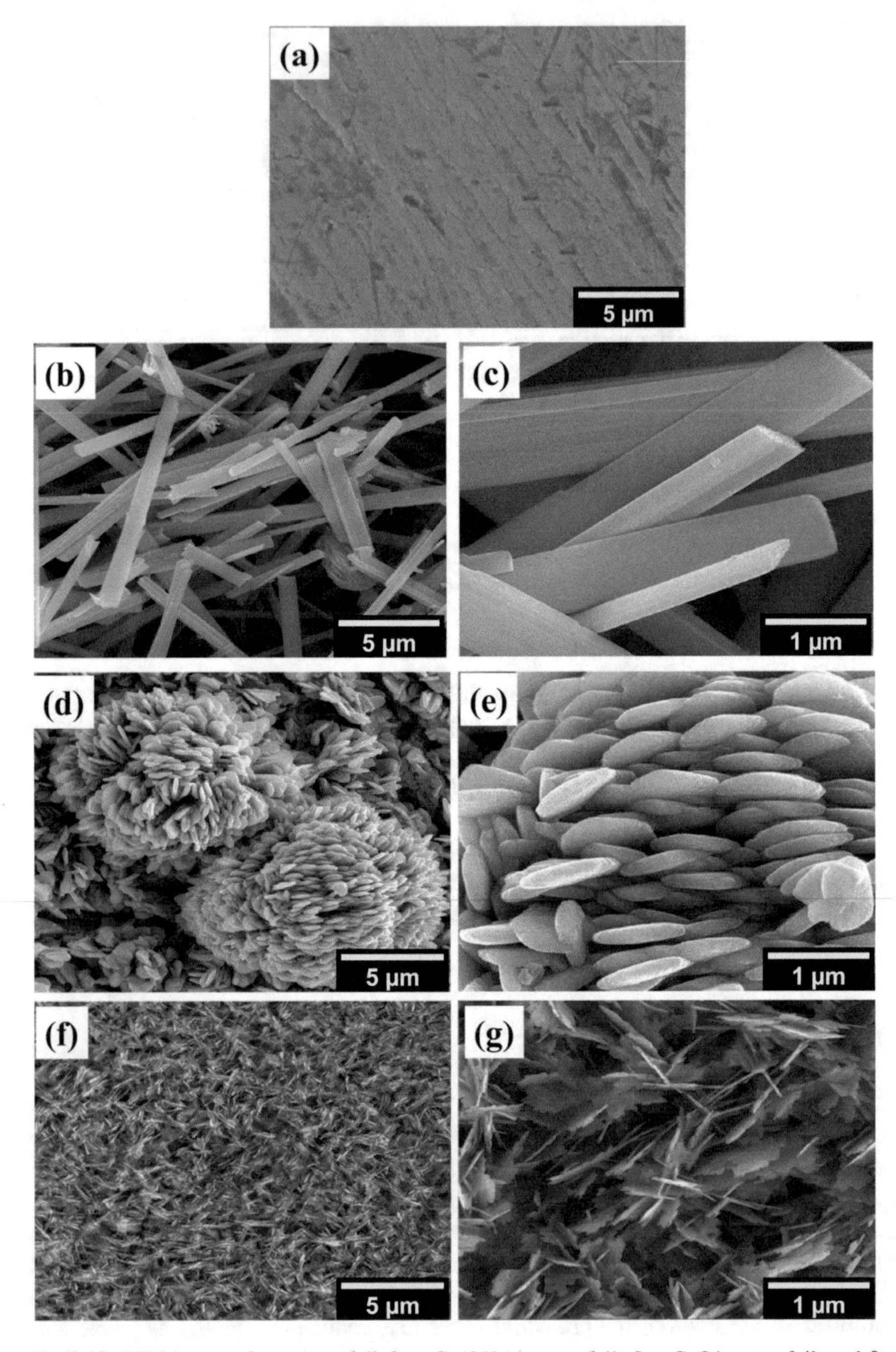

Fig. 8.15 SEM image of **a** copper foil, **b**, **c** $Cu(OH_2)$/copper foil, **d**, **e** CuO/copper foil, and **f**, **g** STA-CuO/Cu-foil (redrawn and reprinted with permission from [45])

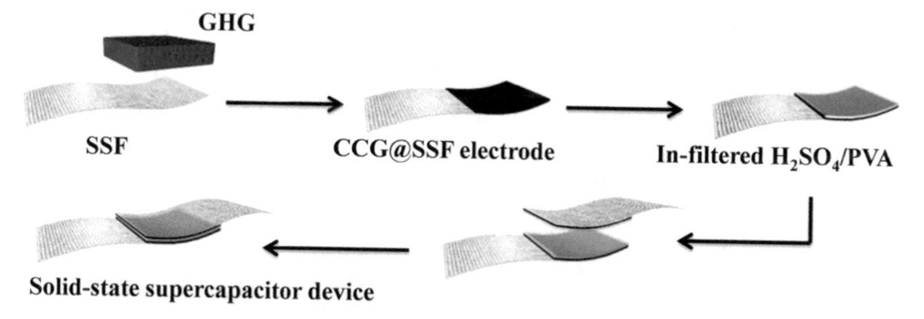

Fig. 8.16 Schematic illustration for the fabrication of flexible supercapacitors using stainless steel fibers as the current collector. (CGC = chemically converted graphene as active material) (SSF = stainless steel fibers) (redrawn and reprinted with permission from [51])

perchlorovinyl, plasticizer, solvent (acetone, n-butyl acetone, toluene), dispersant, wetting agent (surfactant), and the antifoaming agent is coated as a protecting layer for the alloy [46].

Stainless Steel

Flexible electronics grooming rapidly nowadays require a flexible supercapacitor device. To develop a flexible supercapacitor, a flexible current collector is required. Stainless steel fabrics preferably used as a current collector in flexible supercapacitors. The most commonly used flexible current collector that is mentioned in many pieces of literature is carbon fabrics because of high flexibility and good mechanical strength but due to its limited conductivity, carbon fibers restrict its application. Other examples of flexible current collectors are metallic foil and polymer film coated with a conductive material. Ti and Ni show excellent conductivity, but these materials are not used due to its rigidity. Also, the polymer film is not much stable in the acidic and alkaline electrolyte, which decreases the mechanical strength and stability [39, 50]. In this regard, a stainless steel substrate is one of the best alternatives that can be used as a current collector in flexible electronics. Figure 8.16 shows a flexible supercapacitor using a stainless steel current collector. Stainless steel fabrics have very good mechanical strength, cost-effective, very good flexibility, and stability in an acidic medium made it suitable for the flexible current collector in commercial use [51].

Carbon-Based Materials

Carbon-based materials are lightweighted materials, which show good electric conductivity, high thermal, chemical, and electrochemical stability, high flexibility, and good mechanical strength due to the presence of strong C-C bonds networks suitable for efficient current collector materials. Ultra-thin sheets are developed

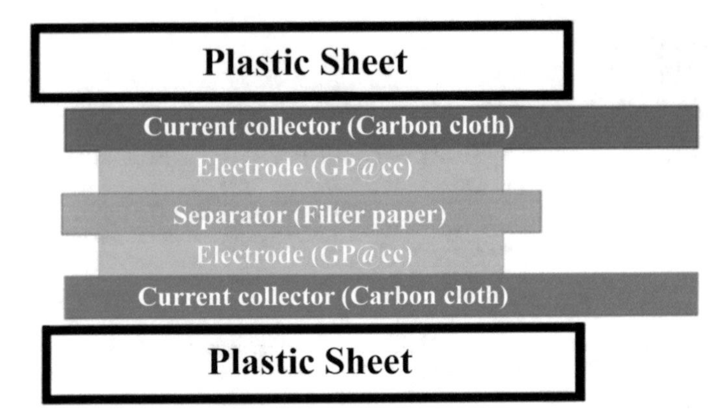

Fig. 8.17 Illustration of the structure of flexible supercapacitor made by carbon cloth current collector, (GP = Graphene–polyaniline, CC = Carbon cloth), (redrawn and reprinted with permission from [52])

by carbon-based materials for stretchable and wearable supercapacitors. Types of carbon-based materials are carbon fibers, carbon cloths, carbon nanotubes, ultra-thin graphite foam [39, 41]. The flexible and binder-free supercapacitor can be designed on graphene—polyaniline at carbon cloth. Figure 8.17 shows carbon cloth current collector-based flexible supercapacitor using graphene polyaniline electrode material.

Carbon cloth can be used as a current collector without the addition of any binder. Graphene–polyaniline composite soaked in H_2SO_4 is used as the electrolyte. Both electrolytes are separated by filter paper known as a separator. The whole supercapacitor is arranged as a "sandwich" structure as shown in Fig. 8.17 [52]. Current collectors made by different carbon materials, and the combination with different electrolytes and electrodes materials are described in Table 8.3 [53].

In order to improve charge transferability and minimize the contact resistance between the electrode and current collector, another strategy is used to grow or nucleate the electrode material onto the current collector. This improves the adhesion property and decreases the resistance. Figure 8.18 shows the fabrication of $NiCo_2O_4$ nanoneedles array around 40–50 nm diameter on a carbon-based current collector to fabricate the $NiCo_2O_4$/CC electrode. Carbon nanotubes integrated carbon fibers are used for this purpose [53].

8.3 Effect of Current Collector Thickness

The resistance of the current collector is not taken into consideration nowadays because everyone is considering electrodes and electrolytes as the main focus area. However, the thickness of the current collector can increase the resistance of the device drastically and that is why it should be considered as a critical parameter.

Table 8.3 Presenting different combinations of current collectors made by carbon-based materials with different electrolytes and, electrodes

Current collector	Electrode materials	Electrolyte	Ref.
Carbon fibers	$NiCo_2O_4$ decorated on PAN/lignin-based carbon fiber	KOH	[54]
Carbon cloth	Pyridinic-nitrogen-doped nanotubular carbon arrays on carbon cloths	KOH	[55]
Co_3O_4 on graphene	Co_3O_4 on graphene	KOH	[56]
PAN carbon fiber	PAN carbon fiber	KOH	[57]
Electrochemically exfoliated 3D graphene	Graphene	KOH-H_2SO_4	[58]
Cotton fabric	CNT-RGO and polypyrrole on cotton fabric	Na_2SO_4	[59]
CNT film	Adenine-based metal organic framework-derived carbon	H_3PO_4-PVA	[60]
Carbon cloth	Polyaniline MOF composite	PVA/H_2SO_4	[61]
Carbon fabric	Carbon nanoparticles/MnO_2 nanorod hybrid	PVA/H_3PO_4	[62]
Carbon fabric	Carbon nanoparticles/MnO_2 nonfoods	Polyvinyl alcohol/H_3PO_4	[62]
Carbon cloth	Hydrogenated ZnO core–shell nanocables	LiCl/PVA Gel electrolyte	[63]
Graphene/CNFs/MnO_2	Graphene/CNFs/MnO_2	Na_2SO_4	[64]
Carbon cloth	Ni-Co-Mn multicomponent metal oxides	KOH/PVA gel electrolyte	[65]
Carbon fabric	WO_{3-x}@Au@MnO_2 core-shell nanowires	PVA/H_3PO_4	[61]
Carbon	NiO/MnO_2 core-shell nanoflakes	Na_2SO_4	[66]

The thickness of the current collector and potential drop is inversely proportional to each other and this relationship valid up to thickness 500 μm, above this thickness relationship, is almost constant [68].

If the thickness of the current collector is increased, then it will reduce the specific and volumetric performance of the device. To reduce the volume of the current collector, it is important to maintain minimal thickness. The traditional method of using metal foil as a current collector can be changed by some advanced methods like direct sputtering of Al on the electrode surface. The interfaces made by the sputtering of Al on electrolyte are very firm and tight. This also improves other electrochemical properties of supercapacitor like stability and high power handling. Around 300 and 186% of improvement in specific power and specific energy respectively, is observed in Al sputtered current collector as compared to traditionally used Al current collector [69]. Figure 8.19a shows a sputtered aluminum surface having a thickness of 600 nm on the carbon electrode with a thickness of 200 μm. Ragone plot of carbon and Al-foil

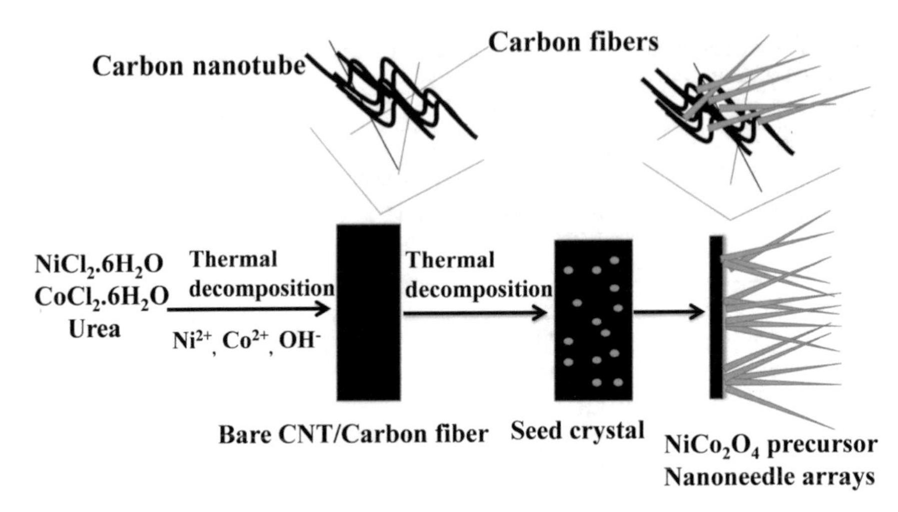

Fig. 8.18 Fabrication of $NiCo_2O_4$ precursor nanoneedles arrays (redrawn and reprinted with permission from [67])

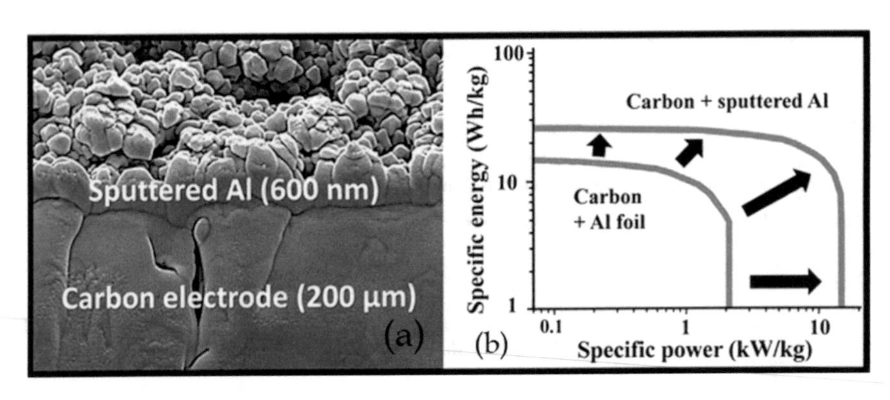

Fig. 8.19 **a** Illustration of sputtered aluminum (Al) surface on carbon electrode (left) and **b** performance comparison using Ragone plot in carbon + Al-foil and carbon + sputtered Al current collector (redrawn and reprinted with permission from [69])

current collector with carbon and sputtered Al current collector-based supercapacitor is shown in Fig. 8.19b. It shows that sputtered aluminum-based supercapacitor has higher energy density and power density [69].

8.4 Effect of Temperature

Generally, current collectors have good thermal conductivity and stability. So, the increase in temperature does not affect the stability of the current collector. However, at high temperatures, the oxidation process may occur at the current collector surface and the oxide layer can also between the surface of the current collector and electrodes. This can deteriorate the device performance by increasing the internal resistance and ESR value [70].

8.5 Effect of Electrolytes

Commonly used electrolytes are acidic or basic in nature. These electrolytes are corrosive in nature for the current collector. If the electrolyte is acidic like H_2SO_4, then high corrosion-resistant material like gold is used traditionally. Nowadays, some other materials like conducting polymers, carbon-based materials, indium tin oxides (ITO) are being used to reduce the cost of the current collector in a strongly acidic medium. Indium tin oxides (ITO) current collectors may also get corroded in HCl media so that it is suggested that carbon-based material more suitable material in this environment and for flexible electronics. For alkaline electrolytes, titanium (Ti), and platinum (Pt) are more efficient compared to nickel (Ni), where the oxides/hydroxide layer is on the nickel (Ni) surface. Other materials like stainless steel and nonporous gold (Au) are also used in alkaline electrolytes to minimize corrosion. The corrosion effect of neutral electrolytes is very less on the current collector so that a very wide variety of materials are available like nickel (Ni), titanium (Ti), copper (Cu), TiO_2, CNT, etc [71].

8.6 Dimension of Current Collector

Supercapacitor dimensions depend upon required electrical specifications that are operating voltage and capacitance so that size can be varied according to materials used to fabricate supercapacitor components to meet requirements [46, 72, 73]. Different case sizes can be used to fabricate supercapacitors of range 6.8–1000 mF capacitance and 3.6–16 V voltage ratings [73].

- case size 1 = 28 × 17 mm
- case size 2 = 48 × 30 mm
- case size 3 = 20 × 15 mm
- case size 4 = 17 × 15 mm.

According to lead configuration, it can be further classified in

- N-Style: Two terminal planar mount
- S-Style: Three terminal planar mount
- L-Style: Four terminal planar mount
- A Style: Through-hole mount
- H-Style: Extended stand-off through-hole mount

Table 8.4 Voltage (V) and capacitance (mF) available for different case sizes [73]

Voltage (V)	3.6				4.5				5.5				9.0			
Capacitance (mF)	50	280	–	–	33	–	22	15	30	200	15	–	22	120	–	–
	70	560	–	–	–	–	47	–	50	400	33	–	33	–	–	–
	100	–	–	–	–	–	–	–	60	1000	100	–	–	–	–	–
	140	–	–	–	–	–	–	–	100	–	–	–	–	–	–	–
Case size	1	2	3	4	1	2	3	4	1	2	3	4	1	2	3	4

Voltage (V)	12				16				20			
Capacitance (mF)	15	90	10	–	6.8	120	–	–	6.8	–	4.7	–
	22	–	–	–	–	–	–	–	–	–	–	–
	–	–	–	–	–	–	–	–	–	–	–	–
	–	–	–	–	–	–	–	–	–	–	–	–
Case size	1	2	3	4	1	2	3	4	1	2	3	4

- W-Style: Wire lead mount.

Different combinations of case sizes with voltage and capacitance are shown in Table 8.4. For example, case size 1 can be selected for 3.6 V and 50, 70, 100, 140 mF capacitance and for 4.5 V and 33 mF capacitance, and so on [73].

Some examples of current collectors described in Table 8.5 showing dimensions of current collector with different compositions of the conductive basis of the current collector, active mass, and electrolyte.

Table 8.5 Specifications of supercapacitors with current collector dimensions

Ex.	Current collector dimensions (mm)	Protective coating thickness (μm)	Compositions of conductive basis of current collector	Active mass	Electrolyte	Electrical capacitance (F)	Coulomb capacity (A h)	Ref.
1	135 × 72 × 0.2	50	97% Lead and 3% tin	Activated carbon materials	Sulfuric acid with density 1.26 g/cm^3	8500	2.45	[46]
2	135 × 72 × 0.2	50	99% Lead and 1% tin	Activated carbon materials	Sulfuric acid with density 1.26 g/cm^3	1400	0.22	[46]
3	50 × 100 × 0.05	–	Aluminum	Activated carbon materials	1 M solution of $LiPF_6$ in a 2:1 mixture of ethylene carbonate	19.2	3.5	[72]

8.7 Current Collector Material Selection for Supercapacitors

Supercapacitor performance not only depends upon electrode and electrolyte materials but also other components like a current collector and separator materials. The role of the current collector in the performance of the cell is to provide low electric and contact resistance with the electrode material. The current collector also provides mechanical strength to the electrode material. Ideal current collector material should be strong enough to bear the external pressure or small shocks and should be flexible. It should be chemically inert and thermally stable. Other important properties of current collector material include low cost and environmentally friendly. Commonly used current collector materials for commercial purposes are listed in Table 8.7.

8.8 Objectives for Current Collector Material Selection

For the selection of the appropriate current collector material, some specific parameters that are necessary to examine according to the need of a supercapacitor cell. The objectives of the current collector material selection are divided into three categories according to the priorities as tabulated in Table 8.6. Primary objectives are the most required properties of the current collector material. Secondary objectives are also essential, but these properties do not directly affect the performance of supercapacitor cells. Tertiary objectives are concern about the economic and environmental effects of a supercapacitor. Table 8.7 shows a list of various properties of different types of current collector materials.

Table 8.6 Primary, secondary and tertiary objectives for selection of a suitable current collector [39]

Primary objective	Secondary objective	Tertiary objective
High electrical conductivity	High electrochemical stable	Low cost
High mechanical flexibility	High chemical stability	Environmental friendly
Low contact resistance	Low density	Easy to fabricate
High thermal conductivity	High thermal stability	
High corrosion resistance	High compression strength	

Table 8.7 Current collector materials list with electrical, mechanical, and thermal properties [39]

Current collector	Electrical conductivity (S/m) × 10^6	ESR (Ω)	Tensile strength (MPa)	Compression strength (MPa)	Young's modulus (GPa)	Thickness (mm)	Thermal stability (°C)	Density (g/cm^3)	Thermal conductivity W/mK	Ref.
Nickel foam	15.80	0.07	–	0.25	200	0.5–40	1455	0.1–0.62	90.9	[74, 75]
Nickel foil	14.43	0.28	570	–	200	0.04	1455	8.9	90.9	[74, 76, 77]
Aluminum foam	39.2	–	3–15	3–17	70	4-50	660-800	0.25-0.75	205	[78–80]
Aluminum foil	35	3.3	105-120	–	70	0.016	660	2.70	205	[47, 81–84]
Stainless steel foam	–	–	2–3	0.25	190	1-25	1371-1537	7.65–7.94	15.1	[85, 86]
Stainless steel foil	1.4	–	1000–1200	–	190	0.15–0.36	–	7.98	15.1-19	[87, 88]
Copper foam	59.6	1.6	–	–	110-128	1-25	1085	0.7–1.1	401	[89]
Copper foil	59.6	0.39	318	–	110-128	0.25	1083	8.96	401	[90–92]
Titanium foil	2.38	0.94	140	–	116	0.25	1668	4.54	21.9	[93, 94]
Platinum foil	9.43	–	125–165	–	168	0.025–3	1772	21.45	71.6	[95]
Tantalum foil	7.63	–	–	–	186	–	3017	16.69	57.5	[96]
Carbon fiber	0.071	–	950-985	–	228	0.005	900-1500	2.267	119–165	[97–101]

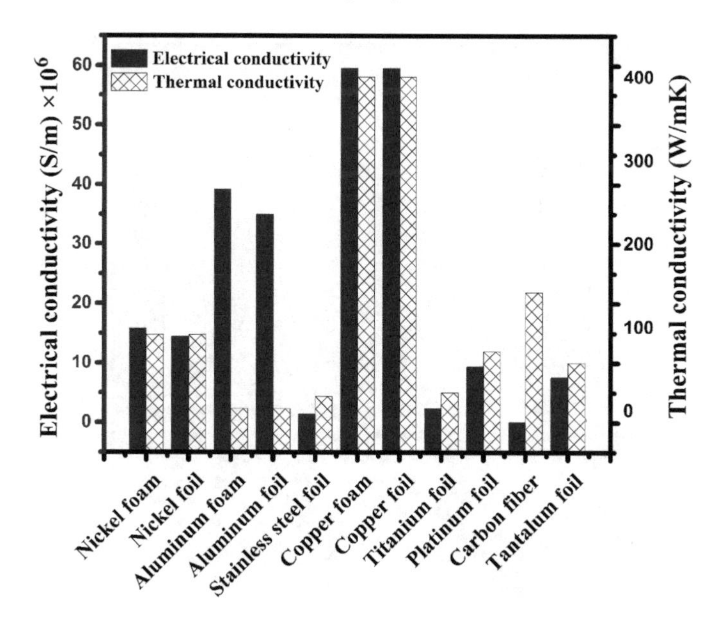

Fig. 8.20 Electrical conductivity and thermal conductivity of different current collector materials

8.8.1 Screening Using Constraints

The primary objective for material selection of a current collector is to have a combination of extraordinary electrical and thermal conductivity, mechanical flexibility, and corrosion resistance properties. Figure 8.20 demonstrates the electrical and thermal conductivity of different current collectors used in energy storage applications. Copper foam and foil display the highest electrical and thermal conductivity. Nickel and aluminum deliver good electrical conductivity. Although carbon fiber has the lowest electrical conductivity, its lightweight is its major advantage for supercapacitor applications. Figure 8.21 represents the elastic modulus and density bars for some common current collectors. In this figure nickel, stainless steel, tantalum, and platinum materials are showing high modulus, but they also contain high density, which will increase the device weight. Copper, carbon fiber, and aluminum have adequate elastic modulus and are lightweight material.

8.8.2 Governing Equations

8.8.2.1 Electrical Conductivity

The electrical conductivity of material indicates the ability to pass electricity in the material. For the high power density and rate capability of the supercapacitor,

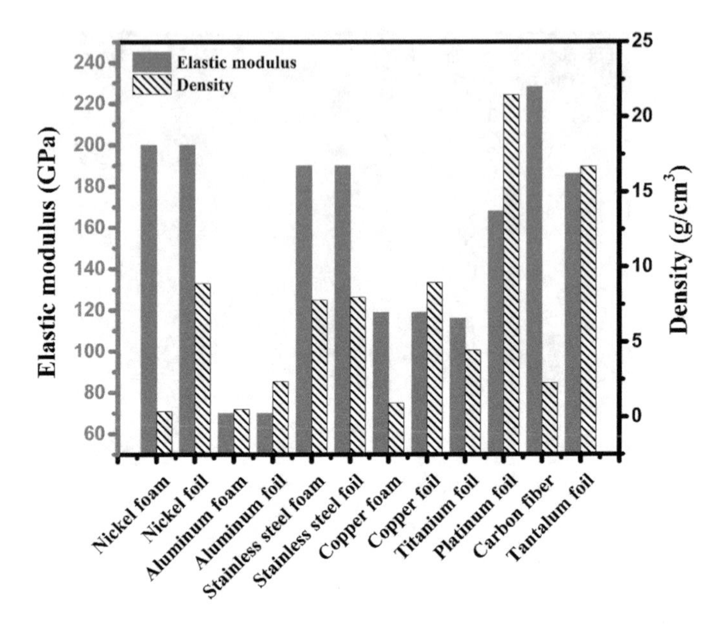

Fig. 8.21 Elastic modulus and density of different current collector materials

the electrical conductivity of the current collector should be high. The electrical conductivity of the current collector will be calculated from the following equations

$$R = \frac{\rho \times l}{A} \tag{8.1}$$

Resistivity and conductivity are related as

$$\rho = \frac{1}{\sigma} \tag{8.2}$$

where R (ohm) denotes resistance, l (m) is the thickness of the current collector, A (m^2) is the cross-sectional area of the current collector, ρ (Ω m) is resistivity, and σ (S/m) represents the conductivity.

8.8.2.2 Thermal Conductivity

The thermal conductivity of material shows the ability to pass the heat across the cross section of material. During the charging and discharging cycle of the supercapacitor, heat is generated due to the repeated movement of electrolyte ions. This is very important for a supercapacitor to maintain thermal stability with any deformation. The high thermal conductivity of current collector is necessary for the assembly of a thermally stable supercapacitor. Thermal conductivity will be calculated using the following equation

$$Q = kA\frac{\Delta T}{l} \tag{8.3}$$

where Q(W) denotes heat flux, k(W/mK) indicates thermal conductivity, $A(m^2)$ is the cross-sectional area, ΔT shows the temperature difference, and l(m) is the thickness of the current collector.

8.8.2.3 Tensile Strength

The tensile strength of the material is the ability of the material to resist the external load. The strength of the current collector should be high enough to bear the external load. Tensile strength will be calculated using the following equation

$$S = \frac{F}{A_0} \tag{8.4}$$

where $S(N/m^2)$ denotes strength, F(N) is axial force, and $A_0(m^2)$ represents the cross-sectional area.

8.8.2.4 Mass

One of the major issues with supercapacitor is lower energy density in comparison to batteries. For high energy density, a supercapacitor should be lightweight. So the current collector should also have lightweight. Mass will be calculated using the following equation

$$M = \rho_0 \times l \times A \tag{8.5}$$

where M(kg) indicates mass, $\rho_0(kg/m^3)$ stands for density, l(m) represents thickness, and $A(m^2)$ stands for the area.

8.8.2.5 Bending Strength

Bending strength is the strength of a material to resist the deformation under bending load. For flexible supercapacitors, materials should have high bending strength for high restoration. The bending strength of the material will be calculated using the following equation

$$\frac{M}{I} = \frac{\sigma_0}{y} = \frac{E}{R} \tag{8.6}$$

where M(N/m) represents the bending moment, I(m^4) is the moment of inertia, σ_0(N/m^2) indicates bending stress, E(GPa) stands for elastic modulus, and R(m) is the radius of curvature.

8.8.3 *Material Index*

The material index is the way to characterize the performance of the material and according to the attributes, materials get ranked.

8.8.3.1 Electrical Conductivity and Thermal Conductivity

The material index of electrical conductivity and thermal conductivity shows tentative constant relation between them. Here we can see that electrical conductivity is directly proportional to the thermal conductivity because metal electron is only cause to conduct heat and electricity.

From the Eqs. 8.1 and 8.2

$$R = \frac{\rho \times l}{A}$$

and

$$Q = kA\frac{\Delta T}{l}$$

There is a relationship between the electrical and thermal conductivity of the material. According to the Wiedemann–Franz law, the ratio of thermal and electrical conductivity is proportional to temperature [60].

$$\frac{k}{\sigma} = LT \tag{8.7}$$

where k(W/mK) is thermal conductivity, σ(S/m) represents electrical conductivity, T(K) stands for the temperature, and L(m) is proportionality constant (Lorenz number).

Hence, the material index of electrical and thermal conductivity is given as

$$M_1 = \frac{k}{\sigma} \tag{8.8}$$

8.8.3.2 Young's Modulus and Density

Material index of Young's modulus and density guides for the selection of lightweight and stiff current collector plates. For calculating the material index, following equations are used.

From Eqs. 8.5 and 8.6

$$M = \rho_0 \times l \times A$$

and

$$\frac{M}{I} = \frac{\sigma_0}{y} = \frac{E}{R}$$

The flexural modulus of the current collector can be calculated using the Eq. 8.6

$$E = \frac{l^3 F}{4wh^3 d}$$

or

$$E^{1/3} = \frac{l F^{1/3} W^{2/3}}{4^{1/3} d^{1/3} A} \tag{8.9}$$

Here E is flexural strength, l is the length of the plate, F is the applied force, W is the width of the plate, h is the height of plate, A is the area of plate, and d is the deflection in the plate due to load F.

From Eqs. 8.5 and 8.9

For minimum weight design of stiff plates [61]

$$E^{1/3}/\rho_0 = C$$

Hence, the material index for elastic modulus and density will be given as

$$M_2 = E^{1/3}/\rho_0 \tag{8.10}$$

8.8.3.3 Final Material Index

Final material index of material will be multiplication of M_1 and M_2 (it can be in any other form like $M_3 = M_1 \times\pm\div M_2 \times\pm\div \mathrm{M}_3$ or $a_1M_1 \times\pm\div a_2M_2 \times\pm\div a_3M_3$, where a_1, a_2, and a_3 are any integers; M_3 is another material index).

So,

$$M_3 = M_1 \times M_2 \tag{8.11}$$

8.8.3.4 Materials Property Chart

Ashby chart provides information on the materials with its specific properties within a specified range. The material index line decides the choice of an appropriate material for a particular application.

8.8.3.5 Ashby Chart for Material Index M_1

From Eq. 8.8

$$M_1 = \frac{k}{\sigma}$$

Taking log both sides

$$\log M_1 = \log k - \log \sigma$$

or

$$\log k = \log \sigma + \log M_1 \tag{8.12}$$

In the Ashby chart, the governing line of thermal and electrical conductivity will have a positive unit slope as shown in Fig. 8.22. The most appropriate current collector materials will stay over the material index line as indicated in Fig. 8.22.

Figure 8.22 demonstrates different materials' thermal and electrical conductivity properties. Cost is also an important factor for the material selection of the current collector. It is well known that gold and silver show high thermal and electrical conductivity. Also, these materials are noble in nature, which cannot get corroded in harsh environments. However, its high cost limits the application as current collector. Along with this, mechanical strength and density also should be taken into consideration. So, there should be other alternative materials to be selected as a current

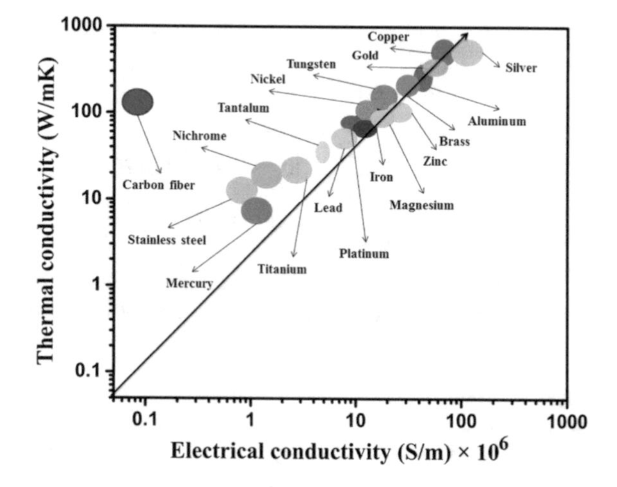

Fig. 8.22 Ashby chart of thermal and electrical conductivity properties of different materials ([102–104])

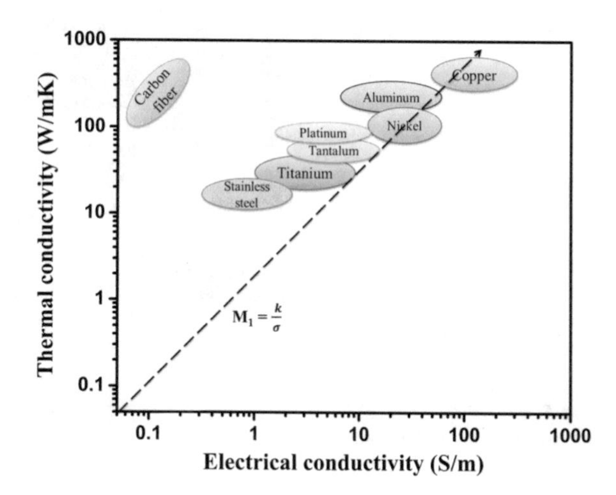

Fig. 8.23 Ashby chart of thermal and electrical conductivity

collector. The appropriate materials for the selection of current collectors are shown in Fig. 8.23.

For the high performance of supercapacitors, we need high thermal and electrical conductivity of the current collector. High electrical conductivity indicates low resistivity and hence high power density. During the charge–discharge cycle, the temperature of the device gets increased, which affects the performance metrics. According to the Ashby chart, the appropriate materials are copper, titanium, platinum, nickel, stainless steel, and tantalum.

8.8.3.6 Ashby Chart for Material Index M_2

From Eq. 6.10

$$M_2 = E^{1/3}/\rho_0$$

Taking log both side

$$\log M_2 = \frac{1}{3}\log E - \log \rho_0$$
$$\log E = 3\log \rho_0 + 3\log M_2 \quad (8.13)$$

The Ashby chart of Young's modulus and density governing line will have a slope angle of 71.56° as shown in Figs. 8.24 and 8.25. Appropriate material for the current

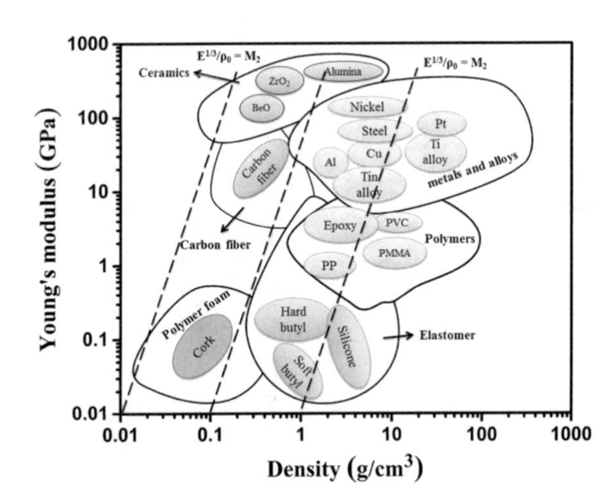

Fig. 8.24 Ashby chart of Young's modulus and density properties of different materials [102]

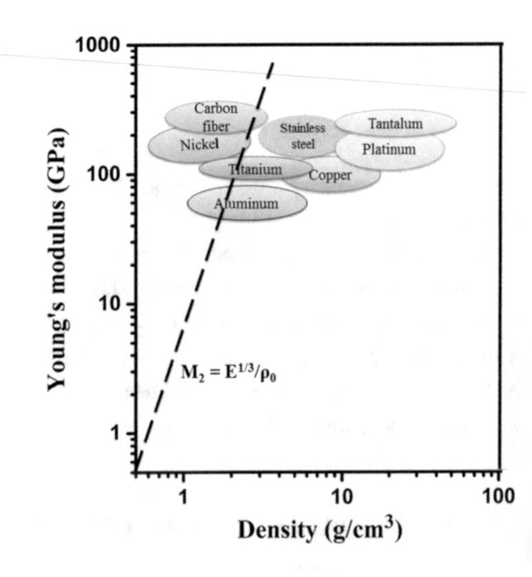

Fig. 8.25 Ashby charts of elastic modulus and density

Table 8.8 List of materials with a material index value

Current collector material	M_1	M_2	$M_3 = M_1 \times M_2$
Nickel	5.75	16.24	93.38
Copper	6.72	5.46	36.69
Aluminum	5.85	8.24	48.20
Titanium	9.20	1.09	10.028
Platinum	7.59	0.25	1.89
Stainless steel	12.17	0.73	8.88
Carbon fiber	2000	2.70	5400
Tantalum	7.53	0.34	2.56

collector will lie above the material index line because the material index should be high for these materials.

The primary condition for the current collector is that it should be electrically conductive. In Fig. 8.24, different materials are placed appropriately in terms of their Young's modulus and density in the material property bobble chart. Carbon fiber, metal, and metal alloys are materials that contain high electrical conductivity. The appropriate materials for the current collector are demonstrated in Fig. 8.25.

For high performance of supercapacitor, high elastic modulus and low density of current collectors are required. According to Fig. 8.25, appropriate materials for supercapacitor applications are nickel, carbon fiber, copper, and aluminum.

8.8.4 List of Material Index

Material indexes M_1 and M_2 are calculated from Table 8.7 data. The final material index M_3 value will decide appropriate materials for current collectors (Table 8.8). For the high performance of supercapacitors, the M_3 value should be high.

According to the high material index (M_3), suitable current collector material is listed in Table 8.9.

8.9 Concluding Remarks

The application of supercapacitors is increasing day by day and has a wide scope in research and development. New innovations in carbon and metal oxide-based nanostructured material and flexible electronics made supercapacitors more efficient by using such advanced material as parts of a supercapacitor. This had enhanced the performance by reducing internal resistance and improving flexibility, making energy-efficient and long-life devices. As a critical part of the supercapacitor, the

Table 8.9 List of suitable current collector materials with properties, advantages, and disadvantages

Current collector material	Properties and advantages	Disadvantages
Carbon fiber	High thermal conductivity high flexibility lightweight high thermal stability long service life	Low electrical conductivity high cost create etching in the human body difficult to fabricate
Nickel	High electrical conductivity high mechanical strength lightweight high bendability low contact resistance high electrochemical stability	Get oxidized at high temperature affect the capacitance of the electrode material high cost
Aluminum	High electrical conductivity low cost lightweight high flexibility	Low thermal stability form oxide layer coating at high temperature low thermal conductivity
Copper	High electrical conductivity high thermal conductivity high flexibility high corrosion resistance low contact resistance low cost	Get oxidized at high and above 150°C low electrochemical stability.
Titanium	High thermal stability high thermal conductivity high mechanical strength durable high corrosion resistance high electrical conductivity	High cost low chemical stability high production cost Reactive at high temperature.
Stainless steel	High corrosion resistance low cost electrochemically stable high mechanical strength	Poor cycle stability high weight
Tantalum	High mechanical strength high thermal stability high corrosion resistance	High cost high density
Platinum	High mechanical strength high thermal stability high electrochemical stability high thermal conductivity	High cost high density

current collector also needs updating its materials and manufacturing process. The current collector has traveled a path from simple aluminum foil to the nanostructured and flexible current collector. However, more focus on optimization in its characteristics is required by applying new materials and techniques. Laser-treated surface modification is a very good example that shows how existing materials could be more efficient by using existing technology. To optimize the characteristics of the current collector, the thickness of current collector, size and shape, and available surface area should be improved, which can provide low internal resistance, good stability by eliminating delamination between current collector and electrode. Effective current collector materials for supercapacitor applications are carbon fiber, nickel (Ni), copper (Cu), stainless steel, aluminum (Al), titanium (Ti), platinum (Pt), and tantalum (Ta).

Acknowledgements The authors acknowledge the financial support provided by the Science and Engineering Research Board, Department of Science and Technology, India (SR/WOS-A/ET-48/2018) for carrying out this research work.

References

1. J. Tahalyani, J. Akhtar, J. Cherusseri, K.K. Kar, Characteristics of capacitor: fundamental aspects, in *Handbook of Nanocomposite Supercapacitor Materials I Characteristics* edited by K.K. Kar. (Springer, Berlin, Heidelberg, 2020). https://doi.org/10.1007/978-3-030-43009-2_1
2. S. Banerjee, P. Sinha, K.D. Verma, T. Pal, B. De, J. Cherusseri, P.K. Manna, K. K. Kar, Capacitor to supercapacitor, in *Handbook of Nanocomposite Supercapacitor Materials I Characteristics* edited by K.K. Kar. (Springer, Berlin, Heidelberg, 2020). https://doi.org/10.1007/978-3-030-43009-2_2
3. A. Sani, S. Siahaan, N. Mubarakah, Suherman, IOP Conf. Ser. Mater. Sci. Eng. **309**, 012078 (2018)
4. P. Sinha, K.K. Kar Introduction to supercapacitors, in *Handbook of Nanocomposite Supercapacitor Materials II Performance* edited by K.K. Kar. (Springer, Berlin, Heidelberg, 2020). https://doi.org/10.1007/978-3-030-52359-6_1
5. Woodford, Chris, Supercapacitors (2020), https://www.explainthatstuff.com/how-supercapacitors-work.html. Accessed 22 June 2020
6. P. Sinha, K.K. Kar, Characteristics of supercapacitors, in *Handbook of Nanocomposite Supercapacitor Materials II Performance* edited by K.K. Kar. (Springer, Berlin, Heidelberg, 2020). https://doi.org/10.1007/978-3-030-52359-6_3
7. B.K. Kim, S. Sy, A. Yu, J. Zhang, *Electrochemical Supercapacitors for Energy Storage and Conversion* (Wiley, Chichester, UK, 2015)
8. C. Zhong, Y. Deng, W. Hu, J. Qiao, L. Zhang, J. Zhang, Chem. Soc. Rev. **44**, 7484 (2015)
9. K.D. Verma, P. Sinha, S. Banerjee, K.K. Kar, Characteristics of electrode materials for supercapacitors, in *Handbook of Nanocomposite Supercapacitor Materials I Characteristics* edited by K.K. Kar. (Springer, Berlin Heidelberg, 2020). https://doi.org/10.1007/978-3-030-43009-2_9
10. Z.S. Iro, C. Subramani, S.S. Dash, Int. J. Electrochem. Sci. **11**, 10628 (2016)
11. K. Yu, Z. Wen, H. Pu, G. Lu, Z. Bo, H. Kim, Y. Qian, E. Andrew, S. Mao, J. Chen J. Mater. Chem. A **1**, 188–193 (2013)
12. P. Sinha, S. Banerjee, K.K. Kar, Activated carbon as electrode materials for supercapacitors, in *Handbook of Nanocomposite Supercapacitor Materials II Performance* edited by K.K. Kar. (Springer, Berlin, Heidelberg, 2020). https://doi.org/10.1007/978-3-030-52359-6_5
13. B. De, S. Banerjee, K.D. Verma, T. Pal, P.K. Manna, K.K. Kar, Carbon nanofiber as electrode materials for supercapacitors, in *Handbook of Nanocomposite Supercapacitor Materials II Performance* edited by K.K. Kar. (Springer, Berlin, Heidelberg, 2020). https://doi.org/10.1007/978-3-030-52359-6_7

14. B. De, S. Banerjee, K.D. Verma, T. Pal, P.K. Manna, K.K. Kar, Carbon Nanotube as Electrode Materials for Supercapacitors, in *Handbook of Nanocomposite Supercapacitor Materials II Performance* edited by K.K. Kar. (Springer, Berlin, Heidelberg, 2020). https://doi.org/10.1007/978-3-030-52359-6_9
15. B. De, S. Banerjee, T. Pal, K.D. Verma, P.K. Manna, K.K. Kar, Graphene/reduced graphene oxide as electrode materials for supercapacitors, in *Handbook of Nanocomposite Supercapacitor Materials II Performance* edited by K.K. Kar. (Springer, Berlin, Heidelberg, 2020). https://doi.org/10.1007/978-3-030-52359-6_11
16. B. De, S. Banerjee, K.D. Verma, T. Pal, P.K. Manna, K.K. Kar, Transition metal oxides as electrode materials for supercapacitors, in *Handbook of Nanocomposite Supercapacitor Materials II Performance* edited by K.K. Kar. (Springer, Berlin, Heidelberg, 2020). https://doi.org/10.1007/978-3-030-52359-6_4
17. A. Tyagi, S. Banerjee, J. Cherusseri, K.K. Kar, Characteristics of transition metal oxides, in *Handbook of Nanocomposite Supercapacitor Materials I Characteristics* edited by K.K. Kar. (Springer, Berlin, Heidelberg, 2020). https://doi.org/10.1007/978-3-030-43009-2_3
18. C. An, Y. Zhang, H. Guo, Y. Wang, Nanoscale Adv. **1**, 4644–4658 (2019)
19. B. De, S. Banerjee, T. Pal, K.D. Verma, A. Tyagi, P.K. Manna, K.K. Kar, Transition metal oxide/electronically conducting polymer composites as electrode materials for supercapacitors, in *Handbook of Nanocomposite Supercapacitor Materials II Performance* edited by K.K. Kar. (Springer, Berlin, Heidelberg, 2020). https://doi.org/10.1007/978-3-030-52359-6_14
20. B. De, S. Banerjee, T. Pal, K.D. Verma, A. Tyagi, P.K. Manna, K.K. Kar, Transition metal oxide-/carbon-/electronically conducting polymer-based ternary composites as electrode materials for supercapacitors, in *Handbook of Nanocomposite Supercapacitor Materials II Performance* edited by K.K. Kar. (Springer, Berlin, Heidelberg, 2020). https://doi.org/10.1007/978-3-030-52359-6_15
21. P. Sinha, S. Banerjee, K.K. Kar, Characteristics of activated carbon, in *Handbook of Nanocomposite Supercapacitor Materials I Characteristics* edited by K.K. Kar. (Springer, Berlin, Heidelberg, 2020). https://doi.org/10.1007/978-3-030-43009-2_4
22. R. Sharma, K.K. Kar, Characteristics of carbon nanofibers, in *Handbook of Nanocomposite Supercapacitor Materials I Characteristics* edited by K.K. Kar. (Springer, Berlin, Heidelberg, 2020). https://doi.org/10.1007/978-3-030-43009-2_7
23. S. Banerjee, K.K. Kar, Characteristics of carbon nanotubes, in *Handbook of Nanocomposite Supercapacitor Materials I Characteristics* edited by K.K. Kar. (Springer, Berlin, Heidelberg, 2020). https://doi.org/10.1007/978-3-030-43009-2_6
24. P. Chamoli, S. Banerjee, K.K. Raina, K.K. Kar, Characteristics of graphene/reduced graphene oxide, in *Handbook of Nanocomposite Supercapacitor Materials I Characteristics* edited by K.K. Kar. (Springer, Berlin, Heidelberg, 2020). https://doi.org/10.1007/978-3-030-43009-2_5
25. P. Sinha, S. Banerjee, K.K. Kar, Transition metal oxide/activated carbon-based composites as electrode materials for supercapacitors, in *Handbook of Nanocomposite Supercapacitor Materials II Performance* edited by K.K. Kar. (Springer, Berlin, Heidelberg, 2020). https://doi.org/10.1007/978-3-030-52359-6_6
26. B. De, S. Banerjee, K.D. Verma, T. Pal, P.K. Manna, K.K. Kar, Transition metal oxide/carbon nanofiber composites as electrode materials for supercapacitors, in *Handbook of Nanocomposite Supercapacitor Materials II Performance* edited by K.K. Kar. (Springer, Berlin, Heidelberg, 2020). https://doi.org/10.1007/978-3-030-52359-6_8
27. B. De, S. Banerjee, T. Pal, K.D. Verma, A. Tyagi, P.K. Manna, K.K. Kar, Transition metal oxide/carbon nanotube composites as electrode materials for supercapacitors, in *Handbook of Nanocomposite Supercapacitor Materials II Performance* edited by K.K. Kar. (Springer, Berlin, Heidelberg, 2020). https://doi.org/10.1007/978-3-030-52359-6_10
28. B. De, P. Sinha, S. Banerjee, T. Pal, K.D. Verma, A. Tyagi, P.K. Manna, K.K. Kar, Transition metal oxide/graphene/reduced graphene oxide composites as electrode materials for supercapacitors, in *Handbook of Nanocomposite Supercapacitor Materials II Performance* edited by K.K. Kar. (Springer, Berlin, Heidelberg, 2020)https://doi.org/10.1007/978-3-030-52359-6_12

29. S. Banerjee, K.K. Kar, Conducting polymers as electrode materials for supercapacitors, in *Handbook of Nanocomposite Supercapacitor Materials II Performance* edited by K.K. Kar. (Springer, Berlin, Heidelberg, 2020). https://doi.org/10.1007/978-3-030-52359-6_13
30. T. Pal, S. Banerjee, P.K. Manna, K.K. Kar, Characteristics of conducting polymers, in *Handbook of Nanocomposite Supercapacitor Materials I Characteristics* edited by K.K. Kar. (Springer, Berlin, Heidelberg, 2020). https://doi.org/10.1007/978-3-030-43009-2_8
31. D.P. Dubal, N.R. Chodamkar, Do-H. Kim, P.G. Romero, Chem. Soc. Rev. **47**, 2065–2129 (2018)
32. K.D. Verma, S. Banerjee, K.K. Kar, Characteristics of electrolytes, in *Handbook of Nanocomposite Supercapacitor Materials I Characteristics* edited by K.K. Kar. (Springer, Berlin, Heidelberg, 2020). https://doi.org/10.1007/978-3-030-43009-2_10
33. L. Liudvinavicius, *Supercapacitors: Theoretical and Practical Solutions* (Intech Open, London, United Kingdom, 2017)
34. K.D. Verma, P. Sinha, S. Banerjee, K.K. Kar, M.K. Ghorai, Characteristics of separator materials for supercapacitors, in *Handbook of Nanocomposite SupercapacitorMaterials I Characteristics* edited by K.K. Kar. (Springer, Berlin, Heidelberg, 2020). https://doi.org/10.1007/978-3-030-43009-2_11
35. H. Yu, Q. Tang, J. Wu, Y. Lin, L. Fan, M. Huang, J. Lin, Y. Li, F. Yu, J. Power Sources **206**, 463–468 (2012)
36. B. Szubzda, A. Szmaja, M. Ozimek, S. Mazurkiewicz, Appl. Phys. A **117**, 1801–1809 (2014)
37. Z. Zhul, S. Tang, J. Yuan, X. Qin, Y. Deng, R. Qu, G.M. Haarberg, Int. J. Electrochem. Sci. **11**, 8270–8279 (2016)
38. D. Bresser, D. Buchholz, A. Moretti, A. Varzi, S. Passerini, Energ. Environ. Sci. **11**, 3096–3127 (2018)
39. K.D. Verma, P. Sinha, S. Banerjee, K.K. Kar, Characteristics of current collector materials for supercapacitors, in *Handbook of Nanocomposite Supercapacitor Materials I Characteristics* edited by K.K. Kar. (Springer, Berlin, 2020). https://doi.org/10.1007/978-3-030-43009-2_12
40. L. Liu, H. Zhao, Y. Wang, Y. Fang, J. Xie, Y. Lei, Adv. Funct. Mater. **28**(6), 1–9 (2018)
41. C. Zhong, Y. Deng, W. Hu, D. Sun, X. Han, J. Qiao, J. Zhang, *Electrolytes for Electrochemical Supercapacitors*, 1st ed. (CRC Press, 2016)
42. A.K. Samantara, S. Ratha, *Components of Supercapacitor* (Springer, Singapore, 2017)
43. H.R. Kim, W.M. Choi, Scr. Mater **146**, 100–104 (2018)
44. S. Pilathottathil, M.S. Thayyil, M.P. Pillai, A.P. Jemshihas, J. Electron. Mater. **48**(9), 5835–5842 (2019)
45. H.N. Miankushki, A. Sedghi, S. Baghshahi, J. Energ. Storage **19**, 201–212 (2018)
46. S.A. Kazaryan, V.P. Nedoshivin, V.A. Kazarov, G.G. Kharisov, S.V. Litvinenko, S.N. Razumov, US Patent 7,446, 998 (2008)
47. Y. Huang, Y. Li, Q. Gong, G. Zhao, G. Zhao, P. Zhang, J. Bai, J. Gan, M. Zhao, Y. Shao, D. Wang, L. Liu, G. Zou, D. Zhuang, J. Liang, H. Zhu, C. Nan, A.C.S. Appl, Mater. Interfaces **10**, 16572–16580 (2018)
48. Y. Li, Q. Wang, Y. Wang, M. Bai, J. Shao, H. Ji, H.H. Feng, J. Zhang, X. Ma, W. Zhao, Dalton Trans. **48**, 7659–7665 (2019)
49. Z. Ren, Y. Li, J. Yu, Science **9**, 138–148 (2018)
50. M. Kumar, P. Sinha, T. Pal, K.K. Kar, Materials for supercapacitors, in *Handbook of Nanocomposite Supercapacitor Materials II Performance* edited by K.K. Kar. (Springer, Berlin, Heidelberg, 2020). https://doi.org/10.1007/978-3-030-52359-6_2
51. J. Yu, J. Wu, H. Wang, A. Zhou, C. Huang, H. Bai, L. Li, A.C.S. Appl, Mater. Interfaces **8**, 4724–4729 (2016)
52. L. Wen, K. Li, J. Liu, Y. Huang, F. Bu, B. Zhao, Y. Xu, RSC Adv. **7**, 7688–7693 (2017)
53. S. Rajendran, M. Naushad, S. Balakumar, *Nanostructured Materials for Energy Related Applications*. (Springer Nature Switzerland AG, 2019)
54. D. Lei, X.D. Li, M.K. Seo, M.S. Khil, H.Y. Kim, B.S. Kim, Polym. Polym. **132**, 31–40 (2017)
55. R. Li, X. Li, J. Chen, J. Wang, H. He, B. Huang, Y. Liu, Y. Zhou, G. Yang, Nanoscale **10**, 3981–3989 (2018)

56. D. Xiong, X. Li, Z. Bai, J. Li, H. Shan, L. Fan, C. Long, D. Li, X. Lu, Electrochim. Acta **259**, 338–347 (2018)
57. J. Tan, Y. Han, L. He, Y. Dong, X. Xu, D. Liu, H. Yan, Q. Yu, C. Huang, L. Mai, J. Mater. Chem. A **5**, 23620–23627 (2017)
58. N.P. Sari, D. Dutta, A. Jamaluddin, J.K. Chang, C.Y. Su, Phys. Chem. Chem. Phys. **19**(45), 30381–30392 (2017)
59. Y. Liang, W. Weng, J. Yang, L. Liu, Y. Zhang, L. Yang, X. Luo, Y. Cheng, M. Zhu, RSC Adv. **7**, 48934–48941 (2017)
60. H. Li, D. Fu, X.M. Zhang, R. Soc, Open Sci. **5**, 171028 (2018)
61. L. Shao, Q. Wang, Z. Ma, Z. Ji, X. Wang, D. Song, Y. Liu, N. Wang, J. Power Sour. **379**, 350–361 (2018)
62. L. Yuan, X.H. Lu, X. Xiao, T. Zhai, J. Dai, F. Zhang, B. Hu, X. Wang, L. Gong, J. Chen, C. Hu, Y. Tong, J. Zhou, Z.L. Wang, ACS Nano **6**, 656–661 (2012)
63. P. Yang, X. Xiao, Y. Li, Y. Ding, P. Qiang, X. Tan, W. Mai, Z. Lin, W. Wu, T. Li, H. Jin, P. Liu, J. Zhou, C.P. Wong, Z.L. Wang, ACS Nano **7**, 2617–2626 (2013)
64. Y. He, W. Chen, X. Li, Z. Zhang, J. Fu, C. Zhao, E. Xie, ACS Nano **7**, 174–182 (2013)
65. P. He, Q. Huang, B. Huang, RSC Adv. **7**, 24353–24358 (2017)
66. S. Xi, Y. Zhu, Y. Yang, S. Jiang, Z. Tang, Nanoscale Res. Lett. **12**, 171 (2017)
67. S. Wu, Q. Liu, F. Xue, M. Wang, S. Yang, H. Xu, F. Jiang, J. Wang, J. Mater. Sci.Mater. Electron **28**, 11615–11623 (2017)
68. J.M. Campillo-Robles, X. Artetxe, K.D.T. Sánchez, Appl. Phys. Lett. **111**, 093902 (2017)
69. J. Busom, A. Schreiber, A. Tolosa, N. Jäckel, I. Grobelsek, N.J. Peter, V. Presser, J. Power Sour. **329**, 432–440 (2016)
70. R. Kotz, P.W. Ruch, D. Cericola, J. Power Sour. **195**, 923–928 (2010)
71. C. Zhong, D. Sun, Y. Deng, W. Hu, J. Qiao, J. Zhang, (2016), Compatibility of Electrolytes with Inactive Components of Electrochemical Supercapacitors, In: *Electrolytes for Electrochemical Supercapacitors*, CRC Press, Boca Raton, https://doi.org/10.1201/b21497-4
72. G. G. Amatucci, A. D. Pasquier, J. M. Tarascon, Supercapacitor structure and method of making same, Patent, WO2000017898A1, 30 March 2000
73. AVX a kyocera group company. http://catalogs.avx.com/BestCap.pdf. Accessed 30 July 2020
74. J.H. Kim, K.H. Shin, C.S. Jin, D.K. Kim, Y.G. Kim, Y.G. Kim, J.H. Park, Y.S. Lee, Y.S. Joo, K.H. Lee, Electrochemistry **69**, 853 (2001)
75. Nickel Foam. https://www.americanelements.com/nickel-foam-7440–02-0. Accessed 2 August 2019
76. Electroformed nickel foil.http://www.specialmetals.com/assets/smc/documents/alloys/other/electroformed-nickel-foil.pdf . Accessed 5 August 2019
77. Nickel Foill AMERICAN ELEMENTS. https://www.americanelements.com/nickel-foil-7440–02-. Accessed 27 July 2019
78. Y. Feng, H. Zheng, Z. Zhu, F. Zu, Mater. Chem. Phys. **78**, 196 (2003)
79. D. Puspitasari, F.K.H. Rabie, T.L. Ginta, J.C. Kurnia, M. Mustapha, MATEC Web Conf. **225**, 01006 (2018)
80. Aluminum Foam. https://www.americanelements.com/aluminum-foam-7429–90-5. Accessed 30 July 2019
81. J. Butt, H. Mebrahtu, H. Shirvani, Prog. Addit. Manuf. **1**, 93 (2016)
82. S. Tóth, M. Füle, M. Veres, J.R. Selman, D. Arcon, I. Pócsik, M. Koós, Thin Solid Films **482**, 207 (2005)
83. R. Vicentini, L. H. Costa, W. Nunes, O. Vilas Boas, D. M. Soares, T. A. Alves, C. Real, C. Bueno, A. C. Peterlevitz, H. Zanin, J. Mater. Sci. Mater. Electron. **29**, 10573 (2018)
84. S. Banerjee, K.K. Kar, Polymer J. **109**, 176 (2017)
85. Y. Li, X. Zhang, L. Nie, Y. Zhang, X. Liu, J. Power Sources **245**, 520 (2014)
86. Stainless Steel Foam. https://www.americanelements.com/stainless-steel-foam-65997–19-5. Accessed 29 July 2019
87. B. Yao, J. Zhang, T. Kou, Y. Song, T. Liu, Y. Li, Adv. Sci. **4**, (2017)

88. Aluminum Versus Steel Conductivity. https://sciencing.com/aluminum-vs-steel-conductivity-5997828.html. Accessed 29 July 2019
89. H. Feng, Y. Chen, Y. Wang, Procedia Eng. **215**, 136 (2018)
90. M. Valvo, M. Roberts, G. Oltean, B. Sun, D. Rehnlund, D. Brandell, L. Nyholm, T. Gustafsson, K. Edström, J. Mater. Chem. A **1**, 9281 (2013)
91. K.N. Kang, I.H. Kim, A. Ramadoss, S.I. Kim, J.C. Yoon, J.H. Jang, Phys. Chem. Chem. Phys. **20**, 719 (2018)
92. X.Q. Yin, L.J. Peng, S. Kayani, L. Cheng, J.W. Wang, W. Xiao, L.G. Wang, G.J. Huang, Rare Met. **35**, 909 (2016)
93. R. Quintero, D.Y. Kim, K. Hasegawa, Y. Yamada, A. Yamada, S. Noda, RSC Adv. **4**, 8230 (2014)
94. Titanium Foil, https://www.americanelements.com/titanium-foil-7440–32-6, accessed 31 July 2019
95. Platinum (Pt) - Properties, Applications, https://www.azom.com/article.aspx?Article ID = 9235, accessed 31 July 2019
96. C. Shi, J. Dai, C. Li, X. Shen, L. Peng, P. Zhang, D. Wu, D. Sun, J. Zhao, Polymers **9**, 10 (2017)
97. Carbon Fiber. https://www.americanelements.com/carbon-fiber-7440–44-0, Accessed 31 July 2019
98. D. Wentzel, I. Sevostianov, Int. J. Eng. Sci. **130**, 129 (2018)
99. G. E. Mostovoi, L. P. Kobets, Frolov, V. I., Highly Conduct. One-Dimensional Solids 15, 247 (1979)
100. J. Maria, F. De Paiva, S. Mayer, M. Cerqueira **9**, 83 (2006)
101. S.K. Singh, M.J. Akhtar, K.K. Kar, A.C.S. Appl, Mater. Interfaces **10**, 24816 (2018)
102. M. F. Ashby, *Materials Selection in Mechanical Design*, 3rd edition, by M.F. Ashby (Elsevier-Butterworth Heinemann, Oxford, 2005)
103. Thermal Conductivity for all the elements in the Periodic Table, https://periodictable.com/Properties/A/ThermalConductivity.an.html, Accessed 13 Aug 2019
104. Table of Electrical Resistivity and Conductivity. https://www.thoughtco.com/table-of-electrical-resistivity-conductivity-608499. Accessed 13 Aug 2019

Chapter 9
Integrated Energy Storage System

Ravi Nigam and Kamal K. Kar

Abstract Intelligent energy storage systems utilize information and communication technology with energy storage devices. Energy management systems help in energy demand management and the effective use of energy storage devices. Supercapacitor management systems have been developed for supercapacitor usage during demand within safe operating limits. Supercapacitors and batteries are used together with the help of hybrid energy management configurations. Rule-based, optimization-based, and artificial intelligence-based energy management strategies are deployed for hybrid energy storage systems. The main parameters are adaptability, reliability, and robustness. Computational complexity is a driving parameter for using these techniques in online or offline mode.

9.1 Introduction

Supercapacitors are an electrochemical energy storage device, which is used to provide power back-up and used with batteries in hybrid energy storage systems. These energy storage devices have a basic configuration known as a cell that is combined to form cell strings, modules, and packs. The combined supercapacitor pack is protected, managed and used with the help of management systems utilizing sensors, electronics, algorithms, and programming. Various consumer devices are used in a controlled environment for energy efficiency. The main power supply from the grid is also managed. Integrated energy storage systems are the term for a

R. Nigam · K. K. Kar (✉)
Advanced Nanoengineering Materials Laboratory, Materials Science Programme, Indian Institute of Technology Kanpur, Kanpur 208016, India
e-mail: kamalkk@iitk.ac.in

R. Nigam
e-mail: ravinigam09@gmail.com

K. K. Kar
Advanced Nanoengineering Materials Laboratory, Department of Mechanical Engineering, Indian Institute of Technology Kanpur, Kanpur 208016, India

K. K. Kar (ed.), *Handbook of Nanocomposite Supercapacitor Materials III*,
Springer Series in Materials Science 313,
https://doi.org/10.1007/978-3-030-68364-1_9

combination of energy management of main power supply, energy storage devices, energy storage management devices, and energy management aspects for consumer general applications like billing, controlling appliances through a portal. The integrated energy storage system lowers the capital cost, energy consumption losses, and increase energy efficiency. An example of an integrated energy storage system is in the vehicle to grid or home systems.

9.1.1 Energy Security as a Component of National Security

National security is the concept of the state to protect and defend its citizen. It has two views, viz. military and non-military. The non-military aspects are political security, economic security, energy and natural resources security, homeland security, cyber-security, human security, and environmental security. Energy security is the uninterrupted availability or usage of energy sources at an affordable price. Foreign entities or internal conflict should not be able to compromise this ability. Energy fuels are oil, coal, natural gas, hydro, nuclear, solar, and wind. India is a net oil importer with huge expenditure and heavy reliance on West Asia for meeting import demands. The sea lines of communications are limited in the Indian Ocean region. The oil imports can be adversely affected during the conflict with independent states.

The solutions to energy security are diversification of fuel, reduce energy consumption, and harness renewable sources of energy. The depleting natural resources-based fuel combined with intermittent renewable energy sources result in an important role of energy storage in energy security.

9.1.2 Energy Storage and Energy Security

Energy storage technologies are vital for energy security. A secure grid has energy storage provisions, which transmit electricity with more control. Solar and wind energy can be stored and used during the night or as per need. The electricity supply for essential services is maintained during grid failure, extreme weather, or unseen circumstances. Energy storage mechanisms can decrease pollution and also have cost–benefit. They are essential for the mass utilization of clean energy sources. The vulnerability to cyber-attacks is reduced due to energy storage systems acting as independent power sources in critical services and infrastructure. Electric transportation requires energy storage. The national mission on transformative mobility and battery storage has been approved in India in 2019.

9.1.3 Supercapacitor and Energy Security

Supercapacitors are electrochemical energy storage devices [1, 2]. These can act as an energy buffer to absorb and supply power, which stabilizes output. Maxwell Technologies Inc., with the US Department of Energy and California Energy Commission, was using supercapacitors in a microgrid for a constant output from a wind turbine and a hydroelectric plant and power bridging to the end-user [3]. Supercapacitors are used as an additional power source. They can prevent a blackout, fault ride-through, and blackout recovery. It can be used as an emergency source of power for critical services. They can be used with batteries in a hybrid system and protect the battery during extreme peak loads. USA, Japan, China, Canada, and Europe, along with France, Poland, Italy, and Germany have done substantial work on supercapacitors. Asian countries, Korea, Taiwan, and India, are doing considerable research on supercapacitors [4].

9.1.4 Information Communication Technology (ICT) and Energy Infrastructure

ICT in the context of energy infrastructure is the connected communication network between all parts of the electricity transmission and distribution network ranging from operations, service providers, and customers. There is an interaction between machines, between users, and between users and machines done through ICT software and hardware algorithms and programming. ICT can include sensors for remote measurements, integrated circuits for monitoring, smart meters, grid management systems, software for measuring demand, and low maintenance and management system for supercapacitors or batteries. This leads to energy efficiency and economic growth. It facilitates the easy incorporation of non-polluting energy sources and storage. It can also do real-time forecasting of intermittent renewable energy sources that can lead to an adjustment in smart appliances, meters, smart homes, and electric vehicles [5, 6]. With household solar panels and other technology, users can sell excess energy to the grid, which can be done through ICT. The data through ICT can be used for better energy management programs by governments and regulatory requirements.

9.2 Intelligent Energy Management System

Smart systems can sense, communicate, operate, and control based on the analysis of the surroundings. They can describe the situational parameters, predict, and make decisions through collected data. The devices and appliances are developed smart, and they can communicate with each other or users wirelessly, remote access, and automation. This can reduce the peak energy demands with benefits to both users and the grid.

Smart systems can support demand-side energy management. The users can have lower electricity bills due to time of use tariffs and load shifting to cheaper time periods. Service providers and grid operators can have bi-directional energy transfer to and fro between grid and smart systems. There is real-time data transmission due to data analytics providing efficient control on energy management. The other benefits are convenience, security due to monitoring, and environmental protection [8].

Sensors embedded within the device gather information on the current or voltage consumed by the utilities. This is feedback provided to the users in the form of energy consumption real-time information and further decisions to be taken by the user or system itself. An example of a smart system may be an air conditioner, which is switched on. Figure 9.1 shows a smart home as a source of information. The user has forgotten to switch off the AC. The sensors will detect the temperature and humidity of the room. In case no user is present for a long time, the AC will automatically switch off saving energy. This is like a computer going to sleep mode or switched off. The same concept is extended in the smart system to take decisions remotely based on user preferences.

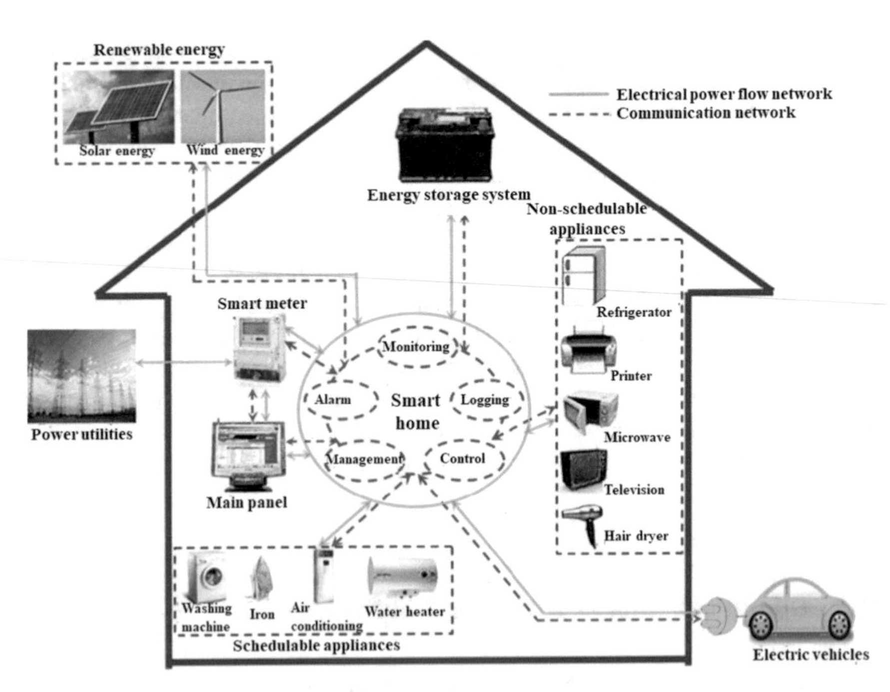

Fig. 9.1 Smart home as a source of information (Reprinted and redrawn with permission from [7])

9.2.1 Intelligent Sensors Network

The intelligent energy storage systems work on the data obtained from sensors. A smart sensor is defined as a combination of the sensor with digital circuitry like analog to digital converter in one housing. An intelligent sensor is different from a smart sensor in that it is self-adaptive, self-testing, predictive, and decisive based on user control. This may be performed with a help of a microprocessor or microcontroller. These sensors are technologically merged and integrated with other components in a single chip. These sensors receive and analyze the signals. They even increase signal quality.

Multiple sensors are required connected in a network to perform energy management of a facility. These can be wireless. The sensors require both hardware and software. There are sensor nodes and a command node. The sensor nodes are clustered around the target to gather, process, and produce information. Various internationally accepted protocols like HomeRF, Bluetooth, IEEE 802.11X, and IrDA have been developed [9].

9.2.2 Integrated Energy Management Portal

The intelligent energy management system can receive data from various devices through embedded sensors. Sensors collect data and transmit it to users in a meaningful form through software. The settings can be done that the smart system decides on own or the user himself directly involves in energy management. The user acts and controls energy transmission or savings or storage through an interface portal. The web-based portals can be accessed locally or remotely from anywhere in the world as shown in Fig. 9.2. The portals can be used for macro-management of energy consumption of the whole facility to micromanagement of individual devices and track their usage performing billing like functions too. The functions that can be performed through the portal are automated electricity readings, auto-billing, automated demand response management, real-time electricity pricing monitoring, time of use electricity pricing scheme, electricity supplier selection, outage detection, remote electrical assistance, distributed energy resources interface, and control and energy management [10].

With the availability of electric vehicles, energy storage devices, and small solar panels, there are provisions for individual households or companies to sell electricity to the grid, which can be controlled and monitored through portals. Consumers can check their energy consumption by utilities individually and collectively and billing information to control energy consumption by shifting demand to time periods of low cost or efficient usage of utilities. Multiple companies are providing these solutions. Auto grid systems energy data platform incorporated weather data along with energy management solutions.

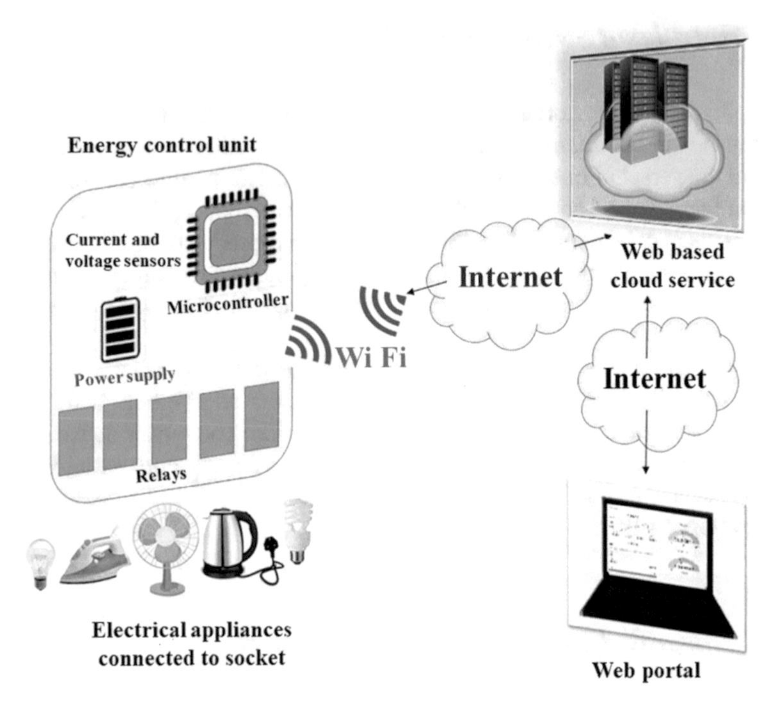

Fig. 9.2 Portal for macro-energy and micro-energy management through remote access. *Source* Reprinted and redrawn with permission from [11]

9.3 Supercapacitor Management System

The intelligent energy storage management system should maintain the proper state of charge and health of supercapacitors and batteries as per specifications. A supercapacitor cell is a basic unit consisting of electrodes [12], electrolyte [13], separator [14] and current collector [15]. Various materials are used to make these components of supercapacitors. These are transition metal oxides [16, 17], activated carbons [18, 19], graphene/reduced graphene oxides [20, 21], carbon nanotubes [22, 23], carbon nanofibers [24, 25], conducting polymers [26, 27] and their composites [28–33]. The connection of the supercapacitor in series, which generates higher voltage, is a supercapacitor cell string. The supercapacitor module, pack, rack, or stack is the connection of supercapacitors in series or parallel configuration as shown in Fig. 9.3. A supercapacitor management system is a supervisory system through which control, monitoring, balancing, and protective functions of the supercapacitor system are performed. It influences safety, performance, and supercapacitor service life. A supercapacitor system is a set of supercapacitor packs, management systems, ancillary devices, and protective components.

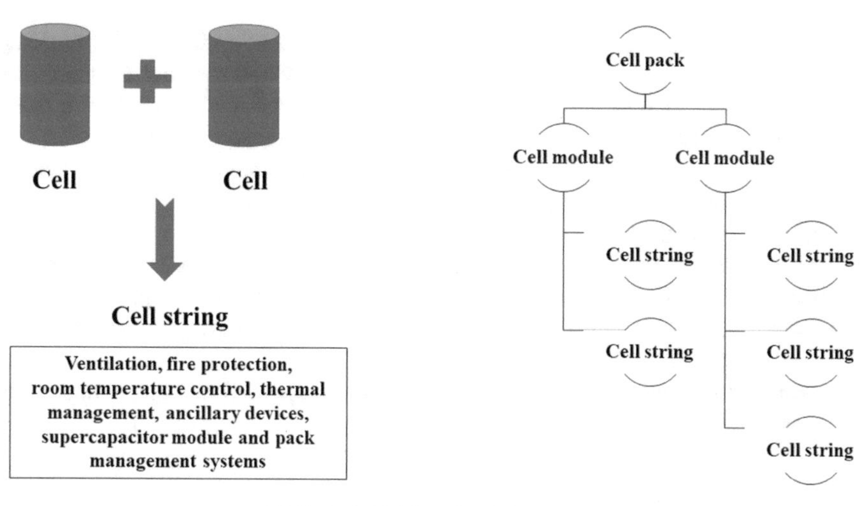

Fig. 9.3 Components of a supercapacitor room

The supercapacitor management system performs cell balancing. Cell balancing is a non-instantaneous process that results in an identical voltage of all supercapacitor cells. The importance of cell balancing is the protection of cell modules from overcharging or undercharging of individual supercapacitor cells. The cell balancing is achieved through a balancing circuit. The principle of equalization is the transfer of energy from a higher voltage pack to a lower voltage pack by using another balancing supercapacitor stack as shown in Fig. 9.4. The process of equalization is repeated from a no. of cycles until individual stacks have identical voltages [34].

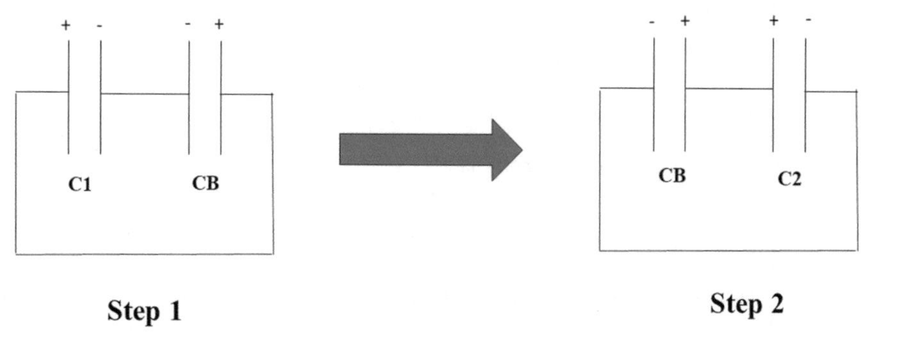

Fig. 9.4 Principle of equalization

The other functions of the supercapacitor management system are monitoring supercapacitor cell voltage, cell temperature, supercapacitor string current, and module voltage. It informs the failure of the normal power supply to the user. The supercapacitor management system protects supercapacitor against cell over-voltage, high surrounding temperature or cell over temperature, cell voltage unbalances, supercapacitor module/pack ground fault, connection failure or error with the power management system of which supercapacitor management system is a part, electrical isolation tripping, failure of supercapacitor system or any module or pack, reverse electrical connections, bad health or poor state of charge of the supercapacitor system, ventilation failure, short circuits, disable over-charging and uniform load sharing between supercapacitor cells and modules.

A simple C++ code to inform of operating voltage range and operating temperature range is as follows:

```
#include <iostream>
using namespace std;

int main()
{
  int voltage=0, temperature=0;

  cout << "Input voltage\n";
  cin>>voltage;
  cout<<"Input temperature\n";
  cin>>temperature;

  //Checking supercapacitor operational parameters with inputs. Taking a decision
based on it.
  //Let supercapacitor cell operating voltage be 2.75 V
  //Let supercapacitor operating temperature be -40 C to 70 C

  if(voltage>2.75)
  {
    cout<<"Overvoltage\n";
  }
  else
  {
    cout<<"Supercapacitor operating\n";
  }

  if((temperature>-40) && (temperature<70))
  {
    cout<<"Permissible operating temperature";
  }
  else
  {
    cout<<"Supercapacitor can get damaged";
  }

  return 0;
}
```

9.4 Energy Management Strategy

The hybrid energy storage systems utilize both batteries and supercapacitors. The various topologies are passive topology, semi-active topology, and active topology. The cost, adaptability, efficiency, and performance are the parameters that determine the usage of a particular topology in an application. The energy management strategy is used to improve the efficiency, reliability, and lifetime of the systems meeting power requirements. The goal of the energy management strategy is to use the high-power density of supercapacitors and the high energy density of batteries [35]. Supercapacitors act as a power buffer device in a hybrid energy storage system. Figure 9.5 shows different energy management techniques for hybrid energy storage systems.

9.4.1 Rule-Based Control Strategy

The rule-based control strategy for a hybrid energy storage system is based on heuristic and empiric experiences. The performance of this strategy may not be good in real applications as it is a pre-determined approach. The benefits are it is simple and easy to compute, robust, and reliable [37].

(i) Deterministic rule control: The deterministic rule control strategy is based on the time constant and peak power duration of supercapacitor and battery in a

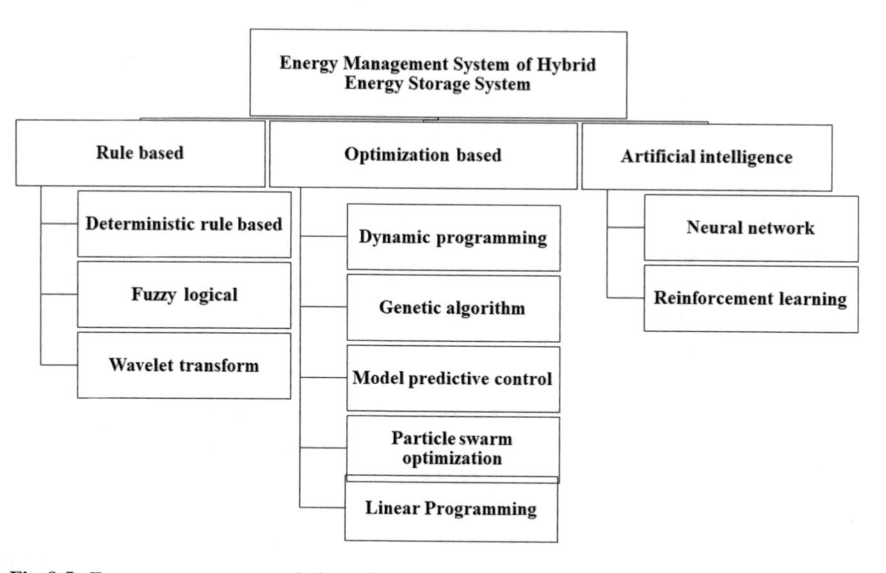

Fig. 9.5 Energy management techniques for a hybrid energy storage system (Redrawn and reprinted with permission from [36])

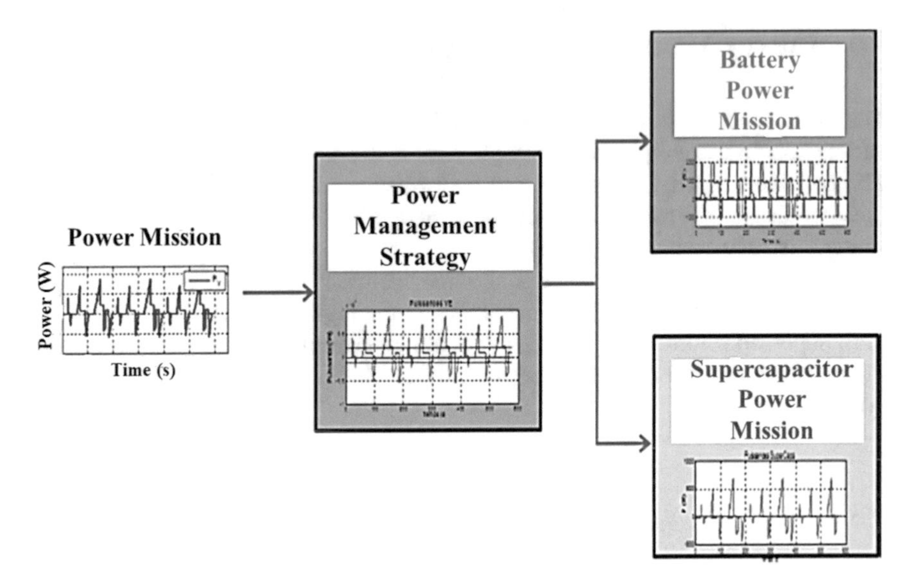

Fig. 9.6 Schematic showing power management strategy utilized for battery and supercapacitor power profiles (Redrawn and reprinted with permission from [38])

hybrid energy storage system [39]. The state of charge or minimum threshold charge of battery to function is taken as initial parameters. Based on these parameters, the energy management strategy is made to use a battery or a supercapacitor in HESS. Filters are also used that utilize supercapacitors in high-frequency demand and low-frequency current is provided by batteries. Supercapacitors can be utilized for high-frequency current because they have electrostatic energy for charge storage. Batteries are slow and have phase changes during charge storage. Figure 9.6 shows the division of power for battery and supercapacitor power management profiles. The supercapacitor is shown to meet peak power demands.

(ii) Fuzzy logic control: Fuzzy logic control is adaptable and robust. They do not depend on exact values "0" or "1" as in digital computation. The need for pre-determined and exact mathematical relation is not there. They can work in different operating conditions. Algorithms are made based on experience and this technique is used with different control strategies for improved performance [40].

(iii) Wavelet transform: Wavelet transform strategy is used in signal processing in both the time and frequency domain. Various base functions can be used. An example is the Haar wavelet function used on a signal $x(t)$ as shown in (9.1).

$$W_s x(t) = x(t).\Phi_S(t) = \frac{1 \int_{-\infty}^{\infty} S(t)\Phi\left(\frac{t-u}{s}\right)\mathrm{d}u}{s} \tag{9.1}$$

where s is the scale factor, $\Phi(t)$ is the scaled signal

Wavelet transform can divide the power into high-frequency and low-frequency components. This strategy can be deployed to efficiently use battery and supercapacitor in a hybrid energy storage system [41].

9.4.2 Optimization-Based Control Strategy

The optimization-based strategy is not widely used but considerable research is underway. This strategy is complex, not robust but the advantage is it helps in framing rules and designing other control-based strategies [42].

(i) Dynamic programming: Dynamic programming is of two types, viz. stochastic dynamic programming and deterministic programming. This can provide optimal results for complex models. Battery lifetime under various temperatures and costs have been optimized by researchers using this technology [43].
(ii) Genetic algorithm: Genetic algorithm is based on the simulation of Charles Darwin's theory of natural evolution. It can be nonlinear and versatile with robust in wide applications. It is adaptable and self-learning. The method has been applied to reduce the charge-discharge rate of a battery in a hybrid energy storage system. This method is on the stochastic principle, which results in non-optimal solutions in real-time scenarios [44]. It has been used with fuzzy-based technique to reduce cost.
(iii) Model predictive control: Model predictive control is based on feedback principles along with predictive analysis and online optimization. All three aspects can be covered with different processes like the Markov process [45]. Quadratic programming can be done for optimization control.
(iv) Particle swarm optimization: Particle swarm optimization is based on a natural bird flocking or metaheuristic approach. The members of the group share information with each other, and this is used to determine the collective group and individual best performance. The algorithm is used to have both stochastic and deterministic approaches [46].
(v) Linear programming: This is a simple technique used in linear processes. There are linear constraints and linear objectives. The objective is to maximize or minimize the values of the variables of the objective [47].

9.4.3 Artificial Intelligence-Based Control Strategy

Artificial intelligence-based algorithms are the latest techniques for energy management [48]. These are used for vehicle density and transport and efficient use of a hybrid energy storage system. The values are predicted by the algorithm of the variables based on the past values and modeling.

(i) Neural network: The human brain is made of neurons. The neural network is based on the same concept, which is trained by a large set of data. The main features are parallelism, nonlinear global role, fault tolerance, and self-adaptability [49].
(ii) Reinforcement learning: Reinforcement learning technique is also like a feedback technique. The current values of the control system and the trial-and-error method are deployed to predict the unknown variables and values [50].

Table 9.1 shows the summary of the energy management systems with merits, demerits, and application modes. Optimization-based techniques are computationally heavy and are used offline. Their advantage is adaptability and nonlinear systems.

Table 9.1 Summary of the energy management systems (reprinted with permission from [36])

S. No.	Energy management system	Merits	Demerits	Application mode
1	Rule-based control strategy Application: Optimization of energy storage device lifetime [37]	Low computational complexity, simple, robust, reliable	Poor adaptive control of the algorithm	Online
2	Dynamic programming Application: Hybrid energy storage system control [54]	Global optimality, optimization helpful for non-linear problems	Heavy calculation burden	Offline
3	Genetic algorithm Application: Energy management system of hybrid energy storage system [55]	Global optimality	Difficult to handle	Offline
4	Model predictive control Application: Distribute the power of hybrid energy storage system [56]	Accurate	Heavy calculation	Online
5	Neural network Application: Predict the usage of supercapacitor [57]	Low computational complexity	A large amount of demand training data, instability	Online
6	Reinforcement learning Application: Improve efficiency of real-time energy management systems [50]	Robust	Large calculation burden	Online

The other rule-based techniques and artificial intelligence-based techniques are used online. The rule-based technique is pre-deterministic and poor adaptability. Artificial intelligence-based techniques are in the development phase. The different energy management techniques are used together for optimal results and comparison of data from each model.

Supercapacitors have been developed with self-healing, shape memory, thermally chargeable, and electrochromic properties making them useful in wearable electronics and portable electronic devices. Biological supercapacitors power medical implants. These devices are also used in transportation [51, 52], industrial applications for peak power supply, and power electronics [53].

9.5 Concluding Remarks

Energy storage is an important aspect of national security to fulfill the uninterruptible energy demand of the country by utilizing renewable sources or other emerging concepts. Supercapacitors are electrochemical energy storage devices used with batteries in a hybrid system or even replacing batteries due to high power density and improved energy density. They have a long usable life. Proper use of energy is important and made possible through intelligent energy storage management systems using algorithms, software, and hardware. The two main functions of the management system are to reduce energy consumption by scheduling the demand or reducing wastage through constant monitoring and control. This is made possible by a machine–user interface like an integrated energy management portal. The second function of the energy management system is to increase the life of energy storage devices by protecting against over-voltage, temperature, or other specifications. Supercapacitor management systems increase the reliability and efficient use of supercapacitors. The supercapacitors are used with batteries in various circuit configurations like passive, semi-active, or fully active with each having its advantages and disadvantages. There are various energy management techniques based on algorithms, modeling, and simulation. Rule-based techniques, optimization-based techniques, and artificial intelligence-based techniques are used together. These are further divided into various sub-categories having their advantages and disadvantages.

Acknowledgements The authors acknowledge the financial support provided by the Science and Engineering Research Board, Department of Science and Technology, India (SR/WOS-A/ET-48/2018) for carrying out this research work.

References

1. K.K. Kar (ed.), *Handbook of Nanocomposite Supercapacitor Materials I Characteristics* (Springer, Cham, 2020). https://doi.org/10.1007/978-3-030-43009-2
2. K.K. Kar (ed.), *Handbook of Nanocomposite Supercapacitor Materials II Performance* (Springer, Cham, 2020). https://doi.org/10.1007/978-3-030-52359-6
3. P. Diwan, A.N. Sarkar (ed.), *Energy Security* (Pentagon Press, 2009)
4. F. Lufrano, P. Staiti, Int. J. Electrochem. Sci. **4**, 173 (2009)
5. Z. Arshad, M. Robaina, A. Botelho, Environ. Sci. Pollut. Res. **27**, 32913 (2020)
6. V. Cirimele, M. Diana, F. Bellotti, R. Berta, N.E. Sayed, A. Kobeissi, P. Guglielmi, R. Ruffo, M. Khalilian, A.L. Ganga, J. Colussi, A.D. Gloria, IEEE Trans. Veh. Technol. **69**(3) (2020)
7. B. Zhou, W. Li, K.W. Chan, Y. Cao, Y. Kuang, X. Liu, X. Wang, Renew. Sustain. Energ. Rev. **61**, 30 (2016)
8. R. Ford, M. Pritoni, A. Sanguinetti, B. Karlin, Build. Environ. **123**, 543 (2017)
9. M. Staroswiecki, IEEE Trans. Ind. Informatics **1**(4), 238 (2005)
10. I. Sudit, J. Longbottom, N. Morita, U.S. Patent 2007/0060171 A1, 15 March 2007
11. S.A. Hashmi, C.F. Ali, S. Zafar, Int. J. Energ. Res. **1** (2020)
12. K.D. Verma, P. Sinha, S. Banerjee, K.K. Kar, in *Handbook of Nanocomposite Supercapacitor Materials I Characteristics*, ed. by K.K. Kar (Springer, Cham, 2020), p. 269. https://doi.org/10.1007/978-3-030-43009-2_9
13. K.D. Verma, S. Banerjee, K.K. Kar, in *Handbook of Nanocomposite Supercapacitor Materials I Characteristics*, ed. by K.K. Kar (Springer, Cham, 2020), p. 287. https://doi.org/10.1007/978-3-030-43009-2_10
14. K.D. Verma, P. Sinha, S. Banerjee, K.K. Kar, M.K. Ghorai, in *Handbook of Nanocomposite Supercapacitor Materials I Characteristics*, ed. by K.K. Kar (Springer, Cham, 2020), p. 315. https://doi.org/10.1007/978-3-030-43009-2_11
15. K.D. Verma, P. Sinha, S. Banerjee, K.K. Kar, in *Handbook of Nanocomposite Supercapacitor Materials I Characteristics*, ed. by K.K. Kar (Springer, Cham, 2020), p. 327. https://doi.org/10.1007/978-3-030-43009-2_12
16. A. Tyagi, S. Banerjee, J. Cherusseri, K.K. Kar, in *Handbook of Nanocomposite Supercapacitor Materials I Characteristics*, ed. by K.K. Kar (Springer, Cham, 2020), p. 91. https://doi.org/10.1007/978-3-030-43009-2_3
17. B. De, S. Banerjee, K.D. Verma, T. Pal, P.K. Manna, K.K. Kar, in *Handbook of Nanocomposite Supercapacitor Materials II Performance*, ed. by K.K. Kar (Springer, Cham, 2020), p. 89. https://doi.org/10.1007/978-3-030-52359-6_4
18. P. Sinha, S. Banerjee, K.K. Kar, in *Handbook of Nanocomposite Supercapacitor Materials II Performance*, ed. by K.K. Kar (Springer, Cham, 2020), p. 113. https://doi.org/10.1007/978-3-030-52359-6_5
19. P. Sinha, S. Banerjee, K. K. Kar, in *Handbook of Nanocomposite Supercapacitor Materials I Characteristics*, ed. by K.K. Kar (Springer, Cham, 2020), p. 125. https://doi.org/10.1007/978-3-030-43009-2_4
20. P. Chamoli, S. Banerjee, K.K. Raina, K.K. Kar, in *Handbook of Nanocomposite Supercapacitor Materials I Characteristics*, ed. by K.K. Kar (Springer, Cham, 2020), p. 155. https://doi.org/10.1007/978-3-030-43009-2_5
21. B. De, S. Banerjee, T. Pal, K.D. Verma, P.K. Manna, K.K. Kar, in *Handbook of Nanocomposite Supercapacitor Materials II Performance*, ed. by K.K. Kar (Springer, Cham, 2020), p. 271. https://doi.org/10.1007/978-3-030-52359-6_11
22. S. Banerjee, K.K. Kar, in *Handbook of Nanocomposite Supercapacitor Materials I Characteristics*, ed. by K.K. Kar (Springer, Cham, 2020), p. 179
23. B. De, S. Banerjee, K.D. Verma, T. Pal. P.K. Manna, K.K. Kar, in *Handbook of Nanocomposite Supercapacitor Materials II Performance*, ed. by K.K. Kar (Springer, Cham, 2020), p. 229. https://doi.org/10.1007/978-3-030-52359-6_9

24. R. Sharma, K.K. Kar, in *Handbook of Nanocomposite Supercapacitor Materials I Characteristics*, ed. by K.K. Kar (Springer, Cham, 2020), p. 215. https://doi.org/10.1007/978-3-030-43009-2_7
25. B. De, S. Banerjee, K.D. Verma, T. Pal. P.K. Manna, K.K. Kar, in *Handbook of Nanocomposite Supercapacitor Materials II Performance*, ed. by K.K. Kar (Springer, Cham, 2020), p. 179. https://doi.org/10.1007/978-3-030-52359-6_7
26. T. Pal, S. Banerjee, P.K. Manna, K.K. Kar, in *Handbook of Nanocomposite Supercapacitor Materials I Characteristics*, ed. by K.K. Kar (Springer, Cham, 2020), p. 247. https://doi.org/10.1007/978-3-030-43009-2_8
27. S. Banerjee, K.K. Kar, in *Handbook of Nanocomposite Supercapacitor Materials II Performance*, ed. by K.K. Kar (Springer, Cham, 2020), p. 333. https://doi.org/10.1007/978-3-030-52359-6_13
28. P. Sinha, S. Banerjee, K.K. Kar, in *Handbook of Nanocomposite Supercapacitor Materials II Performance*, ed. by K.K. Kar (Springer, Cham, 2020), p. 145. https://doi.org/10.1007/978-3-030-52359-6_6
29. B. De, S. Banerjee, K.D. Verma, T. Pal, P.K. Manna, K.K. Kar, in *Handbook of Nanocomposite Supercapacitor Materials II Performance*, ed. by K.K. Kar (Springer, Cham, 2020), p. 201. https://doi.org/10.1007/978-3-030-52359-6_8
30. B. De, S. Banerjee, T. Pal, A. Tyagi, K.D. Verma, P.K. Manna, K.K. Kar, in *Handbook of Nanocomposite Supercapacitor Materials II Performance*, ed. by K.K. Kar (Springer, Cham, 2020), p. 245. https://doi.org/10.1007/978-3-030-52359-6_10
31. B. De, P. Sinha, S. Banerjee, T. Pal, K.D. Verma, A. Tyagi, P.K. Manna, K.K. Kar, in *Handbook of Nanocomposite Supercapacitor Materials II Performance*, ed. by K.K. Kar (Springer, Cham, 2020), p. 297. https://doi.org/10.1007/978-3-030-52359-6_12
32. B. De, S. Banerjee, T. Pal, K.D. Verma, A. Tyagi, P.K. Manna, K.K. Kar, in *Handbook of Nanocomposite Supercapacitor Materials II Performance*, ed. by K.K. Kar (Springer, Cham, 2020), p. 353. https://doi.org/10.1007/978-3-030-52359-6_14
33. B. De, S. Banerjee, T. Pal, K.D. Verma, A. Tyagi, P.K. Manna, K.K. Kar, in *Handbook of Nanocomposite Supercapacitor Materials II Performance*, ed. by K.K. Kar (Springer, Cham, 2020), p. 387. https://doi.org/10.1007/978-3-030-52359-6_15
34. R. Lu, C. Zhu, L. Tian, Q. Wang, IEEE Trans. Magn. **43**(1) (2007)
35. P. Sinha, K.K. Kar, in *Handbook of Nanocomposite Supercapacitor Materials II Performance*, ed. by K.K. Kar (Springer, Cham, 2020), p. 71. https://doi.org/10.1007/978-3-030-52359-6_3
36. R. Xiong, H. Chen, C. Wang, F. Sun, J. Cleaner Prod. **202**, 1228 (2018)
37. A. Castaings, W. Lhomme, R. Trigui, A. Bouscayrol, Appl. Energ. **163**, 190 (2016)
38. N. Rizoug, T. Mesbahi, R. Sadoun, P. Bartholomeüs, P.L. Moigne, Energ. Efficiency **11**, 823 (2018)
39. Z.Y. Song, H. Hofmann, J.Q. Li, J. Hou, X. Han, M. Ouyang, Appl. Energ. **134**, 321 (2014)
40. Y.Z. Wang, W. Wang, Y. Zhao, L. Yang, W. Chen, Energies **9**(1), 1 (2016)
41. O. Erdinc, B. Vural, M. Uzunoglu, J. Power Sour. **194**(1), 369 (2009)
42. O. Gomozov, J.P. Trovao, X. Kestelyn, M.R. Dubois, IEEE Trans. Veh. Technol. **66**(7), 5520 (2017)
43. A. Santucci, A. Sorniotti, C. Lekakou, J. Power Sour. **258**, 395 (2014)
44. J. McCall, J. Comput. Appl. Math. **184**(1), 205 (2005)
45. F. Zhou, F. Xiao, C. Chang, Y. Shao, C. Song, Energies **10**(7), 1 (2017)
46. D. Wang, D. Tan, L. Liu, Soft. Comput. **22**, 387 (2018)
47. N. Karmarkar, Combinatorica **4**(4), 373 (1984)
48. P. Xu, Q. Yin, Y. Huang, Y.Z. Song, Z.Y. Ma, L. Wang, T. Xiang, W.B. Kleijn, J. Guo, Neurocomputing **278**, 75 (2018)
49. V.I. Herrera, H. Gaztanaga, A. Milo, A. Saez-de-Ibarra, I.E. Otadui, T. Nieva, IEEE Trans. Ind. Appl. **52**(4), 3367 (2016)
50. T. Liu, Y. Zou, D.X. Liu, F.C. Sun, IEEE Trans. Ind. Electron. **62**(12), 7837 (2015)
51. S. Banerjee, B. De, P. Sinha, J. Cherusseri, K.K. Kar, in *Handbook of Nanocomposite Supercapacitor Materials I Characteristics*, ed. by K.K. Kar (Springer, Cham, 2020), p. 341. https://doi.org/10.1007/978-3-030-43009-2_13

52. R. Kumar, S. Sahoo, E. Joanni, R.K. Singh, K. Maegawa, W.K. Tan, G. Kawamura, K.K. Kar, A. Matsuda, Mater. Today **39**, 47 (2020)
53. R. Kumar, S. Sahoo, E. Joanni, R.K. Singh, W.K. Tan, K.K. Kar, A. Matsuda, Prog. Energ. Combust. Sci. **75**, 100786 (2019)
54. Z. Song, H. Hofmann, J. Li, X. Zhang, M. Ouyang, Appl. Energ. **159**, 576 (2015)
55. M. Wieczorek, M. Lewandowski, Appl. Energ. **192**, 222 (2017)
56. K. Zhang, C. Mao, J. Lu, D. Wang, X. Chen, J. Zhang, IET Renew. Power Gener. **8**(1), 58 (2014)
57. R. Nigam, K.D. Verma, T. Pal, K.K. Kar, in *Handbook of Nanocomposite Supercapacitor Materials II Performance*, ed. by K.K. Kar (Springer, Cham, 2020), p. 463. https://doi.org/10.1007/978-3-030-52359-6_17

Chapter 10
Global Trends in Supercapacitors

Dylan Lasrado, Sandeep Ahankari, and Kamal K. Kar

Abstract The global supercapacitor market is expected to grow at a rapid rate in the coming years owing to the rising demand for supercapacitors in various applications. These supercapacitors are available in varying sizes, capacitances, voltage ranges, etc., and are sometimes tailor-made for certain applications. The markets in the Asia—Pacific region—are expected to grow at the highest rates with China being at the forefront. At present it is currently dominated by a few major players such as Murata Technology, Maxwell Technologies, Eaton Corporation, Nippon Chemi-Con, Nesscap among others. These major players are focussing immensely on research and development of supercapacitors to meet further demands as well as maintain their competitive advantage over the others. The current chapter deals with the trends in the supercapacitor market and also sheds light on the properties of supercapacitors cells and modules manufactured by key market players.

10.1 Introduction

The supercapacitor market is expected to grow at a compound annual growth rate (CAGR) of about 30% to US$8.3 billion by the year 2025 (Fig. 10.1) [1]. Major changes in the different categories of supercapacitors were observed in early 2010 due to the introduction of global renewable energy applications into the market. Among these various types of supercapacitors, board-mounted supercapacitors, which held the largest market share in 2014 are expected to grow at the highest CAGR of 22.7%

D. Lasrado · S. Ahankari (✉)
School of Mechanical Engineering, VIT University, Vellore 632014, India
e-mail: asandeep.s@vit.ac.in

D. Lasrado
e-mail: lasradodylan@gmail.com

K. K. Kar
Advanced Nanoengineering Materials Laboratory, Department of Mechanical Engineering, Materials Science Programme, Indian Institute of Technology Kanpur, Kanpur 208016, India
e-mail: kamalkk@iitk.ac.in

K. K. Kar (ed.), *Handbook of Nanocomposite Supercapacitor Materials III*,
Springer Series in Materials Science 313,
https://doi.org/10.1007/978-3-030-68364-1_10

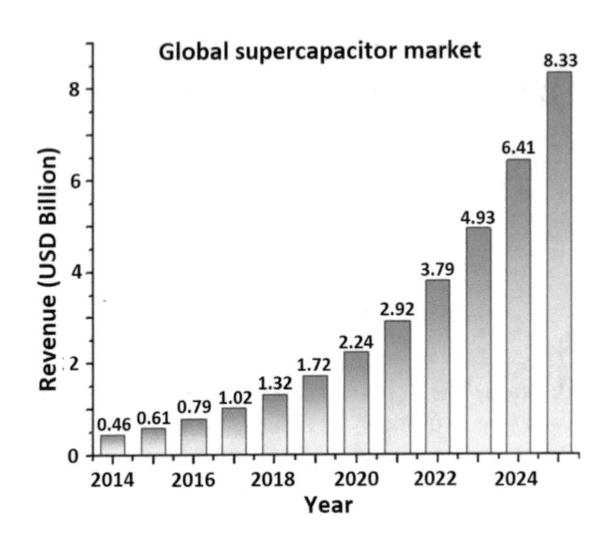

Fig. 10.1 Estimated value of the global supercapacitor market for the period 2014–2025

for the period of 2015–2022. This growth rate is attributed to the increasing demand for these supercapacitors in automotive applications such as regenerative braking, start-stop systems for a reduction in fuel consumption [2, 3]. Hybrid supercapacitors are also predicted to have an exponential growth rate in the coming years [4].

The global supercapacitor market is categorized based on product type, module type, region, etc. Based on product type, the market is further divided into a double-layer capacitor, pseudo capacitor, and hybrid capacitors. The characteristics of capacitors [5–7] and capacitors to supercapacitors are reported elsewhere [8]. Pseudo capacitors have dominated the supercapacitor markets in terms of revenue until 2019 and this trend is expected to continue [9]. Electric double-layer capacitors on the other hand have shown potential to replace conventional batteries in applications such as laptops, computers, smart wearables [2, 3]. The market is also segmented based on modules such as less than 10 V modules, 10–25 V modules, 25–50 V modules, 50–100 V modules, 100 V, and greater modules. Forecasts suggest that although modules of 10–25 V dominated the market share for 2019, modules of 100 V and greater will have the highest growth rate from 2020 to 2027 [3]. The tremendous growth in the market has been observed in certain countries of various geographical regions. Among these regions, the supercapacitor market is projected to grow at a high CAGR of 23.2% in the Asia Pacific region from 2015 to 2023 owing to industrialization as well as the development of infrastructure [2]. Among the various countries in this region, China is expected to have a dominating market share during the forecast period. This growth of the market is being facilitated by the tech-based companies that are coming up in growing economies such as India, China [10]. Investments by these companies in emerging technologies have led to a sharp fall in the price of supercapacitors. This has led to greater acceptance of supercapacitor technology by consumers. Figure 10.2 shows the supercapacitor market share in different regions for the year 2018 [11]. The countries contributing to the

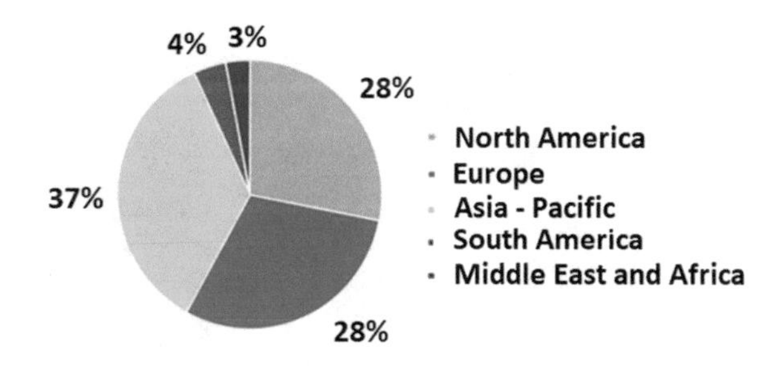

Fig. 10.2 Market share of supercapacitors classified based on regions

growth of the global supercapacitor market in different geographical regions are mainly classified as

(a) China, Korea, Japan, and India in the Asia—Pacific region
(b) USA and Canada in the North American region
(c) Brazil and Argentina in the South American region
(d) Germany, Italy, France, and the UK in Europe.

In the past, the demand for supercapacitors was strictly restricted to providing battery backup. Lower awareness levels and high initial costs were some of the other factors that had inhibited the growth of the supercapacitor industry. However, the recent growing demand for supercapacitors is being fuelled by a broad variety of factors. The automotive and transportation sector, for instance, is expected to be the fastest-growing market for supercapacitors and is projected to reach US$2.7 billion by 2026. The global demand for supercapacitors in this sector is anticipated to compound annually at 16.6% during 2019–2026 [12]. Global awareness about the menace caused by greenhouse gasses and climate change on our ecosystem has led many major players in the automotive sector to shift towards using the environmentally friendly material and cutting down on their CO_2 emissions. Energy conservation has also gained prime importance in this sector. These factors have led to the development of hybrid automobiles and electric vehicles, wherein supercapacitors are widely being used [13, 14]. The rise in demand for hybrid vehicles has stimulated the growth of the supercapacitor industry, as increasing research is being carried out on various fronts such as different materials that could be used for the fabrication of supercapacitor electrodes and separators, operating conditions under which stable functioning of supercapacitors can be obtained, etc., [15]. Manufacturers of electric vehicles are also considering shifting from battery-powered electric vehicles to supercapacitor powered electric vehicles primarily because of their wider operating conditions and their scalability [16]. Another contributor to the growing global demand for supercapacitors is the field of renewable energy, wherein the demand for supercapacitors in solar lighting applications and solar PV panels is on the rise. It is estimated that the demand for supercapacitors in this sector by the year 2020 would be about 416,000 units [17]. Apart from the automobile sector and the renewable energy sector, fields, such as consumer electronics, energy harvesting applications, military applications, artificial intelligence, medical applications, have also significantly contributed to the rise in the demand for supercapacitors [3, 18, 19]. Figure 10.3 sheds light on the

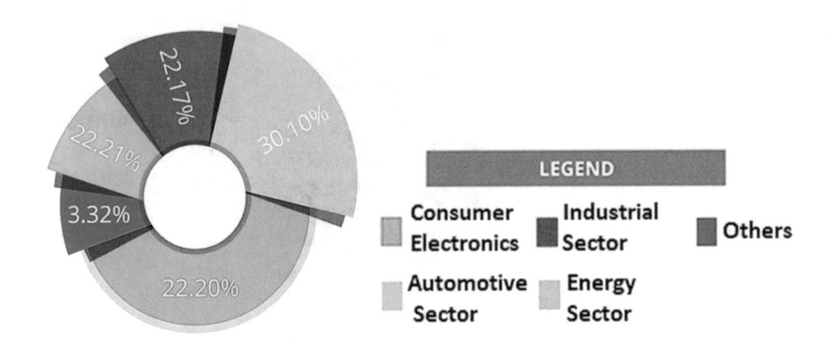

Fig. 10.3 The market share of supercapacitors in different domains

market share of supercapacitors in various applications as of 2016 [20].

Initially, supercapacitors were fabricated using carbon derived materials [21–30] and were symmetric in nature (same material used for the fabrication of both the electrodes). The symmetric nature of the supercapacitor electrodes severely limited the working potential window of supercapacitors. Tremendous research has been carried out in the field of supercapacitor technology over the last few years [31]. Advanced materials, i.e., conducting polymers [32, 33], transition metal oxides [34, 35], and their composites, are now being studied to determine whether the fabrication of supercapacitor electrodes, separators [36], current collectors [37], etc. from these materials is feasible. Organic electrolytes and other alternative electrolytes [38] with wider operating voltages as well as active composite electrode materials [39–44] with higher capacitance capabilities have also been developed. Another major advancement in the fabrication of asymmetric supercapacitors results in supercapacitors having enhanced working potential windows [45]. Recently developed supercapacitors have superior properties in terms of power and energy densities, capacitance range, voltage, rapid charge-discharge rates, etc. as compared to their predecessors. These advancements in supercapacitor technology have attracted the attention of global consumers to consider supercapacitors as a possible replacement to normal batteries in various applications. The ability of supercapacitors to operate under harsh climatic conditions has worked in the favour of supercapacitors. Figure 10.4 clearly shows the ability of supercapacitors to bridge the gap between capacitors and batteries in terms of power density and energy density [46, 47]. However, major challenges in the field of supercapacitor technology still exist, which have restricted consumers from preferring supercapacitors over batteries such as higher cost, lower energy densities, and durability issues related to supercapacitors [48].

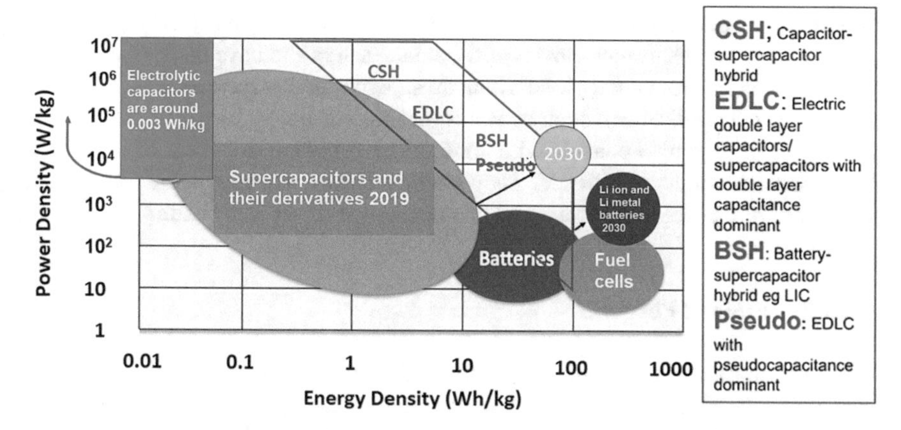

Fig. 10.4 Ragone plot of energy storage devices

10.2 Key Global Players in the Supercapacitor Market

The supercapacitor market is a very competitive market with few major players dominating the market share. General Electric was the first company to patent electrochemical capacitors based on the structure of double layers in 1957 [49]. However, they did not commercialize these supercapacitors. The first commercialized supercapacitors were introduced by Nippon Electric Corporation in 1978 based on designs patented by Standard Oil of Ohio [50]. These supercapacitors offered capacitances of 1 F and were rated as 5.5 V; used primarily to provide backup power for volatile CMOS (complementary metal-oxide-semiconductor) computer memory. Various companies have since entered the market with a wide range of supercapacitors of different sizes, capacitance ranges, rated voltages, etc. customized according to the needs of different applications [2, 51, 52]. Some of the major players in the supercapacitor market include Murata Technology, Maxwell Technologies, Eaton Corporation, Nippon Chemi-Con, Nesscap among others. Many of these companies have focused on expanding their customer base and widening their product lines to attract more customers. Given below is an analysis of the variable ranges of supercapacitor cells and modules, which are manufactured by some of the top companies in the global supercapacitor market.

10.2.1 Kamcap

10.2.1.1 Company Profile

Kamcap, which was established in the year 1999, is currently one of the major developers and manufacturers of supercapacitors in China [53]. Their supercapacitors are currently being used in a range of applications such as consumer electronics, green energy, car electronics, power grids, smart meters and have their customers in over 40 countries around the globe. Kamcap supercapacitors have a market share of around 90% in applications such as water meters, electric meters, and other smart meters. Kamcap is also considered as one of the founding members of the China Supercapacitor Industry Alliance. They have their own research and development centre, where it develops and produces its own range of supercapacitors.

10.2.1.2 Product Portfolio

Kamcap manufactures a variety of supercapacitor cells such as coin type, winding type, hybrid, combined type, and high-temperature supercapacitor cells. These supercapacitor cells are widely being used in energy storage, backup power, and high-power support applications. The energy storage applications mainly include electricity and gas meters, security equipment, small appliances, flashlights, etc., while backup power applications include digital cameras, clock circuits, car recorders, etc. The high-power support applications include the starting of automobiles, electric vehicles, etc. The following are some of the specifications of supercapacitor cells manufactured by Kamcap (Table 10.1).

Table 10.1 Specifications of supercapacitor cells manufactured by Kamcap [53]

Type	Series	Capacitance range (F)	Rated voltage (V)	Operating temperature	Application
Coin	–	0.1–1.5	5.5	−25 to 70 °C	Energy storage
Winding	High energy series	0.35–25	2.7	−25 to 70 °C	Energy storage
	High-power series	0.1–3000	2.7	−40 to 70 °C	Energy storage and high-power support
		1–500	3	−40 to 65 °C	
Combined series	Standard power series	0.1–8	5.5	−40 to 70 °C	Backup power
		0.33–8	7.5		
	Standard energy series	0.33–7.5	5.5	−25 to 70 °C	Energy storage
		0.33–3.3	7.5		
High-temperature series		1–100	2.7	−40 to 85 °C	Backup power
		0.5–10	5.5		
Hybrid series		10–10,000	2.3	−25 to 60 °C	Solar lighting, LED lighting, handheld electronics, medical equipment

Table 10.2 Specifications of supercapacitor modules manufactured by Kamcap [53]

Series	Capacitance range (F)	Rated voltage (V)	Operating temperature	Application
MK	10–200	5.4	−40 to 65 °C	Backup power and high current operations
	0.5–117	10		
	3–500	15		
	1–250	30		
	1–150	50		
	1–75	100		
	1–62	125		

Apart from the single cells, Kamcap also manufactures supercapacitor modules also known as the MK module series. These modules have good stability and high conversion efficiency. These modules have found wide-ranging applications in backup power and high current operation applications such as in rail powered systems, hybrid vehicles, smart grid control. Given below are some of the specifications of the supercapacitor modules manufactured by Kamcap (Table 10.2).

10.2.2 *Skeleton Technologies*

10.2.2.1 Company Profile

Skeleton Technologies is one of Europe's largest manufacturers of ultracapacitors, which are used in a vast number of areas such as transportation, maritime, industrial, manufacturing, and other sectors [54]. They began the development of its patented "curved graphene" based industrial ultracapacitors in 2009, which is currently known as SkelCap ultracapacitors. Skeleton Technology currently has the financing of over 50 million euros, which it plans to invest in manufacturing as well as research and development to improve their competitive advantage.

10.2.2.2 Product Portfolio

Skeleton Technologies manufactures SkelCap ultracapacitors, which claim to have four times higher energy densities and two times higher power densities as compared to their competitors. These SkelCap ultracapacitor cells are of five different types, namely SCA0500, SCA0750, SCA1200, SCA1800, and SCA3200 with its capacitance ranging from 500 F to 3200 F. The SkelCap ultracapacitor cells have attractive properties such as they possess a high-power density, high-temperature resistance, and high cycle life (>100,000 cycles). Apart from these, the SkelCap ultracapacitor cells also generate a low amount of heat at high-power profiles and have a long

Table 10.3 Specifications of ultracapacitor cells manufactured by Skeleton Technologies [54]

Series	Capacitance (F)	Rated voltage (V)	Specific energy (Wh/kg)	Specific power (kW/kg)	Thermal resistance (°C/W)	Operating temperature	Application
SCA0300	300	2.85	5.3	32	10.8	−40 to 65 °C	Automotive, Industrial, Oil and Gas, Medical and other sectors
SCA0500	500		5.1	80	7.1	−40 to 85 °C	
SCA0750	750		5.8	66	6.6		
SCA1200	1200		5.4	73	5.7		
SCA1800	1800		6.0	46.4	4.3		
SCA3200	3200		6.8	34.6	3.0		

Table 10.4 Specifications of ultracapacitor modules manufactured by Skeleton Technologies [54]

Series	Capacitance (F)	Rated voltage (V)	Max. series voltage (V)	No. of cells	Max. energy (Wh)	Nominal power (kW)	Applications
SKELMOD 17 V	533	17	750	6	21.3	65.7	Wind turbine pitch control, intralogistics shuttles, and AGV's. Uninterruptable power supply for medical applications
SKELMOD 51 V	177	51	850	18	63.9	197	KERS, hybrid and peak shaving applications in rail transportation, etc. Engine start/stop for hybrid electric busses
SKELMOD 102 V	88	102	1260	36	127.1	419.5	Power quality applications in grid and industrial. Modular, fits 19" racks, building block for SkelGrid system. Peak power supply and load levelling in industrial application
SKELMOD 131 V	6.7	131	400	46	15.9	69.2	Medical machinery applications, intralogistics shuttles, and AGV's Kinetic energy recovery system in elevators, etc.

(continued)

Table 10.4 (continued)

Series	Capacitance (F)	Rated voltage (V)	Max. series voltage (V)	No. of cells	Max. energy (Wh)	Nominal power (kW)	Applications
SKELMOD 170 V	53	170	920	60	212.7	701.5	Kinetic energy recovery system in transportation sector

application lifetime of about 15 years. These cells are used as a source of power backup and regeneration of brake energy in automotive applications. Similarly, they are also used for peak shaving, generator gradient control in industrial as well as oil and gas applications. Some of the specifications of SkelCap ultracapacitor cells are mentioned in Table 10.3.

Skeleton Technologies also manufacture ultracapacitor modules known as SkelMod, which are characterized by their high-power density and high-power output. These ultracapacitor modules have an operating temperature range of – 40 to 65 °C, a life cycle of 1,000,000 cycles, and shelf life of 10 years. Apart from these, Skeleton Technology also manufactures SkelStart Engine Start ultracapacitor modules, which are specifically meant for engine starting of large diesel engines. These ultracapacitor modules are of 2 types (12 and 24 V) and can be used in busses, heavy machinery, generators, maritime, etc. The advantage of this ultracapacitor module is that it improves the life of the battery and ensures the engine starting even with a dead battery. Given below are some of the specifications of ultracapacitor modules manufactured by Skeleton Technologies (Table 10.4).

10.2.3 SPSCAP

10.2.3.1 Company Profile

Supreme Power Solutions Co., Ltd. (SPS), which is currently one of the largest ultra-capacitor manufacturing companies in the world, was established in China in 2002 [55]. The ultracapacitor cells and modules manufactured by SPSCAP are widely used in industrial applications such as trams, subways, heavy-duty machinery, industrial equipment, etc., across 26 countries.

10.2.3.2 Product Portfolio

Currently, SPSCAP has 10 product series and 2 production lines of supercapacitors. The supercapacitors manufactured by SPSCAP have an activated carbon coating and

Table 10.5 Specifications of supercapacitor cells manufactured by SPSCAP [55]

Series	Capacitance (F)	Rated Volt. (V)	Maximum stored energy (Wh)	Specific energy (Wh/kg)	Thermal resistance (°C/W)	ESR DC (mΩ)	Operat. Temp.	Application
SCE-SHZ	360–600	2.7	0.36–0.61	4.7–5.2	7.6–10.9	3–3.1	−40 to 65 °C	Electronic tools and police flashlights, UPS and backup power supply, vehicle power supply
SCE-SPD	350–800		0.35–0.81	4.9–6.5	6.3–10.1	1.3–1.8		
SCE-SPZ	100–180		0.1–0.18	5.1–5.7	19–26	15–16.5		
SCE-TCZ	450–700		0.46–0.71	5.8	7.4–19.5	3–12		
SCP-STA	650–5000		0.66–5.06	3.0–6.3	2.3–6.5	0.28–0.8		Locomotive engine start-up, heavy-duty machinery, electric vehicles
SCP-WLH	650–5000		0.66–5.06	3.1–6.4	2.3–6.5	0.28–0.8		

Table 10.6 Specifications of supercapacitor modules manufactured by SPSCAP [55]

Series	Capacitance (F)	Rated voltage (V)	No. of cells	ESR (mΩ)	Max. energy (Wh)	Specific power (W/kg)	Thermal resistance (°C/W)	Applications
MCE	60	15	6	25	1.88	2250	4	Energy storage systems
	36	75	90	50	28.1	710	0.15	Heavy-duty machinery and wind turbines
	10	90	36	120	11.2	1010	0.5	
	5.8	160	60	200	20.6	2560	1.1	
MCP	500	16		1.8	17.8	2994	0.8	Automotive sector
	83	48	18	9	26.5	2898	0.4	Railway transportation and heavy-duty machinery
	165	48		5	52.8	3736	0.3	
	125	64	24	8	71.1	3072	0.2	

an organic electrolyte system. These supercapacitors have low leakage and low resistance and a high-power capacity. The SCE supercapacitor series have a higher power density and are smaller in size and can provide an instant power output. These are mainly used in vehicle power supply, industrial backup power, electronic tools, etc. The SCP series on the other hand has a larger capacitance and large charge-discharge currents. These are used in rail transportation, heavy-duty machinery, electric vehicles, etc. The SCE and SCP series supercapacitors have a cycle life of 1,000,000 cycles. Table 10.5 shows some of the specifications of the SCP and SCE series of supercapacitors.

Supreme Power Solutions Co., Ltd. (SPS) also manufactures two series of supercapacitor modules, namely the MCE series and the MCP series. These modules are suitably designed for different power and energy requirements. The MCE series is known for its high-power density and strong charge. The MCP series on the other hand has a high energy density and a high energy storage capacity. These supercapacitor modules have an operating temperature range of −40 to 65 °C, shelf life of 4 years, and cycle life of 1,000,000 cycles. Given below are some of the specifications of supercapacitor modules manufactured by SPS (Table 10.6).

10.2.4 Yunasko

10.2.4.1 Company Profile

Yunasko, which is one of the leading developers of energy storage devices, was established in 2010 in the UK [56]. They also have a research and development laboratory in Kiev, which explores the use of new materials as well as helps them maintain a competitive advantage over their competitors.

10.2.4.2 Product Portfolio

Yunasko's product portfolio includes both Li-ion capacitors and ultracapacitors. The ultracapacitor cells, developed by Yunasko are classified into two series, namely energy series and power series depending on their properties. These ultracapacitor cells are prismatic in shape and have a low internal resistance as well as excellent power and energy characteristics owing to which they have found wide-ranging applications in energy harvesting and recovery systems such as in automobiles, busses, trains, aerospace, and defence sectors. Yunasko ultracapacitors have a shelf life of 4 years and a cycle life of 1,000,000 cycles. Table 10.7 shows the specifications of ultracapacitor cells manufactured by Yunasko.

Yunasko also manufactures ultracapacitor modules whose voltage range varies from 2.7 V (single-cell) to 750 V (modules). These ultracapacitor modules are classified into two series namely the power series and the energy series. The ultracapacitor modules have an operating temperature range of −40 to 65 °C, shelf life of 4 years,

Table 10.7 Specifications of ultracapacitor cells manufactured by Yunasko [56]

Series	Capacitance range (F)	Rated voltage (V)	Energy density (Wh/kg)	Power density (W/kg)	ESR (mΩ)	Operating temperature	Application
Power cells	400	2.7	4.8	41	0.3	−40 to 65 °C	Energy harvesting and recovery systems
	1200		4.2	38	0.15		
Energy cells	1200		5.7	11.3	0.43		
	3000		6.2	7.1	0.28	−40 to 60 °C	

Table 10.8 Specifications of ultracapacitor modules manufactured by Yunasko [56]

Series	Capacitance range (F)	Rated voltage (V)	ESR (mΩ)	Max. energy (Wh)	Specific power (kW/kg)	Application
Power	200	16	1.5	7.1	34	High-power defence and aerospace applications
Energy	500	16	2.1	17.8	2.8	
	165	48	6.2	53	13	
	13	90	80	14.6	12	Wind turbine blade pitch control

and cycle life of 1,000,000 cycles. The specifications of supercapacitor modules manufactured by Yunasko are mentioned in Table 10.8.

10.2.5 Ioxus

10.2.5.1 Company Profile

Ioxus Inc. designs and manufactures ultracapacitor cells and modules in North America [57]. These ultracapacitors are designed specifically for various applications, where they have to operate under harsh mechanical, electrical, and other environmental conditions. Ioxus ultracapacitor cells and modules are used in various applications such as Hybrid drive trains, windmill pitch control, automotive voltage stabilization, starting and stopping of automobiles, backup power.

10.2.5.2 Product Portfolio

Ioxus Inc. manufactures ultracapacitor cells, ultracapacitor modules, and other ultracapacitor-based products. These lightweight ultracapacitor cells known as the

Table 10.9 Specifications of ultracapacitor cells manufactured by IOXUS [57]

Series	Capacitance (F)	Rated voltage (V)	Specific energy (Wh/kg)	Specific power (kW/kg)	Thermal resistance (°C/W)	Operating temperature	Application
Titan	1200	2.85	4.6	26	5	−40 to 65 °C	Engine starting, backup power, etc.
	2000		5.8	24	4.5		
	3000		6.6	20	4		

Table 10.10 Specifications of ultracapacitor modules manufactured by IOXUS [57]

Series	Capacitance (F)	Rated voltage (V)	DC ESR (mΩ)	Max. energy stored (Wh)	Applications
iMOD X series	200	16.2	1.6	7.2	Wind turbine pitch control, capture of regenerative energy, hybrid energy storage, Transportation sector
	500		1.2	18.2	
	100	32.4	3.2	14.5	
	250		2.4	36.4	
	66	48.6	4.8	21.6	
	166		3.6	54.4	
	50	64.8	6.4	29.1	
	125		4.8	72.9	
	40	81	8.1	36.4	
	100		6.0	91.1	
	28	113.4	11.3	50	
	25	129.6	15.3	58.3	
iMOD AGV	62	54	18	25	Automated guided vehicles and robotics
	72	59	20	36	
iMOD UPS	20.8	162	45	76	Short-term UPS for hospitals and other critical systems
	41.5	81	22	38	
iMOD Energy	62	54	17	25	Automated guided vehicles, industrial equipment, and robotics
iMOD Power	2	216	89	13	Short-term UPS for hospitals, industries, and other critical systems
	11.4	108	16	18	

Titan series can operate at elevated temperatures of 85 °C and have a wide operating range. The stand out properties of these ultracapacitor cells are that they are maintenance-free and have a long shelf life. These products also do not have any side effects on the environment and are hence considered to be an example of green technology. Table 10.9 shows the specifications of ultracapacitor cells manufactured by Ioxus.

Ioxus also manufactures ultracapacitor modules such as the iMOD modules. The iMOD modules are further categorized into various series based on their applications. These series range from 16 V to 128 V and are used in wide-ranging applications as they have a high voltage range, long-life cycle and can operate over a wide temperature range and vast climatic conditions. Some of the specifications of iMOD modules are given in Table 10.10.

Apart from ultracapacitor cells and modules, Ioxus Inc. also manufactures ultracapacitor-based products such as uStart. The uStart is specifically designed for vehicles in order to improve their battery life, vehicle voltage health, and sustainability. These modules store a small amount of energy but is capable of generating engine cranking current. These are perfectly suited for start/stop applications. The uStart has a cycle life of 1,000,000 cycles, design life of 15 years, and can function over wider operating temperatures (−40 to 85 °C) as compared to a battery (−40 to 60 °C).

10.2.6 LS Ultracapacitor

10.2.6.1 Company Profile

LS Ultracapacitor is a Korean company, which began the mass production of ultracapacitors in 2007 [58]. Its product range includes both ultracapacitor cells and modules. LS Ultracapacitors have their own manufacturing technology when it comes to electrodes, cell, and packaging.

10.2.6.2 Product Portfolio

The LSUC ultracapacitors are mainly categorized on the basis of their voltages. These ultracapacitors have a wide operating temperature and a shelf life of 4 years, and found applications in windmills, heavy-duty vehicles, smart meters, toys, power tools, hybrid busses, etc. Some of the specifications of LS ultracapacitor cells are mentioned in Table 10.11.

LS Ultracapacitors also manufactures ultracapacitor modules. These ultracapacitor modules have an operating temperature range of −40 to 65 °C, shelf life of 4 years, and cycle life of 500,000 cycles, and found applications in energy storage and power delivery requirements. Some of the specifications of ultracapacitor modules manufactured by LS Ultracapacitors are given in Table 10.12

Table 10.11 Specifications of ultracapacitor cells manufactured by LS Ultracapacitor [58]

Series	Shape	Capacitance range (F)	Rated voltage (V)	Maximum stored energy (Wh)	Operating temperature	Application
LSUC 2.7 V	Cylindrical	650	2.7	0.65	−40 to 65 °C	Backup power and auxiliary power units, instantaneous power, and peak power compensation for various applications
		1500		1.51		
LSUC 2.8 V	Radial	100	2.8	0.1		
		400		0.43		
		600		0.65		
	Prismatic	3000		3.26		
LSUC 3.0 V	Radial	100	3	0.12		
		480		0.6		
	Cylindrical	3000		3.75		

Table 10.12 Specifications of ultracapacitor modules manufactured by LS Ultracapacitor [58]

Series	Capacitance (F)	Rated voltage (V)	Max. series voltage (V)	ESR (mΩ)	Max. energy (Wh)	Specific power (W/kg)
LSUM	2.5	380.8	408	650	50.4	1400
	58	16.8	750	22	2.3	2100
	5.8	168		240	22.7	2100
	166	51.3		5	60.7	5200
	166	48.6		5	54.5	4000
	250	16.2		2	9.1	4000
	500	16.2		1.7	18.2	3300
	250	32.4		3.3	36.5	3800
	93	86.4		11.3	96.4	2900
	15.5	162.4		110	56.7	1500
	62	129.6	1500	13.2	144.6	2700

10.2.7 VINATech

10.2.7.1 Company Profile

VINATech is a Korean company, which was established in the year 1999 [59]. They focus research and development on the use of supercapacitors and fuel cells in new fields with the help of modern technology. Their research and development in supercapacitors mainly focus on their use in regenerative energy and solar energy systems.

10.2.7.2 Product Portfolio

VINATech manufactures a range of products such as lithium-ion capacitors, fuel cell materials, and supercapacitor cells and modules. They began the mass production of supercapacitors in 2005. The supercapacitor cells manufactured by VINATech have found applications in various fields such as automobiles, energy, IoT, industrial and communication equipment. The supercapacitor cell product range includes Hy-Cap, Hy-Cap NEO, Hy-Cap NEO VET, and Hy-Cap Hybrid supercapacitors. These supercapacitor cells can be used in extreme environments over a wide temperature range. They have a shelf life of 2 years and a cycle life of 500,000 cycles. Table 10.13 shows the specifications of supercapacitor cells manufactured by Vinatech.

VINATech also manufactures two series of supercapacitor modules one with a rated voltage of 5.4 V and the other with a rated voltage of 6 V. Apart from these two series, VINATech also manufactures supercapacitor modules, which are customized

Table 10.13 Specifications of supercapacitor cells manufactured by VINATech [59]

Type	Series	Capacitance range (F)	Rated voltage (V)	Operating temperature	Application
Lead terminal	Hy-Cap	1	2.7	−40 to 65 °C	Automotive vehicles, communication equipment, industrial equipment, IoT, etc.
			3		
		5	2.7		
			3		
		10	2.7		
			3		
		50	2.7		
			3		
Snap in		100	2.7		
			3		
		400	2.7		
			3		
		500	2.7		
			3		
–	Hy-Cap NEO VET	3.3	2.7	−40 to 85 °C	High temperature and high humidity applications
		10			
		15			
Lead terminal	Hy-Cap Hybrid	10	2.3	−25 to 60 °C	High energy density-based applications
		50			
Snap in		300			
		800			

Table 10.14 Specifications of supercapacitor modules manufactured by VINATech [59]

Series	Capacitance range (F)	Rated voltage (V)	ESR (mΩ)	Max. Energy (Wh)	Application
Customized	40	28	40	5	Automotive and industrial, Transportation, wind turbines
	36	27	40	4.5	
	12	27	65	1.5	
	10	28	65	1.25	
	10	100	118	16.2	
	4.15	60	154	2.43	
	7.5	134.4	154	21.6	
	120	16.8	10	4.7	
	80	15	20	2.5	
	16.6	90	100	18.75	
Hy-Cap	0.5	5.4	265	–	–
	1.5		115		
	3.5		115		
	5		65		
	0.5	6	295	–	–
	1.5		145		
	3.5		135		
	5		75		

according to the needs of the customers. Based on customer applications these supercapacitor modules are optimized with respect to shape, capacitance, volume, protection level, capacitance, etc. These supercapacitor modules have an operating temperature range of −40 to 65 °C, shelf life of 2 years, and cycle life of 500,000 cycles. The specifications of supercapacitor modules manufactured by VINATech are given in Table 10.14

10.2.8 *Eaton Corporation*

10.2.8.1 Company Profile

Eaton is a power management company that began operations in 1911 and now does business in over 175 countries [60]. It has one of the strongest distribution networks in North America. Eaton manufactures a range of energy efficient, sustainable, reliable products that have wide-ranging applications. Eaton's revenues for the year 2019 was found to be $21.4 billion.

Table 10.15 Specifications of supercapacitor cells manufactured by Eaton Corporation [60]

Type	Series	Capacitance range (F)	Rated voltage (V)	DC ESR (mΩ)	Max. energy (Wh)	Operating temperature	Application
Cylindrical type	B series	0.22–2.2	2.5	200–2000	–	−25 to 70 °C	Backup power
	HB series	3–110	2.5	20–160	0.0026–0.095	−25 to 70 °C	Electric, gas smart meters, solar capture, RF radio power
	HV series	1–100	2.7	12–200	0.001–0.101	−40 to 85 °C	
	Hybrid	30–220	3.8	100–550	0.04–0.293	−25 to 70 °C	Medical and industrial backup power, IOT energy storage
	M series	1–9	2.5	20–210	–	−40 to 85 °C	Pulse power and backup power
	XV series	300–600	2.7	2.6–4.5	0.3–0.6	−40 to 85 °C	Hybrid battery and fuel cell systems
Coin cell type	KR series	0.1–1.5	5.5	30,000–75,000	–	−25 to 70 °C	Utility meters, HVAC controls, computers and peripherals
	KW series	0.1–1	5.5	30,000–50,000	–	−40 to 85 °C	Programmable logic controllers, solar inverters

Table 10.16 Specifications of supercapacitor modules manufactured by Eaton Corporation [60]

Series	Capacitance range (F)	Rated voltage (V)	No. of cells	ESR (mΩ)	Stored energy (Wh)	Peak power (kW)	Thermal resistance (°C/W)	Application
XLM module	130	62.1	23	6.7	69.6	143.9	0.5	Hybrid power systems, healthcare UPS, etc.
XLR module	166	48	18	5	54	118	0.4	Hybrid and electric vehicles, automated guided vehicles, etc.
	188	51			68.7	131.6		
XLR—LV module	500	16	6	1.7	18.2	38.6	–	Engine starting, backup power, remote power for sensors, etc.
XTM—18 module	61.7	18	6	22	2.8	3.7	1.5	Soft shutdown for industrial robots, industrial computers, etc.
XVM—PCBA module	4.17	259		310	38.9	54.2	–	Material handling systems
	6.25			250	58.3	67.2		
XVM module	65	16		22	2.4	3	1.5	Battery assisted engine starting in automobiles

10.2.8.2 Product Portfolio

Eaton offers a wide range of products such as electrical products, hydraulic products, aerospace products, and vehicle products. The electrical products include wiring devices, sensors, electronic components such as transformers, inductors and supercapacitors, enclosures. These electrical products are used across industries, institutional buildings, mines, data centres, etc. Eaton's hydraulic products include pumps, motors, generators, cylinders while their aerospace products include fuel systems, pumps, actuators, heat exchangers, etc. They have developed a wide range of supercapacitor cells specifically tailored for various applications ranging from a few microamps to several hundred amps per second. These supercapacitor cells have a high-power density, low ESR, and a lifetime of around 20 years, and have found wide-ranging applications based on their properties. Table 10.15 shows the specifications of supercapacitor cells manufactured by Eaton.

Apart from supercapacitor cells, Eaton also manufactures supercapacitor modules. These supercapacitor modules have a life of 20 years and a wide operating temperature. These are cost-effective, have a low operating cost, and do not require maintenance. The modules can be used in a variety of applications based on their properties. Some of the specifications of supercapacitor modules manufactured by Eaton are given in Table 10.16

10.2.9 Maxwell Technologies

10.2.9.1 Company Profile

Maxwell Technologies is an American manufacturing company that mainly focuses on the development of energy storage devices to be used in various fields such as renewable energy, wireless communication, heavy-duty vehicles. Maxwell technologies was recently acquired by Tesla in 2019 [61].

10.2.9.2 Product Portfolio

The supercapacitor cells manufactured by Maxwell Technologies are classified under three major series namely the Standard series, the XP series, and the Dura Blue Series. The Standard series supercapacitor cells are specially designed for applications operating under normal environmental conditions that need fast charge/discharge capabilities. The XP series on the other hand is designed for applications operating under hot and humid conditions and has a wider operating temperature range as compared to the Standard series. The Dura Blue series of supercapacitors are designed for applications such as transportation and industries, which experience high amounts of shock and vibrations. All these supercapacitor cells have a shelf life of 4 years and a projected cycle life of 500,000 cycles. Maxwell Technologies has about 65

Table 10.17 Specifications of supercapacitor cells manufactured by Maxwell Technologies [61]

Series	Capacitance range (F)	Rated voltage (V)	DC ESR (mΩ)	Max. energy (Wh)	Specific power (kW/kg)	Operating temperature	Application
Standard	3	2.7	70	0.003	8.9	−40 to 65 °C	Emergency lighting, security equipment
	50		16	0.0506	4.4		
	100		15	0.101	2.5		Smart meters, wireless transmitters
	360		3.2	0.36	3.8		Actuators, backup systems, UPS systems
XP	5	2.7	45	0.005	9.2	−40 to 85 °C	Smoke detectors, advanced metering, emergency lighting
	25		25	0.0253	5.1		
	50		16	0.0506	4.4		
Dura Blue	650–3000	2.7	0.29–0.80	0.66–3.04	5.8–6.8	−40 to 65 °C	Automotive subsystems, hybrid vehicles
	3400		0.23	3.66	7.4		Environments with high shock and vibrations
	3500	2.85	0.28	3.95	8.5		

Table 10.18 Specifications of supercapacitor modules manufactured by Maxwell Technologies [61]

Capacitance range (F)	Rated voltage (V)	DC ESR (mΩ)	No. of cells	Max. energy (Wh)	Specific power (kW/kg)	Application
1.5	5	130	–	0.0052	6.7	Actuators, emergency lighting, smoke detectors, UPS systems
2.5		85		0.0086	7.0	
58	16	22	6	2.1	2.2	Small UPS systems
500	16	2.1	6	18	2.7	Industrial equipment and transportation
9	24	139	–	0.72	1.2	Source of backup power for robots
165	48	6.3	18	53	3.3	Hybrid vehicles
94	75	13	32	73	2.1	Pitch control of wind turbines
63	125	18	48	140	1.7	Mining equipment, cranes, busses, electric vehicles

million supercapacitors cells in use in various applications today. The following are some of the specifications of the supercapacitor cells manufactured by Maxwell Technologies.

Maxwell Technologies also manufactures a wide range of supercapacitor modules. These supercapacitor modules have a shelf life of 4 years and a projected cycle life of 1,000,000 cycles. These can also operate under a wide operating temperature range of −40 to 65 °C. All these supercapacitor modules are used in a range of applications based on their properties. Table 10.18 shows the specifications of the supercapacitor modules manufactured by Maxwell Technologies (Table 10.17).

10.2.10 Nippon Chemi-Con Corporation

10.2.10.1 Company Profile

Nippon Chemi-Con Corporation is a Japan-based company that was established in the year 1931. They mainly manufacture mechanical and electronic components including a wide range of capacitors [62]. Nippon Chemi-Con Corporation is currently the largest producer of aluminium electrode foil, which is one of the primary materials used in capacitors. They were the first to commercialize electrolyte capacitors in Japan in 1931 and currently has the largest share of aluminium electrolyte capacitors in the world. Nippon Chemi-Con Corporation announced that the total revenue they generated for the financial year 2018 amounted to 140,951 million yen.

10.2.10.2 Product Portfolio

The product portfolio of Nippon Chemi-Con Corporation includes aluminium electrolytic capacitors, multilayer ceramic capacitors, film capacitors, camera modules, EDLCs, etc. The EDLCs manufactured by Nippon Chemi-Con Corporation are known as DLCAP supercapacitors and are cylindrical in shape. They are further classified into four series namely DKA, DXF, DXG, and DXE series. These supercapacitors have varied applications based on their properties. The specifications of supercapacitors manufactured by Nippon Chemi-Con Corporation are mentioned in Table 10.19.

Nippon Chemi-Con Corporation also manufactures supercapacitor modules known as DLCAP modules, which are suitable for high voltage applications and have the ability to store a large amount of charge. These supercapacitor modules can function over a wide operating temperature range of −40 to 70 °C and are

Table 10.19 Specifications of supercapacitor cells manufactured by Nippon Chemi-Con Corporation [62]

Series	Type	Capacitance range (F)	Rated voltage (V)	Max. energy (Wh)	Operating temperature	Application
DKA	High-power type	50	2.5	0.04	−40 to 75 °C	
DXF	High voltage type	3150–3500	2.8	3.5	−40 to 60 °C	Battery assistance, electricity storage
DXG	High-temperature type	300–330	2.5	0.3	−40 to 85 °C	Low temperature engine cranking application, electricity storage
		590–650		0.6		
		910–1000		0.8		
DXE	High-power type	400	2.5	0.4	−40 to 70 °C	Kinetic energy recapturing, start and stop application in automobiles
		800		0.7		
		1200		1.1		
		1400		1.3		

Table 10.20 Specifications of supercapacitor modules manufactured by Nippon Chemi-Con Corporation [62]

Series	Capacitance range (F)	Rated voltage (V)	ESR (mΩ)	Max. energy (Wh)	Application
DLCAP	133	7.5	6.6	1.1	Energy saving, renewable energy, emergency applications
	400		2.7	3.2	
	466		3.6	3.7	

widely used in renewable energy-based applications such as stabilization of windmill power, energy-saving applications such as effective recapture of kinetic energy, and other emergency applications. Some of the specifications of DLCAP supercapacitor modules are mentioned in Table 10.20.

10.2.11 CAP-XX

10.2.11.1 Company Profile

CAP-XX is an Australian company that specializes in the design and manufacture of thin, flat supercapacitors, and other electronic devices [63]. They have been developing supercapacitors for the past 20 years. CAP-XX has recently acquired Murata Manufacturing's supercapacitor production line and will now manufacture these supercapacitors in Australia.

10.2.11.2 Product Portfolio

These supercapacitors manufactured by CAP-XX have a high power and energy density, long cycle life, and wide operating temperatures. The supercapacitors are mainly of two types, namely prismatic and cylindrical. The prismatic supercapacitors are suitable to be used in space-constrained. The cylindrical supercapacitors on the other hand are used in peak power support for locks, actuators, GSM/GPRS transmissions, and portable drug delivery systems. Some of the specifications of cylindrical and prismatic supercapacitors are given in Table 10.21.

Table 10.21 Specifications of supercapacitor cells manufactured by CAP-XX [63]

Shape	Capacitance range (F)	Rated voltage (V)	ESR (mΩ)	Operating temperature	Application
Prismatic	0.17–2.4	2.5	15–50	−40 to 70 °C	Tablet computers, digital cameras, mobile phones, micro-hybrid vehicles
		2.75	25–90	−40 to 90 °C	
Cylindrical	1–100	2.7	10–240	−40 to 65 °C	Energy harvesting for wireless sensors, peak power support for portable drug delivery systems
	1–850	2.7	6–120		
	1–50	3	18–180		
	650–3000	2.7	0.22–0.7		Engine cranking, regenerative energy capture, grid smoothing

10.2.12 Murata Manufacturing

10.2.12.1 Company Profile

Murata Manufacturing is a Japan-based company that was established in the year 1950 [64]. They initially began with the production of titanium oxide ceramic capacitors and gradually expanded their product range. Today, Murata Manufacturing is one of the industrial leaders in the design and manufacture of electronic components, which are used in varied applications such as household appliances, automotive applications, mobile phones, healthcare devices. The production lines of Murata Manufacturing supercapacitors have recently been acquired by CAP-XX, an Australia-based supercapacitor manufacturer.

10.2.12.2 Product Portfolio

The product portfolio of Murata Manufacturing includes lithium-ion batteries, piezoelectric sensors, and buzzers, inductors, short-range wireless communication modules, multilayer ceramic capacitors, ceramic filters, etc. Murata manufacturing also manufactured three series of supercapacitor cells, namely DMF, DMT, and DMH. The DMF supercapacitor cells are used to meet power needs. The DMT supercapacitor cells are used in the set and forget long-life applications (10–15 years) such

Table 10.22 Specifications of supercapacitor cells manufactured by Murata manufacturing [64]

Series	Capacitance range (F)	Rated voltage (V)	ESR (mΩ)	Operating temperature	Application
DMF	0.47	5.5	45	−40 to 70 °C	Power applications
	1		40		
DMT	0.22	4.2	300	−40 to 85 °C	Building and industrial sensors
	0.47		130		
DMH	0.035	4.5	300	−40 to 85 °C	Smart cards, wearables,

as smart cards, sensors, etc. The DMH supercapacitor cells are ultrathin supercapacitor cells (20 mm × 20 mm × 0.4 mm) and are used in applications that require sleek designs. Table 10.22 shows some of the specifications of supercapacitor cells manufactured by Murata manufacturing.

10.2.13 AVX Corporation

10.2.13.1 Company Profile

AVX Corporation was established in the year 1972 and currently is one of the leading manufacturers and suppliers of electronic equipment with 29 manufacturing facilities spread over 16 countries [65]. These components have been widely used in the industrial, medical, military, transportation, commination sectors for over 50 years. AVX Corporation provides its customers with customized products that have a competitive advantage over others.

10.2.13.2 Product Portfolio

The product portfolio of AVX Corporation includes antennas, capacitors, diodes, filters, inductors, thermistors, etc. AVX Corporation also has various series of supercapacitor cells such as BestCap—BZ series, SCC series, and SCP series. These supercapacitor cells offer a very high capacitance and a very low ESR. They have a shelf life of 2 years and a cycle life of 500,000 cycles and can be used in wide-ranging applications. The BestCap—BZ series supercapacitor cells have a unique proton polymer membrane and are considered as direct competition to devices having organic electrolytes. Given below are some of the specifications of the supercapacitor cells manufactured by AVX Corporation (Table 10.23).

Table 10.23 Specifications of supercapacitor cells manufactured by AVX Corporation [65]

Series	Capacitance range (F)	Rated voltage (V)	ESR DC (mΩ)	Operating temperature	Max energy (Wh)	Power density (W/kg)	Application
SCC series	1–3000	2.3 and 2.7	Very low	−40 to 85 °C	0.0010–3.0375	1842–6033	Camera flash systems, wireless alarms, scanners, toys and games, remote metering, UPS to industries, etc.
SCM series	0.47–7.5	5 and 5.4	160–1000	−40 to 65 °C	0.0016–0.0304	1429–3069	
	0.47–15	5.5 and 6	50–3000		0.0020–0.0630	864–4235	
	0.33–1	7.5	360–900		0.0026–0.0078	2419–4076	
	0.33–1	8.1 and 9	840–2850		0.0030–0.0113	869–2338	
SCP series	1–11	2.1	39–485	−55 to 65 °C	0.0006–0.0067	535–2804	Power peripherals, Bluetooth keyboard, battery assistance, wearables, tablets, etc.
BestCap—BZ series	50×10^{-3} to 140×10^{-3}	3.6	70–100 (at 1 kHz)	−40 to 75 °C	–	–	–
	30×10^{-3} to 100×10^{-3}	5.5	80 to 160 (at 1 kHz)				
	15×10^{-3} to 22×10^{-3}	12	350 (at 1 kHz)				
	6.8×10^{-3}	20	400 (at 1 kHz)				

10.2.14 Nichicon Corporation

10.2.14.1 Company Profile

Nichicon Corporation is a Japanese company, which was established in the year 1950. They initially began with the production of aluminium electrolytic capacitors and gradually expanded their product range. Nichicon Corporation's network of customers spread across Asia, Europe, and the USA. They registered a net sale of 119,675 million yen as of March 2020.

10.2.14.2 Product Portfolio

Nichicon Corporation manufactures a wide range of products such as small Li-ion rechargeable batteries, aluminium electrolyte capacitors, plastic film capacitors, thermistors, public and industrial power storage systems. Nichicon also manufactures EDLCs known as EVerCAP supercapacitors. These EVerCAP supercapacitors are further divided into various series based on several factors such as energy density, power density, lifetime, capacitance range. These are used in wide-ranging applications widely as an emergency source of backup power for medical equipment and lighting including regenerative power systems of hybrid electric vehicles. The specifications of these supercapacitors are given in Table 10.24.

Table 10.24 Specifications of supercapacitor cells manufactured by Nichicon Corporation [66]

Series	Capacitance range (F)	Rated voltage (V)	ESR DC (Ω)	Operating temperature	Applications
JUM (standard)	1–47	2.7	0.06–3	−25 to 70 °C	Quick charge–discharge applications
JJC (standard)	56–200	2.5	0.03–0.07	−25 to 60 °C	
JUW (high capacitance)	1–82	2.7	0.06–1.8	−25 to 70 °C	
JUK (lower resistance)	6.8–27	2.5	0.040–0.075	−40 to 70 °C	Smart meters
JJD (high energy density type)	1000–2500	2.5	–	−25 to 60 °C	Electric power storage
JJL (high-power density type)	700–2000	2.5	–	−25 to 60 °C	Regeneration and UPS applications
JUA (lower resistance and longer life)	2.5–4.7	2.7	0.1–0.15	−40 to 70 °C	

10.2.15 KEMET Corporation

10.2.15.1 Company Profile

KEMET, which was established in 1919, manufactures products whose applications range from spacecraft to devices inside the human body [67]. It is one of the largest manufacturers of supercapacitors and ships more than 50 billion components each year to over 180,000 customers across the globe. KEMET was recently acquired by Yageo Corporation and the combined organization has an aim to increase their global footprint.

10.2.15.2 Product Portfolio

KEMET manufactures a variety of products such as capacitors, sensors, transformers, inductors, piezoelectric devices, varistors, relays. These devices have found applications in various sectors such as defence, aerospace, telecommunication, medical,

Table 10.25 Specifications of supercapacitor cells manufactured by KEMET [67]

Series	Capacitance range (F)	Rated voltage (V)	ESR at 1 kHz (Ω)	Operating temperature	Application
FA	0.047–1	5.5	2.5–20	−25 to 70 °C	Power source of toys, LED, buzzers, etc.
	0.022–0.47	11	4–20		
FE	0.047–1.5	5.5	0.6–14	−40 to 70 °C	Applications needing high current supply for short periods of time such as gas igniters, actuators
FG	1.5	3.5	65	−40 to 85 °C	CMOS micro-computer
	0.010–4.7	5.5	25–300		
FM	0.047–0.22	3.5	50–200	−25 to 70 °C	CMOS micro-computer
	0.010–0.33	5.5	20–300		
	0.047	6.5	200		
FR	0.022–1	5.5	60–220	−40 to 85 °C	Automotive power source
FS	0.022–1	5.5	7–60	−25 to 70 °C	Camera, projector, printer, micro-computer, microwave oven, etc.
	0.47–1	11	7		
	1–5	12	4–7.5		
HV	50	2.5	50×10^{-3}	−25 to 70 °C	Power supply for street signs, display lights, UPS
	1–200	2.7	30×10^{-3} to 300×10^{-3}		

automotive, computing among others. The supercapacitor cells manufactured by KEMET are either surface mount or radial in shape. These cells are lead-free, leak-proof, and have a shelf life of up to 1 year. They have found applications in low voltage devices such as an embedded microprocessor with flash memory and in capturing energy during regenerative braking. Given below are some of the specifications of KEMET supercapacitors (Table 10.25).

10.2.16 Elna

10.2.16.1 Company Profile

Elna Co. Ltd. is a Japanese company, which was established in 1929 and manufactures and sells various electronic components [68]. These electronic components mainly include capacitors such as aluminium electrolytic capacitors, solid conductive polymer electrolytic capacitors, and electric double-layer capacitors. Elna Co. Ltd. also has several associated companies all around the globe as well as has a capital of 100 million Japanese yen as of March 2020.

Table 10.26 Specifications of supercapacitor cells manufactured by Elna Co. Ltd [68]

<table>
<tr><th>Series</th><th>Capacitance range (F)</th><th>Rated voltage (V)</th><th>Operating temperature</th><th>Application</th></tr>
<tr><td>DVN</td><td>0.047–0.33</td><td rowspan="3">5.5</td><td>−25 to 70 °C</td><td rowspan="3">Smart meters and momentary power assistance to batteries</td></tr>
<tr><td>DVL</td><td>0.047–0.22</td><td>−40 to 85 °C</td></tr>
<tr><td>DH</td><td>0.047–1.0</td><td>−25 to 85 °C</td></tr>
<tr><td>DBS</td><td rowspan="2">0.047–1.0</td><td>3.6</td><td>−25 to 85 °C</td><td rowspan="6">IC’s of camera, micro-computers, RAMs</td></tr>
<tr><td>DBJ</td><td rowspan="5">5.5</td><td>−10 to 85 °C</td></tr>
<tr><td>DX</td><td>0.047–1.5</td><td>−25 to 70 °C</td></tr>
<tr><td>DXJ</td><td rowspan="2">0.047–1.0</td><td>−10 to 85 °C</td></tr>
<tr><td>DHC</td><td>−25 to 85 °C</td></tr>
<tr><td>DB</td><td>0.047–1.5</td><td>−25 to 70 °C</td></tr>
<tr><td>DC</td><td rowspan="2">0.2</td><td>2.5</td><td>−25 to 70 °C</td><td rowspan="2">Backing up of pager, solar watches and calculators, cameras, etc.</td></tr>
<tr><td>DCK</td><td>3.3</td><td>−10 to 60 °C</td></tr>
<tr><td>DZ</td><td>1–200</td><td>2.5 and 2.7</td><td>−25 to 70 °C</td><td rowspan="3">Power supplies of LED displays and personal wireless items</td></tr>
<tr><td>DZP</td><td>0.47–4.7</td><td>5</td><td>−25 to 70 °C</td></tr>
<tr><td>DZH</td><td>22–300</td><td>2.5</td><td>−25 to 60 °C</td></tr>
<tr><td>DZN</td><td>1–200</td><td>2.5 and 2.7</td><td>−25 to 70 °C</td><td rowspan="2">Actuators in motors and electromagnetic coil drives</td></tr>
<tr><td>DU</td><td>1–50</td><td>2.7</td><td>−40 to 65 °C</td></tr>
</table>

10.2.16.2 Product Portfolio

The supercapacitors manufactured by Elna Co. Ltd. have found wide-ranging applications in numerous fields performing important roles such as memory backup and power storage, and are classified under different series based on their properties. Some of the specifications of the supercapacitors manufactured by Elna Co. Ltd are reported in Table 10.26.

10.3 Concluding Remarks

The growing awareness about global warming and the looming energy crisis we might face in the near future due to the rapid depletion of fossil fuel reserves has led to many major technology-based companies shifting to renewable and lasting energy sources to satisfy their energy demands. Supercapacitors/ultracapacitors play a vital role in storing and recycling this energy that is generated through the cyclical nature of wind, solar, and wave type systems.

Supercapacitors/ultracapacitors have found tremendous applications in the automotive and transportation sector, renewable energy sector, consumer electronic applications, etc. It is also predicted that the demand for supercapacitors for applications in industrial equipment, material handling equipment, smart meters, and UPS shall rise sharply in the coming years. The global supercapacitor market is expected to grow at a CAGR of 23.9% to about $16.95 billion by the year 2027. This growth rate is being driven by countries such as India and China in the Asia Pacific region, where the CAGR of the supercapacitor market is predicted to be 23.2% for the period 2015–2023. Tremendous research is being carried out on materials from which supercapacitor electrodes and separators can be fabricated, operating conditions under which these supercapacitors can perform, etc. The global supercapacitor market is currently dominated by a few players that supply to customers all over the globe. These companies invest a lot of their resources in research and development of supercapacitors in order to customize their supercapacitors according to the needs of a specific application as well as gain a competitive advantage over their rivals.

Major challenges still exist, which have impeded the growth of the supercapacitor market. Though supercapacitors have the potential to replace traditional batteries in various applications, challenges still exist with regard to the high price of supercapacitors. The comparison between a supercapacitor and a traditional battery begins with the price. Supercapacitors can be considered an economically viable technology only if its price is at par with lead-acid batteries. Supercapacitors have thus not been able to achieve the kind of economies of scale as that of batteries and reduction of the price of supercapacitors is the need of the hour.

Acknowledgements The authors acknowledge the financial support provided by the Department of Science and Technology, India (DST/TMD/MES/2K16/37(G)) for carrying out this research work.

References

1. K.V.G. Raghavendra, R. Vinoth, K. Zeb, C.V. V. Muralee Gopi, S. Sambasivam, M.R. Kummara, I.M. Obaidat, H.J. Kim, J. Energy Storage **31**, 101652 (2020)
2. R. Nigam, K.D. Verma, T. Pal, K.K. Kar, in *Handbook of Nanocomposite Supercapacitor Materials II Performance*, ed. by K.K. Kar (Springer International Publishing, Cham, 2020), pp. 463–483
3. S. Banerjee, B. De, P. Sinha, J. Cherusseri, K.K. Kar, in *Handbook of Nanocomposite Supercapacitor Material I* (2020), pp. 341–350
4. L. Zhang, D.P. Wilkinson, Z. Chen, J. Zhang, in *Lithium-Ion Supercapacitors Fundam. Energy Appl.* (Taylor & Francis Group, 2018), pp. 129–164
5. J. Tahalyani, M.J. Akhtar, J. Cherusseri, K.K. Kar, in *Handbook of Nanocomposite Supercapacitor Materials I characteristics* (2020), pp. 1–51
6. P. Sinha, K.K. Kar, in *Handbook of Nanocomposite Supercapacitor Materials II Performance*, ed. by K.K. Kar (Springer International Publishing, Cham, 2020), pp. 1–28
7. P. Sinha, K.K. Kar, in *Handbook of Nanocomposite Supercapacitor Materials II Performance.* (2020), pp. 71–87
8. S. Banerjee, P. Sinha, K.D. Verma, T. Pal, B. De, J. Cherusseri, P.K. Manna, K.K. Kar, in *Handbook of Nanocomposite Supercapacitor Materials I Characteristics* (2020), pp. 53–89
9. *Supercapacitor Market by Product Type, Module Type, Material and Application: Global Opportunity Analysis and Industry Forecast, 2020–2027* (2020)
10. *Supercapacitor Market Study 2019—World Market to Grow from $498.6 Million in 2018 to $1.29 Billion by 2024—ResearchAndMarkets.Com.* (2019)
11. *Supercapacitors Market, Size, Share, Opportunities and Forecast, 2020–2027* (2020)
12. *Global Supercapacitors Market—Products, Modules and Applications* (2020)
13. J. Cherusseri, R. Sharma, K.K. Kar, in *Handbook of Polymer Nanocomposites. Processing, Performance* and *Application* (Springer Berlin Heidelberg, Berlin, Heidelberg, 2015), pp. 479–510
14. G. Wang, L. Zhang, J. Zhang, Chem. Soc. Rev. **41**, 797 (2012)
15. R.B. Choudhary, S. Ansari, B. Purty, J. Energy Storage **29**, 101302 (2020)
16. *Global Supercapacitor Market (2020 to 2025)—Growth in Global Demand for Electric Vehicles Is Driving the Market—ResearchAndMarkets.Com* (2020)
17. *Supercapacitors Market—Growth, Trends, and Forecast (2020–2025)* (2020)
18. P. Sharma, V. Kumar, J. Electron. Mater. **49**, 3520 (2020)
19. D.P. Dubal, N.R. Chodankar, D.-H. Kim, P. Gomez-Romero, Chem. Soc. Rev. **47**, 2065 (2018)
20. *Global Supercapacitor Market 2018–2022* (2018)
21. B. De, S. Banerjee, T. Pal, K.D. Verma, P.K. Manna, K.K. Kar, in *Handbook of Nanocomposite Supercapacitor Materials II Performance* (2020), pp. 271–296
22. B. De, S. Banerjee, K.D. Verma, T. Pal, P.K. Manna, K.K. Kar, in *Handbook of Nanocomposite Supercapacitor Materials II Performance.* (2020), pp. 229–243
23. B. De, S. Banerjee, K.D. Verma, T. Pal, P.K. Manna, K.K. Kar, in *Handbook of Nanocomposite Supercapacitor Materials II Performance* (2020), pp. 179–200
24. P. Sinha, S. Banerjee, K.K. Kar, in *Handbook of Nanocomposite Supercapacitor Materials II Performance* (2020), pp. 113–144
25. M. Kumar, P. Sinha, T. Pal, K.K. Kar, in *Handbook of Nanocomposite Supercapacitor Materials II Performance.* (2020), pp. 29–70
26. K.D. Verma, P. Sinha, S. Banerjee, K.K. Kar, in *Handbook of Nanocomposite Supercapacitor Materials I Characteristics* (2020), pp. 269–285
27. R. Sharma, K.K. Kar, in *Handbook of Nanocomposite Supercapacitor Materials I Characteristics* (2020), pp. 215–245
28. S. Banerjee, K.K. Kar, in *Handbook of Nanocomposite Supercapacitor Materials I Characteristics* (2020), pp. 179–214
29. P. Chamoli, S. Banerjee, K.K. Raina, K.K. Kar, in *Handbook Nanocomposite Supercapacitor Materials I Characteristics* (2020), pp. 155–177

30. P. Sinha, S. Banerjee, K.K. Kar, in *Handbook of. Nanocomposite Supercapacitor Materials I Characteristics*(2020), pp. 125–154
31. P. Sinha, B. De, S. Banerjee, K.D. Verma, T. Pal, P.K. Manna, K.K. Kar, in *Handbook of Nanocomposite Supercapacitor Materials II Performance* (2020), pp. 435–461
32. S. Banerjee, K.K. Kar, in *Handbook of Nanocomposite Supercapacitor Materials II Performance* (2020), pp. 333–352
33. T. Pal, S. Banerjee, P.K. Manna, K.K. Kar, in *Handbook of Nanocomposite Supercapacitor Materials I Characteristics* (2020), pp. 247–268
34. A. Tyagi, S. Banerjee, J. Cherusseri, K.K. Kar, in *Handbook of Nanocomposite Supercapacitor Materials I Characteristics* (2020), pp. 91–123
35. B. De, S. Banerjee, K.D. Verma, T. Pal, P.K. Manna, K.K. Kar, in *Handbook of Nanocomposite Supercapacitor Materials II Performance.* (2020), pp. 89–111
36. K D. Verma, P. Sinha, S. Banerjee, K.K. Kar, M.K. Ghorai, in *Handbook of Nanocomposite Supercapacitor Materials I Characteristics* (2020), pp. 315–326
37. K.D. Verma, P. Sinha, S. Banerjee, K.K. Kar, in *Handbook of Nanocomposite Supercapacitor Materials I Characteristics* (2020), pp. 327–340
38. K.D. Verma, S. Banerjee, K.K. Kar, in *Handbook of Nanocomposite Supercapacitor Materials I Characteristics* (2020), pp. 287–314
39. P. Sinha, S. Banerjee, K.K. Kar, in *Handbook of Nanocomposite Supercapacitor Materials II Performance.* (2020), pp. 145–178
40. B. De, S. Banerjee, K.D. Verma, T. Pal, P.K. Manna, K.K. Kar, in *Handbook of. Nanocomposite Supercapacitor Materials II Performance.* (2020), pp. 201–227
41. B. De, S. Banerjee, T. Pal, A. Tyagi, K.D. Verma, P.K. Manna, K.K. Kar, in *Handbook of Nanocomposite Supercapacitor Materials II Performance.* (2020), pp. 245–270
42. B. De, P. Sinha, S. Banerjee, T. Pal, K.D. Verma, A. Tyagi, P.K. Manna, K.K. Kar, in *Handbook of Nanocomposite Supercapacitor Materials II Performance* (2020), pp. 297–331
43. B. De, S. Banerjee, T. Pal, K. D. Verma, A. Tyagi, P.K. Manna, K.K. Kar, in *Handbook of Nanocomposite Supercapacitor Materials II Performance.* (2020), pp. 353–385
44. B. De, S. Banerjee, T. Pal, K. D. Verma, A. Tyagi, P.K. Manna, K.K. Kar, in *Handbook of Nanocomposite Supercapacitor Materials II Performance.* (2020), pp. 387–434
45. A.K. Samantara, S. Ratha, in *Materials Dev. Act. Components a Supercapacitor* (Springer Singapore Pte. Limited, 2017), p. 47
46. B.K. Kim, S. Sy, A. Yu, J. Zhang, in *Handbook of Clean Energy Systems* (Wiley, Chichester, UK, 2015), pp. 1–25
47. *Supercapacitor Markets, Technology Roadmap, Opportunities 2021–2041* (2020)
48. Y. Aiping, V. Chabot, Z. Jiujun, Z. Jiujun, in *Electrochemical Supercapacitors for Energy Storage and Delivery* (Taylor & Francis Group, 2013), pp. 335–344
49. H. Becker, US2800616A (1957)
50. R. Rightmire, US3288641A (1966)
51. F. Beguin, E. Frackowiak, M. Lu, in *Supercapacitors Materials Syst. Appl.* (Wiley, 2013), pp. 510–525
52. P. Sharma, T.S. Bhatti, Energy Convers. Manag. **51**, 2901 (2010)
53. https://www.kamcappower.com/
54. http://www.skeletontech.com/
55. https://www.spscap.com/
56. https://yunasko.com/
57. https://ioxus.com/english
58. https://www.ultracapacitor.co.kr:8001/
59. https://www.vina.co.kr/eng/
60. https://www.eaton.com/us/en-us.html
61. https://www.maxwell.com/
62. https://www.chemi-con.co.jp/e/
63. https://www.cap-xx.com/
64. https://www.murata.com/en-sg

65. http://www.avx.com/
66. https://www.nichicon.co.jp/english/index.html
67. https://www.kemet.com/en/us.html
68. http://www.elna.co.jp/en/capacitor/index.html

Chapter 11
Applications of Supercapacitors

T. P. Sumangala, M. S. Sreekanth, and Ariful Rahaman

Abstract Supercapacitors exhibit large power density, fast charge and discharge capability, and long cycle stability. These characteristics find applications in transportation, energy and utilities, aerospace, military, electronics, industrial, and medical fields. Supercapacitors are currently used as one of the most efficient energy storage systems replacing batteries in many applications. In the transportation and aerospace sector, supercapacitor-based hybrid energy storage systems are widely utilized for improved efficiency. The use of supercapacitors in various sectors such as automotive, energy, medicine, electronics, aerospace, and defense is presented with consideration of the various products offered by manufacturers. The application of supercapacitor in portable and wearable electronics and medical sectors is discussed in detail. In this chapter, most of the possible application areas of supercapacitors along with manufacturers are discussed in detail.

11.1 Introduction

Supercapacitors are one of the emerging energy storage devices with the capability of bridging the gap between high energy density batteries and high power density conventional capacitors. Supercapacitors evolved as the most efficient energy convention and storage systems in sustainable and renewable-based energy storage systems due to large power density, fast charge, and discharge capability, higher capacitance,

T. P. Sumangala
Department of Physics, School of Advanced Sciences, Vellore Institute of Technology, Vellore, Tamil Nadu 632014, India
e-mail: sumangala.tp@vit.ac.in

M. S. Sreekanth (✉) · A. Rahaman
Department of Manufacturing Engineering, School of Mechanical Engineering, Vellore Institute of Technology, Vellore, Tamil Nadu 632014, India
e-mail: sreekanth.ms@vit.ac.in

A. Rahaman
e-mail: arahaman@vit.ac.in

K. K. Kar (ed.), *Handbook of Nanocomposite Supercapacitor Materials III*,
Springer Series in Materials Science 313,
https://doi.org/10.1007/978-3-030-68364-1_11

and long cycle stability [1, 2]. Conventionally, energy storage systems can be divided into two major classes namely batteries and supercapacitors. Supercapacitors have longer cycle stability compared to batteries due to less chemical phase changes in electrodes during continuous charge and discharge operations. Supercapacitors are generally used for harvesting energy and delivering high pulse power for short periods [3, 4]. The use of these capacitors is found in electronic devices like mobile phones, wearable device, health monitors, sensors, LED display, bar code scanners, and GPS chips [5, 6]. Depending upon the amount and extent of charge storage, the electrode materials can be utilized in various applications such as hybrid vehicles, military weapons, telecommunications, and spaceships [7]. Today, there are 40 major supercapacitor producers across the globe. This number is growing day by day. The distribution of the major players is represented in Fig. 11.1. The statistics of these companies in terms of the continents are represented in Fig. 11.2.

The main contribution is from Asia followed by North America and Europe. In Asia, China is leading the manufacturing process. This is followed by Japan, Taiwan,

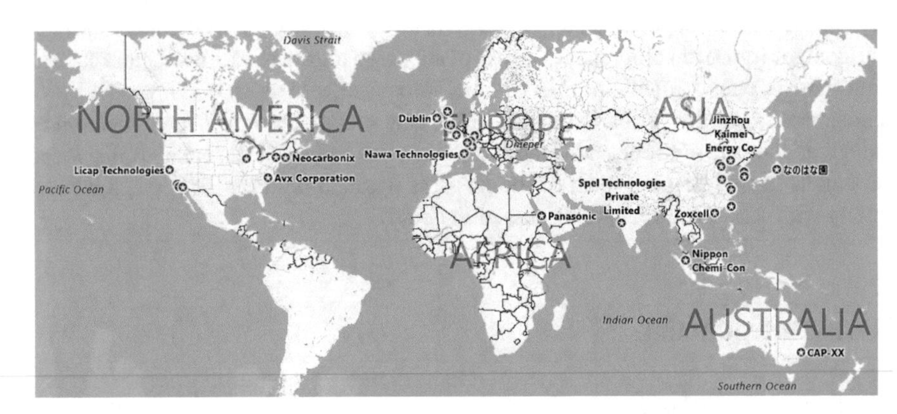

Fig. 11.1 Distribution of major supercapacitor manufacturers across the globe (data obtained from [8–19])

Fig. 11.2 Statistics of the distribution of supercapacitor manufacturers across various continents (data obtained from [8–19])

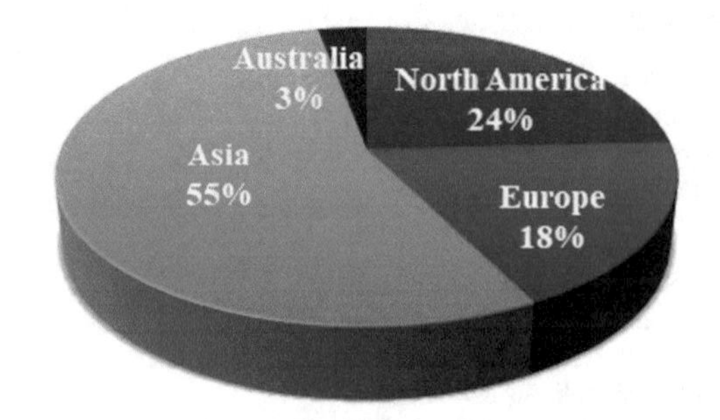

South Korea, and India. In Europe, various countries like Germany, France, Switzerland, and the UK contribute to the manufactures. Recently, countries like Russia, India, etc. have made strong initiatives on supercapacitors. Globally, supercapacitor growth rate is currently more than 39%.

The supercapacitor companies are highly volatile and dynamic. Many acquisitions are happening in the case of supercapacitor companies. The major and most relevant one is the acquisition of Maxwell Technologies by Tesla, Inc. (NASDAQ: TSLA) in May 2019 [14]. In addition to it, in April 2020, XS Power Batteries® finished the acquisition of IOXUS® [15]. In December 2019, Sydney-based supercapacitor manufactures, CAP-XX, reported the acquisition of production lines of Murata [16]. A few other acquisitions worth mentioning are that of Nesscap by Maxwell technology and that of EPCOS by TDK electronics [17, 18]. Most of the leading players in the supercapacitor industry are companies with more than 100 years in the energy/electronic section, sparing a few. Skeleton Technologies is Estonia/Germany-based Startup Company started in 2009. Now, it is the leading supercapacitor manufacturer in Europe with a net investment of 60 million euros. The net production per year is over 3 million cells a year [19]. The major companies in the US are Maxwell Technologies, Eaton Corporation, KEMET, and AVXCorporation. Their net income from the 2019 financial year is \$90.5 million, \$ 21.39 Billion, \$206.6 million, and \$1792 million, respectively. These companies are having a series of products other than supercapacitors.

The global market for supercapacitors in 2020 has reached approximately US\$92.3 billion [20], according to Bosch's 2007–2022 Research Report on the Current Situation and Investment Prospects of China's Supercapacitor Market (Fig. 11.3). The growth rate has shown an initial increase during the beginning of the last decade. However, this showed a constant decrease for the later four years. However, after 2013, the growth rate of the supercapacitor market showed a continuous increase. The constant increase from 2013 could be attributed to the widespread of electric vehicles, including railways and hybrid buses, power recuperation, and dynamic brake energy.

Energy storage devices are emerging areas for the scientific community due to changes in trends of the global market. In this regard, various energy storage

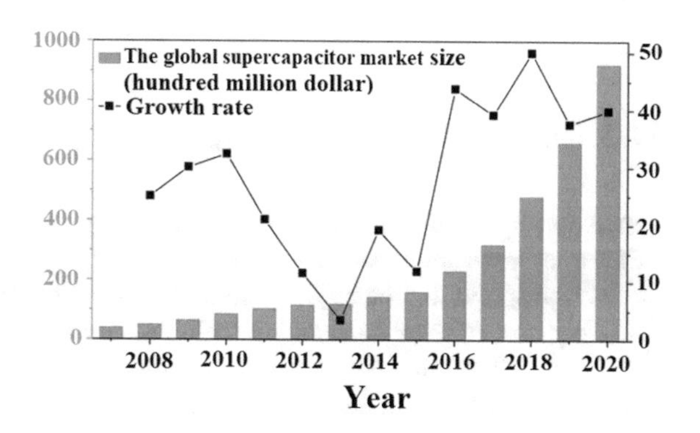

Fig. 11.3 Global supercapacitor market size and growth rate (redrawn and reprinted under Creative Commons license from [20])

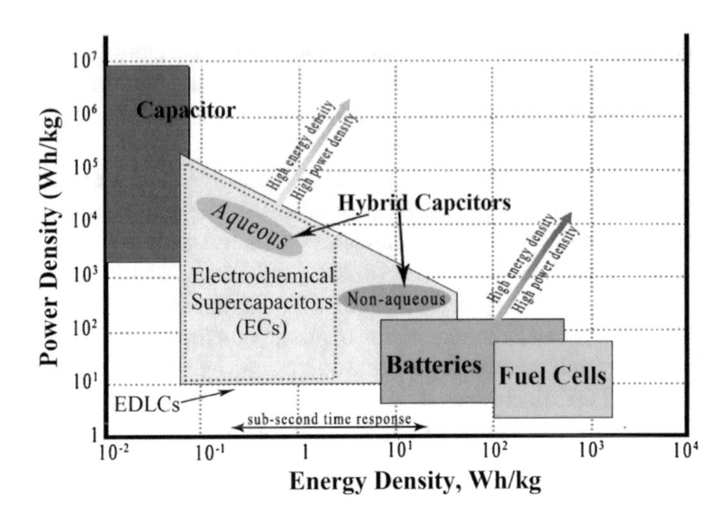

Fig. 11.4 Ragone plot of performance range of various energy storage systems (redrawn and reprinted with permission from [21])

devices were being developed for higher efficiencies. Supercapacitors were developed with higher surface area electrode materials, and thin dielectrics to produce improved capacitance values compared to conventional capacitors. Supercapacitors and batteries are used to store and release energy for various systems and the major difference is in specific application areas. Batteries are used to provide constant values of DC voltage, whereas supercapacitors provide quick charge and discharge capabilities for high power requirements. Supercapacitors have attracted more attention due to higher power capabilities and longer cycle times for efficient hybrid energy storage systems. Figure 11.4 shows the Ragone plot of the performance range of various energy storage devices. Supercapacitors technology provides a bridge between batteries and conventional capacitors in terms of both power density and energy density applications [21]. One of the emerging areas in energy storage systems is the use of hybridization of the electrode to increase capacitance values, which lead to the development of higher energy density hybrid supercapacitors (supercapacitor and battery).

11.2 Application of Supercapacitors

Supercapacitor application spans over a wide range from basic electronic gadgets like LEDs to defense and medical devices. Supercapacitors are extensively used in automotive/transportation, energy, electronics, aerospace, medical, industry, and other fields because of their outstanding features. Growth projected by industry experts in 2014 and 2020 for many sectors is shown in Fig. 11.5 [22]. The picture gives an idea of the change in the usage of supercapacitors. The electronic industry holds

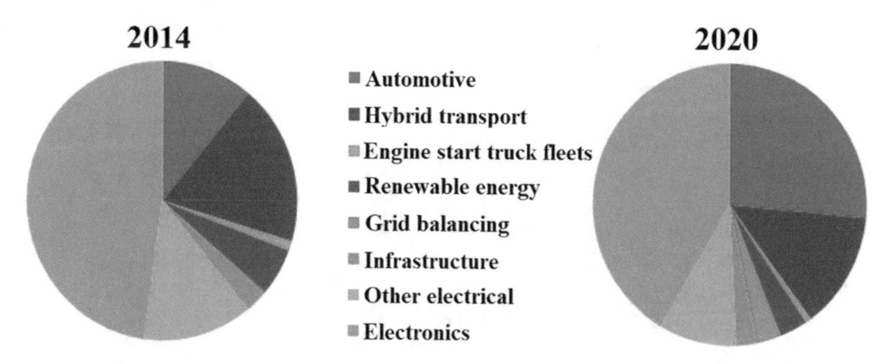

Fig. 11.5 Supercapacitor demand projections for many applications (redrawn and reprinted with permission from [22])

the major share and had a marginal decrease from 48% to 43% from 2014 to 2020. The automotive industry is the one that had undergone major growth in the years under consideration. The automotive part accounted for 16.5% in 2014. It showed an increase to around 27% in 2020. Hybrid transport, engine start truck fleets, renewable energy, and other electrical industries suffered from a decline in these years. Infrastructure showed almost the same response. One of the notable features is the entry of grid balancing into the market.

Commercialization of supercapacitors for the transportation sector evolved through various stages as shown in Fig. 11.6. The initial development started in 1996, in which Russia developed supercapacitor powered electric cars [20].

The first supercapacitor bus was developed in the year 2004. Shanghai Zhangiang high-tech park has developed world's first supercapacitor bus with fast-charging station. Also, Shanghai 11th has become the first commercial supercapacitor bus line in the world in the year 2006. Nanotexture developed supercapacitor for hybrid electric vehicle in the year 2009. In 2010, German manufacturer MAN commercialized hybrid electric vehicle. Later in 2014, Chariot e-bus used ultra capacitor of Aowei for the first time in Sofia and Bulgaria. In 2018, Russia and Finland jointly developed flexible supercapacitor for various applications. The United States developed a new type of rechargeable supercapacitor in 2019. Various developments and commercialization of supercapacitor are going on in almost all the sectors.

11.2.1 Automotive/Transportation

Electric vehicles are getting attention in many countries because of zero pollution. The commercialization of electric vehicles (EV) is one of the thrust areas in the current scenario. Most of the automotive applications are presently based on fuel

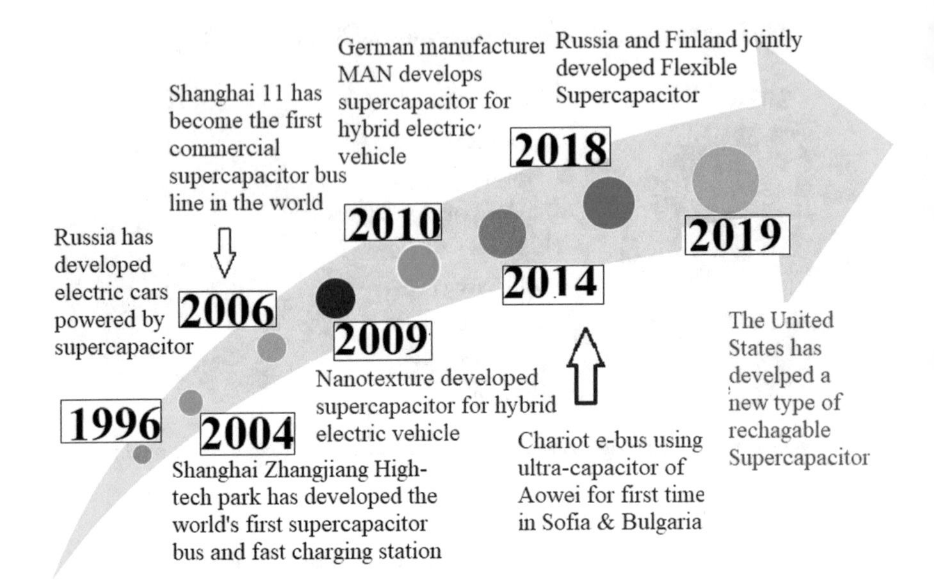
German manufacturer MAN develops supercapacitor for hybrid electric vehicle
Russia and Finland jointly developed Flexible Supercapacitor
Shanghai 11 has become the first commercial supercapacitor bus line in the world
2018
2010
2019
Russia has developed electric cars powered by supercapacitor
2006
2014
2009
The United States has develped a new type of rechagable Supercapacitor
Nanotexture developed supercapacitor for hybrid electric vehicle
1996
2004
Chariot e-bus using ultra-capacitor of Aowei for first time in Sofia & Bulgaria
Shanghai Zhangjiang High-tech park has developed the world's first supercapacitor bus and fast charging station

Fig. 11.6 Evolution of supercapacitor for transport application (data adapted from [20])

cells as the main source of energy. The major issue of fuel cell technology is slow dynamics to meet fuel shortage problems and to improve performance and cycle stability (life). The use of a supercapacitor as an auxiliary power source will solve the issues in an electric vehicle. The development of supercapacitor-based electric vehicles is an efficient approach to solve the pollution problem affected by public transportation. Optimization of power distribution between fuel cells and supercapacitor is a promising issue for efficient hybrid energy storage systems. Various elements can be located in a real vehicle with a hybrid energy system such as a DC–DC converter, supercapacitor battery, fuel cell, and inverter. The DC–DC converters control power and maintain constant voltage DC output against load variations, which is the input requirement for electric vehicle inverter. DC–DC converters are used as propulsion systems by adjusting the voltage to the required levels. DC bus has a capacitor, which provides traction wheel drives with the help of a three-phase inverter to convert the DC power into AC power. The system gives torque control for each wheel in a controlled manner. Supercapacitor battery-based hybrid system plays an important role in providing high power efficiencies [23].

The supercapacitors market is segmented by different features in automotive applications such as start-stop, light, ignition, electric, and hybrid vehicle drive support. Railways and hybrid buses show a significant role in the supercapacitor market. It benefits from increasing demand among operators and consumers, and rising vehicle availability at lower prices. Veneri et al. [24] reported the performance of a hybrid energy system consisting of ZEBRA batteries and supercapacitor, which has been

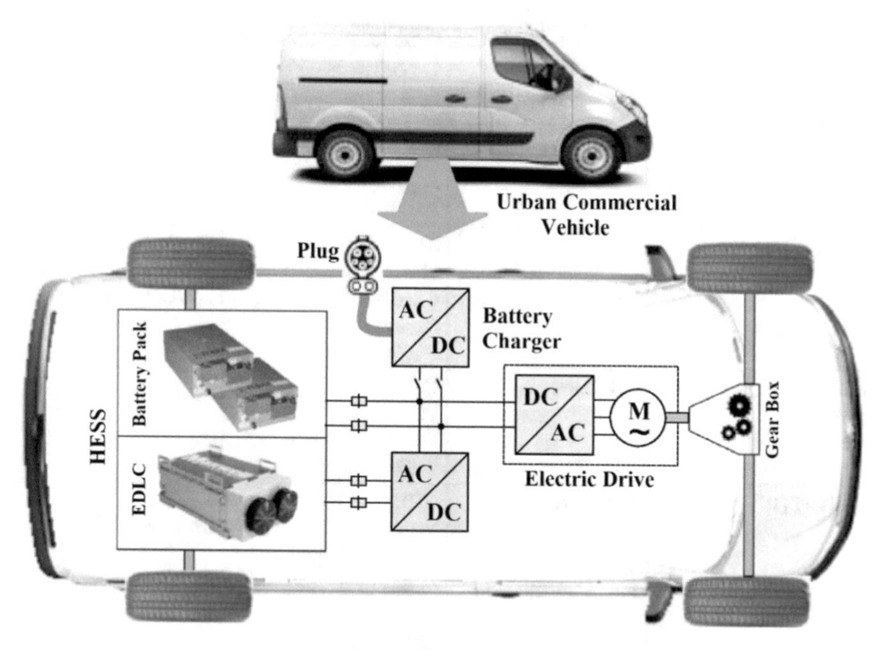

Fig. 11.7 Supercapacitor used in hybrid energy storage system for urban commercial vehicle (redrawn and reprinted with permission from [24])

tested on a laboratory scale to simulate real road urban commercial vehicle operations. Figure 11.7 shows the schematic diagram of the vehicle propulsion system using hybrid energy storage.

Maxwell technologies [25] deal with supercapacitor modules with capacitance range between 65 and 3400 F. This company has designed supercapacitors connected across a smaller lead-acid battery by replacing conventional vehicle battery. Supercapacitor products in automotive applications from various industries are presented in Table 11.1.

Supercapacitor-based electric vehicle has already been demonstrated. According to [27], AFS Trinity is preparing an extreme hybrid drive vehicle. In this vehicle, there is two energy storage systems: lithium-ion batteries combined with a supercapacitor. The existence of a high power density supercapacitor will protect the lithium-ion battery from these high current events and prolong battery life up to ten years. Recently, Skeleton Technologies have received an order from Škoda Ltd. for supplying supercapacitors for electric trams (Fig. 11.8) [26]. In addition to this, Skeleton Technologies is providing specialized services for various sectors. These are summarized in Fig. 11.9.

Table 11.1 Supercapacitor products in automotive applications from various industries

Company name	Supercapacitor	Specification	Application
ZoxCell [8]	2.7 V 3000 F	C: 3000 F, V: 2.7 V, E_{max}: 4.1–7.4 Wh/kg, P_{max} = 12,000–14,000 W/kg	Electric vehicles
KAM [9]	MK module	C: 1–500 F, V: 3.3–120 V, E_{max}: 4.1–7.4 Wh/kg, P_{max} = 12,000–14,000 W/kg	Electrified railway, hybrid vehicles, rail power system
Skeleton technologies [26]	SCA0300	C: 300 F, V: 2.85 V, Energy: 0.34 Wh, Power = 1.3 kW	Electric vehicles
EATON [10]	B supercapacitor	C: 0.2–2.2 F, V: 2.5 V, ESR (mΩ): 2000	Safety system, asset tracking
	HB supercapacitor	C: 3–110 F, V: 2.5 V, ESR (mΩ): 155–18	
Maxwell technologies [25]	K2 series	C: 65–3400 F, V: 2.7–2.85 V, E_{max}: 4.1–7.4 Wh/kg, P_{max} = 12,000–14,000 W/kg	Mass transit bus, fleet vehicles, trains
	125 V modules	C: 63 F, V: 125 V, E_{max}: 2.3 Wh/kg, P_{max} = 3600 W/kg	

11.2.2 Energy and Utilities

There has been a great interest in using supercapacitors in energy harvesting. The various areas of energy application by supercapacitors include UPS/power backup, power generation, transmission and distribution, uninterrupted elevator, pitch control, and wireless sensors. Supercapacitor products in energy applications from various industries are presented in Table 11.2.

Figure 11.10 shows supercapacitor-based efficient renewable energy storage system [30]. For higher voltage and power requirements of the renewable power system, hundreds of supercapacitors need to be connected in series to form a module. A power smoothening strategy is used for DC/DC converter with a sliding mode technique [31]. Generally, a supercapacitor module supplies energy to regulate DC bus energy and a fuel cell supplies energy to the supercapacitor to charge.

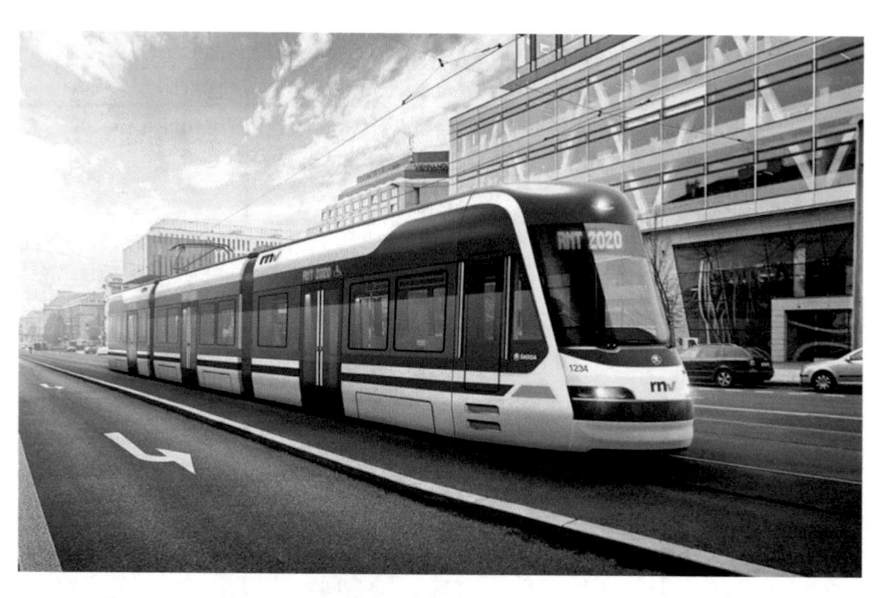

Fig. 11.8 Škoda electric trams to begin using ultra capacitors produced by Skeleton technologies (reprinted with permission from [28])

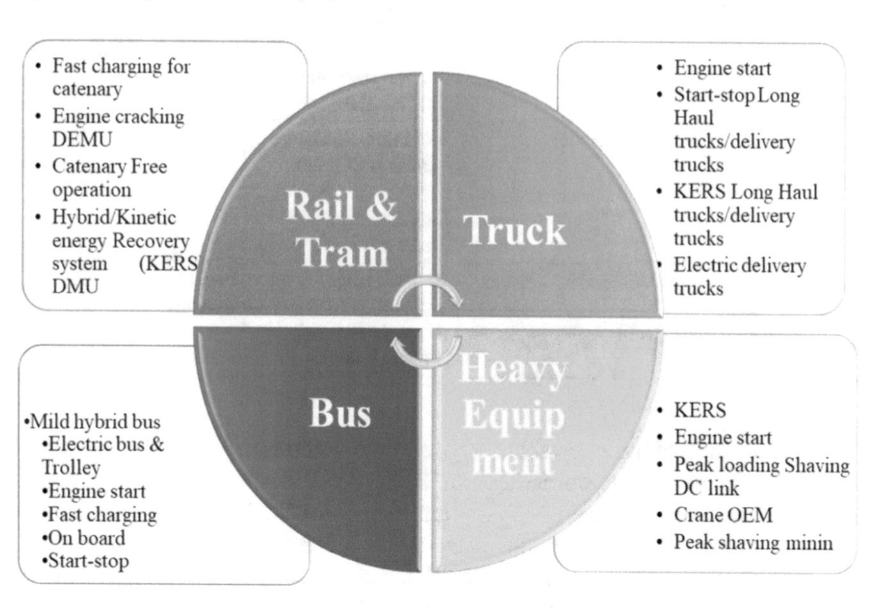

Fig. 11.9 Various transport services offered by skeleton technologies

Table 11.2 Supercapacitor products in energy applications from various industries

Company name	Supercapacitor	Specification	Application
Skeleton technologies [26]	SkelMod 17	C: 533 F, V: 17 V, Energy: 21.3 Wh, Power = 65.7 kW	Wind pitch control
EATON [10]	B supercapacitor	C: 0.2–2.2 F, V: 2.5 V, ESR (mΩ): 2000	Metering
	HB supercapacitor	C: 3–110 F, V: 2.5 V, ESR (mΩ): 155–18	
CAP-XX Supercapacitor [29]	GY13RO series	C: 1–50 F, V: 3 V, ESR_{max} @ 1 kHz (mΩ): 180	Energy harvesting for wireless sensor
	125 V modules	C: 63 F, V: 125 V, E_{max}: 2.3 Wh/kg, P_{max} = 3600 W/kg	
Newcell technologies [12]	LSC14250	Maximum discharge current (mA): 0.8, V: 4 V, Internal resistance (mΩ): <250	Automatic meter reading, advanced metering infrastructure
	LSC14505	Maximum discharge current (mA): 2.0, V: 4 V, Internal resistance (mΩ): <150	
Goldencell technologies [13]	WPSM5R5474Z-L	C: 0.47–1.5 F, V: 5.5 V, Internal resistance (mΩ): <900	Wind turbine pitch, Backup power

Currently, most of the renewable energy storage systems for higher efficiencies are based on the hybrid energy storage system (HESS) [32], which combines supercapacitors for quick dynamic power regulation and battery for durable energy management as shown in Fig. 11.11. The major components are renewable energy generators, supercapacitors, batteries, inverter, controller, and load. Supercapacitors are directly connected in parallel to the battery. In this system, batteries provide continuous energy for a long duration and supercapacitor gives peak power smoothening. The hybrid energy storage systems are designed not only for stable output power but also to meet peak power output and load requirements.

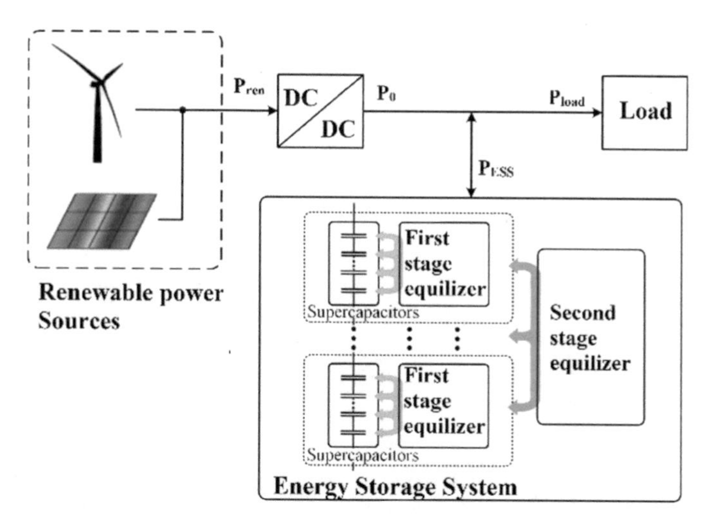

Fig. 11.10 Supercapacitor energy storage system from renewable sources (redrawn and reprinted under Creative Commons license [30])

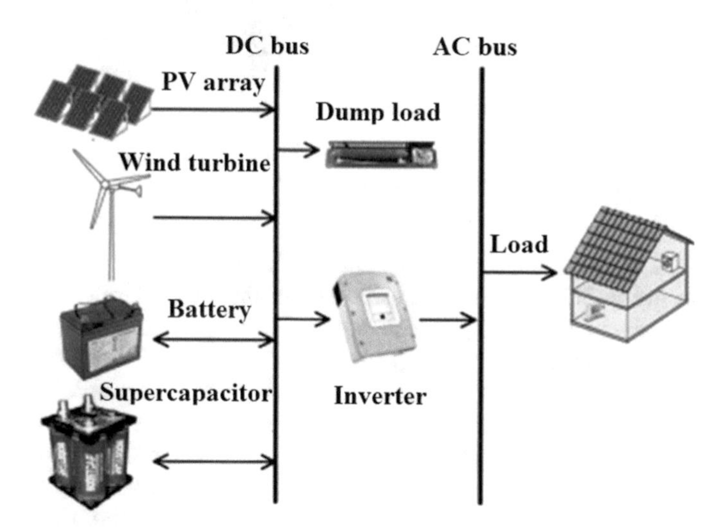

Fig. 11.11 Hybrid energy storage using supercapacitor and batteries for industrial applications (redrawn and reprinted with permission from [32])

11.2.3 Electronics

The mainstream of supercapacitor income in the consumer electronics section is produced by TV devices. Supercapacitor products in electronic applications from industry is presented in Table 11.3.

Table 11.3 Supercapacitor products in electronic applications from various industries

Company name	Supercapacitor	Specification	Application
Goldencell technologies [13]	WPSM5R5474Z-L	C: 0.47–1.5 F, V: 5.5 V, Internal resistance (mΩ): <900	Consumer electronics

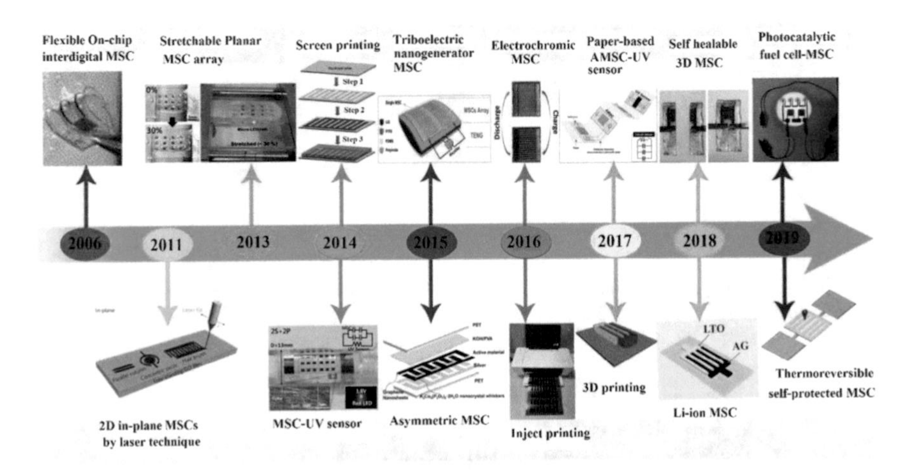

Fig. 11.12 Timeline of the advancement of flexible MSC in terms of features and fabrication techniques (redrawn and reprinted with permission from [33])

There has been a huge advancement in the research on flexible micro-supercapacitors (MSC). Figure 11.12 gives the timeline of the development of flexible supercapacitors. One of the first studies on flexible supercapacitors was using photolithography and electrochemical polymerization in 2006 [33]. Following that, various techniques have been explored for flexible micro-supercapacitor production. These include laser technique, screen printing, and inkjet printing for two-dimensional device formation, and 3D printing. In addition to the advancement in the fabrication process, specific features such as stretchable, triboelectric nanogenerator, asymmetric, electrochromic, paper-based, self-heat able, photocatalytic, and thermoreversible self-protection have been incorporated in MSC.

There has been a lot of studies on supercapacitors based on wearable fabrics. These studies have been developed with preferences on coated energy textiles, fiber, and yarn-based electrodes, and woven/knitted textiles with engineering materials. The most commonly studied materials on the coated textile are single-walled CNT, activated carbon fabric, metal oxides like Zn_2SnO_4/MnO_2 and, $LiFePO_4/Li_4Ti_5O_{10}$ [34]. Fiber and yarn-based electrodes are yet another approach with flexibility in materials. Use of Kevlar fibers with gold and ZnO, graphitic pen ink electrode material in plastic casing, core-sheath geometries based on graphene fiber and graphene oxide core, PEDOT, CNT disc rolled yarn, and CNT-Cotton yarn is studied for their potential application in wearable fabric. In addition to the textile-based fabric, knitted/woven fabric could also be developed with the functionalities of wearable electronics. This has also been demonstrated based on yarns of Lithium-based polymer strips. The presence of various electronic elements in wearable electronics using a supercapacitor for energy storage is shown in Fig. 11.13 [34]. Wearable fabric with an on-chip micro-supercapacitor is represented in Fig. 11.14 [33].

In addition to wearable electronics, supercapacitors are found useful in a variety of other electronic applications. The most significant advancement in supercapacitor research is the demonstration of the use of building bricks as a supercapacitor material to generate power [35]. The presence of 8 wt% α-Fe_2O_3, the porosity, and the mechanical robustness make it an ideal substrate for supercapacitors. The electrodes are formed by a Nano fibrillar coating of PEDOT on these substrates. Electrolytes are in the gel phase and epoxy is added to improve the waterproofing nature of

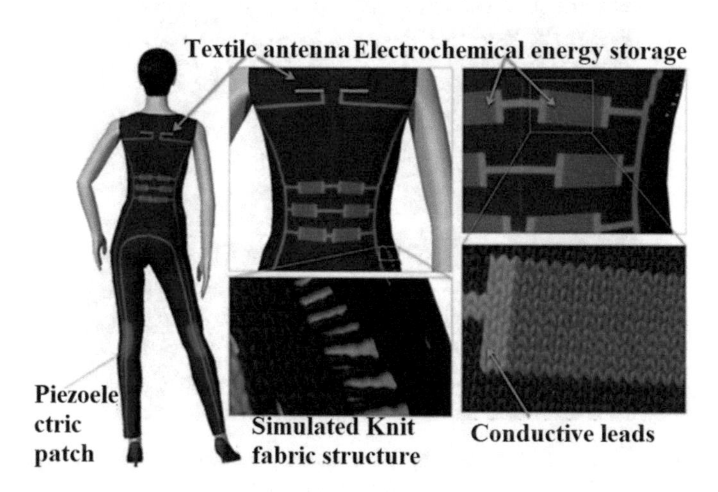

Fig. 11.13 Wearable fabric with various electronic elements integrated using supercapacitors for energy storage (redrawn and reprinted with permission from [34])

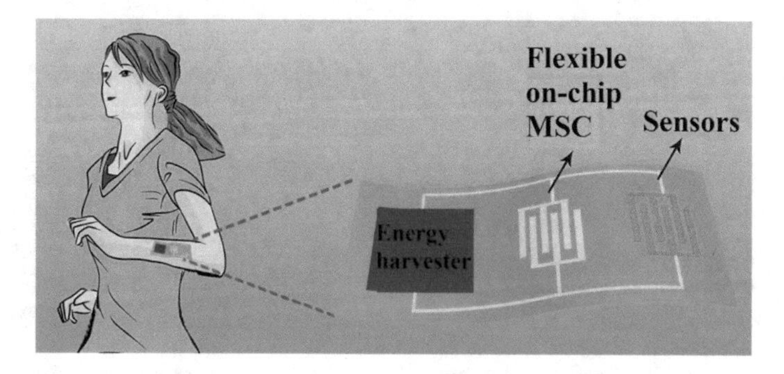

Fig. 11.14 Schematic representation of energy harvesting using flexible on-chip micro-supercapacitor (MSC) (redrawn and reprinted with permission from [33])

the supercapacitor material. The stacking of these bricks makes them function as modules. These supercapacitors have 90% capacitance retention and are stable up to 10,000 cycles. Figure 11.15 shows the glowing of the LED bulb using a PEDOT based supercapacitor in a brick substrate. The module is made by connecting three supercapacitors in series.

A representation of the various applications of supercapacitors is given in Fig. 11.16. A fully solid-state asymmetric supercapacitor using fish swim bladder as the separator and as a nanogenerator can be seen. Power generation is by utilizing bio-generated piezoelectricity. The charging can be done mechanically or electrically. Eight cells can be connected for the powering of small electronic gadgets

Fig. 11.15 Lighting of a LED bulb using supercapacitor module developed on brick substrate (redrawn and reprinted under Creative Commons license [35])

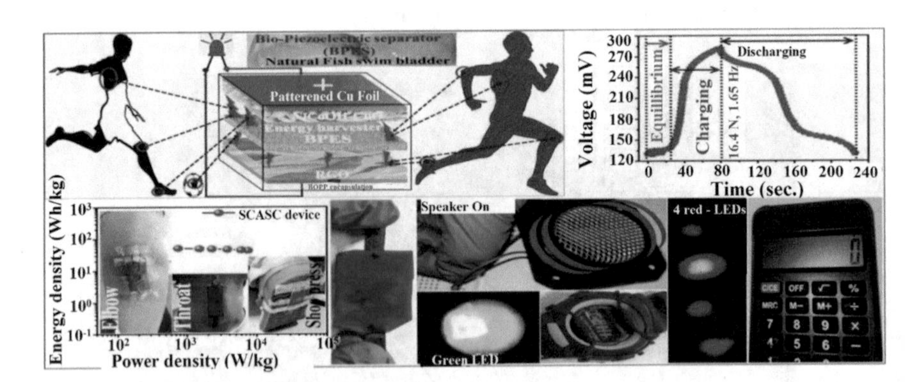

Fig. 11.16 Natural separator-based supercapacitor for electronic applications (redrawn and reprinted with permission from [36])

like watches, speakers, calculators, and various LEDs. The supercapacitor could be attached to various parts of the human body like the elbow, throat, and show press to charge the cell [36].

Another important breakthrough in the use of supercapacitors in electronics is the paper-based mechanical sensor [37]. The fact that the supercapacitor is sensitive to the area between deformable electrolytes and electrodes is being exploited to use it as a mechanical sensor. Using a very simple paper-based fabrication process without the need for a clean room, these sensors could measure the normal and shear components of force. These sensors can be developed with various geometries of the electrolyte. This is illustrated in Fig. 11.17.

The incorporation of asymmetric stretchable supercapacitors in fibers for preparing wearable fabric has been demonstrated and is shown in Fig. 11.18. In this

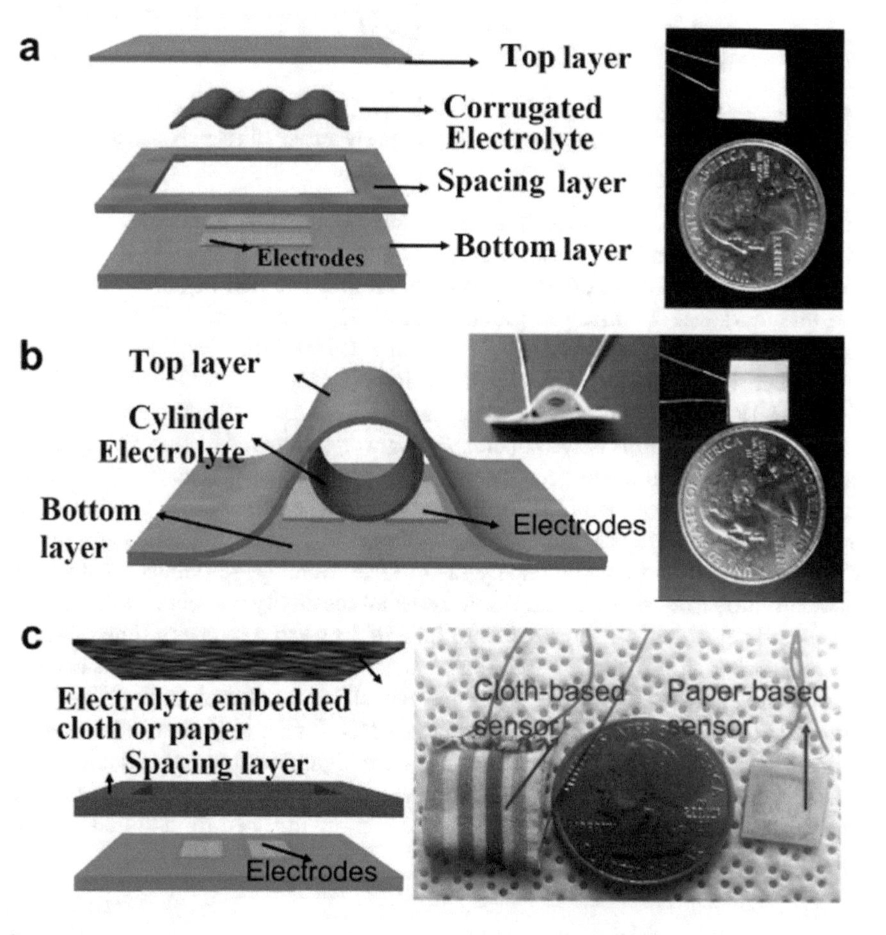

Fig. 11.17 Schematic and photograph of **a** corrugated electrolyte **b** cylinder electrolyte and **c** cloth and paper-based sensor (redrawn and reprinted under Creative Commons license [37])

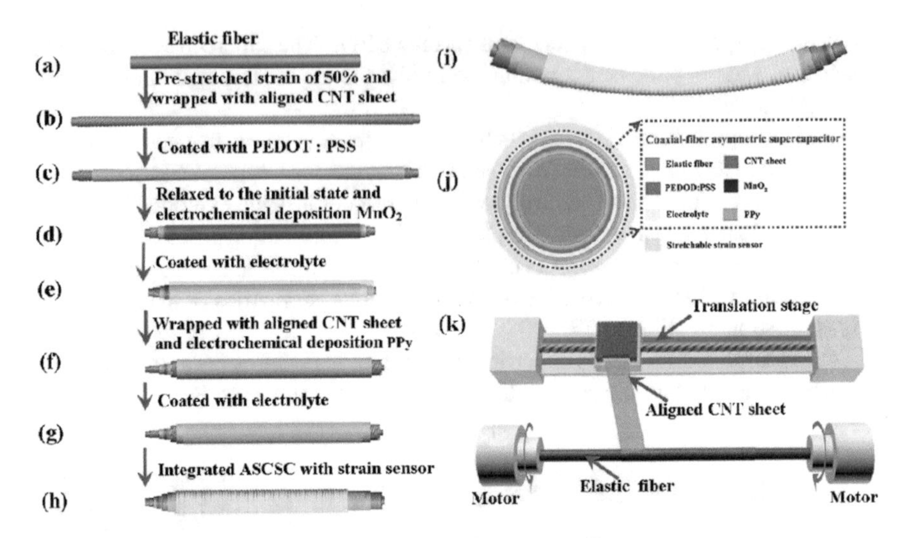

Fig. 11.18 Schematic representation of the fabrication of coaxial fiber strain sensor incorporated with asymmetric stretchable supercapacitor (redrawn and reprinted with permission from [38])

study, the supercapacitor and a strain sensor are incorporated simultaneously into a coaxial strain sensor. This sensor is made of elastic fiber. Manganese dioxide and polypyrrole deposited on CNT are taken as the positive and negative electrodes. The unique device architecture provides an energy density of 1.42 mWh cm^{-3}. Capacitance retention after 6000 cycles is also obtained [38].

Another interesting use of supercapacitors is for gas sensor applications. Figure 11.19 shows the use of a supercapacitor array for a self-driven ethanol gas sensor. Polypyrrole films based supercapacitor arrays prepared by photolithography and electrodeposition techniques were used to power ethanol sensor deposited on the same flexible substrate [39]. A concentric circle geometry was employed for the supercapacitor arrays to increase the areal capacitance. A high areal capacitance of 47.42 mF/cm^2 and energy density of 0.185 mW/cm^2 were obtained using this supercapacitor array. A room temperature ethanol sensitivity was obtained using this self-driven device with a short response time of 13 s and a recovery time of 4.5 s. The response of the graphene-based sensor integrated with the micro-supercapacitor array is also shown in Fig. 11.19. The supercapacitor consisted of multiwalled CNT wrapped in polyaniline. The sensor could sense NO_2 for about 50 min on integrating with the supercapacitor array. These sensors were highly flexible and operating even with a uniaxial stretching of 50%. These supercapacitor array-integrated sensors are one step in the direction of wearable devices without the need for additional cords for external power supply [40].

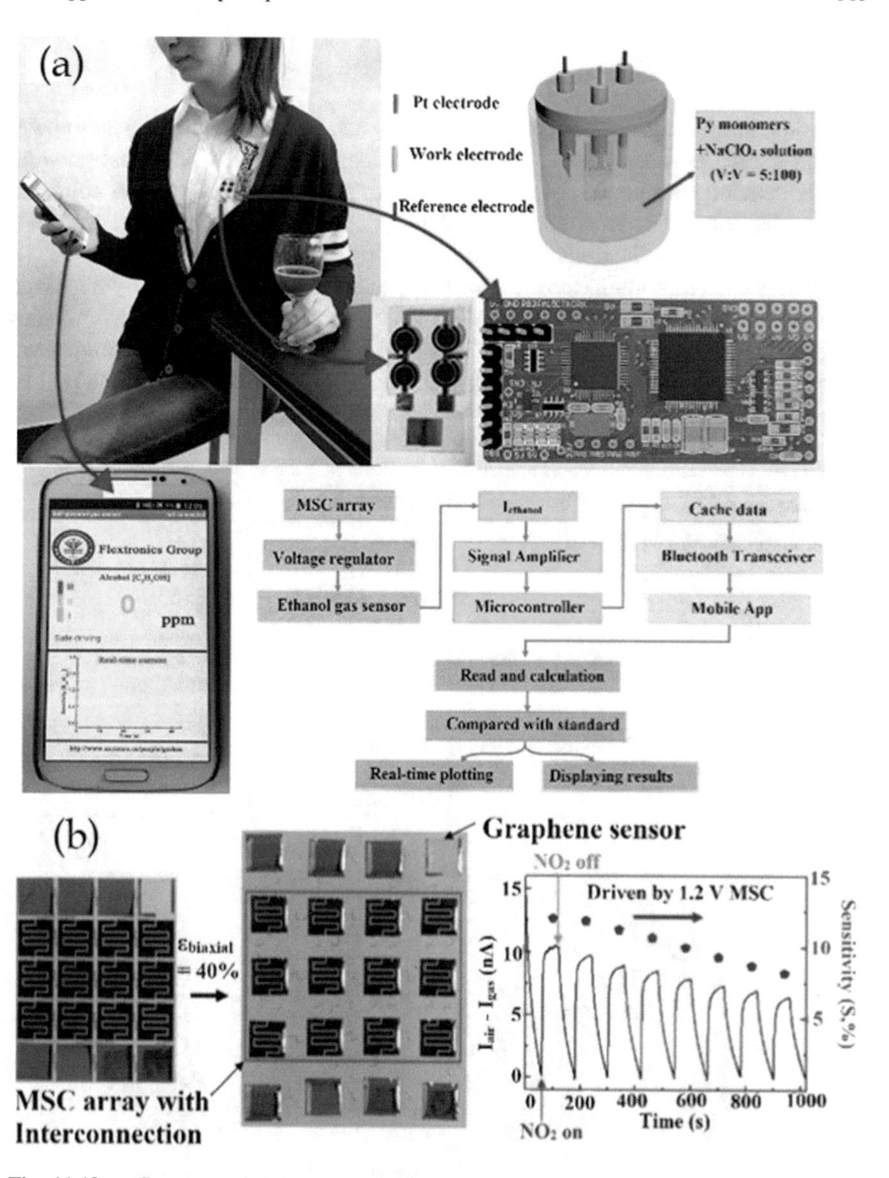

Fig. 11.19 **a** Gas sensor integrated with supercapacitor array for self-driven wearable devices (reprinted with permission from [39]) and **b** on-chip NO_2 detection using graphene sensor (redrawn and reprinted with permission from [40])

11.2.4 Industrial

For industrial applications, supercapacitors are generally used as an emergency power backup source because of their immediate discharging ability. The fast-growing market demand for energy-efficient light emitting diode (LED) lighting solutions and results in rapid expansion of LED lighting applications. Supercapacitor products in industrial applications from various industries are presented in Table 11.4. The power consumption in the elevator is expected to reduce by 50% by the usage of the kinetic energy recovery system (KERS), a product of Skeleton technologies using graphene-based ultracapacitor. This is given in Fig. 11.20. The company in collaboration with a Spanish company, EPIC Power attempts to reduce the cost of

Table 11.4 Supercapacitor products in industrial applications from various industries

Company name	Supercapacitor	Specification	Application
Skeleton technologies [26]	SkelMod 131 V	C: 6 F, V: 131 V, Energy: 15.9 Wh, Power = 56.7 kW	Kinetic energy recovery for the elevator industry
EATON [10]	B supercapacitor	C: 0.2–2.2 F, V: 2.5 V, ESR (mΩ): 2000	Building and home control
	HB supercapacitor	C: 3–110 F, V: 2.5 V, ESR (mΩ): 155–18	

Fig. 11.20 Kinetic energy recovery system (KERS) by Skeleton technologies (reprinted with permission from [41])

elevator operation [41]. There have been many studies on the use of supercapacitors for elevators. Various approaches including bi-directional energy flow with soft communication [42], bi-directional DC-DC converter [43], use of supercapacitor bank for storing braking energy [44], and the use of a multi-channel buck-boost converter is being studied.

11.2.5 Aerospace and Defense

The aerospace and defense industry needs several requirements for technologies, including light-weight and high-performance materials. The first high power supercapacitor was developed for military applications by a research institute in the USA in 1982. Supercapacitor products in aerospace and defense applications from various industries are presented in Table 11.5.

In aerospace systems, supercapacitors are widely used to power various electronic systems and devices which require either more or less energy to be managed continuously and also for delivering impulsive energy for a short duration. It is widely used in satellites and aircraft systems for efficient and consistent power delivery solutions.

For any spacecraft, electronic systems constitute control systems, sensors, and functional tools. For a satellite, the reliability of electronic components is completely critical in terms of materials selection, design, and performance reliability for longer durations. Thus, self-healing electronic system includes supercapacitors, photodiodes, transistors, and other elements such as self-healing alloys and self-healing conductive ink to ensure higher reliability and performance. Figure 11.21 shows advanced materials for satellite applications [45]. Supercapacitors are used with ultra-thin solar cells to form an efficient power system.

Table 11.5 Supercapacitor products in aerospace applications from various industries

Company name	Supercapacitor	Specification	Application
Newcell technologies [12]	LSC14250	Maximum discharge current (A): 0.5, V: 4 V, Internal resistance (mΩ): <200	Defense industry
	LSC14505	Maximum discharge current (mA): 2.0, V: 4 V, Internal resistance (mΩ): <150	
Goldencell technologies [13]	WPSM5R5474Z-L	C: 0.47–1.5 F, V: 5.5 V, Internal resistance (mΩ): <900	Military

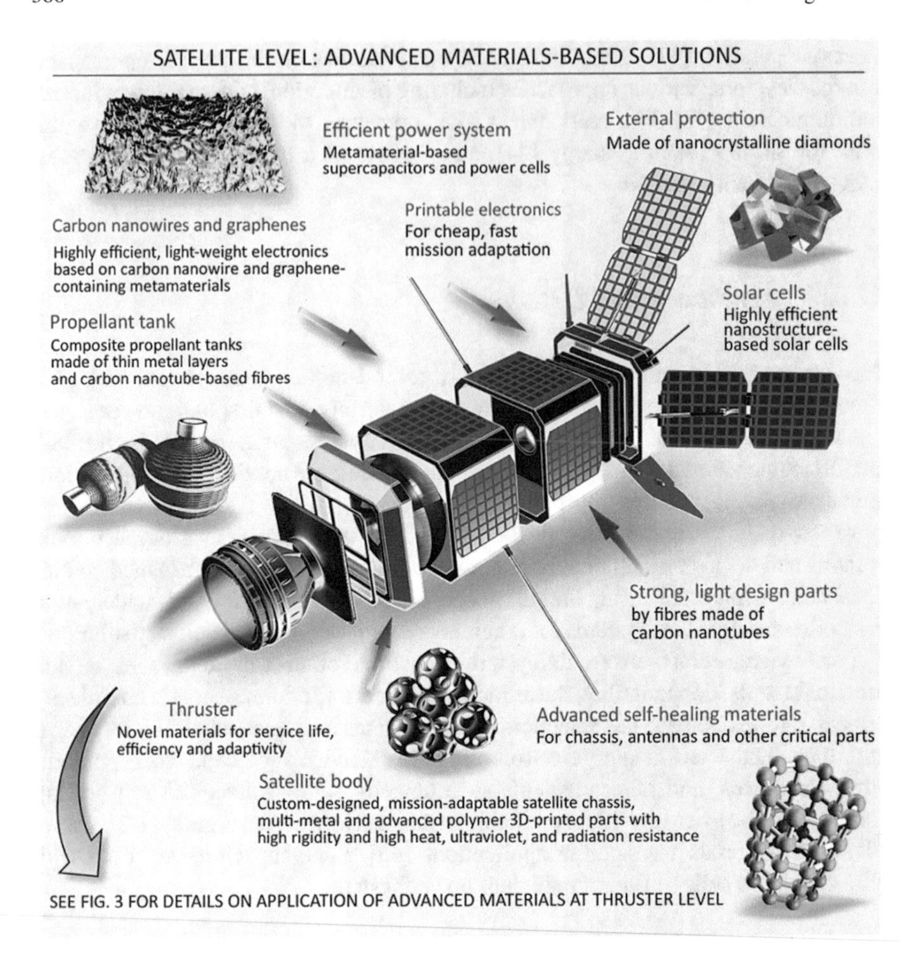

Fig. 11.21 Supercapacitor in satellites (reprinted with permission from [45])

The latest development for manned electric aircraft by Siemens and Rolls Royce is the use of supercapacitor hybrid energy storage systems. Batteries manufactured by Maxwell technologies and Panasonic corporation are widely used to power the aircraft [46]. The battery cells are also produced by LGChem with about 260 Wh/kg capabilities.

Unmanned aerial vehicle (UAV) started developing in recent years and integration of hybrid energy system plays a major role in design and development of high performance aircraft. Figure 11.22 shows UAV under flight test condition. The supercapacitor delivers load smoothing for fuel cell during dynamic working conditions [47]. Complete hybrid power integration system with fuel cell, supercapacitor, and battery and associated control systems are shown in Fig. 11.23. Hybrid power systems using multiple devices are integrated systematically to have higher efficiency and performance as per requirements. Supercapacitors and batteries acts as energy sources in

Fig. 11.22 Unmanned aerial vehicle (UAV) during flight test (reprinted with permission from [47])

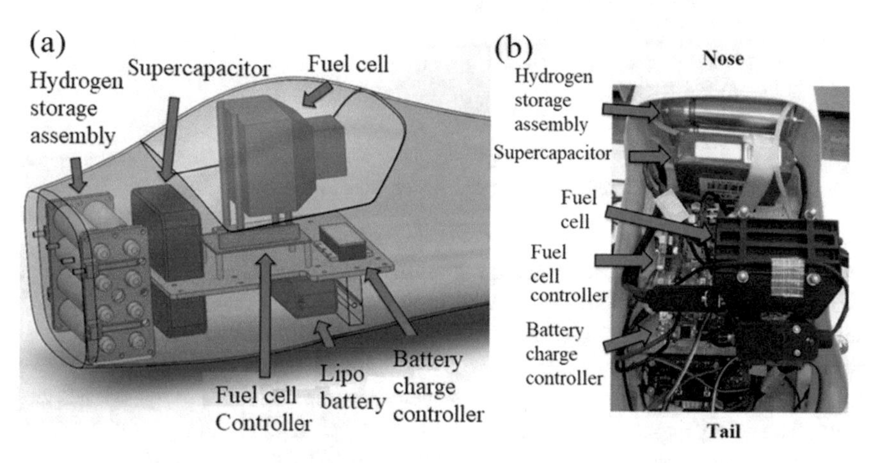

Fig. 11.23 **a** Complete hybrid vehicle integration system and **b** hybrid power integration control system in aircraft (redrawn and reprinted with permission from [47])

high power and dynamic conditions, whereas fuel cells satisfies high specific energy and lower specific power requirement. The hybrid power integration control system provides energy for smooth functioning of aircraft.

Military application of supercapacitors mainly includes laser weapons, electromagnetically accelerated projectiles, radar systems, propeller of an aircraft carrier, and vehicle driven trains. Many of these military systems are manufactured by various manufacturers [48]. Tecate group manufacture various military and aerospace system [49] such as radar antena, avionics, munitions, and unmanned aerial vehicles (UAVs) (Fig. 11.24).

Recently, high power laser beams were developed by the USA and Russia to destroy enemy aircraft and ballistic missiles [50]. In the case of a fast-moving target, it is not possible to hold a laser beam on the target, thus the energy that can destroy

Fig. 11.24 Structure of a HEMTT ProPulse, A-generator, B-diesel engine, C-supercapacitors, D-AC engines (redrawn and reprinted with permission from [50])

the target must be delivered as single radiation of millisecond duration. Thus, supercapacitors are the only solution for accumulating electric power for a few seconds and delivering it in a matter of milliseconds. Electrically driven vehicles are available with hybrid as well as pure storage systems. Batteries are used in supply for constant energy requirement, whereas supercapacitor delivers energy for varying loads. The heavy expanded mobility tactical truck (HEMTT) is an eight-wheel drive, diesel-powered truck used by the military [50]. HEMTT A3 is a military truck with a 470LE Cummins-based diesel engine, which drives both the wheels and also a 340 kW generator, which charges a series of supercapacitors and exhibits 1.9 MJ nominal capacitance, which drives four AC engines of 480 V through an inverter. The HEMTT requires only 20% lesser fuel than diesel driven vehicle of the same specification. With the use of supercapacitor, HEMTT Propulse provides efficient military requirements such as communication systems, radar, and laser weapons systems.

11.2.6 Medical

Nowadays modern healthcare facilities pay more attention to high-tech equipment and the quality of services they offer. Furthermore, an oversized electrical system makes it almost impossible to offer mobile medical services, such as mobile MRI.

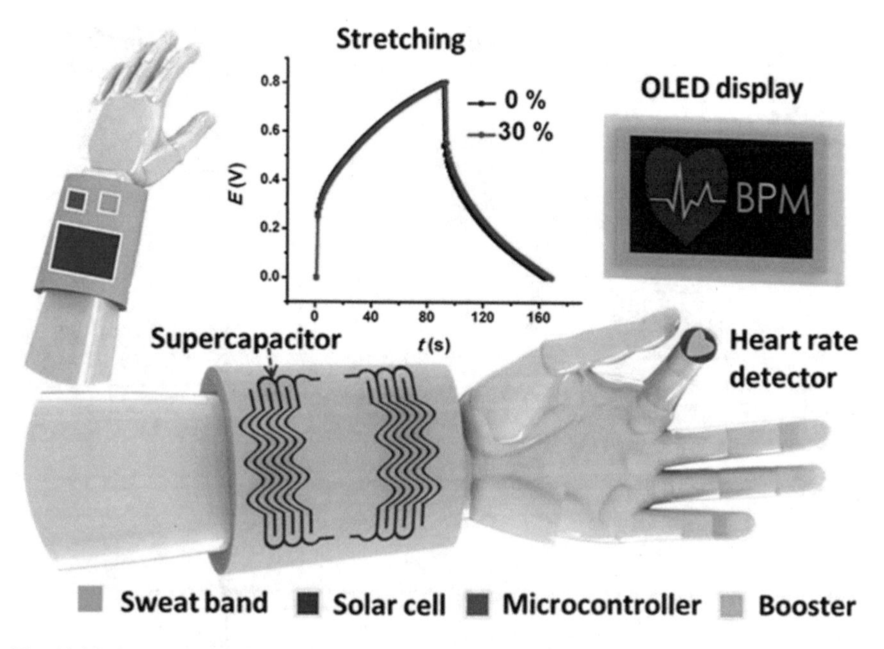

Fig. 11.25 Schematic illustration of ISSC-bonded stretchable sweatband (reprinted with permission from [51])

There are several biomedical implanted devices such as cardiac and gastric pacemakers, bladder, bone stimulators, and automated drug delivery systems, where supercapacitors can offer significant advantages. Rajendran et al. [51] conducted a study on a flexible solid-state supercapacitor for self-powered wearable electronics. In this study, they have prepared a CNT-PANI-based interdigitated solid-state supercapacitor (ISSC) and integrated ISSC on a wrist-worn stretchable sweatband for powering ability checking devices. Figure 11.25 shows a schematic illustration of the ISSC-bonded stretchable sweatband.

Park et al. [52] developed a printing technique for solid-state supercapacitor for soft, smart contact lenses. The contact lens contains an integrated wireless supercapacitor array for continuous operation. Mosa et al. [53] have successfully developed a biological supercapacitor, which is capable of storing electrical energy inside the body. For the development of next-generation implanted devices, they have designed an ultra-thin supercapacitor using human biofluids as electrolytes for the first time, as shown in Fig. 11.26.

Supercapacitor products in medical applications from various industries are presented in Table 11.6.

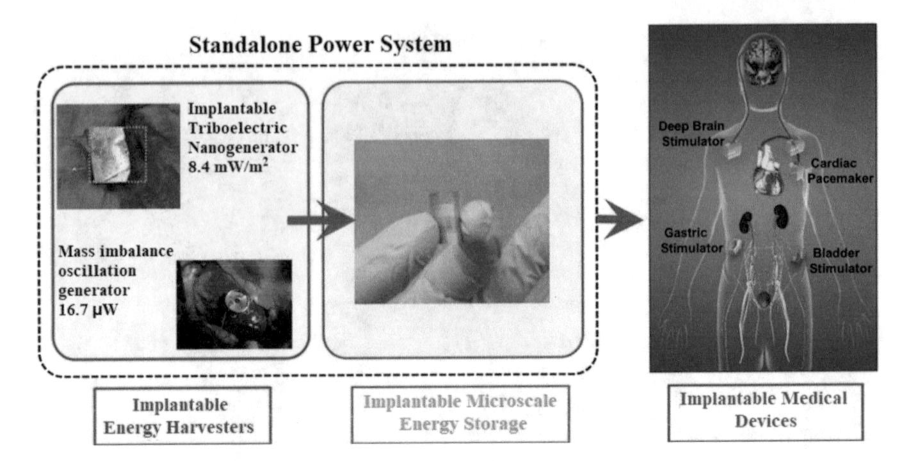

Fig. 11.26 Schematic illustration of supercapacitor based implantable medical devices (reprinted with permission from [53])

Table 11.6 Supercapacitor products in medical applications from various industries

Company name	Supercapacitor	Specification	Application
KAM [9]	2.3 V hybrid	C: 10–10,000 F, V: 2.3 V, E_{max}: 0.0059 m Wh, $P_{max} = 0.0024$ W, Operating temp. range: −25–60 °C	Medical equipment
Skeleton technologies [26]	SkelMod 17 V	C: 533 F, V: 17 V, Energy: 21.3 Wh, Power = 65.7 kW	Medical UPS
	SkelMod 131 V	C: 6 F, V: 131 V, Energy: 15.9 Wh, Power = 56.7 kW	MRI power quality
EATON [10]	B supercapacitor	C: 0.2–2.2 F, V: 2.5 V, ESR (mΩ): 2000	Personal and hospital systems
	HB supercapacitor	C: 3–110 F, V: 2.5 V, ESR (mΩ): 155–18	

11.3 Conclusions

Supercapacitors are utilised in a wide range of applications due to their feature like large power density, fast charging/discharging, high capacitance, robustness, and long life cycles. Supercapacitor finds application in various applications like automobiles/transportation, energy and utilities, aerospace military, electronics, industrial, and medical. The products and services offered by various companies and the range of products with the specification are provided. Supercapacitor manufacturers are categorized continent wise and a few selected commercial supercapacitor is reviewed. The supercapacitor market has showed a steady state increase from 2013 onward. Due to the policy across the globe for a pollution-free environment and saving fossil fuels, supercapacitor-based hybrid vehicles are becoming an alternative to the existing transportation system. The main market of supercapacitors is electronics which is highly dependent on the large power density. Wearable and self-charging devices are becoming a groundbreaking development in the use of supercapacitors in day-to-day life. A hybrid energy system is very convenient for rendering a solution to energy production and transmission issues. It also provides a backbone to the utilization of renewable energy sources. These hybrid energy systems are also shown to be useful in industries and smart homes. The recovery of kinetic energy from braking is reducing the operational cost of elevators and metro trains. Supercapacitors are used in spacecraft and military applications which require high power density, high voltage, and longer cycle life as compared to conventional storage systems. Supercapacitors are also useful in a variety of biomedical applications wherein they can power sweatband for continuous monitoring and is useful in wireless contact lenses. In summary, the horizon of supercapacitors is expanding and the application is only limited to the imagination of the users.

References

1. *Handbook of Nanocomposite Supercapacitor Materials I, Book Subtitle: Characteristics*, ed. by K.K. Kar (eBook ISBN:978-3-030-52359-6, Hardcover ISBN:978-3-030-52358-9)
2. *Handbook of Nanocomposite Supercapacitor Materials II, Book Subtitle: Performance*, ed. by K.K. Kar (eBook ISBN: 978-3-030-43009-2, Hardcover ISBN: 978-3-030-43008-5). https://doi.org/10.1007/978-3-030-43009-2
3. A. Burke, Ultracapacitors: why, how, and where is the technology. J. Power Sources **91**, 37–50 (2000)
4. L.L. Zhang, X.S. Zhao, Carbon-based materials as supercapacitor electrodes. Chem. Soc. Rev. **38**, 2520–2531 (2009)
5. S. Banerjee, B. De, C. Sinha, C. Jayesh, K.K. Kar, in *Applications of Supercapacitors*, ed. by K.K. Kar. Handbook of Nanocomposite Supercapacitor Materials I Characteristics, (Springer, Berlin, Heidelberg, 2020) Chapter 13, pp. 341–350. (Print ISBN 978-3-030-43008-5, Online ISBN 978-3-030-43009-2). https://doi.org/10.1007/978-3-030-43009-2_13
6. R. Nigam, K.D. Verma, T. Pal, K.K. Kar, in *Applications of Supercapacitors*, ed. by K.K. Kar. Handbook of Nanocomposite Supercapacitor Materials II Performance (Springer Nature, Berlin, Heidelberg, 2020, Chapter 17). https://doi.org/10.1007/978-3-030-52359-6
7. K. Kotz, M. Carlen, Principles and applications of electrochemical capacitors. Electrochim. Acta **45**, 2483–2498 (2000)
8. https://www.zoxcell.com/. Accessed 01 Oct 2020
9. https://www.kamcappower.com/. Accessed 01 Oct 2020
10. https://www.eaton.com/us/en-us/products/electronic-components/supercapacitors.html. Accessed 01 Oct 2020

11. http://www.aowei.com/en/program/application.html. Accessed 01 Oct 2020
12. http://www.newcell-supercapacitor.com/. Accessed 01 Oct 2020
13. http://www.goldencellbattery.com/product/41/. Accessed 01 Oct 2020
14. https://ir.tesla.com/press-release/tesla-completes-acquisition-maxwell-technologies. Accessed 01 Oct 2020
15. https://ioxus.com/english/media/press-releases/. Accessed 01 Oct 2020
16. https://www.cap-xx.com/wp-content/uploads/2019/12/CAP-XX-Press-Release-CAP-XX-acquires-Muratas-supercapacitor-production-lines_final.pdf. Accessed 01 Oct 2020
17. https://www.maxwell.com/maxwell-nesscap#:~:text=Maxwell%20Technologies%20and%20Nesscap%20are,and%20most%20innovative%20ultracapacitor%20solutions. Accessed 01 Oct 2020
18. https://eu-japan.com/2008/07/tdk-epcos/#:~:text=EPCOS%20becomes%20part%20of%20100,thus%20acquiring%20100%25%20of%20EPCOS. Accessed 01 Oct 2020
19. https://www.skeletontech.com/investors. Accessed 01 Oct 2020
20. S. Huang, X. Zhu, S. Sarkar, Y. Zhao, Challenges and opportunities for supercapacitors. APL Mater. **7**, 100901 (2019)
21. A. Afif, S.M.H. Rahman, A.T. Azad, J. Zaini, M.A. Islan, A.K. Azad, Advanced materials and technologies for hybrid supercapacitors for energy storage—a review. J. Energy Storage **25**, 100852 (2019)
22. https://www.idtechex.com/de/research-article/supercapacitors-market-to-achieve-30-cagr-over-the-next-decade/6502. Accessed 01 Oct 2020
23. B. Allaoua, K. Asnoune, B. Mebarki, Energy management of PEM fuel cell/ supercapacitor hybrid power sources for an electric vehicle. Int. J. Hydrogen Energy **42**, 21158–21166 (2017)
24. O. Veneri, C. Capasso, S. Patalano, Experimental investigation into the effectiveness of a supercapacitor based hybrid energy storage system for urban commercial vehicles. Appl. Energy **227**, 312–323 (2018)
25. https://www.maxwell.com/products/ultracapacitors/cells. Accessed 01 Oct 2020
26. https://www.skeletontech.com/. Accessed 01 Oct 2020
27. http://afstrinity.com/xh/tech.htm. Accessed 01 Oct 2020
28. https://www.skoda.cz/. Accessed 24 Nov 2020
29. https://www.cap-xx.com/. Accessed 01 Oct 2020
30. L. Li, Z. Huang, H. Li, H. Lu, A high-efficiency voltage equalization scheme for supercapacitor energy storage system in renewable generation applications. Sustainability **8**, 548 (2016)
31. J. Pegueroles-Queralt, F.D. Bianchi, O. Gomis-Bellmunt, A power smoothing system based on supercapacitors for renewable distributed generation. IEEE Trans. Ind. Electron. **62**, 343 (2015)
32. T. Ma, H. Yang, L. Lu, Development of hybrid battery–supercapacitor energy storage for remote area renewable energy systems. Appl. Energy **153**, 56–62 (2015)
33. R. Jia, G. Shen, F. Qu, D. Chen, Flexible on-chip micro-supercapacitors: Efficient power units for wearable electronics. Energy Storage Mater. **27**, 169–186 (2020)
34. K. Jost, G. Dion, Y. Gogotsi, Textile energy storage in perspective. J. Mater. Chem. A **2**, 10776–10787 (2014)
35. H. Wang, Y. Diao, Y. Lu, H. Yang, Q. Zhou, K. Chrulski, J.M. D'Arcy, Energy storing bricks for stationary PEDOT supercapacitors. Nat. Commun. **11**, 3882 (2020)
36. A. Maitra, S.K. Karan, S. Paria, A.K. Das, R. Bera, L. Halder, S.K. Si, A. Bera, B.B. Khatua, Fast charging self-powered wearable and flexible asymmetric supercapacitor power cell with fish swim bladder as an efficient natural bio-piezoelectric separator. Nano Energy **40**, 633–645 (2017)
37. Y. Zhang, S. Sezen, M. Ahmadi, X. Cheng, R. Rajamani, Paper-Based Supercapacitive Mechanical Sensors. Sci. Rep. **8**, 816284 (2018)
38. Z. Pan, J. Yang, L. Li, X. Gao, L. Kang, Y. Zhang, Q. Zhang, Z. Kou, T. Zhang, L. Wei, Y. Yao, J. Wang, All-in-one stretchable coaxial-fiber strain sensor integrated with high-performing supercapacitor. Energy Storage Mater. **27**, 124–130 (2020)

39. L. Li, C. Fu, Z. Lou, S. Chen, W. Han, K. Jiang, D. Chen, G. Shen, Flexible planar concentric circular micro-supercapacitor arrays for wearable gas sensing application. Nano Energy **41**, 261–268 (2017)
40. J. Yun, Y. Lim, G.N. Jang, D. Kim, S.J. Lee, H. Park, S.Y. Hong, G. Lee, G. Zi, J.S. Ha, Stretchable patterned graphene gas sensor driven by integrated micro-supercapacitor array. Nano Energy **19**, 401–414 (2016)
41. https://www.skeletontech.com/news/press-release-graphene-cuts-elevator-energy-consumption-in-half. Accessed 01 Nov 2020
42. A. Rufer, P. Philippe, A supercapacitor-based energystorage system for elevators with soft commutated interface. IEEE Trans. Ind. Appl. **38**, 1151–1159 (2002)
43. N. Jabbour, C. Mademlis, Supercapacitor-based energy recovery system with improved power control and energy management for elevator applications. IEEE Trans. Power Electron. **32**, 9389–9399 (2017)
44. P. Gao, W. Niu, Z. Quanji, Y. Yang, Y. Lv, Elevator regenerative energy feedback technology. Adv. Comput. Sci. Res. **63**, 168–175 (2016)
45. I. Levchenko, K. Bazaka, T. Belmonte, M. Keidar, Advanced materials for next-generation spacecraft. Adv. Mater. **30**, 1802201 (2018)
46. https://www.aircraftinteriorsinternational.com/industry-opinion/the-impact-of-tesla-on-the-aviation-market.html. Accessed 14 Nov 2020
47. A. Gong, J.L. Palmer, D. Verstraete, in Flight test of a fuel-cell/battery/ supercapacitor triple hybrid Uav propulsion system, in *31st Congress of the International Council of the Aeronautical Sciences*, **0864**, 1–10 (2018)
48. A. Muzaffar, M.B. Ahamed, K. Deshmukh, J. Thirumalai, A review on recent advances in hybrid supercapacitors: design, fabrication and applications. Renew. Sustain. Energy Rev. **101**, 123–145 (2019)
49. https://www.tecategroup.com/markets/?market=Military-Aerospace. Accesses 14 Nov 2020
50. Z. Végvári, Supercapacitors and their Military Applicability. Hungarian Defence Rev. **147**, 38 (2019)
51. V. Rajendran, A.M. Vinu Mohan, M. Jayaraman, T. Nakagawa, All-printed, interdigitated, freestanding serpentine interconnects based flexible solid state supercapacitor for self powered wearable electronics. Nano Energy **65**, 104055 (2019)
52. J. Park, D.B. Ahn, J. Kim, E. Cha, B.S. Bae, S.Y. Lee, J.U. Park, Printing of wirelessly rechargeable solid-state supercapacitors for soft, smart contact lenses with continuous operations. Sci. Adv. **5**, 1–8 (2019)
53. I.M. Mosa, A. Pattammattel, K. Kadimisetty, P. Pande, M.F. El-Kady, G.W. Bishop, M. Novak, R.B. Kaner, A.K. Basu, C.V. Kumar, J.F. Rusling, Ultrathin graphene-protein supercapacitors for miniaturized bioelectronics. Adv. Energy Mater. **7**, 700358 (2017)

Index

K. K. Kar (ed.), *Handbook of Nanocomposite Supercapacitor Materials III*, Springer Series in Materials Science 313,
https://doi.org/10.1007/978-3-030-68364-1

F

G

Printed in the United States
by Baker & Taylor Publisher Services